Student Solutions Manual

for

Technical Calculus with Analytic Geometry

Fourth Edition

Peter Kuhfittig
Milwaukee School of Engineering

With the assistance of:

Christopher Schroeder
Morehead State University

THOMSON

BROOKS/COLE

Australia • Brazil • Canada • Mexico • Singapore • Spain • United Kingdom • United States

Printed in the United States of America
   2   3   4   5   6   7   09   08   07

Printer: Thomson/West
ISBN-13: 978-0-495-10545-9
ISBN-10: 0-495-10545-7
Cover Image Credit: Getty Images

Thomson Higher Education
10 Davis Drive
Belmont, CA 94002-3098
USA

For more information about our products,
contact us at:
Thomson Learning Academic Resource Center
1-800-423-0563

For permission to use material from this text or product, submit a request online at
http://www.thomsonrights.com.
Any additional questions about permissions can be submitted by email to **thomsonrights@thomson.com.**

Contents

1 Introduction to Analytic Geometry   **1**

    1.1   The Cartesian Coordinate System 1

    1.2   The Slope 3

    1.3   The Straight Line 4

    1.4   Curve Sketching 7

    1.5   Curves with Graphing Utilities 18

    1.7   The Circle 20

    1.8   The Parabola 23

    1.9   The Ellipse 27

    1.10 The Hyperbola 32

    1.11 Translation of Axes; Standard Equations of the Conics 37

    Chapter 1 Review 43

2 Introduction to Calculus: The Derivative   **51**

    2.1   Functions and Intervals 51

    2.2   Limits 53

    2.4   The Derivative by the Four-Step Process 56

    2.5   Derivatives of Polynomials 59

    2.6   Instantaneous Rates of Change 60

    2.7   Differentiation Formulas 62

    2.8   Implicit Differentiation 69

    2.9   Higher Derivatives 72

    Chapter 2 Review 74

3 Applications of the Derivative   **79**

    3.1   The First-Derivative Test 79

    3.2   The Second-Derivative Test 83

    3.3   Exploring with Graphing Utilities 93

    3.4   Applications of Minima and Maxima 97

    3.5   Related Rates 104

    3.6   Differentials 110

    Chapter 3 Review 111

4   The Integral   **117**

   4.1   Antiderivatives 117

   4.2   The Area Problem 118

   4.3   The Fundamental Theorem of Calculus 119

   4.5   Basic Integration Formulas 119

   4.6   Area Between Curves 123

   4.7   Improper Integrals 129

   4.8   The Constant of Integration 133

   4.9   Numerical Integration 139

   Chapter 4 Review 142

5   Applications of the Integral   **147**

   5.1   Means and Root Mean Squares 147

   5.2   Volumes of Revolution: Disk and Washer Methods 148

   5.3   Volumes of Revolution: Shell Method 153

   5.4   Centroids 161

   5.5   Moments of Inertia 173

   5.6   Work and Fluid Pressure 179

   Chapter 5 Review 189

6   Derivatives of Transcendental Functions   **193**

   6.1   Review of Trigonometry 193

   6.2   Derivatives of Sine and Cosine Functions 195

   6.3   Other Trigonometric Functions 198

   6.4   Inverse Trigonometric Functions 202

   6.5   Derivatives of Inverse Trigonometric Functions 204

   6.6   Exponential and Logarithmic Functions 207

   6.7   Derivative of the Logarithmic Function 210

   6.8   Derivative of the Exponential Function 212

   6.9   L'Hospital's Rule 215

   6.10 Applications 216

   6.11 Newton's Method 221

   Chapter 6 Review 223

7   Integration Techniques   **227**

   7.1   The Power Formula Again 227

   7.2   The Logarithmic and Exponential Forms 229

   7.3   Trigonometric Forms 233

   7.4   Further Trigonometric Forms 236

   7.5   Inverse Trigonometric Forms 242

   7.6   Integration by Trigonometric Substitution 245

   7.7   Integration by Parts 251

   7.8   Integration of Rational Functions 254

   7.9   Integration by Use of Tables 259

   Chapter 7 Review 261

8 Parametric Equations, Vectors, and Polar Coordinates   **267**

     8.1   Vectors and Parametric Equations 267

     8.2   Arc Length 271

     8.3   Polar Coordinates 273

     8.4   Curves in Polar Coordinates 276

     8.5   Areas in Polar Coordinates 279

     Chapter 8 Review 286

9 Three Dim. Space; Partial Derivatives; Multiple Integrals   **289**

     9.1   Surfaces in Three Dimensions 289

     9.2   Partial Derivatives 296

     9.3   Applications of Partial Derivatives 300

     9.4   Curve Fitting 307

     9.5   Iterated Integrals 309

     9.6   Volumes by Double Integration 314

     9.7   Mass, Centroids, and Moments of Inertia 320

     9.8   Volumes in Cylindrical Coordinates 327

     Chapter 9 Review 330

10 Infinite Series   **337**

     10.1   Introduction to Infinite Series 337

     10.2   Tests for Convergence 338

     10.3   Maclaurin Series 342

     10.4   Operations with Series 345

     10.5   Computations with Series; Applications 346

     10.6   Fourier Series 351

     Chapter 10 Review 358

11 First-Order Differential Equations   **361**

     11.1   What is a Differential Equation? 361

     11.2   Separation of Variables 363

     11.3   First-Order Linear Differential Equations 368

     11.4   Applications of First-Order Differential Equations 373

     11.5   Numerical Solutions 381

     Chapter 11 Review 383

12 Higher-Order Linear Differential Equations   **389**

     12.1   Higher-Order Homogeneous Differential Equations 389

     12.2   Auxiliary Equations with Repeating or Complex Roots 392

     12.3   Nonhomogeneous Equations 396

     12.4   Applications of Second-Order Equations 406

     Chapter 12 Review 414

13 The Laplace Transform   419

Sections 13.1-13.3 419

13.4 Solution of Linear Equations by Laplace Transforms 425

Chapter 13 Review 434

Chapter 1

Introduction to Analytic Geometry

1.1   The Cartesian Coordinate System

1. Let $(x_2, y_2) = (2, 4)$ and $(x_1, y_1) = (5, 2)$. From the distance formula

$$d = \sqrt{(x_2 - x_1)^2 + (y_2 - y_1)^2}$$

we get

$$d = \sqrt{(2 - 5)^2 + (4 - 2)^2} = \sqrt{(-3)^2 + 2^2} = \sqrt{9 + 4} = \sqrt{13}.$$

3. Let $(x_2, y_2) = (-3, -6)$ and $(x_1, y_1) = (5, -2)$. Then

$$d = \sqrt{(-3 - 5)^2 + [-6 - (-2)]^2} = \sqrt{(-8)^2 + (-4)^2} = \sqrt{64 + 16} = \sqrt{16 \cdot 5} = 4\sqrt{5}.$$

5. Let $(x_2, y_2) = (\sqrt{3}, 4)$ and $(x_1, y_1) = (0, 2)$. Then

$$d = \sqrt{(\sqrt{3} - 0)^2 + (4 - 2)^2} = \sqrt{3 + 4} = \sqrt{7}.$$

7. $d = \sqrt{[1 - (-1)]^2 + (-\sqrt{2} - 0)^2} = \sqrt{4 + 2} = \sqrt{6}$

9. $d = \sqrt{[-10 - (-12)]^2 + (-2 - 0)^2} = \sqrt{2^2 + (-2)^2} = \sqrt{8} = \sqrt{2 \cdot 4} = 2\sqrt{2}$

11b. x/y is negative whenever x and y have opposite signs: quadrants II and IV.

13a. Any point on the y-axis has coordinates of the form $(0, y)$.

15. Let $A = (-2, 5), B = (3, 5)$ and $C = (-2, 2)$. After calculating $AB = 5, AC = 3$ and $BC = \sqrt{34}$, we observe that

$$(AB)^2 + (AC)^2 = (BC)^2.$$

17. The points $(12, 0), (-4, 8)$ and $(-1, -13)$ are all $5\sqrt{5}$ units from $(1, -2)$.

19. Distance from $(-1, -1)$ to $(2, 8)$:

$$\sqrt{(-1-2)^2 + (-1-8)^2} = \sqrt{9+81} = \sqrt{90} = \sqrt{9 \cdot 10} = 3\sqrt{10}$$

Distance from $(2, 8)$ to $(5, 17)$:

$$\sqrt{(5-2)^2 + (17-8)^2} = \sqrt{90} = 3\sqrt{10}$$

Distance from $(-1, -1)$ to $(5, 17)$:

$$\sqrt{6^2 + 18^2} = \sqrt{360} = 6\sqrt{10}$$

Total distance $6\sqrt{10} = 3\sqrt{10} + 3\sqrt{10}$, the sum of the other two distances.

21. Distance from (x, y) to y-axis: x units

Distance from (x, y) to $(2, 0)$: $\sqrt{(x-2)^2 + (y-0)^2} = \sqrt{(x-2)^2 + y^2}$

By assumption,

$$\begin{aligned} \sqrt{(x-2)^2 + y^2} &= x \\ (x-2)^2 + y^2 &= x^2 \qquad \text{squaring both sides} \\ x^2 - 4x + 4 + y^2 &= x^2 \\ y^2 - 4x + 4 &= 0 \end{aligned}$$

23. Let $(x_1, y_1) = (-2, 6)$ and $(x_2, y_2) = (2, -4)$. Then from the midpoint formula

$$\left(\frac{x_1 + x_2}{2}, \frac{y_1 + y_2}{2} \right)$$

we get

$$\left(\frac{-2+2}{2}, \frac{6 + (-4)}{2} \right) = (0, 1).$$

25. Let $(x_1, y_1) = (5, 0)$ and $(x_2, y_2) = (9, 4)$. Then from the midpoint formula

$$\left(\frac{x_1 + x_2}{2}, \frac{y_1 + y_2}{2} \right)$$

we get

$$\left(\frac{5+9}{2}, \frac{0+4}{2} \right) = (7, 2).$$

27. $\left(\dfrac{-3 + (-6)}{2}, \dfrac{10 + 0}{2} \right) = \left(-\dfrac{9}{2}, 5 \right)$

1.2 The Slope

1. Let $(x_2, y_2) = (1, 7)$ and $(x_1, y_1) = (2, 6)$. Then, by formula (1.4),

$$m = \frac{y_2 - y_1}{x_2 - x_1}$$

we get

$$m = \frac{7 - 6}{1 - 2} = \frac{1}{-1} = -1.$$

3. Let $(x_1, y_1) = (0, 2)$ and $(x_2, y_2) = (-4, -4)$. Then

$$m = \frac{-4 - 2}{-4 - 0} = \frac{-6}{-4} = \frac{3}{2}.$$

5. Let $(x_2, y_2) = (7, 8)$ and $(x_1, y_1) = (-3, -4)$. Then

$$m = \frac{8 - (-4)}{7 - (-3)} = \frac{8 + 4}{7 + 3} = \frac{12}{10} = \frac{6}{5}.$$

7. $m = \dfrac{-8 - 0}{-4 - 0} = 2$

9. $m = \dfrac{-5 - 4}{3 - 3} = \dfrac{-9}{0}$ (undefined)

11. $m = \dfrac{-3 - (-3)}{9 - 5} = \dfrac{0}{4} = 0$

13. $m = \dfrac{-6 - 0}{0 - 8} = \dfrac{-6}{-8} = \dfrac{3}{4}$

15. See answer section of book.

17. Slope of given line is $\dfrac{1 - (-5)}{-7 - 6} = \dfrac{6}{-13} = -\dfrac{6}{13}$. Slope of perpendicular is given by the negative reciprocal and is therefore $-\dfrac{1}{(-6/13)} = \dfrac{13}{6}$.

19. Slope of line through $(-4, 6)$ and $(6, 10)$: $\dfrac{6 - 10}{-4 - 6} = \dfrac{-4}{-10} = \dfrac{2}{5}$

Slope of line through $(6, 10)$ and $(10, 0)$: $\dfrac{10 - 0}{6 - 10} = \dfrac{10}{-4} = -\dfrac{5}{2}$

Since the slopes are negative reciprocals, the lines are perpendicular.

21. Slope of line through $(0, -3)$ and $(-2, 3)$: $\dfrac{-3-3}{0-(-2)} = \dfrac{-6}{2} = -3$

    Slope of line through $(7, 6)$ and $(9, 0)$: $\dfrac{6-0}{7-9} = \dfrac{6}{-2} = -3$

    Slope of line through $(-2, 3)$ and $(7, 6)$: $\dfrac{3-6}{-2-7} = \dfrac{-3}{-9} = \dfrac{1}{3}$

    Slope of line through $(0, -3)$ and $(9, 0)$: $\dfrac{-3-0}{0-9} = \dfrac{1}{3}$

    Since -3 and $\frac{1}{3}$ are negative reciprocals, adjacent sides are perpendicular and opposite sides are parallel.

23. Midpoint: $\left(\dfrac{-3+9}{2}, \dfrac{-2+0}{2}\right) = (3, -1)$

    Slope of line through $(5, 6)$ and $(3, -1)$: $\dfrac{6-(-1)}{5-3} = \dfrac{7}{2}$

25. Similar to *Ex.* 23. One median is the line segment CD, where D is the midpoint of $AB = (-3/2, 1)$.

    Slope of $CD = \dfrac{-3-1}{2-(-3/2)} = \dfrac{-4}{7/2} = -\dfrac{8}{7}$

27. Slope of line through $(-1, -1)$ and $(3, -5)$: $\dfrac{-1-(-5)}{-1-3} = \dfrac{4}{-4} = -1$

    Slope of line through $(x, 2)$ and $(4, -6)$: $\dfrac{2+6}{x-4} = \dfrac{8}{x-4}$

    Since the two slopes must be equal, we have:

$$\begin{aligned}
\dfrac{8}{x-4} &= -1 \\
8 &= -x + 4 \qquad \text{multiplying both sides by } x - 4 \\
x &= -4
\end{aligned}$$

1.3   The Straight Line

1. Since $(x_1, y_1) = (-7, 2)$ and $m = 1/2$, we get

$$\begin{aligned}
y - 2 &= \tfrac{1}{2}(x + 7) & y - y_1 = m(x - x_1) \\
2y - 4 &= x + 7 & \text{clearing fractions} \\
0 &= x - 2y + 7 + 4 \\
x - 2y + 11 &= 0
\end{aligned}$$

3.
$$\begin{aligned}
y - y_1 &= m(x - x_1) \\
y + 4 &= 3(x - 3) & (x_1, y_1) = (3, -4); \; m = 3 \\
y + 4 &= 3x - 9 \\
3x - y - 13 &= 0
\end{aligned}$$

5.
$$\begin{aligned}
y - y_1 &= m(x - x_1) \\
y - 0 &= -\tfrac{1}{3}(x - 0) & (x_1, y_1) = (0, 0); \; m = -1/3 \\
3y &= -x \\
x + 3y &= 0
\end{aligned}$$

7. The line $y = 1 = 0x + 1$ has slope 0.

$$\begin{aligned} y - y_1 &= m(x - x_1) \\ y - 0 &= 0(x + 4) \qquad (x_1, y_1) = (-4, 0);\ m = 0 \\ y &= 0 \qquad\qquad\quad x\text{-axis} \end{aligned}$$

9. First determine the slope using $m = \dfrac{y_2 - y_1}{x_2 - x_1}$ to get $m = \dfrac{4 - (-6)}{-3 - 3} = \dfrac{10}{-6} = -\dfrac{5}{3}$.

   Then let $(x_1, y_1) = (-3, 4)$ to get

$$\begin{aligned} y - 4 &= -\tfrac{5}{3}(x + 3) \qquad y - y_1 = m(x - x_1) \\ 3y - 12 &= -5x - 15 \qquad \text{multiplying by 3} \\ 5x + 3y + 3 &= 0 \end{aligned}$$

11. $m = \dfrac{-4 - 0}{9 - 5} = -1$

$$\begin{aligned} y - 0 &= -1(x - 5) \qquad \text{choosing } (x_1, y_1) = (5, 0) \\ x + y - 5 &= 0 \end{aligned}$$

13. $m = \dfrac{3 - 4}{2 - (-6)} = \dfrac{-1}{8} = -\dfrac{1}{8}$

$$\begin{aligned} y - 3 &= -\tfrac{1}{8}(x - 2) \qquad (x_1, y_1) = (2, 3) \\ 8y - 24 &= -x + 2 \qquad \text{multiplying by 8} \\ x + 8y - 26 &= 0 \end{aligned}$$

15. $m = \tan 45° = 1$

$$\begin{aligned} y - 10 &= 1(x - 0) \qquad (x_1, y_1) = (0, 10);\ m = 1 \\ x - y + 10 &= 0 \end{aligned}$$

17. $(x_1, y_1) = (0, -2),\ m = -\dfrac{1}{3}$

$$\begin{aligned} y + 2 &= -\tfrac{1}{3}(x - 0) \\ 3y + 6 &= -x \\ x + 3y + 6 &= 0 \end{aligned}$$

19. $$\begin{aligned} 6x + 2y &= 5 \\ 2y &= -6x + 5 \\ y &= -3x + \tfrac{5}{2} \qquad y = mx + b \end{aligned}$$

   $m = -3$, y-intercept $= \frac{5}{2}$; see graph in answer section of book.

21. Since $2x = 3y$, $y = \frac{2}{3}x$. From the form $y = mx + b$, $m = \frac{2}{3}$ and $b = 0$. The line passes through the origin and has slope $\frac{2}{3}$. See graph in answer section of book.

23. $$\begin{aligned} 2y - 7 &= 0 \\ y &= 0x + \tfrac{7}{2} \qquad y = mx + b \end{aligned}$$

   $m = 0$, y-intercept $= \frac{7}{2}$; see graph in answer section of book.

25. $2x - 3y = 1$ $4x - 6y + 3 = 0$

$-3y = -2x + 1$ $-6y = -4x - 3$

$y = \frac{2}{3}x - \frac{1}{3}$ $y = \frac{4}{6}x + \frac{3}{6}$

$y = \frac{2}{3}x + \frac{1}{2}$

From the form $y = mx + b$, $m = \frac{2}{3}$ in both cases, so that the lines are parallel.

27. $3x - 4y = 1$ $3y - 4x = 3$

$-4y = -3x + 1$ $3y = 4x + 3$

$y = \frac{3}{4}x - \frac{1}{4}$ $y = \frac{4}{3}x + 1$

The lines are neither parallel nor perpendicular.

29. $x + 3y = 5$ $y - 3x - 2 = 0$

$3y = -x + 5$ $y = 3x + 2$

$y = -\frac{1}{3}x + \frac{5}{3}$

The slopes are $-\frac{1}{3}$ and 3, respectively. Since the slopes are negative reciprocals, the lines are perpendicular.

31. $3x - 4y = 7$

$-4y = -3x + 7$

$y = \frac{3}{4}x - \frac{7}{4}$ slope $= 3/4$

$y - 1 = \frac{3}{4}(x + 1)$ $y - y_1 = m(x - x_1);\ (x_1, y_1) = (-1, 1)$

$4y - 4 = 3x + 3$

$3x - 4y + 7 = 0$

33. To find the coordinates of the point of intersection, solve the equations simultaneously:

$2x - 4y = 1$

$\underline{3x + 4y = 4}$

$5x = 5$ adding

$x = 1$

From the second equation, $3(1) + 4y = 4$, and $y = \frac{1}{4}$. So the point of intersection is $(1, \frac{1}{4})$.

From the equation $5x + 7y + 3 = 0$, we get

$7y = -5x - 3$

$y = -\frac{5}{7}x - \frac{3}{7}$ slope$= -5/7$

Thus $(x_1, y_1) = (1, \frac{1}{4})$ and $m = -\frac{5}{7}$. The desired line is $y - \frac{1}{4} = -\frac{5}{7}(x - 1)$. To clear fractions, we multiply both sides by 28:

$28y - 7 = -20(x - 1)$

$28y - 7 = -20x + 20$

$20x + 28y - 27 = 0$

35. See graph in answer section of book.

37. From $F = kx$, we get $3 = k \cdot \frac{1}{2}$. Thus $k = 6$ and $F = 6x$.

39. $$\begin{aligned} F &= mC + b \\ 212 &= m(100) + b \qquad &F = 212, \; C = 100 \\ 32 &= m(0) + b \qquad &F = 32, \; C = 0 \\ \hline b &= 32 \qquad &\text{second equation} \\ 212 &= m(100) + 32 \qquad &\text{substituting into first equation} \\ \hline \end{aligned}$$

$$m = \frac{180}{100} = \frac{9}{5}$$

Solution: $F = \frac{9}{5}C + 32$

41. $$\begin{aligned} R &= aT + b \\ 51 &= a \cdot 100 + b \qquad &R = 51, \; T = 100 \\ 54 &= a \cdot 400 + b \qquad &R = 54, \; T = 400 \\ \hline -3 &= -300a \qquad &\text{subtracting} \\ a &= \frac{-3}{-300} = 0.01 \end{aligned}$$

From the first equation, $51 = a \cdot 100 + b$, we get

$$\begin{aligned} 51 &= (0.01)(100) + b \qquad (a = 0.01) \\ b &= 50 \end{aligned}$$

So the formula $R = aT + b$ becomes $R = 0.01T + 50$.

1.4   Curve Sketching

3. <u>Intercepts.</u> If $x = 0$, then $y = -9$. If $y = 0$, then

$$\begin{aligned} 0 &= x^2 - 9 \\ x^2 &= 9 \qquad &\text{solving for } x \\ x &= \pm 3 \qquad &x = 3 \text{ and } x = -3. \end{aligned}$$

<u>Symmetry.</u> If x is replaced by $-x$, we get $y = (-x)^2 - 9$, which reduces to the given equation $y = x^2 - 9$. The graph is therefore symmetric with respect to the y-axis. There is no other type of symmetry.

<u>Asymptotes.</u> Since the equation is not in the form of a quotient with a variable in the denominator, there are no asymptotes.

<u>Extent.</u> y is defined for all x.

<u>Graph.</u>

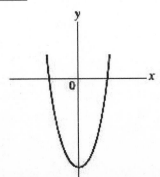

5. <u>Intercepts.</u> If $x = 0$, then $y = 1$. If $y = 0$, then

$$\begin{aligned} 0 &= 1 - x^2 \\ x^2 &= 1 \qquad\qquad \text{solving for } x \\ x &= \pm 1 \qquad\qquad x = 1 \text{ and } x = -1. \end{aligned}$$

<u>Symmetry.</u> If x is replaced by $-x$, we get $y = 1 - (-x)^2$, which reduces to the given equation $y = 1 - x^2$. The graph is therefore symmetric with respect to the y-axis. There is no other type of symmetry.

<u>Asymptotes.</u> Since the equation is not in the form of a quotient with a variable in the denominator, there are no asymptotes.

<u>Extent.</u> y is defined for all x.

<u>Graph.</u>

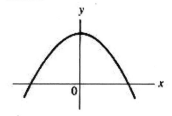

7. <u>Intercepts.</u> If $x = 0$, then $y = 4$. If $y = 0$, then

$$\begin{aligned} 0 &= 4 - 2x^2 \\ x^2 &= 2 \qquad\qquad \text{solving for } x \\ x &= \pm\sqrt{2} \qquad\qquad x = \sqrt{2} \text{ and } x = -\sqrt{2}. \end{aligned}$$

<u>Symmetry.</u> If x is replaced by $-x$, we get $y = 4 - 2(-x)^2$, which reduces to $y = 4 - 2x^2$. The graph is therefore symmetric with respect to the y-axis. There is no other type of symmetry.

<u>Asymptotes.</u> Since the equation is not in the form of a quotient with a variable in the denominator, there are no asymptotes.

<u>Extent.</u> y is defined for all x.

<u>Graph.</u>

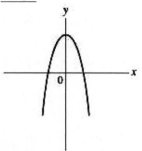

9. <u>Intercepts.</u> If $x = 0$, then $y = 0$, and if $y = 0$, then $x = 0$. So the only intercept is the origin.

<u>Symmetry.</u> If we replace x by $-x$, we get $y^2 = -x$, which does not reduce to the given equation. So there is no symmetry with respect to the y-axis.

If y is replaced by $-y$, we get $(-y)^2 = x$, which reduces to $y^2 = x$, the given equation. It follows that the graph is symmetric with respect to the x-axis.

To check for symmetry with respect to the origin, we replace x by $-x$ and y by $-y$: $(-y)^2 = -x$.

The resulting equation, $y^2 = -x$, does not reduce to the given equation. So there is no symmetry with respect to the origin.

Asymptotes. Since the equation is not in the form of a fraction with a variable in the denominator, there are no asymptotes.

Extent. Solving the equation for y in terms of x, we get

$$y = \pm\sqrt{x}.$$

Note that to avoid imaginary values, x cannot be negative. It follows that the extent is $x \geq 0$.

Graph.

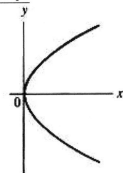

11. Intercepts. If $x = 0$, then $y = \pm 1$. If $y = 0$, then $x = -1$.

Symmetry. If we replace x by $-x$ we get $y^2 = -x + 1$, which does not reduce to the given equation. So there is no symmetry with respect to the y-axis.

If y is replaced by $-y$, we get $(-y)^2 = x + 1$, which reduces to $y^2 = x + 1$, the given equation. It follows that the graph is symmetric with respect to the x-axis.

The graph is not symmetric with respect to the origin.

Asymptotes. Since the equation is not in the form of a fraction with a variable in the denominator, there are no asymptotes.

Extent. Solving the equation for y, we get $y = \pm\sqrt{x + 1}$. To avoid imaginary values, we must have $x + 1 \geq 0$ or $x \geq -1$. Therefore the extent is $x \geq -1$.

Graph.

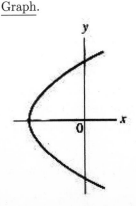

13. <u>Intercepts</u>. If $x = 0$, then $y = (0 - 3)(0 + 5) = -15$. If $y = 0$, then

$$0 = (x - 3)(x + 5)$$
$$x - 3 = 0 \qquad x + 5 = 0$$
$$x = 3 \qquad x = -5.$$

<u>Symmetry</u>. If x is replaced by $-x$, we get $y = (-x - 3)(-x + 5)$, which does not reduce to the given equation. So there is no symmetry with respect to the y-axis. Similarly, there is no other type of symmetry.

<u>Asymptotes</u>. Since the equation is not in the form of a quotient with a variable in the denominator, there are no asymptotes.

<u>Extent</u>. y is defined for all x.

<u>Graph</u>.

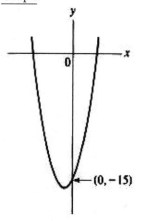

15. <u>Intercepts</u>. If $x = 0$, then $y = 0$. If $y = 0$, then

$$0 = x(x + 3)(x - 2)$$
$$x = 0, -3, 2.$$

<u>Symmetry</u>. If x is replaced by $-x$, we get $y = -x(-x + 3)(-x - 2)$, which does not reduce to the given equation. So the graph is not symmetric with respect to the y-axis. There is no other type of symmetry.

<u>Asymptotes</u>. Since the equation is not in the form of a quotient with a variable in the denominator, there are no asymptotes.

<u>Extent</u>. y is defined for all x.

<u>Graph</u>.

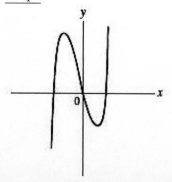

17. <u>Intercepts.</u> If $x = 0$, $y = 0$; if $y = 0$, then

$$
\begin{aligned}
x(x-1)(x-2)^2 &= 0 \\
x &= 0, 1, 2.
\end{aligned}
$$

<u>Symmetry.</u> If x is replaced by $-x$, we get $y = -x(-x-1)(-x-2)^2$, which does not reduce to the given equation. So there is no symmetry with respect to the y-axis. Similarly, there is no other type of symmetry.

<u>Asymptotes.</u> None (the equation does not have the form of a fraction).

<u>Extent.</u> y is defined for all x.

<u>Graph.</u>

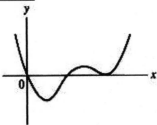

19. <u>Intercepts.</u> If $x = 0$, then $y = 0$. If $y = 0$, then

$$
\begin{aligned}
0 &= x(x-1)^2(x-2) \\
x &= 0, 1, 2.
\end{aligned}
$$

<u>Symmetry.</u> If x is replaced with $-x$, we get $y = -x(-x-1)^2(-x-2)$, which does not reduce to the given equation. Therefore there is no symmetry with respect to the y-axis. There is no other type of symmetry.

<u>Asymptotes.</u> None (the equation does not have the form of a fraction).

<u>Extent.</u> y is defined for all x.

<u>Graph.</u>

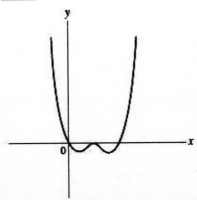

21. <u>Intercepts.</u> If $x = 0$, $y = 1$; if $y = 0$, we have

$$0 = \frac{2}{x+2}.$$

This equation has no solution.

<u>Symmetry.</u> Replacing x by $-x$, we get

$$y = \frac{2}{-x+2}$$

which does not reduce to the given equation. So there is no symmetry with respect to the y-axis. Similarly, there is no other type of symmetry.

<u>Asymptotes.</u> Setting the denominator equal to 0, we get

$$x + 2 = 0 \text{ or } x = -2.$$

It follows that $x = -2$ is a vertical asymptote. Also, as x gets large, y approaches 0. So the x-axis is a horizontal asymptote.

<u>Extent.</u> To avoid division by 0, x cannot be equal to -2. So the extent is all x except $x = -2$.

<u>Graph.</u>

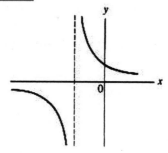

23. <u>Intercepts.</u> If $x = 0$, then $y = 2$. If $y = 0$, then

$$\frac{2}{(x-1)^2} = 0.$$

This equation has no solution.

<u>Symmetry.</u> Replacing x by $-x$, we get

$$y = \frac{2}{(-x-1)^2}$$

which does not reduce to the given equation. There are no other types of symmetry.

<u>Asymptotes.</u> Setting the denominator equal to 0 gives

$$(x-1)^2 = 0 \text{ or } x = 1.$$

It follows that $x = 1$ is a vertical asymptote. Also, as x gets large, y approaches 0. So the x-axis is a horizontal asymptote.

<u>Extent.</u> To avoid division by 0, x cannot be equal to 1. So the extent is the set of all x except $x = 1$.

Graph.

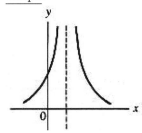

25. Intercepts. If $x = 0$, then $y = 0$. If $y = 0$, then

$$0 = \frac{x^2}{x - 1}.$$

The only solution is $x = 0$.

Symmetry. Replacing x by $-x$ yields

$$y = \frac{(-x)^2}{-x - 1} = \frac{x^2}{-x - 1}$$

which is not the same as the given equation. So the graph is not symmetric with respect to the y-axis. Replacing y by $-y$, we have

$$-y = \frac{x^2}{x - 1}$$

which does not reduce to the given equation. So the graph is not symmetric with respect to the x-axis.

Similarly, there is no symmetry with respect to the origin.

Asymptotes. Setting the denominator equal to 0, we get $x - 1 = 0$, or $x = 1$. So $x = 1$ is a vertical asymptote. There are no horizontal asymptotes.

(Observation: for very large x the 1 in the denominator becomes insignificant. So the graph gets ever closer to $y = \dfrac{x^2}{x} = x$; the line $y = x$ is a slant asymptote.)

Extent. To avoid division by 0, x cannot be equal to 1. So the extent is all x except $x = 1$.

Graph.

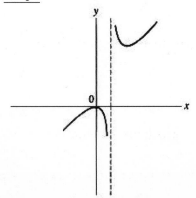

27. <u>Intercepts</u>. If $x = 0$, then $y = -1/2$. If $y = 0$, then

$$0 = \frac{x+1}{(x-1)(x+2)}.$$

The only solution is $x = -1$.

<u>Symmetry</u>. Replacing x by $-x$ yields

$$y = \frac{-x+1}{(-x-1)(-x+2)}$$

which is not the same as the given equation. There are no types of symmetry.

<u>Asymptotes</u>. Setting the denominator equal to 0, we get $(x-1)(x+2) = 0$. So $x = 1$ and $x = -2$ are the vertical asymptotes. As x gets large, y approaches 0, so the x-axis is a horizontal asymptote.

<u>Extent</u>. To avoid division by 0, the extent is all x except $x = 1$ and $x = -2$.

<u>Graph</u>.

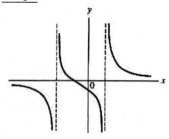

29. <u>Intercepts</u>. If $x = 0$, $y = 0$; if $y = 0$, we have

$$\begin{aligned} x\sqrt{1-x^2} &= 0 \\ x &= 0, \pm 1. \end{aligned}$$

<u>Symmetry</u>. If x and y are replaced by $-x$ and $-y$ respectively, we get

$$-y = -x\sqrt{1-(-x)^2},$$

which reduces to

$$y = x\sqrt{1-x^2}.$$

The graph is therefore symmetric with respect to the origin.

<u>Asymptotes</u>. None (no fractions).

<u>Extent</u>. To avoid imaginary values, the radicand has to be greater than or equal to 0:

$$\begin{aligned} 1 - x^2 &\geq 0 \\ 1 &\geq x^2 \qquad \text{adding } x^2 \text{ to both sides} \\ x^2 &\leq 1 \\ -1 &\leq x \leq 1. \end{aligned}$$

<u>Graph</u>.

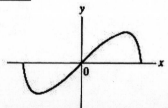

31. <u>Intercepts</u>. If $x = 0$, then $y = 4$. If $y = 0$, then

$$\begin{aligned}
\frac{x^2 - 4}{x^2 - 1} &= 0 \\
x^2 - 4 &= 0 && \text{multiplying by } x^2 - 1 \\
x &= \pm 2. && \text{solution}
\end{aligned}$$

<u>Symmetry</u>. Replacing x by $-x$ reduces to the given equation. So there is symmetry with respect to the y-axis. There is no other type of symmetry.

<u>Asymptotes</u>. Vertical: setting the denominator equal to 0, we have

$$x^2 - 1 = 0 \text{ or } x = \pm 1.$$

Horizontal: dividing numerator and denominator by x^2, the equation becomes

$$y = \frac{1 - \frac{4}{x^2}}{1 - \frac{1}{x^2}}.$$

As x gets large, y approaches 1. So $y = 1$ is a horizontal asymptote.

<u>Extent</u>. All x except $x = \pm 1$ (to avoid division by 0).

<u>Graph</u>.

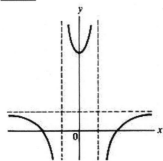

33. <u>Intercepts</u>. If $x = 0$, $y^2 = (-3)(5) = -15$, or $y = \pm\sqrt{15}\,j$, which is a pure imaginary number. If $y = 0$,

$$\begin{aligned}
(x - 3)(x + 5) &= 0 \\
x &= 3, -5.
\end{aligned}$$

<u>Symmetry</u>. Replacing y by $-y$, we get $(-y)^2 = (x - 3)(x + 5)$, which reduces to the given equation. Hence the graph is symmetric with respect to the x-axis.

<u>Asymptotes</u>. None (no fractions).

<u>Extent</u>. From $y = \pm\sqrt{(x - 3)(x + 5)}$, we conclude that $(x - 3)(x + 5) \geq 0$. If $x \geq 3$, $(x - 3)(x + 5) \geq 0$. If $x \leq -5$, $(x - 3)(x + 5) \geq 0$, since both factors are negative (or zero). If $-5 < x < 3$, $(x - 3)(x + 5) < 0$. [For example, if $x = 0$, we get $(-3)(5) = -15$.] These observations are summarized in the following chart.

	test values	$x - 3$	$x + 5$	$(x - 3)(x + 5)$
$x > 3$	4	+	+	+
$-5 < x < 3$	0	−	+	−
$x < -5$	−6	−	−	+

Extent: $x \leq -5$, $x \geq 3$

Graph.

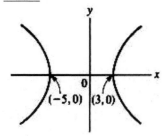

35. Intercepts. If $x = 0$, $y = 0$; if $y = 0$, $x = 0$.

Symmetry. Replacing y by $-y$ leaves the equation unchanged. So there is symmetry with respect to the x-axis. There is no other type of symmetry.

Asymptotes. Vertical: setting the denominator equal to 0, we get

$$(x - 3)(x - 2) = 0 \text{ or } x = 3, 2.$$

Horizontal: as x gets large, y approaches 0 (x-axis).

Extent. From

$$y = \pm\sqrt{\frac{x}{(x - 3)(x - 2)}}$$

we conclude that

$$\frac{x}{(x - 3)(x - 3)} \geq 0.$$

Since signs change only at $x = 0, 2$ and 3, we need to use "test values" between these points. The results are summarized in the following chart.

	test values	x	$x - 2$	$x - 3$	$\dfrac{x}{(x - 3)(x - 2)}$
$x < 0$	-1	$-$	$-$	$-$	$-$
$0 < x < 2$	1	$+$	$-$	$-$	$+$
$2 < x < 3$	$5/2$	$+$	$+$	$-$	$-$
$x > 3$	4	$+$	$+$	$+$	$+$

So the inequality is satisfied for $0 < x < 2$ and $x > 3$. In addition, $y = 0$ when $x = 0$. So the extent is $0 \leq x < 2$ and $x > 3$.

Graph.

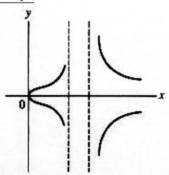

37. <u>Intercepts</u>. If $x = 0$, then $y^2 = \frac{-4}{-1} = 4$, or $y = \pm 2$. If $y = 0$, then

$$0 = \frac{x^2 - 4}{x^2 - 1}$$

which is possible only if $x^2 - 4 = 0$, or $x = \pm 2$.

<u>Symmetry</u>. The even powers on x and y tell us that if x is replaced by $-x$ and y is replaced by $-y$, the resulting equation will reduce to the given equation. The graph is therefore symmetric with respect to both axes and the origin.

<u>Asymptotes</u>. Vertical: setting the denominator equal to 0, we get

$$x^2 - 1 = 0 \text{ or } x = \pm 1.$$

Horizontal: dividing numerator and denominator by x^2, we get

$$y^2 = \frac{1 - \frac{4}{x^2}}{1 - \frac{1}{x^2}}.$$

The right side approaches 1 as x gets large. Thus y^2 approaches 1, so that $y = \pm 1$ are the horizontal asymptotes.

<u>Extent</u>. From

$$y = \pm\sqrt{\frac{x^2 - 4}{x^2 - 1}}$$

we conclude that

$$\frac{x^2 - 4}{x^2 - 1} = \frac{(x - 2)(x + 2)}{(x - 1)(x + 1)} \geq 0.$$

Since the signs change only at $x = 2, -2, 1$, and -1, we need to use arbitrary "test values" between these points. The results are summarized in the following chart.

	test values	$x - 2$	$x - 1$	$x + 1$	$x + 2$	$\dfrac{(x-2)(x+2)}{(x-1)(x+1)}$
$x > 2$	3	+	+	+	+	+
$1 < x < 2$	3/2	−	+	+	+	−
$-1 < x < 1$	0	−	−	+	+	+
$-2 < x < -1$	−3/2	−	−	−	+	−
$x < -2$	−3	−	−	−	−	+

Note that the fraction is positive only when $x > 2$, $-1 < x < 1$ and $x < -2$. Since $y = 0$ when $x = \pm 2$, the extent is $x \geq 2$, $-1 < x < 1$, $x \leq -2$.

<u>Graph</u>.

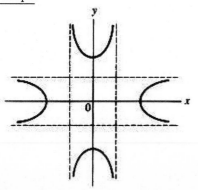

39. $C = \dfrac{10^{-2}C_1}{C_1 + 10^{-2}}$, $C_1 \geq 0$

   The only intercept is the origin. Dividing numerator and denominator by C_1, the equation becomes

   $$C = \frac{10^{-2}}{1 + 10^{-2}/C_1}.$$

   As C_1 gets large, C approaches 10^{-2}; so $C = 10^{-2}$ is a horizontal asymptote.

   See graph in answer section of book.

41. <u>Intercepts</u>. If $t = 0$, $S = 0$; if $S = 0$, we get

   $$\begin{aligned} 0 &= 60t - 5t^2 \\ 0 &= 5t(12 - t) \end{aligned}$$

   or $t = 0, 12$.

   <u>Symmetry</u>. None.

   <u>Asymptotes</u>. None.

   <u>Extent</u>. $t \geq 0$ by assumption.

   <u>Graph</u>. See graph in answer section of book.

43. <u>Extent</u> $L \geq 0$.

   See graph in answer section of book.

1.5   Discussion of Curves with Graphing Utilities

1. If $y = 0$, then

   $$x^2(x - 1)(x - 2) = 0.$$

   Setting each factor equal to 0, we get

   $$x = 0, 1, 2.$$

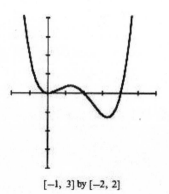

[−1, 3] by [−2, 2]

5. $$\begin{aligned} x^4 - 2x^3 &= 0 \\ x^3(x - 2) &= 0 \\ x &= 0, 2 \end{aligned}$$

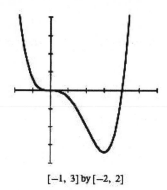

[-1, 3] by [-2, 2]

9. Domain: $x \geq 0$ (to avoid imaginary values).

   Vertical asymptotes: None. (The denominator is always positive, that is, $1 + \sqrt{x} \neq 0$.)

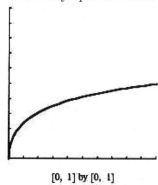

[0, 1] by [0, 1]

13. To find the vertical asymptotes, we set the denominator equal to 0:

$$2x^2 - 3 = 0$$
$$2x^2 = 3$$
$$x^2 = \frac{3}{2} \cdot \frac{2}{2} = \frac{6}{4}$$
$$x = \pm\frac{\sqrt{6}}{2}.$$

   Domain: y is defined for all x except $x = \pm\dfrac{\sqrt{6}}{2}$.

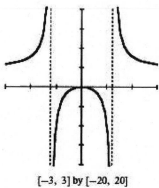

[-3, 3] by [-20, 20]

17. See graph in answer section of book.

21. See graph in answer section of book.

1.7   The Circle

1. Since $(h, k) = (0, 0)$ and $r = 5$, we get from the form

$$(x - h)^2 + (y - k)^2 = r^2$$

the equation

$$x^2 + y^2 = 25.$$

3. The radius of the circle is the distance from the origin to $(-6, 8)$. Hence $r^2 = (0+6)^2 + (0-8)^2 = 100$. From the standard form of the equation of the circle we get

$$\begin{aligned}(x - 0)^2 + (y - 0)^2 &= 100 \qquad \text{center: } (0, 0) \\ x^2 + y^2 &= 100.\end{aligned}$$

5. $$\begin{aligned}(x - h)^2 + (y - k)^2 &= r^2 \\ (x + 2)^2 + (y - 5)^2 &= 1^2 \\ x^2 + y^2 + 4x - 10y + 28 &= 0\end{aligned}$$

7. The radius is the distance from $(-1, -4)$ to the origin:

$$r^2 = (-1 - 0)^2 + (-4 - 0)^2 = 1 + 16 = 17.$$

Hence,

$$\begin{aligned}(x + 1)^2 + (y + 4)^2 &= 17 \qquad (x - h)^2 + (y - k)^2 = r^2 \\ x^2 + 2x + 1 + y^2 + 8y + 16 &= 17 \\ x^2 + y^2 + 2x + 8y &= 0.\end{aligned}$$

9. Diameter: distance from $(-2, -6)$ to $(1, 5)$. Hence

$$r = \frac{1}{2}\sqrt{(-2 - 1)^2 + (-6 - 5)^2} = \frac{1}{2}\sqrt{9 + 121} = \frac{1}{2}\sqrt{130}$$

and thus,

$$r^2 = \frac{1}{4}(130) = \frac{65}{2}.$$

Center: midpoint of the line segment, whose coordinates are

$$\left(\frac{-2 + 1}{2}, \frac{-6 + 5}{2}\right) = \left(-\frac{1}{2}, -\frac{1}{2}\right).$$

Thus

$$\begin{aligned}(x + \tfrac{1}{2})^2 + (y + \tfrac{1}{2})^2 &= \tfrac{65}{2} \\ x^2 + x + \tfrac{1}{4} + y^2 + y + \tfrac{1}{4} &= \tfrac{65}{2} \\ x^2 + y^2 + x + y - 32 &= 0.\end{aligned}$$

11. $$\begin{aligned}x^2 + y^2 - 2x - 2y - 2 &= 0 \\ x^2 - 2x + y^2 - 2y &= 2\end{aligned}$$

We now add to each side the square of one-half the coefficient of x:

$$\begin{aligned}\left[\tfrac{1}{2}(-2)\right]^2 &= 1 \\ x^2 - 2x + \underline{1} + y^2 - 2y &= 2 + \underline{1}.\end{aligned}$$

Similarly, we add 1 (the square of one-half the coefficient of y):

$$(x^2 - 2x + 1) + (y^2 - 2y + \underline{1}) \;=\; 2 + 1 + \underline{1}$$
$$(x - 1)^2 + (y - 1)^2 \;=\; 4.$$

Center: $(h, k) = (1, 1)$; radius: $\sqrt{4} = 2$.

13. $x^2 + y^2 + 4x - 8y + 4 \;=\; 0$

  $x^2 + 4x \;\; + y^2 - 8y \;\;=\; -4$

Since

$$\left(\frac{1}{2} \cdot 4 \right)^2 = 4 \quad \text{and} \quad \left[\frac{1}{2}(-8) \right]^2 = 16$$

we get

$$(x^2 + 4x + \underline{4}) + (y^2 - 8y + \underline{16}) \;=\; -4 + \underline{4} + \underline{16}$$
$$(x + 2)^2 + (y - 4)^2 \;=\; 16.$$

The equation can be written

$$[x - (-2)]^2 + (y - 4)^2 = 4^2.$$

It follows that

$$(h, k) = (-2, 4) \text{ and } r = 4.$$

15. $x^2 + y^2 + 4x + 2y + 2 \;=\; 0$

  $x^2 + 4x + y^2 + 2y \;=\; -2$

We add to each side the square of one-half the coefficient of x: $\left(\dfrac{1}{2} \cdot 4 \right)^2 = 4$ to get

$$x^2 + 4x + \underline{4} + y^2 + 2y = -2 + \underline{4}.$$

Similarly, we add to each side the square of one-half the coefficient of y: $\left(\dfrac{1}{2} \cdot 2 \right)^2 = 1$ to get

$$x^2 + 4x + 4 + y^2 + 2y + \underline{1} \;=\; -2 + 4 + \underline{1}$$
$$(x + 2)^2 + (y + 1)^2 \;=\; 3.$$

Center: $(-2, -1)$; radius: $\sqrt{3}$.

17. $x^2 + y^2 - 4x + y + \frac{9}{4} \;=\; 0$

  $x^2 - 4x + y^2 + y \;=\; -\frac{9}{4}$

We add to each side the square of one-half the coefficient of x: $\left[\dfrac{1}{2}(-4) \right]^2 = 4$. This gives

$$x^2 - 4x + \underline{4} + y^2 + y \;=\; -\frac{9}{4} + \underline{4}.$$

Similarly, we add the square of one-half the coefficient of y: $\left[\dfrac{1}{2} \cdot 1 \right]^2 = \dfrac{1}{4}$. This gives

$$x^2 - 4x + 4 + y^2 + y + \tfrac{1}{4} \;=\; -\tfrac{9}{4} + 4 + \tfrac{1}{4}$$
$$(x - 2)^2 + (y + \tfrac{1}{2})^2 \;=\; -\tfrac{8}{4} + 4$$
$$(x - 2)^2 + (y + \tfrac{1}{2})^2 \;=\; 2.$$

Center: $(2, -\frac{1}{2})$; radius: $\sqrt{2}$.

19. $4x^2 + 4y^2 - 8x - 12y + 9 = 0$

$$x^2 + y^2 - 2x - 3y + \tfrac{9}{4} = 0 \qquad \text{dividing by 4}$$

$$x^2 - 2x + y^2 - 3y = -\tfrac{9}{4}$$

Add to each side: $\left[\tfrac{1}{2}(-2)\right]^2 = 1$ and $\left[\tfrac{1}{2}(-3)\right]^2 = \tfrac{9}{4}$ to get

$$(x^2 - 2x + 1) + (y^2 - 3y + \tfrac{9}{4}) = -\tfrac{9}{4} + 1 + \tfrac{9}{4}$$

$$(x - 1)^2 + (y - \tfrac{3}{2})^2 = 1.$$

Center: $(1, \tfrac{3}{2})$; radius: 1.

21. $x^2 + y^2 + 4x - 2y - 4 = 0$

$$x^2 + 4x + y^2 - 2y = 4$$

Note that

$$\left(\frac{1}{2} \cdot 4\right)^2 = 4 \text{ and } \left[\frac{1}{2}(-2)\right]^2 = 1.$$

Adding 4 and 1, respectively, we get

$$(x^2 + 4x + 4) + (y^2 - 2y + 1) = 4 + 4 + 1$$

$$(x + 2)^2 + (y - 1)^2 = 9.$$

The equation can be written

$$[x - (-2)]^2 + (y - 1)^2 = 3^2.$$

So the center is $(-2, 1)$ and the radius is 3.

23. $x^2 + y^2 - x - 2y + \tfrac{1}{4} = 0$

$$x^2 - x + y^2 - 2y = -\tfrac{1}{4}$$

Add to each side: $\left[\tfrac{1}{2}(-1)\right]^2 = \tfrac{1}{4}$ and $\left[\tfrac{1}{2}(-2)\right]^2 = 1$ to get

$$(x^2 - x + \tfrac{1}{4}) + (y^2 - 2y + 1) = -\tfrac{1}{4} + \tfrac{1}{4} + 1$$

$$(x - \tfrac{1}{2})^2 + (y - 1)^2 = 1.$$

Center: $(\tfrac{1}{2}, 1)$; radius: 1.

25. $x^2 + y^2 - 4x + y + \tfrac{9}{4} = 0$

$$x^2 - 4x + y^2 + y = -\tfrac{9}{4}$$

Note that

$$\left[\frac{1}{2}(-4)\right]^2 = 4 \text{ and } \left(\frac{1}{2} \cdot 1\right)^2 = \frac{1}{4}.$$

Adding 4 and $\tfrac{1}{4}$, respectively, we get

$$(x^2 - 4x + 4) + (y^2 + y + \tfrac{1}{4}) = -\tfrac{9}{4} + 4 + \tfrac{1}{4}$$

$$(x - 2)^2 + (y + \tfrac{1}{2})^2 = 2.$$

The equation can be written

$$(x - 2)^2 + \left[y - \left(-\frac{1}{2}\right)\right]^2 = (\sqrt{2})^2.$$

Center: $(2, -\tfrac{1}{2})$; radius: $\sqrt{2}$.

27. $\quad 4x^2 + 4y^2 + 12x + 16y + 5 = 0$

$\qquad\quad x^2 + y^2 + 3x + 4y + \frac{5}{4} = 0$

$\qquad\quad x^2 + 3x \;\; + y^2 + 4y = -\frac{5}{4}$

Add to each side: $\left[\frac{1}{2} \cdot 3\right]^2 = \frac{9}{4}$ and $\left(\frac{1}{2} \cdot 4\right)^2 = 4$ to get

$\qquad x^2 + 3x + \frac{9}{4} + y^2 + 4y + 4 = -\frac{5}{4} + \frac{9}{4} + 4$

$\qquad\qquad (x + \frac{3}{2})^2 + (y + 2)^2 = 5.$

Center: $(-\frac{3}{2}, -2)$; radius: $\sqrt{5}$.

29. $\quad 4x^2 + 4y^2 - 20x - 4y + 26 = 0$

$\qquad\quad x^2 + y^2 - 5x - y + \frac{26}{4} = 0 \qquad\qquad$ dividing by 4

$\qquad\quad x^2 - 5x \;\; + y^2 - y = -\frac{26}{4}$

$\qquad x^2 - 5x + \frac{25}{4} + y^2 - y + \frac{1}{4} = -\frac{26}{4} + \frac{25}{4} + \frac{1}{4}$

$\qquad\qquad (x - \frac{5}{2})^2 + (y - \frac{1}{2})^2 = 0$

Locus is the single point $(\frac{5}{2}, \frac{1}{2})$.

31. $\qquad\qquad x^2 + y^2 - 6x + 8y + 25 = 0$

$\qquad\qquad x^2 - 6x \;\; + y^2 + 8y = -25$

$\qquad (x^2 - 6x + 9) + (y^2 + 8y + 16) = -25 + 9 + 16$

$\qquad\qquad\qquad (x - 3)^2 + (y + 4)^2 = 0$

Locus is the single point $(3, -4)$.

33. $\qquad\quad x^2 + y^2 - 6x - 8y + 30 = 0$

$\qquad\qquad x^2 - 6x \;\; + y^2 - 8y = -30$

$\qquad x^2 - 6x + 9 + y^2 - 8y + 16 = -30 + 9 + 16$

$\qquad\qquad (x - 3)^2 + (y - 4)^2 = -5 \qquad\qquad$ (imaginary circle)

35. $x^2 + y^2 = (2.00)^2 = 4.00;\; x^2 + y^2 = (3.40)^2 = 11.6$

37. The radius is $22,300 + 4000 = 26,300\,\text{mi}$.

1.8 The Parabola

1. Since the focus is on the x-axis, the form is $y^2 = 4px$. Since the focus is at $(3, 0)$, $p = 3$ (positive). Thus $y^2 = 4(3)x$, or $y^2 = 12x$.

3. Since the focus is on the y-axis, the form is $x^2 = 4py$. Since the focus is at $(0, -5)$, $p = -5$ (negative). Thus $x^2 = 4(-5)y$, or $x^2 = -20y$.

5. Since the focus is on the x-axis, the form is $y^2 = 4px$. The focus is on the left side of the origin, at $(-4, 0)$. So $p = -4$ (negative). It follows that $y^2 = 4(-4)x$, or $y^2 = -16x$.

7. Since the directrix is $x = -1$, the focus is at $(1, 0)$. So the form is $y^2 = 4px$ with $p = 1$, and the equation is $y^2 = 4x$.

9. Since the directrix is $x = 2$, the focus is at $(-2, 0)$. So the form is $y^2 = 4px$ with $p = -2$. Thus $y^2 = -8x$.

11. Form: $y^2 = 4px$. Substituting the coordinates of the point $(-2, -4)$, we get

$$(-4)^2 = 4p(-2) \quad \text{and} \quad 4p = -8.$$

Thus $y^2 = -8x$.

13. The form is either

$$y^2 = 4px \text{ or } x^2 = 4py.$$

Substituting the coordinates of the point $(1, 1)$, we get

$$1^2 = 4p \cdot 1 \text{ or } 1^2 = 4p \cdot 1.$$

In either case, $p = \frac{1}{4}$. So the equations are $y^2 = x$ and $x^2 = y$.

15. From $x^2 = 8y$, we have $x^2 = 4(2y)$. Thus $p = 2$ and the focus is at $(0, 2)$.

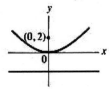

17. From $x^2 = -12y$, we have $x^2 = 4(-3)y$. So $p = -3$ and the focus is at $(0, -3)$.

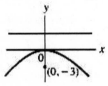

19. $y^2 = 16x = 4(4)x$; $p = 4$ and the focus is at $(4, 0)$.

21. From $y^2 = -4x$, $y^2 = 4(-1)x$. So $p = -1$ and the focus is at $(-1, 0)$.

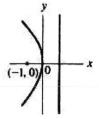

23. $x^2 = 4y = 4(1)y$; $p = 1$ and the focus is at $(0, 1)$.

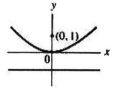

25. From $y^2 = 9x$, $y^2 = 4(\frac{9}{4})x$ (inserting 4). So $p = \frac{9}{4}$ and the focus is at $(\frac{9}{4}, 0)$.

27. $y^2 = -x = 4(-\frac{1}{4})x$; $p = -\frac{1}{4}$ and the focus is at $(-\frac{1}{4}, 0)$.

29. $\begin{aligned}
3y^2 + 2x &= 0 \\
y^2 &= -\frac{2}{3}x \\
y^2 &= 4(-\frac{2}{3} \cdot \frac{1}{4})x \\
y^2 &= 4(-\frac{1}{6})x
\end{aligned}$

So the focus is at $(-\frac{1}{6}, 0)$.

31. $x^2 = 4(3)y$; $p = 3$. Focus: $(0, 3)$; directrix: $y = -3$.

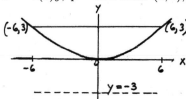

Observe that the points $(6, 3)$ and $(-6, 3)$ lie on the curve because the distance to the focus must be equal to the distance to the directrix.

$\begin{aligned}
\text{Circle:}\quad (x - h)^2 + (y - k)^2 &= r^2 \\
(x - 0)^2 + (y - 3)^2 &= 6^2 \\
x^2 + y^2 - 6y + 9 &= 36 \\
x^2 + y^2 - 6y - 27 &= 0.
\end{aligned}$

33. We need to find the locus of points (x, y) equidistant from $(4, 1)$ and the y-axis. Since the distance from (x, y) to the y-axis is x units, we get

$\begin{aligned}
\sqrt{(x - 4)^2 + (y - 1)^2} &= x \\
(x - 4)^2 + (y - 1)^2 &= x^2 \\
x^2 - 8x + 16 + y^2 - 2y + 1 &= x^2 \\
y^2 - 2y - 8x + 17 &= 0.
\end{aligned}$

35. $\begin{aligned} x^2 &= 4py \\ (-2)^2 &= 4p(-4) \\ 4p &= -1 \end{aligned}$ $\begin{aligned} y^2 &= 4px \\ (-4)^2 &= 4p(-2) \\ 4p &= -8 \end{aligned}$

Equations: $x^2 = -y$ and $y^2 = -8x$.

37. If the origin is the lowest point on the cable, then the top of the right supporting tower is at $(100, 70)$.

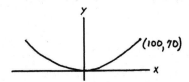

From the equation $x^2 = 4py$, we get
$$\begin{aligned} (100)^2 &= 4p(70) \\ 4p &= \frac{10,000}{70} = \frac{1000}{7}. \end{aligned}$$
The equation is therefore
$$x^2 = \frac{1000}{7}y.$$

To find the length of the cable 30 m from the center, we let $x = 30$:
$$30^2 = \frac{1000}{7}y \quad \text{and} \quad y = \frac{6300}{1000} = 6.3.$$

So the length of the cable is $20 + 6.3 = 26.3$ m.

39.

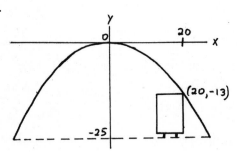

The required minimum clearance of 12 ft yields the point $(20, -13)$ in the figure.
$$\begin{aligned} x^2 &= 4py \\ 20^2 &= 4p(-13) \quad \text{or} \quad 4p = \frac{20^2}{-13} \end{aligned}$$
Equation: $x^2 = -\dfrac{20^2}{13}y$.
When $y = -25$,
$$\begin{aligned} x^2 &= -\frac{20^2}{13}(-25) \\ x &= \sqrt{\frac{20^2 \cdot 25}{13}} = \frac{20 \cdot 5}{\sqrt{13}} = \frac{100}{\sqrt{13}} \\ 2x &= \frac{200}{\sqrt{13}} \approx 55.5 \,\text{ft} \end{aligned}$$

41. We place the vertex of the parabola at the origin, so that one point on the parabola is $(3, -3)$ (from the given dimensions). Substituting in the equation $x^2 = 4py$, we get

$$3^2 = 4p(-3)$$
$$4p = -3.$$

The equation is therefore seen to be $x^2 = -3y$.

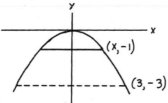

The right end of the beam $2\,\mathrm{m}$ above the base is at $(x, -1)$. To find x, let $y = -1$:

$$x^2 = -3(-1) = 3$$
$$x = \pm\sqrt{3}.$$

Hence the length of the beam is $2|x| = 2\sqrt{3}\,\mathrm{m}$.

43. $x^2 = 4py$. From Figure 1.55, we see that the point $(4, 1)$ lies on the curve: $4^2 = 4p(1)$. So $p = 4\,\mathrm{ft}$.

1.9   The Ellipse

1. The equation is

$$\frac{x^2}{25} + \frac{y^2}{16} = 1.$$

So by (1.16), $a^2 = 25$ and $b^2 = 16$; thus $a = 5$ and $b = 4$. Since the major axis is horizontal, the vertices are at $(\pm 5, 0)$. From $b^2 = a^2 - c^2$,

$$16 = 25 - c^2$$
$$c^2 = 9$$
$$c = \pm 3.$$

The foci are therefore at $(\pm 3, 0)$, on the major axis. Finally, the length of the semi-minor axis is equal to $b = 4$.

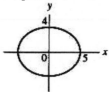

3. The equation is

$$\frac{x^2}{9} + \frac{y^2}{4} = 1.$$

So by (1.16), $a^2 = 9$ and $b^2 = 4$; thus $a = 3$ and $b = 2$. Since the major axis is horizontal, the vertices are at $(\pm 3, 0)$. From $b^2 = a^2 - c^2$,

$$4 = 9 - c^2$$
$$c^2 = 5$$
$$c = \pm\sqrt{5}.$$

The foci are therefore at $(\pm\sqrt{5}, 0)$, on the major axis. Finally, the length of the semi-minor axis is equal to $b = 2$.

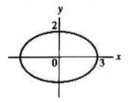

5. The equation is

$$\frac{x^2}{16} + y^2 = 1.$$

By (1.16), $a = 4$ and $b = 1$. From $b^2 = a^2 - c^2$,

$$1 = 16 - c^2$$
$$c^2 = 15$$
$$c = \pm\sqrt{15}.$$

Since the major axis is horizontal, the vertices and foci lie on the x-axis. The vertices are therefore at $(\pm 4, 0)$ and the foci are at $(\pm\sqrt{15}, 0)$. The length of the semi-minor axis is $b = 1$.

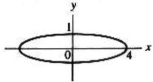

7. $$16x^2 + 9y^2 = 144$$

$$\frac{x^2}{9} + \frac{y^2}{16} = 1$$

So by (1.17), $a^2 = 16$ and $b^2 = 9$; so $a = 4$ and $b = 3$. Since the major axis is vertical, the vertices are at $(0, \pm 4)$. From $b^2 = a^2 - c^2$

$$9 = 16 - c^2$$
$$c = \pm\sqrt{7}.$$

The foci are therefore at $(0, \pm\sqrt{7})$. Semi-minor axis: $b = 3$.

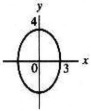

9. $$5x^2 + 2y^2 = 20$$

$$\frac{5x^2}{20} + \frac{2y^2}{20} = 1$$

$$\frac{x^2}{4} + \frac{y^2}{10} = 1$$

By (1.17), $a = \sqrt{10}$ and $b = 2$. Since the major axis is vertical, the vertices are at $(0, \pm\sqrt{10})$. From $b^2 = a^2 - c^2$

$$4 = 10 - c^2$$
$$c = \pm\sqrt{6}.$$

The foci, also on the major axis, are therefore at $(0, \pm\sqrt{6})$, while the length of the semi-minor axis is $b = 2$. (See sketch in answer section of book.)

11. $5x^2 + y^2 = 5$

$$\frac{x^2}{1} + \frac{y^2}{5} = 1 \qquad \text{major axis vertical}$$

Vertices: $(0, \pm\sqrt{5})$, foci: $(0, \pm 2)$. Length of semi-minor axis: $b = 1$. (See sketch in answer section of book.)

13. $x^2 + 2y^2 = 6$

$$\frac{x^2}{6} + \frac{2y^2}{6} = 1$$

$$\frac{x^2}{6} + \frac{y^2}{3} = 1 \qquad \text{major axis horizontal}$$

Thus $a = \sqrt{6}$ and $b = \sqrt{3}$. From $b^2 = a^2 - c^2$,

$$3 = 6 - c^2$$
$$c = \pm\sqrt{3}.$$

Vertices: $(\pm\sqrt{6}, 0)$; foci: $(\pm\sqrt{3}, 0)$. Length of semi-minor axis: $b = \sqrt{3}$. (See sketch in answer section of book.)

15. $15x^2 + 7y^2 = 105$

$$\frac{x^2}{7} + \frac{y^2}{15} = 1 \qquad \text{major axis vertical}$$

Thus $a = \sqrt{15}$ and $b = \sqrt{7}$. From $b^2 = a^2 - c^2$

$$7 = 15 - c^2$$
$$c = \pm\sqrt{8} = \pm 2\sqrt{2}.$$

Vertices: $(0, \pm\sqrt{15})$, foci: $(0, \pm 2\sqrt{2})$. Length of semi-minor axis: $b = \sqrt{7}$.

17. $2x^2 + 5y^2 = 50$

$$\frac{2x^2}{50} + \frac{5y^2}{50} = 1$$

$$\frac{x^2}{25} + \frac{y^2}{10} = 1 \qquad \text{major axis horizontal}$$

Thus $a = 5$ and $b = \sqrt{10}$. From $b^2 = a^2 - c^2$,

$$10 = 25 - c^2$$
$$c^2 = 15.$$

Vertices: $(\pm 5, 0)$; foci: $(\pm\sqrt{15}, 0)$; length of semi-minor axis: $b = \sqrt{10}$.

23. Since the foci $(\pm 3, 0)$ lie along the major axis, the major axis is horizontal. So by (1.16) the form of the equation is

$$\frac{x^2}{a^2} + \frac{y^2}{b^2} = 1.$$

Since the vertices are at $(\pm 4, 0)$, $a = 4$. From $b^2 = a^2 - c^2$ (with $c = 3$), we get $b^2 = 16 - 9 = 7$. So the equation is

$$\frac{x^2}{16} + \frac{y^2}{7} = 1.$$

25. Since the foci $(0, \pm 2)$ lie on the major axis, the major axis is vertical. So by (1.17) the form of the equation is

$$\frac{x^2}{b^2} + \frac{y^2}{a^2} = 1.$$

Since the length of the major axis is 8, $a = 4$. From $b^2 = a^2 - c^2$ with $c = 2$, $b^2 = 16 - 4 = 12$. Hence

$$\frac{x^2}{12} + \frac{y^2}{16} = 1 \quad \text{or} \quad 4x^2 + 3y^2 = 48.$$

27. Since the foci are at $(0, \pm 3)$, $c = 3$, and the major axis is vertical. By (1.17)

$$\frac{x^2}{b^2} + \frac{y^2}{a^2} = 1.$$

Since the length of the minor axis is 6, $b = 3$. From $b^2 = a^2 - c^2$, $9 = a^2 - 9$ and $a^2 = 18$.

$$\text{Equation: } \frac{x^2}{9} + \frac{y^2}{18} = 1 \quad \text{or} \quad 2x^2 + y^2 = 18.$$

29. Since the vertices and foci are on the y-axis, the form of the equation is, by (1.17),

$$\frac{x^2}{b^2} + \frac{y^2}{a^2} = 1.$$

From $b^2 = a^2 - c^2$ with $a = 8$ and $c = 5$, $b^2 = 64 - 25 = 39$. Hence

$$\frac{x^2}{39} + \frac{y^2}{64} = 1.$$

31. Form: $\dfrac{x^2}{a^2} + \dfrac{y^2}{b^2} = 1$; $c = 2\sqrt{3}$, $b = 2$. From $b^2 = a^2 - c^2$, $4 = a^2 - (2\sqrt{3})^2$ and $a^2 = 16$.

$$\text{Equation: } \frac{x^2}{16} + \frac{y^2}{4} = 1 \quad \text{or} \quad x^2 + 4y^2 = 16.$$

33. From the original derivation of the ellipse, $2a = 16$ and $a = 8$. Since the foci are at $(\pm 6, 0)$, $c = 6$. Thus $b^2 = a^2 - c^2 = 64 - 36 = 28$.

By (1.16) the equation is

$$\frac{x^2}{64} + \frac{y^2}{28} = 1.$$

35. $9x^2 + 5y^2 = 45$ or $\dfrac{x^2}{5} + \dfrac{y^2}{9} = 1$; $a = 3$; $b = \sqrt{5}$. From $b^2 = a^2 - c^2$, $5 = 9 - c^2$ and $c = 2$. Thus

$$e = \frac{c}{a} = \frac{2}{3}.$$

37. Distance from (x, y) to $(0, 0)$: $\sqrt{(x-0)^2 + (y-0)^2} = \sqrt{x^2 + y^2}$.

Distance from (x, y) to $(3, 0)$: $\sqrt{(x-3)^2 + (y-0)^2} = \sqrt{(x-3)^2 + y^2}$.

From the given condition:

$$
\begin{aligned}
\sqrt{x^2 + y^2} &= 2\sqrt{(x-3)^2 + y^2} \\
x^2 + y^2 &= 4[(x-3)^2 + y^2] \qquad \text{squaring both sides} \\
x^2 + y^2 &= 4(x^2 - 6x + 9 + y^2) \\
x^2 + y^2 &= 4x^2 - 24x + 36 + 4y^2 \\
0 &= 3x^2 - 24x + 36 + 3y^2 \\
3x^2 + 3y^2 - 24x + 36 &= 0 \\
x^2 + y^2 - 8x + 12 &= 0.
\end{aligned}
$$

The locus is a circle.

39. Since $a = 2$ and $b = \frac{3}{2}$, we get

$$
\frac{x^2}{4} + \frac{y^2}{\frac{9}{4}} = 1 \quad \text{or} \quad \frac{x^2}{4} + \frac{4y^2}{9} = 1,
$$

and $9x^2 + 16y^2 = 36$.

41. Placing the center at the origin, the vertices are at $(\pm 6, 0)$. The road extends from $(-4, 0)$ to $(4, 0)$. Since the clearance is $4\,\text{m}$, the point $(4, 4)$ lies on the ellipse, as shown.

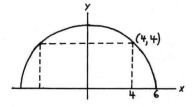

By (1.16),

$$
\begin{aligned}
\frac{x^2}{a^2} + \frac{y^2}{b^2} &= 1 \\
\frac{x^2}{36} + \frac{y^2}{b^2} &= 1.
\end{aligned}
$$

To find b, we substitute the coordinates of $(4, 4)$ in the equation:

$$
\begin{aligned}
\frac{16}{36} + \frac{16}{b^2} &= 1 \\
\frac{16}{b^2} &= \frac{36}{36} - \frac{16}{36} = \frac{20}{36} = \frac{5}{9} \\
\frac{b^2}{16} &= \frac{9}{5} \\
b^2 &= \frac{(9)(16)}{5} \\
b &= \frac{(3)(4)}{\sqrt{5}} = \frac{12}{\sqrt{5}} = \frac{12\sqrt{5}}{5}.
\end{aligned}
$$

So the height of the arch is $\dfrac{12\sqrt{5}}{5} = 5.4\,\text{m}$ to two significant digits.

43. We want the center of the ellipse to be at the origin with the center of the earth at one of the foci. Study the following diagram:

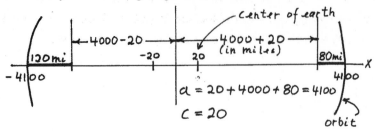

$$b^2 = 4100^2 - 20^2 = 16,809,600.$$

45. Let A = the maximum distance and P = the minimum distance as shown.

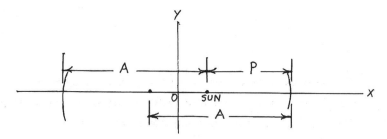

A is also the distance from the left focus to the right vertex. So $A - P$ is the distance between the foci. Therefore $\frac{1}{2}(A - P)$ is the distance from the center to the sun (the focus), or $c = \frac{1}{2}(A - P)$.

Now $a = c + P = \frac{1}{2}(A - P) + P = \frac{1}{2}(A + P)$. So

$$e = \frac{c}{a} = \frac{\frac{1}{2}(A - P)}{\frac{1}{2}(A + P)} = \frac{A - P}{A + P}.$$

In our problem

$$e = \frac{3.285 \times 10^9 - 5.48 \times 10^7}{3.285 \times 10^9 + 5.48 \times 10^7} = 0.967.$$

1.10   The Hyperbola

1. Comparing the given equation,

$$\frac{x^2}{16} - \frac{y^2}{9} = 1$$

to form (1.22), we see that the transverse axis is horizontal, with $a^2 = 16$ and $b^2 = 9$. So $a = 4$ and $b = 3$. From $b^2 = c^2 - a^2$, we get

$$9 = c^2 - 16$$
$$c = \pm 5.$$

So the vertices are at $(\pm 4, 0)$ and the foci are at $(\pm 5, 0)$. Using $a = 4$ and $b = 3$, we draw the auxiliary rectangle and sketch the curve:

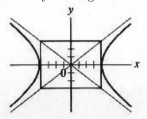

3. $\dfrac{x^2}{9} - \dfrac{y^2}{16} = 1$; by Equation (1.22), the transverse axis is horizontal with $a^2 = 9$ and $b^2 = 16$. So $a = 3$ and $b = 4$. From $b^2 = c^2 - a^2$, we have $16 = c^2 - 9$ or $c = \pm 5$.

It follows that the vertices are at $(\pm 3, 0)$ and the foci are at $(\pm 5, 0)$. Using $a = 3$ and $b = 4$, we draw the auxiliary rectangle and the asymptotes, and then sketch the curve.

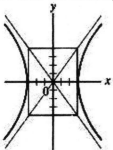

5. By (1.23), $a = 2$ and $b = 2$, transverse axis vertical along the y-axis. From $b^2 = c^2 - a^2$, $4 = c^2 - 4$ and $c = \pm\sqrt{8} = \pm 2\sqrt{2}$. So the vertices are at $(0, \pm 2)$ and the foci are at $(0, \pm 2\sqrt{2})$. Using $a = 2$ and $b = 2$, we draw the auxiliary rectangle and sketch the curve:

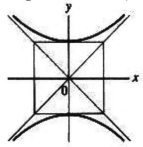

7. $x^2 - \dfrac{y^2}{5} = 1$ \qquad transverse axis horizontal

$a^2 = 1$ and $b^2 = 5$; so $a = 1$ and $b = \sqrt{5}$. From $b^2 = c^2 - a^2$, $5 = c^2 - 1$ and $c = \pm\sqrt{6}$. Vertices: $(\pm 1, 0)$; foci: $(\pm\sqrt{6}, 0)$. Using $a = 1$ and $b = \sqrt{5}$, we draw the auxiliary rectangle and sketch the curve.

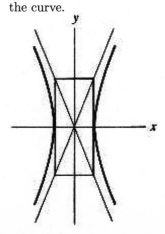

9. $2y^2 - 3x^2 = 24$

$$\frac{2y^2}{24} - \frac{3x^2}{24} = 1$$

$$\frac{y^2}{12} - \frac{x^2}{8} = 1$$

By (1.23), $a = \sqrt{12} = 2\sqrt{3}$ and $b = \sqrt{8} = 2\sqrt{2}$. From $b^2 = c^2 - a^2$, $8 = c^2 - 12$, so that $c = \pm\sqrt{20} = \pm 2\sqrt{5}$. Since the transverse axis lies along the y-axis, the vertices are at $(0, \pm 2\sqrt{3})$ and the foci at $(0, \pm 2\sqrt{5})$. Using $a = 2\sqrt{3}$ and $b = 2\sqrt{2}$, we draw the auxiliary rectangle and sketch the curve:

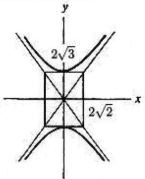

11. $3y^2 - 2x^2 = 6$ or $\dfrac{y^2}{2} - \dfrac{x^2}{3} = 1$

By (1.23) the transverse axis is vertical with $a = \sqrt{2}$ and $b = \sqrt{3}$. From $b^2 = c^2 - a^2$, $3 = c^2 - 2$ or $c = \pm\sqrt{5}$.

So the vertices are at $(0, \pm\sqrt{2})$ and the foci at $(0, \pm\sqrt{5})$. Using $a = \sqrt{2}$ and $b = \sqrt{3}$, we draw the auxiliary rectangle and sketch the curve.

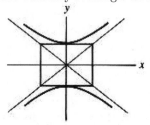

13. Since the foci (and hence the vertices) lie on the x-axis, the transverse axis is horizontal. By (1.22),

$$\frac{x^2}{a^2} - \frac{y^2}{b^2} = 1.$$

Since the length of the transverse axis is 6, $a = 3$, and since the length of the conjugate axis is 4, $b = 2$. It follows that

$$\frac{x^2}{9} - \frac{y^2}{4} = 1 \quad \text{or} \quad 4x^2 - 9y^2 = 36.$$

15. Since the foci lie on the y-axis, the transverse axis is vertical. By (1.23),

$$\frac{y^2}{a^2} - \frac{x^2}{b^2} = 1.$$

Since the length of the conjugate axis is 8, $b = 4$. The foci are at $(0, \pm 5)$, so that $c = 5$. From $b^2 = c^2 - a^2$, we have $16 = 25 - a^2$ and $a^2 = 9$.

Equation: $\dfrac{y^2}{9} - \dfrac{x^2}{16} = 1$ or $16y^2 - 9x^2 = 144$.

17. Since the foci lie in the y-axis, the transverse axis is vertical and by (1.23) the form is

$$\frac{y^2}{a^2} - \frac{x^2}{b^2} = 1.$$

Since the length of the transverse axis is 12, $a = 6$. The foci are $(0, \pm 8)$, so that $c = 8$. To find b: $b^2 = c^2 - a^2 = 64 - 36 = 28$. The equation is therefore

$$\frac{y^2}{36} - \frac{x^2}{28} = 1.$$

19. Since the vertices lie on the x-axis, the transverse axis is horizontal:

$$\frac{x^2}{a^2} - \frac{y^2}{b^2} = 1.$$

Since the vertices are at $(\pm 4, 0)$, $a = 4$. We also have $b = 4$ (since the length of the conjugate axis is 8).

Equation: $\dfrac{x^2}{16} - \dfrac{y^2}{16} = 1$.

21. By (1.22), the from of the equation is $\dfrac{x^2}{a^2} - \dfrac{y^2}{b^2} = 1$. Since $a = 3$ and $c = 6$, $b^2 = 36 - 9 = 27$. Thus

$$\frac{x^2}{9} - \frac{y^2}{27} = 1 \quad \text{or} \quad 3x^2 - y^2 = 27.$$

23. The asymptotes are

$$y = \pm \frac{b}{a} x = \pm 2x$$

which implies that $\dfrac{b}{a} = 2$ and $b = 2a$. Since the vertices are at $(\pm 1, 0)$, it follows that $a = 1$ and $b = 2$. Hence

$$\frac{x^2}{1} - \frac{y^2}{4} = 1 \qquad\qquad \text{transverse axis horizontal}$$

or $4x^2 - y^2 = 4$.

25. By the original derivation of the equation of the hyperbola, $2a = 6$ and $a = 3$. Since $(0, \pm 5)$ are the foci, $c = 5$. Thus $b^2 = 25 - 9 = 16$. By (1.23)

$$\frac{y^2}{a^2} - \frac{x^2}{b^2} = 1$$

$$\frac{y^2}{9} - \frac{x^2}{16} = 1 \quad \text{or} \quad 16y^2 - 9x^2 = 144.$$

27. $\sqrt{(x-1)^2 + (y-2)^2} - \sqrt{(x+3)^2 + (y-2)^2} = \pm 2$

$\sqrt{(x-1)^2 + (y-2)^2} = \sqrt{(x+3)^2 + (y-2)^2} \pm 2$

Squaring both sides

$$(x-1)^2 + (y-2)^2 = (x+3)^2 + (y-2)^2 \pm 4\sqrt{(x+3)^2 + (y-2)^2} + 4$$

which simplifies to

$$-2x - 3 = \pm\sqrt{(x+3)^2 + (y-2)^2}.$$

Squaring both sides again,

$$4x^2 + 12x + 9 = (x+3)^2 + (y-2)^2$$

which reduces to

$$3x^2 - y^2 + 6x + 4y - 4 = 0.$$

29. By (1.23), the equation has the form $\dfrac{y^2}{a^2} - \dfrac{x^2}{b^2} = 1$. Since $a = 12$, we have

$$\frac{y^2}{144} - \frac{x^2}{b^2} = 1.$$

To find b, we substitute the coordinates of $(-1, 13)$ in the last equation:

$$\frac{169}{144} - \frac{1}{b^2} = 1$$

$$-\frac{1}{b^2} = \frac{144}{144} - \frac{169}{144} = -\frac{25}{144}.$$

Thus $b^2 = \dfrac{144}{25}$. The equation is

$$\frac{y^2}{144} - \frac{x^2}{144/25} = 1 \quad \text{or} \quad \frac{y^2}{144} - \frac{25x^2}{144} = 1.$$

31. $\qquad pV = k$

$(12)(3.0) = k \qquad V = 3.0\,\mathrm{m}^3,\ p = 12\,\mathrm{Pa}$

So $pV = 36$. (See graph in answer section of book.)

1.11 Translation of Axes; Standard Equations of the Conics

1. Circle, center at $(1, 2)$, $r = \sqrt{3}$.

3. $(y + 3)^2 = 8(x - 2)$

   $(y + 3)^2 = 4(2)(x - 2)$ $p = +2$

   Vertex at $(2, -3)$, focus at $(2 + 2, -3) = (4, -3)$.

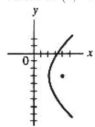

5. $ 2x^2 - 3y^2 + 8x - 12y + 14 = 0$

   $ 2x^2 + 8x - 3y^2 - 12y + 14 = 0$

   $2(x^2 + 4x) - 3(y^2 + 4y) = -14$ factoring 2 and -3

   Note that the square of one-half the coefficient of x and y is $\left(\dfrac{1}{2} \cdot 4\right)^2 = 4$.

   Inserting these values <u>inside</u> the parentheses and balancing the equation, we get

   $$2(x^2 + 4x + \underline{4}) - 3(y^2 + 4y + \underline{4}) = -14 + 2 \cdot \underline{4} - 3 \cdot \underline{4}$$

   $$2(x + 2)^2 - 3(y + 2)^2 = -18$$

   $$\frac{3(y + 2)^2}{18} - \frac{2(x + 2)^2}{18} = 1$$

   $$\frac{(y + 2)^2}{6} - \frac{(x + 2)^2}{9} = 1.$$

   The equation represents a hyperbola with transverse axis vertical. Center: $(-2, -2)$, $a = \sqrt{6}$, $b = 3$.

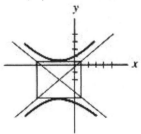

7. $ 16x^2 + 4y^2 + 64x - 12y + 57 = 0$

   $ 16x^2 + 64x + 4y^2 - 12y + 57 = 0$

   $16(x^2 + 4x) + 4(y^2 - 3y) = -57$ factoring 16 and 4

   Note that

   $$\left(\frac{1}{2} \cdot 4\right)^2 = 4 \quad \text{and} \quad \left[\frac{1}{2}(-3)\right]^2 = \frac{9}{4}.$$

   Inserting these values inside the parentheses and balancing the equation, we get

   $$16(x^2 + 4x + 4) + 4(y^2 - 3y + \tfrac{9}{4}) = -57 + 16 \cdot 4 + 4(\tfrac{9}{4})$$

   $$16(x + 2)^2 + 4(y - \tfrac{3}{2})^2 = 16$$

   $$\frac{(x + 2)^2}{1} + \frac{(y - 3/2)^2}{4} = 1.$$

   The equation represents an ellipse with major axis vertical. Center: $(-2, \frac{3}{2})$, $a = 2$, $b = 1$.

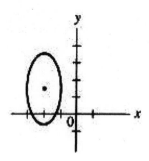

9.
$$\begin{aligned}
x^2 + y^2 + 2x - 2y + 2 &= 0 \\
x^2 + 2x \quad + y^2 - 2y \quad &= -2 \\
(x^2 + 2x + 1) + (y^2 - 2y + 1) &= -2 + 1 + 1 \\
(x+1)^2 + (y-1)^2 &= 0
\end{aligned}$$
Point: $(-1, 1)$.

11.
$$\begin{aligned}
2x^2 - 12y^2 + 60y - 63 &= 0 \\
2x^2 - 12(y^2 - 5y \quad) &= 63
\end{aligned}$$

The square of one-half the coefficient of y is $\left[\frac{1}{2}(-5)\right]^2 = \frac{25}{4}$. Inserting this number <u>inside</u> the parentheses, we get
$$\begin{aligned}
2x^2 - 12(y^2 - 5y + \tfrac{25}{4}) &= 63 - 12(\tfrac{25}{4}) = -12 \\
x^2 - 6(y^2 - 5y + \tfrac{25}{4}) &= -6 \\
x^2 - 6(y - \tfrac{5}{2})^2 &= -6 \\
(y - \tfrac{5}{2})^2 - \frac{x^2}{6} &= 1.
\end{aligned}$$
Hyperbola, center at $(0, \frac{5}{2})$, transverse axis vertical with $a = 1$ and $b = \sqrt{6}$.

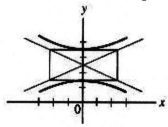

13.
$$\begin{aligned}
64x^2 + 64y^2 - 16x - 96y - 27 &= 0 \\
64x^2 - 16x + 64y^2 - 96y - 27 &= 0 \\
64(x^2 - \tfrac{x}{4} \quad) + 64(y^2 - \tfrac{3y}{2} \quad) &= 27 \\
64(x^2 - \tfrac{x}{4} + \tfrac{1}{64}) + 64(y^2 - \tfrac{3y}{2} + \tfrac{9}{16}) &= 27 + 1 + 36 \\
64(x - \tfrac{1}{8})^2 + 64(y - \tfrac{3}{4})^2 &= 64 \\
(x - \tfrac{1}{8})^2 + (y - \tfrac{3}{4})^2 &= 1.
\end{aligned}$$
Circle of radius 1 centered at $\left(\dfrac{1}{8}, \dfrac{3}{4}\right)$.

15.
$$\begin{aligned}
3x^2 + y^2 - 18x + 2y + 29 &= 0 \\
3x^2 - 18x + y^2 + 2y &= -29 \\
3(x^2 - 6x \quad) + (y^2 + 2y \quad) &= -29
\end{aligned}$$

Observe that $\left[\dfrac{1}{2}(-6)\right]^2 = 9$ and $\left(\dfrac{1}{2}\cdot 2\right)^2 = 1$. Adding these values inside the parentheses and balancing the equation, we get

$$3(x^2 - 6x + 9) + (y^2 + 2y + 1) \;=\; -29 + 3\cdot 9 + 1$$
$$3(x-3)^2 + (y+1)^2 \;=\; -1$$

which is an imaginary locus.

17. $x^2 + 2x - 12y + 25 \;=\; 0$

$$x^2 + 2x \;=\; 12y - 25$$

We add to each side of the equation the square of one-half the coefficient of x, $\left[\dfrac{1}{2}\cdot 2\right]^2 = 1$:

$$x^2 + 2x + \underline{1} \;=\; 12y - 25 + \underline{1}$$
$$(x+1)^2 \;=\; 12y - 24$$
$$(x+1)^2 \;=\; 12(y-2)$$
$$(x+1)^2 \;=\; 4\cdot 3(y-2). \qquad p = +3$$

Vertex at $(-1, 2)$, focus at $(-1, 5)$.

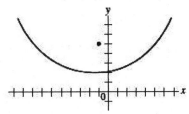

19. $x^2 + 2y^2 + 6x - 4y + 9 \;=\; 0$

$$x^2 + 6x + 2y^2 - 4y \;=\; -9$$
$$(x^2 + 6x \qquad) + 2(y^2 - 2y \qquad) \;=\; -9$$

Adding $\left[\dfrac{1}{2}(6)\right]^2 = 9$ and $\left[\dfrac{1}{2}(-2)\right]^2 = 1$ inside the parentheses and balancing the equation, we have

$$(x^2 + 6x + 9) + 2(y^2 - 2y + 1) \;=\; -9 + 9 + 2$$
$$(x+3)^2 + 2(y-1)^2 \;=\; 2$$
$$\dfrac{(x+3)^2}{2} + (y-1)^2 \;=\; 1.$$

Ellipse, center at $(-3, 1)$ with $a = \sqrt{2}$ and $b = 1$.

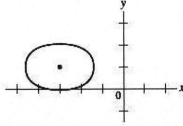

21. $x^2 + 4x + 4y + 16 \;=\; 0$

$$x^2 + 4x \;=\; -4y - 16$$

We add to each side $\left[\dfrac{1}{2}\cdot 4\right]^2 = 4$:

$$\begin{aligned}
x^2 + 4x + \underline{4} &= -4y - 16 + \underline{4} \\
(x+2)^2 &= -4y - 12 \\
(x+2)^2 &= -4(y+3) \\
(x+2)^2 &= 4(-1)(y+3). \qquad p = -1
\end{aligned}$$

Vertex at $(-2, -3)$, focus at $(-2, -4)$.

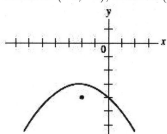

23.
$$\begin{aligned}
x^2 + 2y^2 - 4x + 12y + 14 &= 0 \\
x^2 - 4x \ \ + 2y^2 + 12y &= -14 \\
(x^2 - 4x \ \ ) + 2(y^2 + 6y \ \ ) &= -14 \qquad \text{factoring}
\end{aligned}$$

Observe that $\left[\dfrac{1}{2}(-4)\right]^2 = 4$ and $\left[\dfrac{1}{2}(6)\right]^2 = 9$. Inserting these vales inside the parentheses and balancing the equation, we get

$$\begin{aligned}
(x^2 - 4x + \underline{4}) + 2(y^2 + 6y + \underline{9}) &= -14 + \underline{4} + 2 \cdot \underline{9} \\
(x-2)^2 + 2(y+3)^2 &= 8 \\
\frac{(x-2)^2}{8} + \frac{(y+3)^2}{4} &= 1.
\end{aligned}$$

Ellipse, center at $(2, -3)$.

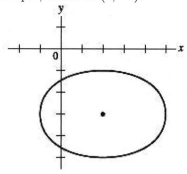

25.

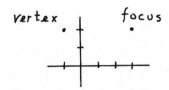

Distance from vertex to focus: $3 - (-1) = 4$. Thus $p = 4$. Since the axis is horizontal, the form of the equation is $(y - k)^2 = 4p(x - h)^2$. Thus

$$\begin{aligned}
(y-2)^2 &= 4 \cdot 4(x+1) \qquad (h, k) = (-1, 2), \ p = 4 \\
(y-2)^2 &= 16(x+1).
\end{aligned}$$

27.

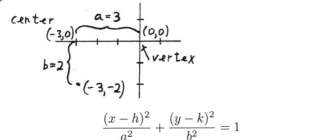

$$\frac{(x-h)^2}{a^2} + \frac{(y-k)^2}{b^2} = 1 \qquad \text{major axis horizontal}$$

Since the center is at $(-3, 0)$, we get for the equation

$$\frac{(x+3)^2}{9} + \frac{y^2}{4} = 1. \qquad a = 3,\ b = 2$$

29.

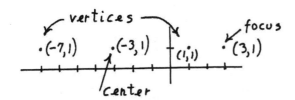

Distance between vertices is 8, so that $a = 4$. Center: $(-3, 1)$ (point midway between vertices). Distance from center to one focus is 6, so that $c = 6$. The transverse axis is horizontal, resulting in the form

$$\frac{(x-h)^2}{a^2} - \frac{(y-k)^2}{b^2} = 1.$$

Since $b^2 = c^2 - a^2 = 36 - 16 = 20$, we get

$$\frac{(x+3)^2}{16} - \frac{(y-1)^2}{20} = 1. \qquad (h, k) = (-3, 1)$$

31.

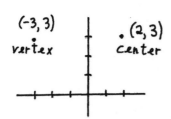

$(h, k) = (2, 3)$; $a = 2 - (-3) = 5$ (distance from center to vertex); $b = 2$ (length of minor axis is 4).

Form: $\dfrac{(x-h)^2}{a^2} + \dfrac{(y-k)^2}{b^2} = 1.$ major axis horizontal

Resulting equation: $\dfrac{(x-2)^2}{25} + \dfrac{(y-3)^2}{4} = 1.$ $(h, k) = (2, 3)$

33. Form: $\dfrac{(x-h)^2}{a^2} - \dfrac{(y-k)^2}{b^2} = 1.$

Distance from center to vertex is 2, so that $a = 2$. From $x - 2y = 1$, $y = \frac{1}{2}x - \frac{1}{2}$. So the slope m of one of the asymptotes is $\frac{1}{2}$. But $m = \frac{b}{a}$. Thus $\frac{1}{2} = \frac{b}{a} = \frac{b}{2}$ or $b = 1$. The equation is

$$\frac{(x-1)^2}{4} - \frac{y^2}{1} = 1. \qquad (h, k) = (1, 0)$$

35.

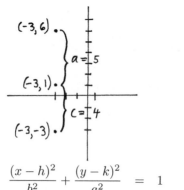

$$b^2 = a^2 - c^2 = 25 - 16 = 9$$

$$\frac{(x-h)^2}{b^2} + \frac{(y-k)^2}{a^2} = 1 \qquad \text{major axis vertical}$$

$$\frac{(x+3)^2}{9} + \frac{(y-1)^2}{25} = 1$$

37. Distance from vertex to focus: $4 - 1 = 3$. Since the focus is to the left of the vertex, $p = -3$.

Form:
$$(y-k)^2 = 4p(x-h) \qquad \text{axis horizontal}$$
$$(y-k)^2 = -12(x-h). \qquad p = -3$$
Equation: $(y+2)^2 = -12(x-4).$ $\qquad\qquad (h,k) = (4,-2)$

39.

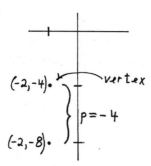

Since the vertex is midway between the focus and directrix, its coordinates are $(-2,-4)$. Since $p = -4$, we have
$$(x+2)^2 = 4(-4)(y+4)$$
$$(x+2)^2 = -16(y+4).$$

41.

$$(-1,3)\cdot \quad \text{vertex}$$
$$(-1,1)\cdot \quad \text{center}$$
$$(-1,-2)\cdot \quad \text{focus}$$

$(h,k) = (-1,1)$; $a = 3 - 1 = 2$ (distance from center to vertex);

$c = 1 - (-2) = 3$ (distance from center to focus); $b^2 = c^2 - a^2 = 9 - 4 = 5$.

Form: $\dfrac{(y-k)^2}{a^2} - \dfrac{(x-h)^2}{b^2} = 1.$ $\qquad\qquad$ transverse axis vertical

Equation: $\dfrac{(y-1)^2}{4} - \dfrac{(x+1)^2}{5} = 1.$ $\qquad\qquad (h,k) = (-1,1)$

Chapter 1 Review

1. Slope of line segment joining $(3, 10)$ and $(7, 4)$: $-\dfrac{3}{2}$.

   Slope of line segment joining $(4, 2)$ and $(7, 4)$: $\dfrac{2}{3}$.

   Since the slopes are negative reciprocals, the line segments are perpendicular.

3. $C = \frac{5}{9}(F - 32)$

   By assumption $C = F$,

   $$
   \begin{aligned}
   F &= \tfrac{5}{9}(F - 32) \\
   F &= \tfrac{5}{9}F - \tfrac{160}{9} \\
   F - \tfrac{5}{9}F &= -\tfrac{160}{9} \\
   \tfrac{4}{9}F &= -\tfrac{160}{9} \\
   F &= -40^\circ.
   \end{aligned}
   $$

5. Slope of line segment joining $(-1, 5)$ and $(3, 9)$: 1.

   Slope of line segment joining $(3, 1)$ and $(7, 5)$: 1.

   Slope of line segment joining $(-1, 5)$ and $(3, 1)$: -1.

   Slope of line segment joining $(3, 9)$ and $(7, 5)$: -1.

   Since opposite sides are parallel, the figure is a parallelogram. Moreover, since the line segment joining $(-1, 5)$ and $(3, 1)$ is perpendicular to the line segment joining $(-1, 5)$ and $(3, 9)$, the figure must be a rectangle. Finally:

   Length of line segment joining $(-1, 5)$ and $(3, 1) = 4\sqrt{2}$.

   Length of line segment joining $(-1, 5)$ and $(3, 9) = 4\sqrt{2}$.

   Thus the figure is a square.

7.
   $$
   \begin{aligned}
   3x + y &= 3 \\
   y &= -3x + 3 \qquad y = mx + b
   \end{aligned}
   $$

   Since $m = -3$, we get

   $$
   \begin{aligned}
   y - 5 &= -3(x + 1) \qquad y - y_1 = m(x - x_1) \\
   3x + y - 2 &= 0.
   \end{aligned}
   $$

9. $r^2 = (1 - 0)^2 + (-2 - 0)^2 = 1 + 4 = 5$. We now get

   $(x - 1)^2 + (y + 2)^2 = 5$ or $x^2 + y^2 - 2x + 4y = 0$.

11.
    $$
    \begin{aligned}
    x^2 + y^2 + 2x + 2y &= 0 \\
    x^2 + 2x + y^2 + 2y &= 0 \\
    (x^2 + 2x + 1) + (y^2 + 2y + 1) &= 1 + 1 \\
    (x + 1)^2 + (y + 1)^2 &= 2
    \end{aligned}
    $$

    Center: $(-1, -1)$; $r = \sqrt{2}$.

13. Ellipse, major axis vertical, $a = 4$, $b = 3$. From $b^2 = a^2 - c^2$,

$$9 = 16 - c^2$$
$$c = \pm\sqrt{7}.$$

Vertices: $(0, \pm 4)$; foci: $(0, \pm\sqrt{7})$. (See sketch in answer section of book.)

15. $\dfrac{y^2}{4} - \dfrac{x^2}{7} = 1$

Hyperbola, transverse axis vertical with $a = 2$ and $b = \sqrt{7}$. From $b^2 = c^2 - a^2$, $7 = c^2 - 4$ and $c = \pm\sqrt{11}$. So the vertices are at $(0, \pm 2)$ and the foci are at $(0, \pm\sqrt{11})$. Using $a = 2$ and $b = \sqrt{7}$, we draw the auxiliary rectangle and sketch the curve.

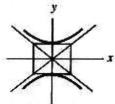

17. Parabola, axis horizontal. From $y^2 = -3x$, we have

$$y^2 = 4\left(-\frac{3}{4}\right)x. \qquad\qquad \text{inserting 4}$$

Thus, $p = -\dfrac{3}{4}$, placing the focus at $\left(-\dfrac{3}{4}, 0\right)$. (See sketch in answer section of book.)

19. $y^2 + 6y + 4x + 1 = 0$

$$y^2 + 6y = -4x - 1$$

Adding $\left[\dfrac{1}{2}(6)\right]^2 = 9$ to each side,

$$y^2 + 6y + 9 = -4x - 1 + 9$$
$$(y + 3)^2 = -4x + 8$$
$$(y + 3)^2 = 4(-1)(x - 2). \qquad p = -1$$

Parabola, vertex at $(2, -3)$, focus at $(2 - 1, -3) = (1, -3)$.

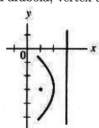

21. $$16x^2 - 64x + 9y^2 + 18y = 71$$
$$16(x^2 - 4x\quad) + 9(y^2 + 2y\quad) = 71 \qquad \text{factoring 16 and 9}$$

Note that $\left[\dfrac{1}{2}(-4)\right]^2 = 4$ and $\left[\dfrac{1}{2}(2)\right]^2 = 1$. Inserting these values inside the parentheses and balancing the equation, we get

$$16(x^2 - 4x + \underline{4}) + 9(y^2 + 2y + \underline{1}) = 71 + 16 \cdot \underline{4} + 9 \cdot \underline{1}$$
$$16(x - 2)^2 + 9(y + 1)^2 = 144$$
$$\frac{(x - 2)^2}{9} + \frac{(y + 1)^2}{16} = 1. \qquad\qquad \text{dividing by 144}$$

Ellipse, center at $(2, -1)$, major axis vertical, $a = 4$, $b = 3$.

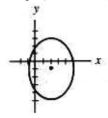

23.
$$x^2 - y^2 - 4x + 8y - 21 = 0$$
$$x^2 - 4x - y^2 + 8y = 21$$
$$(x^2 - 4x \quad) - (y^2 - 8y \quad) = 21$$

Adding $\left[\dfrac{1}{2}(-4)\right]^2 = 4$ and $\left[\dfrac{1}{2}(-8)\right]^2 = 16$ inside the parentheses and balancing the equation we get

$$(x^2 - 4x + 4) - (y^2 - 8y + 16) = 21 + 4 - 1(16)$$
$$(x - 2)^2 - (y - 4)^2 = 9$$
$$\frac{(x - 2)^2}{9} - \frac{(y - 4)^2}{9} = 1.$$

Hyperbola, center at $(2, 4)$.

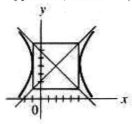

25. Form:

$$(x - h)^2 = 4p(y - k). \qquad \text{axis vertical}$$

Distance from vertex $(1, 3)$ to directrix $y = 0$ is 3, so that $p = 3$. The equation is

$$(x - 1)^2 = 4(3)(y - 3) \qquad (h, k) = (1, 3), \ p = 3$$

or

$$(x - 1)^2 = 12(y - 3).$$

27. Form: $\dfrac{x^2}{b^2} + \dfrac{y^2}{a^2} = 1$; $a = 4$; $c = 3$; $b^2 = a^2 - c^2 = 16 - 9 = 7$.

Equation: $\dfrac{x^2}{7} + \dfrac{y^2}{16} = 1$.

29.

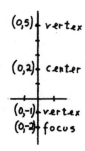

Center: $(0,2)$ (midway between vertices). Distance from center to vertex is 3, so that $a = 3$. Distance from center to focus is 4, so that $c = 4$. Thus $b^2 = c^2 - a^2 = 16 - 9 = 7$. Since the transverse axis is vertical, the form is

$$\frac{(y-k)^2}{a^2} - \frac{(x-h)^2}{b^2} = 1,$$

and the equation is

$$\frac{(y-2)^2}{9} - \frac{x^2}{7} = 1. \qquad\qquad (h,k) = (0,2)$$

31.

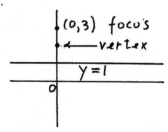

Since the vertex is midway between the focus and directrix, its coordinates are $(0,2)$. It follows that $p = +1$; so the equation is

$$
\begin{aligned}
(x-0)^2 &= 4(1)(y-2) \\
x^2 &= 4(y-2).
\end{aligned}
$$

33.

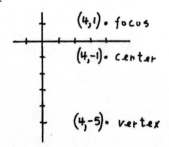

$c = 2$ (distance from center to focus); $a = 4$ (distance from center to vertex); $b^2 = a^2 - c^2 = 16 - 4 = 12$. Form: $\dfrac{(x-h)^2}{b^2} + \dfrac{(y-k)^2}{a^2} = 1.$ \qquad major axis vertical

Equation: $\dfrac{(x-4)^2}{12} + \dfrac{(y+1)^2}{16} = 1.$ \qquad\qquad $(h,k) = (4,-1)$

35. $y = (x + 1)^3$

Intercepts. If $x = 0$, then $y = 1$. If $y = 0$, then $x = -1$.

Symmetry. None: replacing x by $-x$ or y by $-y$ changes the equation.

Asymptotes. None: the equation is not in the form of a fraction with x in the denominator.

Extent. All x and all y.

Graph.

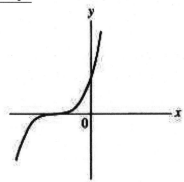

37. Intercepts. If $x = 0$, then $y = 0$; if $y = 0$, then $x(x - 4) = 0$ and thus $x = 0, 4$.

Symmetry. Replacing y by $-y$, we get $(-y)^2 = x(x - 4)$, which reduces to the given equation. So the curve is symmetric with respect to the x-axis.

Asymptotes. None (equation is not in the form of a fraction).

Extent. Solving for y we have

$$y = \pm\sqrt{x(x - 4)}.$$

If $x > 4$, $x(x - 4) > 0$. If $0 < x < 4$, $x(x - 4) < 0$ [for example, if $x = 2$, we get $2(2 - 4) = -4$]. If $x < 0$, $x(x - 4) > 0$, since both factors are negative. So the extent is $x \leq 0$ and $x \geq 4$.

Graph.

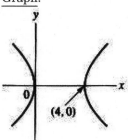

39. $y = \dfrac{x}{x^2 - 4}$

Intercepts. $(0, 0)$

Symmetry. Replacing x by $-x$ and y by $-y$ we get

$$-y = \frac{-x}{(-x)^2 - 4}$$

which reduces to

$$y = \frac{x}{x^2 - 4}.$$

The graph is therefore symmetric with respect to the origin.

Asymptotes. Vertical: setting the denominator equal to 0 results in

$$x^2 - 4 = 0 \quad \text{and} \quad x = \pm 2.$$

If x gets large, y approaches 0, so that the x-axis is a horizontal asymptote.

Extent. All x except $x = 2$ and $x = -2$.

Graph.

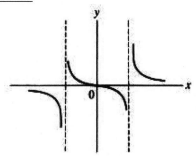

41. Placing the vertex at the origin, one point on the parabola is $(0.90, 0.60)$, as shown in the figure. The form is $y^2 = 4px$. To find p, we substitute the coordinates of the point in the equation:

$$
\begin{aligned}
(0.60)^2 &= 4p(0.90) \\
p &= \frac{(0.60)^2}{(4)(0.90)} = 0.10.
\end{aligned}
$$

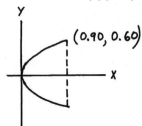

To be at the focus, the light must be placed 0.10 feet from the vertex.

43.

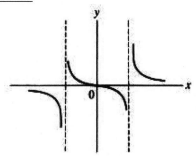

Volume=length×width×height.

$$
\begin{aligned}
V &= (6 - 2x)(6 - 2x)x \\
  &= x(6 - 2x)^2
\end{aligned}
$$

See graph in answer section of book.

45. See Exercise 45, Section 1.9:

$$e = \frac{A - P}{A + P};$$

$$P = 4000 + 119 = 4119\,\text{mi}; \ A = 4000 + 122{,}000 = 126{,}000\,\text{mi};$$

$$e = \frac{126{,}000 - 4119}{126{,}000 + 4119} = 0.94.$$

Chapter 2

Introduction to Calculus: The Derivative

2.1   Functions and Intervals

13. From the given equation $y = x + 2$, we see that y is defined for all x. From $x = y - 2$ it follows that x is defined for all y.

15. To avoid imaginary values, we must have $x - 2 \geq 0$ or $x \geq 2$. We can also say that the domain is the interval $[2, \infty)$.

    When $x = 2$, then $y = 0$. By definition of principal square root, y cannot be negative. So the range is $y \geq 0$ or $[0, \infty)$.

17. For the y-values to be real, the radicand must be positive or zero:
$$
\begin{array}{rcll}
4 - x^2 & \geq & 0 & \\
4 & \geq & x^2 & \text{adding } x^2 \text{ to both sides} \\
x^2 & \leq & 4 & \text{same inequality} \\
-2 \leq x & \leq & 2. & \text{domain}
\end{array}
$$
    If $x = 0$, $y = \sqrt{4} = 2$; $y = 2$ is the largest possible value of y (if $x \neq 0$, then $y < 2$). If $x = \pm 2$, then $y = 0$. So the range is $0 \leq y \leq 2$.

19. $y = x\sqrt{x - 3}$. To avoid imaginary values, we require that $x - 3 \geq 0$ or $x \geq 3$; $x = 0$ is also permitted: $0\sqrt{0 - 3} = 0 \cdot j\sqrt{3} = 0$.

    Range: by definition of principal square root, $\sqrt{x - 3} \geq 0$. So the range is the interval $[0, \infty)$.

21. For $\sqrt{x - 2}$ to be real, we must have $x \geq 2$. If $x = 1$, then $y = (0)(j) = 0$. So the domain is the set of values $x = 1$ and $x \geq 2$. For the x-values under discussion, $y \geq 0$, which is the range.

23. $y = \dfrac{1}{x-1}$. Domain: all x except $x = 1$ (to avoid division by 0).

Range: all y except $y = 0$ because the x-axis ($y = 0$) is a horizontal asymptote. This can also be seen by solving the equation for x in terms of y:

$$x = \frac{y+1}{y}.$$

y cannot be equal to zero.

25. For the y-values to be real, the radicand must be positive or zero:

$$
\begin{aligned}
1 - x^2 &\geq 0 \\
1 &\geq x^2 \qquad \text{adding } x^2 \text{ to both sides} \\
x^2 &\leq 1 \\
-1 \leq x &\leq 1. \qquad \text{domain}
\end{aligned}
$$

If $x = 0$, then $y = 1 + \sqrt{1} = 2$; $y = 2$ is the largest possible value. (Whenever $x \neq 0$, $y < 2$.)

If $x = \pm 1$, then $y = 1$. So the range is the closed interval $[1, 2]$.

27. $f(x) = x^2$; so $f(2) = 2^2 = 4$ and $f(-2) = (-2)^2 = 4$.

29.
$$
\begin{aligned}
f(x) &= 2x \\
f(0) &= 2 \cdot 0 = 0 \qquad & x = 0 \\
f(6) &= 2 \cdot 6 = 12 \qquad & x = 6
\end{aligned}
$$

31. $h(x) = x^2 + 2x$. Leaving a blank space for x, we get $h(\ ) = (\ )^2 + 2(\ )$.

Now fill in the blanks:
$$
\begin{aligned}
h(1) &= 1^2 + 2(1) = 3; \\
h(3) &= 3^2 + 2(3) = 15.
\end{aligned}
$$

33. $f(x) = x^3 + 1$;

$f(0) = 0^3 + 1 = 1$; $f(1) = 1^3 + 1 = 2$; $f(-2) = (-2)^3 + 1 = -7$.

35. $\phi(x) = \dfrac{1}{x}$; $\phi(3) = \dfrac{1}{3}$ and $\phi(a) = \dfrac{1}{a}$.

37. $G(z) = \sqrt{z^2 - 1}$. Leaving a blank space for z, we get $G(\ ) = \sqrt{(\ )^2 - 1}$.

Now fill in the blanks:
$$
\begin{aligned}
G(a^2) &= \sqrt{(a^2)^2 - 1} = \sqrt{a^4 - 1}; \\
G(x - 1) &= \sqrt{(x-1)^2 - 1} = \sqrt{x^2 - 2x}.
\end{aligned}
$$

39. $f(x) = 1 - x^2$; $f(\ ) = 1 - (\ )^2$. Filling in the blanks:

$f(x + \Delta x) = 1 - (x + \Delta x)^2 = 1 - x^2 - 2x\Delta x - (\Delta x)^2$;

$f(x - \Delta x) = 1 - (x - \Delta x)^2 = 1 - x^2 + 2x\Delta x - (\Delta x)^2$.

41. $f(x) = x^2$; $g(x) = x + 1$; $f(\ ) = (\ )^2$; $g(\ ) = (\ ) + 1$.

    Filling in the blanks:

    (a) $f(g(x)) = (x + 1)^2$ $g(x) = x + 1$

    $= x^2 + 2x + 1$;

    (b) $g(f(x)) = (x^2) + 1$ $f(x) = x^2$

    $= x^2 + 1$;

    (c) $f(f(x)) = (x^2)^2$ $f(x) = x^2$

    $= x^4$.

43. See graph in answer section of book.

2.2 Limits

1. Set the calculator in the radian mode and evaluate $\dfrac{\tan x}{x}$ for values near 0:

x :	0.1	0.05	0.01	0.001
$\frac{\tan x}{x}$	1.003	1.0008	1.00003	1.0000003

The values approach 1.

5. Set the calculator in the radian mode and evaluate $\sec x - \tan x$ for values near $\dfrac{\pi}{2} = 1.570796$.

x :	1.5 $(< \frac{\pi}{2})$	1.6 $(> \frac{\pi}{2})$	1.57 $(< \frac{\pi}{2})$	1.571 $(> \frac{\pi}{2})$	1.5705 $(< \frac{\pi}{2})$
$\sec x - \tan x$:	0.04	-0.01	4.0×10^{-4}	-1.0×10^{-4}	1.5×10^{-4}

Based on these calculations, the limit appears to be 0.

7. Since

$$\frac{x^2 + 4x}{x} = \frac{x(x + 4)}{x} = x + 4,$$

the two functions are

$$y = \frac{x^2 + 4x}{x} \quad \text{and} \quad y = x + 4.$$

As $x \to 0$, $y \to 4$.

9. Since

$$\frac{x^2 - 3x - 4}{x - 4} = \frac{(x - 4)(x + 1)}{x - 4} = x + 1,$$

the two functions are

$$y = \frac{x^2 - 3x - 4}{x - 4} \quad \text{and} \quad y = x + 1.$$

As $x \to 4$, $y \to 5$.

11. The function $y = x^2$ is defined for all x. As x approaches 0, y approaches 0.

13. The function $f(x) = x^4 - 3x + 2$ is defined at $x = 1$. As x approaches 1, $f(x)$ approaches $1^4 - 3(1) + 2 = 0$.

15. Since the function $f(x) = \dfrac{x^2 - 4}{x - 2}$ is undefined at $x = 2$, we evaluate the limit as follows:

$$\lim_{x \to 2} \frac{x^2 - 4}{x - 2} = \lim_{x \to 2} \frac{(x - 2)(x + 2)}{x - 2} = \lim_{x \to 2} (x + 2) = 4.$$

(Since x approaches 2 but is never equal to 2, division by 0 has been avoided.)

17. Since the function $f(x) = \dfrac{x^2 - x}{x}$ is undefined at $x = 0$, we evaluate the limit as follows:

$$\lim_{x \to 0} \frac{x^2 - x}{x} = \lim_{x \to 0} \frac{x(x - 1)}{x} = \lim_{x \to 0} (x - 1) = -1.$$

(Since x approaches 0 but is never equal to 0, division by 0 has been avoided.)

19. The function is undefined at $x = -4$.

$$\lim_{x \to -4} \frac{x^2 - 16}{x + 4} = \lim_{x \to -4} \frac{(x - 4)(x + 4)}{x + 4} = \lim_{x \to -4} (x - 4) = -8.$$

21. The function is undefined at $x = 2$.

$$\lim_{x \to 2} \frac{x^2 - 4x + 4}{x - 2} = \lim_{x \to 2} \frac{(x - 2)^2}{x - 2} = \lim_{x \to 2} (x - 2) = 0.$$

23. The function is undefined at $x = 2$.

$$\lim_{x \to 2} \frac{x^2 + x - 6}{x - 2} = \lim_{x \to 2} \frac{(x - 2)(x + 3)}{x - 2} = \lim_{x \to 2} (x + 3) = 5.$$

25. The function is defined at $x = 4$. So we obtain the limit by inspection: as $x \to 4$

$$\frac{x^2 + x - 8}{x + 4} \to \frac{4^2 + 4 - 8}{4 + 4} = \frac{12}{8} = \frac{3}{2}.$$

27. $\displaystyle\lim_{x \to 1} \frac{x^3 - 2x^2 + x}{x - 1} = \lim_{x \to 1} \frac{x(x^2 - 2x + 1)}{x} = \lim_{x \to 1} (x^2 - 2x + 1) = 0.$

29. $\displaystyle\lim_{r \to 3} \frac{r^2 - 9}{3 - r} = \lim_{r \to 3} \frac{(r - 3)(r + 3)}{(-1)(r - 3)} = \lim_{r \to 3} \frac{r + 3}{-1} = -6.$

31. $\displaystyle\lim_{x \to 5} \frac{25 - x^2}{5 - x} = \lim_{x \to 5} \frac{(5 - x)(5 + x)}{5 - x} = \lim_{x \to 5} (5 + x) = 10.$

33. Since the function $f(x) = \dfrac{4x^2 - 2}{x}$ is defined at $x = 1$, we obtain the limit by inspection:
as $x \to 1$

$$\frac{4x^2 - 2}{1} \to 2.$$

35. Since the function is defined at $x = 4$, we obtain the limit by inspection: as $x \to 4$

$$\frac{x^2 + 6x + 4}{x + 4} \to \frac{11}{2}.$$

37. Dividing numerator and denominator by x^3, the highest power of x, we obtain

$$\lim_{x\to\infty} \frac{4 - 2/x^2 + 1/x^3}{5 + 3/x - 1/x^2} = \frac{4 - 0 + 0}{5 + 0 - 0} = \frac{4}{5}.$$

39. Dividing numerator and denominator by x^2, the highest power of x, we get

$$\lim_{x\to\infty} \frac{3 + 2/x}{4 - 3/x^2} = \frac{3 + 0}{4 - 0} = \frac{3}{4}.$$

41. Dividing numerator and denominator by y^2, the highest power of y, we obtain

$$\lim_{y\to\infty} \frac{1 - 4y}{y^2 + 1} = \lim_{y\to\infty} \frac{1/y^2 - 4/y}{1 + 1/y^2} = \frac{0}{1} = 0.$$

(The conclusion also follows from the fact that the denominator is of higher degree than the numerator.)

43. Dividing numerator and denominator by x:

$$\lim_{x\to\infty} \frac{3 + 4/x}{9 - 7/x + 6/x^2} = \frac{3}{9} = \frac{1}{3}.$$

45. Divide numerator and denominator by x^2:

$$\lim_{x\to\infty} \frac{1/x^2 - 12}{2/x^2 + 6/x - 6} = \frac{-12}{-6} = 2.$$

47. $\displaystyle\lim_{x\to\infty} \frac{\sqrt{x^2 + 6}}{x + 1} = \lim_{x\to\infty} \frac{\sqrt{x^2 + 6}}{\sqrt{(x+1)^2}} = \lim_{x\to\infty} \sqrt{\frac{x^2 + 6}{(x+1)^2}} = \lim_{x\to\infty} \sqrt{\frac{x^2 + 6}{x^2 + 2x + 1}} = \sqrt{1} = 1.$

49. Rationalize the numerator:

$$\begin{aligned}
\lim_{x\to\infty} (x - \sqrt{x^2 - 1}) &= \lim_{x\to\infty} \frac{x - \sqrt{x^2 - 1}}{1} \cdot \frac{x + \sqrt{x^2 - 1}}{x + \sqrt{x^2 - 1}} \\
&= \lim_{x\to\infty} \frac{x^2 - (\sqrt{x^2 - 1})^2}{x + \sqrt{x^2 - 1}} = \lim_{x\to\infty} \frac{x^2 - (x^2 - 1)}{x + \sqrt{x^2 - 1}} \\
&= \lim_{x\to\infty} \frac{1}{x + \sqrt{x^2 - 1}} = 0.
\end{aligned}$$

51. (a) As $x \to 0$ from the right (through positive values), $1/x$ gets large beyond all bounds. So

$$\lim_{x\to 0+} 3^{1/x} = \infty. \quad \text{(no limit)}$$

(b) As $x \to 0$ from the left (through negative values), $1/x$ becomes large and negative. As a result

$$\lim_{x\to 0-} 3^{1/x} = 0.$$

53. By the way the function is defined, its value at $x = 2$ is -1, that is $f(2) = -1$. But this value does not match the right-hand limit: $\displaystyle\lim_{x\to 2+} f(x) = 1$.

So the condition $\displaystyle\lim_{x\to a} f(x) = f(a)$ has been violated.

55. $\lim\limits_{x \to -1} \dfrac{x^2 - 1}{x + 1} = \lim\limits_{x \to -1} \dfrac{(x - 1)(x + 1)}{x + 1} = -2.$

But $f(x)$ is undefined at $x = -1$, that is, $f(-1)$ does not exist, even though the limit exists.

This violates the condition

$$\lim_{x \to -1} f(x) = f(-1).$$

2.4   The Derivative by the Four-Step Process

1. To find $f(x + \Delta x)$, write $f(\ ) = 2(\ ) + 1$ and fill in the blanks.

Step 1. $y + \Delta y = f(x + \Delta x) = 2(x + \Delta x) + 1$

Step 2. $\Delta y = 2(x + \Delta x) + 1 - (2x + 1) = 2x + 2\Delta x + 1 - 2x - 1 = 2\Delta x$

Step 3. $\dfrac{\Delta y}{\Delta x} = \dfrac{2\Delta x}{\Delta x} = 2$

Step 4. $f'(x) = \lim\limits_{\Delta x \to 0} \dfrac{\Delta y}{\Delta x} = \lim\limits_{\Delta x \to 0} 2 = 2$

3. To find $f(x + \Delta x)$, write $f(\ ) = 2 - 3(\ )$ and fill in the blanks.

Step 1. $y + \Delta y = f(x + \Delta x) = 2 - 3(x + \Delta x)$

Step 2. $\Delta y = 2 - 3(x + \Delta x) - (2 - 3x) = 2 - 3x - 3\Delta x - 2 + 3x = -3\Delta x$

Step 3. $\dfrac{\Delta y}{\Delta x} = \dfrac{-3\Delta x}{\Delta x} = -3$

Step 4. $f'(x) = \lim\limits_{\Delta x \to 0} \dfrac{\Delta y}{\Delta x} = \lim\limits_{\Delta x \to 0} (-3) = -3$

5. To find $f(x + \Delta x)$, write $f(\ ) = (\ )^2 + 1$ and fill in the blanks.

Step 1. $y + \Delta y = f(x + \Delta x) = (x + \Delta x)^2 + 1$

Step 2. $\Delta y = (x + \Delta x)^2 + 1 - (x^2 + 1) = x^2 + 2x\Delta x + (\Delta x)^2 + 1 - x^2 - 1 = 2x\Delta x + (\Delta x)^2$

Step 3. $\dfrac{\Delta y}{\Delta x} = \dfrac{2x\Delta x + (\Delta x)^2}{\Delta x} = \dfrac{\Delta x(2x + \Delta x)}{\Delta x} = 2x + \Delta x$

Step 4. $f'(x) = \lim\limits_{\Delta x \to 0} \dfrac{\Delta y}{\Delta x} = \lim\limits_{\Delta x \to 0} (2x + \Delta x) = 2x$

7. To find $f(x + \Delta x)$, write $f(\ ) = 2(\ )^2 - (\ )$ and fill in the blanks.

Step 1. $y + \Delta y = f(x + \Delta x) = 2(x + \Delta x)^2 - (x + \Delta x)$

Step 2. $\begin{aligned} \Delta y &= 2(x + \Delta x)^2 - (x + \Delta x) - (2x^2 - x) \\ &= 2x^2 + 4x\Delta x + 2(\Delta x)^2 - x - \Delta x - 2x^2 + x \\ &= 4x\Delta x + 2(\Delta x)^2 - \Delta x \end{aligned}$

Step 3. $\dfrac{\Delta y}{\Delta x} = \dfrac{4x\Delta x + 2(\Delta x)^2 - \Delta x}{\Delta x} = \dfrac{\Delta x(4x + 2\Delta x - 1)}{\Delta x} = 4x + 2\Delta x - 1$

Step 4. $f'(x) = \lim\limits_{\Delta x \to 0} \dfrac{\Delta y}{\Delta x} = \lim\limits_{\Delta x \to 0} (4x + 2\Delta x - 1) = 4x - 1$

9. $f(\ ) = (\ )^3 - 3(\ )^2$

$\underline{\text{Step 1.}}$ $y + \Delta y \;=\; f(x + \Delta x) = (x + \Delta x)^3 - 3(x + \Delta x)^2$

$\underline{\text{Step 2.}}$ $\Delta y \;=\; (x + \Delta x)^3 - 3(x + \Delta x)^2 - (x^3 - 3x^2)$

$\;=\; x^3 + 3x^2\Delta x + 3x(\Delta x)^2 + (\Delta x)^3 - 3x^2 - 6x\Delta x - 3(\Delta x)^2 - x^3 + 3x^2$

$\;=\; 3x^2\Delta x + 3x(\Delta x)^2 + (\Delta x)^3 - 6x\Delta x - 3(\Delta x)^2$

$\;=\; \Delta x[3x^2 + 3x\Delta x + (\Delta x)^2 - 6x - 3\Delta x]$

$\underline{\text{Step 3.}}$ $\dfrac{\Delta y}{\Delta x} \;=\; \dfrac{\Delta x[3x^2 + 3x\Delta x + (\Delta x)^2 - 6x - 3\Delta x]}{\Delta x}$

$\;=\; 3x^2 + 3x\Delta x + (\Delta x)^2 - 6x - 3\Delta x$

$\underline{\text{Step 4.}}$ $f'(x) \;=\; \lim\limits_{\Delta x \to 0}[3x^2 + 3x\Delta x + (\Delta x)^2 - 6x - 3\Delta x]$

$\;=\; 3x^2 - 6x$

11. $\underline{\text{Step 1.}}$ $y + \Delta y = \dfrac{1}{x + \Delta x + 1}$

$\underline{\text{Step 2.}}$ $\Delta y \;=\; \dfrac{1}{x + \Delta x + 1} - \dfrac{1}{x + 1}$

$\;=\; \dfrac{1}{x + \Delta x + 1} \cdot \dfrac{x + 1}{x + 1} - \dfrac{1}{x + 1} \cdot \dfrac{x + \Delta x + 1}{x + \Delta x + 1}$

$\;=\; \dfrac{(x + 1) - (x + \Delta x + 1)}{(x + \Delta x + 1)(x + 1)} = \dfrac{-\Delta x}{(x + \Delta x + 1)(x + 1)}$

$\underline{\text{Step 3.}}$ $\dfrac{\Delta y}{\Delta x} = \dfrac{-\Delta x}{(x + \Delta x + 1)(x + 1)} \cdot \dfrac{1}{\Delta x} = \dfrac{-1}{(x + \Delta x + 1)(x + 1)}$

$\underline{\text{Step 4.}}$ $f'(x) = \lim\limits_{\Delta x \to 0}\dfrac{\Delta y}{\Delta x} = \lim\limits_{\Delta x \to 0}\dfrac{-1}{(x + \Delta x + 1)(x + 1)} = -\dfrac{1}{(x + 1)^2}$

13. $\underline{\text{Step 1.}}$ $y + \Delta y = f(x + \Delta x) = \dfrac{1}{(x + \Delta x)^2}$

$\underline{\text{Step 2.}}$ $\Delta y = \dfrac{1}{(x + \Delta x)^2} - \dfrac{1}{x^2}$

$\underline{\text{Note:}}$ The lowest common denominator is $x^2(x + \Delta x)^2$. We therefore have

$\Delta y \;=\; \dfrac{1}{(x + \Delta x)^2} \cdot \dfrac{x^2}{x^2} - \dfrac{1}{x^2} \cdot \dfrac{(x + \Delta x)^2}{(x + \Delta x)^2}$

$\;=\; \dfrac{x^2 - (x + \Delta x)^2}{x^2(x + \Delta x)^2} = \dfrac{x^2 - x^2 - 2x\Delta x - (\Delta x)^2}{x^2(x + \Delta x)^2}$

$\;=\; \dfrac{\Delta x(-2x - \Delta x)}{x^2(x + \Delta x)^2}.$

$\underline{\text{Step 3.}}$ $\dfrac{\Delta y}{\Delta x} = \dfrac{\Delta x(-2x - \Delta x)}{x^2(x + \Delta x)^2} \cdot \dfrac{1}{\Delta x} = \dfrac{-2x - \Delta x}{x^2(x + \Delta x)^2}$

$\underline{\text{Step 4.}}$ $f'(x) = \lim\limits_{\Delta x \to 0}\dfrac{-2x - \Delta x}{x^2(x + \Delta x)^2} = \dfrac{-2x}{x^2 x^2} = -\dfrac{2}{x^3}$

15. $\underline{\text{Step 1.}}$ $y + \Delta y = \dfrac{1}{1 - (x + \Delta x)^2}$

$\underline{\text{Step 2.}}$ $\Delta y = \dfrac{1}{1 - (x + \Delta x)^2} - \dfrac{1}{1 - x^2} = \dfrac{(1 - x^2) - [1 - (x + \Delta x)^2]}{[1 - (x + \Delta x)^2](1 - x^2)} = \dfrac{2x\Delta x + (\Delta x)^2}{[1 - (x + \Delta x)^2](1 - x^2)}$

$\underline{\text{Step 3.}}$ $\dfrac{\Delta y}{\Delta x} = \dfrac{\Delta x(2x + \Delta x)}{[1 - (x + \Delta x)^2](1 - x^2)} \cdot \dfrac{1}{\Delta x} = \dfrac{2x + \Delta x}{[1 - (x + \Delta x)^2](1 - x^2)}$

$\underline{\text{Step 4.}}$ $f'(x) = \lim\limits_{\Delta x \to 0}\dfrac{\Delta y}{\Delta x} = \lim\limits_{\Delta x \to 0}\dfrac{2x + \Delta x}{[1 - (x + \Delta x)^2](1 - x^2)} = \dfrac{2x}{(1 - x^2)^2}$

17. Step 1. $y + \Delta y = f(x + \Delta x) = \sqrt{x + \Delta x}$

    Step 2. $\Delta y = \sqrt{x + \Delta x} - \sqrt{x}$

    As in Example 4, we rationalize the numerator by multiplying both numerator and denominator by the quantity $\sqrt{x + \Delta x} - \sqrt{x}$:

    $$
    \begin{aligned}
    \Delta y &= \frac{\sqrt{x + \Delta x} - \sqrt{x}}{1} \cdot \frac{\sqrt{x + \Delta x} + \sqrt{x}}{\sqrt{x + \Delta x} + \sqrt{x}} \\
    &= \frac{(\sqrt{x + \Delta x})^2 - (\sqrt{x})^2}{\sqrt{x + \Delta x} + \sqrt{x}} = \frac{x + \Delta x - x}{\sqrt{x + \Delta x} + \sqrt{x}} \\
    &= \frac{\Delta x}{\sqrt{x + \Delta x} + \sqrt{x}}
    \end{aligned}
    $$

    Step 3. $\dfrac{\Delta y}{\Delta x} = \dfrac{\Delta x}{\sqrt{x + \Delta x} + \sqrt{x}} \cdot \dfrac{1}{\Delta x} = \dfrac{1}{\sqrt{x + \Delta x} + \sqrt{x}}$

    Step 4. $f'(x) = \lim\limits_{\Delta x \to 0} \dfrac{1}{\sqrt{x + \Delta x} + \sqrt{x}} = \dfrac{1}{\sqrt{x} + \sqrt{x}} = \dfrac{1}{2\sqrt{x}}$

19. Step 1. $y + \Delta y = \sqrt{1 - (x + \Delta x)}$

    Step 2. $\Delta y = \sqrt{1 - (x + \Delta x)} - \sqrt{1 - x}$

    Step 3. $\dfrac{\Delta y}{\Delta x} = \dfrac{\sqrt{1 - (x + \Delta x)} - \sqrt{1 - x}}{\Delta x} \cdot \dfrac{\sqrt{1 - (x + \Delta x)} + \sqrt{1 - x}}{\sqrt{1 - (x + \Delta x)} + \sqrt{1 - x}}$

    $\qquad = \dfrac{[1 - (x + \Delta x)] - (1 - x)}{\Delta x[\sqrt{1 - (x + \Delta x)} + \sqrt{1 - x}]} = \dfrac{-1}{\sqrt{1 - (x + \Delta x)} + \sqrt{1 - x}}$

    Step 4. $f'(x) = \lim\limits_{\Delta x \to 0} \dfrac{\Delta y}{\Delta x} = \lim\limits_{\Delta x \to 0} \dfrac{-1}{\sqrt{1 - (x + \Delta x)} + \sqrt{1 - x}} = -\dfrac{1}{2\sqrt{1 - x}}$

21. Similar to Exercise 5: $f'(x) = 2x$. Thus $f'(-1) = 2(-1) = -2$; $f'(2) = 4$; $f'(3) = 6$.

23. Step 1. $y + \Delta y = \dfrac{1}{\sqrt{x + \Delta x}}$

    Step 2. $\Delta y = \dfrac{1}{\sqrt{x + \Delta x}} - \dfrac{1}{\sqrt{x}}$

    Step 3. $\dfrac{\Delta y}{\Delta x} = \dfrac{1}{\Delta x} \cdot \dfrac{\sqrt{x} - \sqrt{x + \Delta x}}{\sqrt{x + \Delta x}\sqrt{x}} \cdot \dfrac{\sqrt{x} + \sqrt{x + \Delta x}}{\sqrt{x} + \sqrt{x + \Delta x}}$

    $\qquad = \dfrac{1}{\Delta x} \cdot \dfrac{x - (x + \Delta x)}{\sqrt{x + \Delta x}\sqrt{x}(\sqrt{x} + \sqrt{x + \Delta x})}$

    $\qquad = \dfrac{-1}{\sqrt{x + \Delta x}\sqrt{x}(\sqrt{x} + \sqrt{x + \Delta x})}$

    Step 4. $f'(x) = \lim\limits_{\Delta x \to 0} \dfrac{-1}{\sqrt{x + \Delta x}\sqrt{x}(\sqrt{x} + \sqrt{x + \Delta x})} = \dfrac{-1}{\sqrt{x}\sqrt{x}(\sqrt{x} + \sqrt{x})} = -\dfrac{1}{2x\sqrt{x}}$

    $$f'(4) = -\frac{1}{2 \cdot 4\sqrt{4}} = -\frac{1}{16}; \quad f'(9) = -\frac{1}{2 \cdot 9\sqrt{9}} = -\frac{1}{54}$$

25. First we need to find the slope of the tangent line at the given point. From $f(\ ) = (\ )^2 - 1$:

    <u>Step 1.</u> $y + \Delta y \quad = \quad f(x + \Delta x) = (x + \Delta x)^2 - 1$

    <u>Step 2.</u> $\Delta y \quad = \quad (x + \Delta x)^2 - 1 - (x^2 - 1)$

$$= \quad x^2 + 2x\Delta x + (\Delta x)^2 - 1 - x^2 + 1$$

$$= \quad 2x\Delta x + (\Delta x)^2$$

$$= \quad \Delta x(2x + \Delta x)$$

    <u>Step 3.</u> $\dfrac{\Delta y}{\Delta x} \quad = \quad \dfrac{\Delta x(2x + \Delta x)}{\Delta x} = 2x + \Delta x$

    <u>Step 4.</u> $f'(x) \quad = \quad \lim\limits_{\Delta x \to 0}(2x + \Delta x) = 2x$

So $f'(x) = 2x$ and $f'(2) = 4$. Using the point-slope form $y - y_1 = m(x - x_1)$, we have

$$y - 3 \quad = \quad 4(x - 2) \qquad (x_1, y_1) = (2, 3); \ m = 4$$

$$y \quad = \quad 4x - 5.$$

27. $f'(x) \quad = \quad \lim\limits_{\Delta x \to 0} \dfrac{\frac{2}{x+\Delta x} - \frac{2}{x}}{\Delta x}$

$$= \quad \lim\limits_{\Delta x \to 0} \dfrac{1}{\Delta x} \cdot \dfrac{2x - 2x - 2\Delta x}{(x + \Delta x)x}$$

$$= \quad \lim\limits_{\Delta x \to 0} \dfrac{-2}{(x + \Delta x)x} = \dfrac{-2}{x^2}$$

Thus $f'(x) = \dfrac{-2}{x^2}$ and $f'(2) = -\dfrac{2}{2^2} = -\dfrac{1}{2}$. Using the point-slope form $y - y_1 = m(x - x_1)$, we have

$$y - 1 \quad = \quad -\tfrac{1}{2}(x - 2) \qquad (x_1, y_1) = (2, 1)$$

$$y \quad = \quad \tfrac{1}{2}(4 - x).$$

2.5 Derivatives of Polynomials

9. $y \quad = \quad 5x^3 - 7x^2 + 2$

$$y' \quad = \quad 5(3x^2) - 7(2x) + 0$$

$$= \quad 15x^2 - 14x$$

11. $y \quad = \quad 7x^3 - x^2 - x + 2$

$$y' \quad = \quad 7(3x^2) - 2x - 1 + 0 = 21x^2 - 2x - 1$$

13. $y \quad = \quad \tfrac{1}{3}x^3 + \tfrac{1}{2}x^2 + x$

$$y' \quad = \quad \tfrac{1}{3}(3x^2) + \tfrac{1}{2}(2x) + 1$$

$$= \quad x^2 + x + 1$$

15. $y \quad = \quad \tfrac{1}{2}x^2 - \tfrac{1}{3}x^3$

$$y' \quad = \quad \tfrac{1}{2}(2x) - \tfrac{1}{3}(3x^2) = x - x^2$$

17. $y \quad = \quad 20x^{10} - 24x^6 + 2x^3 - \sqrt{3}$

$$y' \quad = \quad 20(10x^9) - 24(6x^5) + 2(3x^2) + 0 \qquad \text{note that } \sqrt{3} \text{ is a constant}$$

$$= \quad 200x^9 - 144x^5 + 6x^2$$

19. $\quad y \;=\; \frac{1}{5}t^7 - \frac{1}{\sqrt{2}}t^5 + \frac{1}{3}$

$\quad \dfrac{dy}{dt} \;=\; \frac{1}{5}(7t^6) - \frac{1}{\sqrt{2}}(5t^4) + 0$

$\qquad\; =\; \frac{7}{5}t^6 - \frac{5}{\sqrt{2}}t^4$

21. $\quad y \;=\; \frac{1}{6}R^6 + \frac{1}{5}R^4 - \sqrt{3}$

$\quad \dfrac{dy}{dR} \;=\; \frac{1}{6}(6R^5) + \frac{1}{5}(4R^3) - 0$

$\qquad\; =\; R^5 + \frac{4}{5}R^3$

23. $f(x) = 4 - x^2$; $f'(x) = -2x$; $f'(2) = -4$. Slope of tangent line: -4; slope of normal line: $\frac{1}{4}$; point: $(2, 0)$.

$\quad y - y_1 \;=\; m(x - x_1)$

$\quad y - 0 \;=\; -4(x - 2) \qquad$ tangent line

$\quad y - 0 \;=\; \frac{1}{4}(x - 2) \qquad$ normal line

(See graphs in answer section in the back of the book.)

25. $f(x) = \frac{1}{3}x^3 + x$; $f'(x) = x^2 + 1$; $f'(1) = 1^2 + 1 = 2$. Slope of tangent line: 2; slope of normal line: $-\frac{1}{2}$; point: $(1, \frac{4}{3})$.

$\quad y - y_1 \;=\; m(x - x_1)$

$\quad y - \frac{4}{3} \;=\; 2(x - 1) \qquad$ tangent line

$\quad y - \frac{4}{3} \;=\; -\frac{1}{2}(x - 1) \qquad$ normal line

(Simplified forms and graphs are in the answer section in the back of the book.)

27. $y = cu$ where u is a function of x.

$\underline{\text{Step 1.}}$ $y + \Delta y = c(u + \Delta u)$

$\underline{\text{Step 2.}}$ $\Delta y = c(u + \Delta u) - cu = c\Delta u$

$\underline{\text{Step 3.}}$ $\dfrac{\Delta y}{\Delta x} = c\dfrac{\Delta u}{\Delta x}$

$\underline{\text{Step 4.}}$ $\dfrac{dy}{dx} = \lim\limits_{\Delta x \to 0} \dfrac{\Delta y}{\Delta x} = \lim\limits_{\Delta x \to 0} c\dfrac{\Delta u}{\Delta x} = c\dfrac{du}{dx}$

2.6   Instantaneous Rates of Change

1. $\quad s \;=\; 2t^2 \qquad\qquad\qquad\qquad\qquad\quad v = 0$ when $4t = 0$ or $t = 0$.

$\quad v \;=\; \dfrac{ds}{dt} = 2(2t) = 4t$

$\quad a \;=\; \dfrac{dv}{dt} = 4$

3. $\quad s \;=\; t^2 - 2t + 1 \qquad\qquad\qquad\quad v = 0$ when $2t - 2 = 0$ or $t = 1$.

$\quad v \;=\; \dfrac{ds}{dt} = 2t - 2$

$\quad a \;=\; \dfrac{dv}{dt} = 2$

5. $s = 2t - t^2$

$v = \dfrac{ds}{dt} = 2(1) - 2t = 2 - 2t$

$a = \dfrac{dv}{dt} = 0 - 2 = -2$

$v = 0$ when $\quad 2 - 2t = 0$

$\qquad\qquad\qquad -2t = -2$

$\qquad\qquad\qquad\quad t = 1.$

7. $s = 12t - 2t^2$

$v = \dfrac{ds}{dt} = 12 - 4t$

$a = \dfrac{dv}{dt} = -4$

$v = 0$ when $12 - 4t = 0$ or $t = 3.$

9. $s = 3t^3 + 2t^2 + 2$

$v = \dfrac{ds}{dt} = 9t^2 + 4t$

$a = \dfrac{dv}{dt} = 18t + 4$

$v = 0$ when $\quad 9t^2 + 4t = 0$

$\qquad\qquad\qquad t(9t + 4) = 0,$

that is, when $t = 0$ or $\quad 9t + 4 = 0$

$\qquad\qquad\qquad\qquad\qquad\quad t = -\dfrac{4}{9}.$

11. $s = 50t^2$

$v = \dfrac{ds}{dt} = 50(2t) = 100t\Big|_{t=10} = 1000\,\text{m/s}$

13. $y = 50t - 0.83t^2$

$v = \dfrac{dy}{dt} = 50 - 0.83(2t)\Big|_{t=10\,\text{s}} = 33.4\,\text{m/s}$

15. If x is the length of the side, then $A = x^2$;

$$\frac{dA}{dx} = 2x\Big|_{x=2} = 4\,\text{cm}^2 \text{ per cm.}$$

17. The instantaneous rate of change of R with respect to T is $\dfrac{dR}{dT}$:

$R = 20.0 + 0.520T + 0.00973T^2$

$\dfrac{dR}{dT} = 0 + 0.520(1) + 0.00973(2T)$

$\qquad = 0.520 + (0.00973)(2)T\Big|_{T=125}$

$\qquad = 0.520 + (0.00973)(2)(125)$

$\qquad = 2.95\,\dfrac{\Omega}{^\circ\text{C}}.$

19. $P = 15i^2$

$\dfrac{dP}{di} = 30i\Big|_{i=2.1} = 63\,\text{W/A}$

21. The instantaneous rate of change of S with respect to T is

$\dfrac{dS}{dT} = 0 - 0.000085(2T)\Big|_{T=145}$

$\qquad = -0.000085(2)(145) = -0.025\,\text{lb/}^\circ\text{F.}$

23. $\theta = 3t^2$

$\omega = \dfrac{d\theta}{dt} = 6t\Big|_{t=1} = 6\,\text{rad/s}$

$\alpha = \dfrac{d\omega}{dt} = 6\,\text{rad/s}^2$

25. $q = 10.0t^2 + 2.0t$

$$i = \frac{dq}{dt} = 10.0(2t) + 2.0 \Big|_{t=0.01\,s} = 2.2\,A$$

27. $i = C\dfrac{dv}{dt} = 1.0 \times 10^{-2}(20t)\Big|_{t=1} = 0.20\,A$

29. $P = \dfrac{dW}{dt} = \dfrac{d}{dt}(5t^4 + 2t) = 20t^3 + 2\Big|_{t=1} = 22\,W$

31. $V = 6.4P^{-1/2}$; $\dfrac{dV}{dP} = 6.4\left(-\dfrac{1}{2}\right)P^{-3/2}$

$$\beta = -\frac{1}{V}\frac{dV}{dP} = -\frac{1}{6.4P^{-1/2}}(6.4)\left(-\frac{1}{2}\right)P^{-3/2} = \frac{1}{2P}$$

2.7   Differentiation Formulas

Group A

1. $y = 4x^4 - 4x^2 + 8$; $y' = 4(4x^3) - 4(2x) + 0 = 16x^3 - 8x$

3. $y = x^{-1}$; $y' = -1x^{-2} = -\dfrac{1}{x^2}$

5. $y = x^5 - 3x^{-3} + 2x^{-2}$. By (2.4),
$y' = 5x^4 - 3(-3)x^{-4} + 2(-2)x^{-3} = 5x^4 + 9x^{-4} - 4x^{-3}.$

7. $y = x^{5/2} + x^{-1/2}$; $y' = \dfrac{5}{2}x^{3/2} - \dfrac{1}{2}x^{-3/2} = \dfrac{5}{2}x^{3/2} - \dfrac{1}{2x^{3/2}}$

9. $y = \dfrac{2}{x^4} - \dfrac{4}{\sqrt{x}} + \dfrac{1}{\sqrt{2}}$

$y = 2x^{-4} - 4x^{-1/2} + \dfrac{1}{\sqrt{2}}$

By (2.4) and (2.6)

$y' = 2(-4)x^{-5} - 4(-\frac{1}{2})x^{-3/2} + 0$

(recall that the derivative of a constant is zero)

$y' = -\dfrac{8}{x^5} + \dfrac{2}{x^{3/2}}.$

11. $y = (2x^2 - 3)^4$. By the generalized power rule,

$\begin{aligned} y' &= 4(2x^2 - 3)^3\frac{d}{dx}(2x^2 - 3) \\ &= 4(2x^2 - 3)^3(4x) = 16x(2x^2 - 3)^3. \end{aligned}$

13. $y = (4 - 3x^2)^4$. By the generalized power rule

$\begin{aligned} y' &= 4(4 - 3x^2)^3\frac{d}{dx}(4 - 3x^2) \\ y' &= 4(4 - 3x^2)^3(-6x) = -24x(4 - 3x^2)^3. \end{aligned}$

15. $y = (x^{10} + 1)^{10}$. By the generalized power rule,

$$\begin{aligned} y' &= 10(x^{10} + 1)^9 \frac{d}{dx}(x^{10} + 1) = 10(x^{10} + 1)^9(10x^9) \\ &= 100x^9(x^{10} + 1)^9. \end{aligned}$$

17. $y = \dfrac{2}{\sqrt{x^3 - 3x}}$. We avoid the quotient rule by writing the function as follows:

$y = 2(x^3 - 3x)^{-1/2}$. By the generalized power rule,

$$\begin{aligned} y' &= 2\left(-\frac{1}{2}\right)(x^3 - 3x)^{-3/2}\frac{d}{dx}(x^3 - 3x) \\ &= -(x^3 - 3x)^{-3/2}(3x^2 - 3) \\ &= -\frac{3x^2 - 3}{(x^3 - 3x)^{3/2}} = -\frac{3(x^2 - 1)}{(x^3 - 3x)^{3/2}}. \end{aligned}$$

19. $v = \sqrt[3]{t^3 - 3} = (t^3 - 3)^{1/3}$. By the power rule,

$$\begin{aligned} \frac{dv}{dt} &= \frac{1}{3}(t^3 - 3)^{-2/3}\frac{d}{dt}(t^3 - 3) = \frac{1}{3}(t^3 - 3)^{-2/3}(3t^2) \\ &= \frac{t^2}{(t^3 - 3)^{2/3}}. \end{aligned}$$

21. $y = x^3(x + 1)^2$. By the product rule:

$$\begin{aligned} y' &= x^3\frac{d}{dx}(x + 1)^2 + (x + 1)^2\frac{d}{dx}(x^3) \\ &= x^3 \cdot 2(x + 1) + (x + 1)^2(3x^2) \end{aligned}$$

Note that the terms on the right have common factor $x^2(x + 1)$. So

$$\begin{aligned} y' &= x^2(x + 1)[x \cdot 2 + (x + 1)(3)] \\ &= x^2(x + 1)(5x + 3). \end{aligned}$$

23. $y = 2x^4(x + 2)^2$. By the product rule,

$$\begin{aligned} y' &= 2x^4\frac{d}{dx}(x + 2)^2 + (x + 2)^2\frac{d}{dx}(2x^4) \\ &= 2x^4 \cdot 2(x + 2) + (x + 2)^2(8x^3) \\ &= 4x^3(x + 2)[x + 2(x + 2)] \qquad \text{common factor: } 4x^3(x + 2) \\ &= 4x^3(x + 2)(3x + 4). \end{aligned}$$

25. $y = x^2(x^2 - 5)^2$

By the product rule,

$$y' = x^2\frac{d}{dx}(x^2 - 5)^2 + (x^2 - 5)^2\frac{d}{dx}(x^2).$$

Now by the generalized power rule,

$$y' = x^2 \cdot 2(x^2 - 5)(2x) + (x^2 - 5)^2(2x).$$

Factoring $2x(x^2 - 5)$, we get

$$\begin{aligned} y' &= 2x(x^2 - 5)[x^2 \cdot 2 + (x^2 - 5)] \\ &= 2x(x^2 - 5)(3x^2 - 5). \end{aligned}$$

27. $y = 4x(x + 5)^4$. By the product rule,

$$\begin{aligned} y' &= 4x\frac{d}{dx}(x + 5)^4 + (x + 5)^4\frac{d}{dx}(4x) \\ &= 4x \cdot 4(x + 5)^3(1) + (x + 5)^4(4) \\ &= 4(x + 5)^3[4x + (x + 5)] \qquad \text{common factor: } 4(x + 5)^3 \\ &= 4(x + 5)^3(5x + 5) \\ &= 20(x + 5)^3(x + 1). \qquad \text{common factor: } 5 \end{aligned}$$

29. $y = 2x^3(x - 2)^3$. By the product rule,

$$\begin{aligned}
y' &= 2x^3 \frac{d}{dx}(x-2)^3 + (x-2)^3 \frac{d}{dx}(2x^3) \\
&= 2x^3 \cdot 3(x-2)^2 + (x-2)^3 \cdot 6x^2.
\end{aligned}$$

Factoring $6x^2(x-2)^2$, we get

$$\begin{aligned}
y' &= 6x^2(x-2)^2[x + (x-2)] = 6x^2(x-2)^2(2x-2) \\
&= 6x^2(x-2)^2 \cdot 2(x-1) = 12x^2(x-2)^2(x-1).
\end{aligned}$$

31. $y = \dfrac{x}{x-1}$. By the quotient rule,

$$\begin{aligned}
y' &= \frac{(x-1)\frac{d}{dx}(x) - x\frac{d}{dx}(x-1)}{(x-1)^2} = \frac{x-1-x}{(x-1)^2} \\
&= \frac{-1}{(x-1)^2}.
\end{aligned}$$

33. $P = \dfrac{t-2}{t^2+4}$. By the quotient rule:

$$\begin{aligned}
\frac{dP}{dt} &= \frac{(t^2+4)\frac{d}{dt}(t-2) - (t-2)\frac{d}{dt}(t^2+4)}{(t^2+4)^2} \\
&= \frac{(t^2+4)(1) - (t-2)(2t)}{(t^2+4)^2} \\
&= \frac{t^2+4-2t^2+4t}{(t^2+4)^2} = \frac{-t^2+4t+4}{(t^2+4)^2}.
\end{aligned}$$

35. $y = x^2\sqrt{x+1} = x^2(x+1)^{1/2}$. By the product rule,

$$\begin{aligned}
y' &= x^2 \frac{d}{dx}(x+1)^{1/2} + (x+1)^{1/2}\frac{d}{dx}(x^2) \\
&= x^2 \cdot \tfrac{1}{2}(x+1)^{-1/2} + (x+1)^{1/2} \cdot 2x \\
&= \frac{x^2}{2(x+1)^{1/2}} + 2x(x+1)^{1/2}.
\end{aligned}$$

The common denominator is $2(x+1)^{1/2}$:

$$\begin{aligned}
y' &= \frac{x^2}{2(x+1)^{1/2}} + \frac{2x(x+1)^{1/2}}{1} \cdot \frac{2(x+1)^{1/2}}{2(x+1)^{1/2}} \\
&= \frac{x^2 + 4x(x+1)}{2(x+1)^{1/2}} && a^{1/2} \cdot a^{1/2} = a \\
&= \frac{5x^2 + 4x}{2\sqrt{x+1}}.
\end{aligned}$$

37. $y = x(x^2+2)^{1/2}$. By the product rule:

$$\begin{aligned}
y' &= x\frac{d}{dx}(x^2+2)^{1/2} + (x^2+2)^{1/2}\frac{d}{dx}(x) \\
&= x \cdot \tfrac{1}{2}(x^2+2)^{-1/2}(2x) + (x^2+2)^{1/2}(1) && \text{power rule} \\
&= \frac{x^2}{(x^2+2)^{1/2}} + (x^2+2)^{1/2}.
\end{aligned}$$

Since the common denominator is $(x^2+2)^{1/2}$, we need to write the second fraction as

$$\frac{(x^2+2)^{1/2}}{1} \cdot \frac{(x^2+2)^{1/2}}{(x^2+2)^{1/2}}.$$

So

$$\begin{aligned}
y' &= \frac{x^2}{(x^2+2)^{1/2}} + \frac{(x^2+2)^{1/2}}{1} \cdot \frac{(x^2+2)^{1/2}}{(x^2+2)^{1/2}} \\
&= \frac{x^2 + (x^2+2)^{1/2}(x^2+2)^{1/2}}{(x^2+2)^{1/2}} = \frac{x^2 + (x^2+2)}{(x^2+2)^{1/2}} && a^{1/2} \cdot a^{1/2} = a \\
&= \frac{2x^2+2}{(x^2+2)^{1/2}} = \frac{2(x^2+1)}{\sqrt{x^2+2}}.
\end{aligned}$$

39. $R = \dfrac{s^2 - 3}{s - 2}$. By the quotient rule,

$$\frac{dR}{ds} = \frac{(s - 2)(2s) - (s^2 - 3)(1)}{(s - 2)^2} = \frac{s^2 - 4s + 3}{(s - 2)^2}.$$

Group B

1. $y = \sqrt{1 - x} = (1 - x)^{1/2}$. By the power rule,
$$y' = \tfrac{1}{2}(1 - x)^{-1/2}\frac{d}{dx}(1 - x) = \tfrac{1}{2}(1 - x)^{-1/2}(-1) = -\frac{1}{2(1 - x)^{1/2}} = -\frac{1}{2\sqrt{1 - x}}.$$

3. $y = (1 - x^2)^{1/3}$. By the power rule,
$$y' = \tfrac{1}{3}(1 - x^2)^{-2/3}(-2x) = -\frac{2x}{3(1 - x^2)^{2/3}}.$$

5. $y = 4(x - x^2)^{-1/4}$. By the generalized power rule,
$$\begin{aligned}
y' &= 4(-\tfrac{1}{4})(x - x^2)^{-5/4}\frac{d}{dx}(x - x^2) \\
&= -(x - x^2)^{-5/4}(1 - 2x) \\
&= -\frac{1 - 2x}{(x - x^2)^{5/4}} = \frac{2x - 1}{(x - x^2)^{5/4}}.
\end{aligned}$$

7. $n = \dfrac{m^3 + 2m}{m^2 - 8}$. By the quotient rule,
$$\begin{aligned}
\frac{dn}{dm} &= \frac{(m^2 - 8)\frac{d}{dm}(m^3 + 2m) - (m^3 + 2m)\frac{d}{dm}(m^2 - 8)}{(m^2 - 8)^2} \\
&= \frac{(m^2 - 8)(3m^2 + 2) - (m^3 + 2m)(2m)}{(m^2 - 8)^2}.
\end{aligned}$$
After collecting terms,
$$\frac{dn}{dm} = \frac{m^4 - 26m^2 - 16}{(m^2 - 8)^2}.$$

9. $y = x\sqrt{x^2 - 1} = x(x^2 - 1)^{1/2}$. By the product rule,
$$\begin{aligned}
y' &= x\frac{d}{dx}(x^2 - 1)^{1/2} + (x^2 - 1)^{1/2}\frac{d}{dx}(x) \\
&= x(\tfrac{1}{2})(x^2 - 1)^{-1/2}(2x) + (x^2 - 1)^{1/2} \qquad \text{generalized power rule} \\
&= \frac{x^2}{\sqrt{x^2 - 1}} + \sqrt{x^2 - 1}.
\end{aligned}$$
Since the common denominator is $\sqrt{x^2 - 1}$, we need to write the second term as
$$\frac{\sqrt{x^2 - 1}}{1} \cdot \frac{\sqrt{x^2 - 1}}{\sqrt{x^2 - 1}}.$$

So
$$\begin{aligned}
y' &= \frac{x^2}{\sqrt{x^2 - 1}} + \frac{\sqrt{x^2 - 1}}{1} \cdot \frac{\sqrt{x^2 - 1}}{\sqrt{x^2 - 1}} \\
&= \frac{x^2 + (x^2 - 1)}{\sqrt{x^2 - 1}} = \frac{2x^2 - 1}{\sqrt{x^2 - 1}}. \qquad \sqrt{A}\sqrt{A} = A
\end{aligned}$$

11. $y = 2x(1 - x)^{1/2}$. By the product rule,
$$\begin{aligned}
y' &= 2x \cdot \tfrac{1}{2}(1 - x)^{-1/2}(-1) + 2(1 - x)^{1/2} \\
&= \frac{-x}{(1 - x)^{1/2}} + \frac{2(1 - x)^{1/2}}{1} \cdot \frac{(1 - x)^{1/2}}{(1 - x)^{1/2}} \\
&= \frac{-x + 2(1 - x)}{(1 - x)^{1/2}} = \frac{2 - 3x}{\sqrt{1 - x}}.
\end{aligned}$$

13. $T = \theta^3(\theta + 7)^{1/2}$. By the product rule:

$$
\begin{aligned}
\frac{dT}{d\theta} &= \theta^3 \frac{d}{d\theta}(\theta + 7)^{1/2} + (\theta + 7)^{1/2}\frac{d}{d\theta}(\theta^3) \\
&= \theta^3 \cdot \tfrac{1}{2}(\theta + 7)^{-1/2} + (\theta + 7)^{1/2}(3\theta^2) \\
&= \frac{\theta^3}{2(\theta + 7)^{1/2}} + 3\theta^2(\theta + 7)^{1/2}.
\end{aligned}
$$

Since the common denominator is $2(\theta + 7)^{1/2}$, we need to write the second fraction as

$$
\frac{3\theta^2(\theta + 7)^{1/2}}{1} \cdot \frac{2(\theta + 7)^{1/2}}{2(\theta + 7)^{1/2}}.
$$

So

$$
\begin{aligned}
\frac{dT}{d\theta} &= \frac{\theta^3}{2(\theta + 7)^{1/2}} + \frac{3\theta^2(\theta + 7)^{1/2}}{1} \cdot \frac{2(\theta + 7)^{1/2}}{2(\theta + 7)^{1/2}} \\
&= \frac{\theta^3 + 3\theta^2(\theta + 7)^{1/2} \cdot 2(\theta + 7)^{1/2}}{2(\theta + 7)^{1/2}} \\
&= \frac{\theta^3 + 6\theta^2(\theta + 7)}{2(\theta + 7)^{1/2}} \\
&= \frac{\theta^3 + 6\theta^3 + 42\theta^2}{2(\theta + 7)^{1/2}} = \frac{7\theta^3 + 42\theta^2}{2(\theta + 7)^{1/2}} \\
&= \frac{7\theta^2(\theta + 6)}{2\sqrt{\theta + 7}}.
\end{aligned}
$$

15. $y = \dfrac{\sqrt{x}}{x - 4} = \dfrac{x^{1/2}}{x - 4}$. By the quotient rule,

$$
\begin{aligned}
y' &= \frac{(x - 4)\frac{d}{dx}(x^{1/2}) - x^{1/2}\frac{d}{dx}(x - 4)}{(x - 4)^2} \\
&= \frac{(x - 4) \cdot \frac{1}{2}x^{-1/2} - x^{1/2}}{(x - 4)^2} = \frac{\frac{x-4}{2x^{1/2}} - x^{1/2}}{(x - 4)^2} \\
&= \frac{\frac{x-4}{2x^{1/2}} - \frac{x^{1/2}}{1} \cdot \frac{2x^{1/2}}{2x^{1/2}}}{(x - 4)^2} \\
&= \frac{\frac{x-4-2x}{2x^{1/2}}}{(x - 4)^2} = \frac{-x - 4}{2x^{1/2}} \cdot \frac{1}{(x - 4)^2} \\
&= -\frac{x + 4}{2(x - 4)^2\sqrt{x}}
\end{aligned}
$$

$\qquad$ common denominator: $2x^{1/2}$

17. $y = \dfrac{x^2}{\sqrt{x + 1}} = \dfrac{x^2}{(x + 1)^{1/2}}$. By the quotient rule:

$$
\begin{aligned}
y' &= \frac{(x + 1)^{1/2}\frac{d}{dx}(x^2) - x^2\frac{d}{dx}(x + 1)^{1/2}}{[(x + 1)^{1/2}]^2} \\
&= \frac{(x + 1)^{1/2}(2x) - x^2(\frac{1}{2})(x + 1)^{-1/2}}{x + 1} = \frac{2x\sqrt{x + 1} - \frac{x^2}{2\sqrt{x+1}}}{x + 1}.
\end{aligned}
$$

To combine the expressions on top, note that the common denominator is $2\sqrt{x + 1}$:

$$
\begin{aligned}
&= \frac{\frac{2x\sqrt{x+1}}{1} \cdot \frac{2\sqrt{x+1}}{2\sqrt{x+1}} - \frac{x^2}{2\sqrt{x+1}}}{x + 1} = \frac{\frac{4x(x+1)}{2\sqrt{x+1}} - \frac{x^2}{2\sqrt{x+1}}}{x + 1} \\
&= \frac{\frac{4x^2+4x-x^2}{2\sqrt{x+1}}}{x + 1} = \frac{3x^2 + 4x}{2\sqrt{x + 1}} \cdot \frac{1}{x + 1} = \frac{3x^2 + 4x}{2(x + 1)\sqrt{x + 1}}.
\end{aligned}
$$

The expression

$$
\frac{2x\sqrt{x + 1} - \frac{x^2}{2\sqrt{x+1}}}{x + 1}
$$

can also be simplified by clearing fractions: multiply numerator and denominator by $2\sqrt{x + 1}$ to reduce the complex fraction to an ordinary fraction. Thus

$$y' = \frac{2x\sqrt{x+1} - \frac{x^2}{2\sqrt{x+1}}}{x+1} \cdot \frac{2\sqrt{x+1}}{2\sqrt{x+1}}$$

$$= \frac{2x\sqrt{x+1} \cdot 2\sqrt{x+1} - x^2}{(x+1) \cdot 2\sqrt{x+1}}$$

$$= \frac{4x(x+1) - x^2}{2(x+1)\sqrt{x+1}} = \frac{3x^2 + 4x}{2(x+1)\sqrt{x+1}}.$$

19. $y = \dfrac{(x^2-1)^{1/2}}{x^2}$. By the quotient rule,

$$y' = \frac{x^2 \frac{d}{dx}(x^2-1)^{1/2} - (x^2-1)^{1/2}\frac{d}{dx}(x^2)}{x^4}$$

$$= \frac{x^2 \cdot \frac{1}{2}(x^2-1)^{-1/2}(2x) - (x^2-1)^{1/2}(2x)}{x^4}$$

$$= \frac{\frac{x^3}{(x^2-1)^{1/2}} - 2x(x^2-1)^{1/2}}{x^4}.$$

To clear fractions, we multiply numerator and denominator by $(x^2-1)^{1/2}$:

$$y' = \frac{\frac{x^3}{(x^2-1)^{1/2}} - 2x(x^2-1)^{1/2}}{x^4} \cdot \frac{(x^2-1)^{1/2}}{(x^2-1)^{1/2}}$$

$$= \frac{\frac{x^3}{(x^2-1)^{1/2}}(x^2-1)^{1/2} - 2x(x^2-1)^{1/2}(x^2-1)^{1/2}}{x^4(x^2-1)^{1/2}}$$

$$= \frac{x^3 - 2x(x^2-1)}{x^4(x^2-1)^{1/2}} = \frac{2x - x^3}{x^4\sqrt{x^2-1}}$$

$$= \frac{2 - x^2}{x^3\sqrt{x^2-1}}.$$

21. $y = \dfrac{x^2\sqrt{x}}{x^2+3} = \dfrac{x^2 x^{1/2}}{x^2+3} = \dfrac{x^{5/2}}{x^2+3}$. By the quotient rule:

$$y' = \frac{(x^2+3)(\frac{5}{2})x^{3/2} - x^{5/2}(2x)}{(x^2+3)^2}$$

$$= \frac{x^{3/2}[(\frac{5}{2})(x^2+3) - x(2x)]}{(x^2+3)^2}. \qquad \text{common factor } x^{3/2}$$

Now we multiply numerator and denominator by 2 (to clear fractions):

$$y' = \frac{x^{3/2}[\frac{5}{2}(x^2+3) - x(2x)]}{(x^2+3)^2} \cdot \frac{2}{2}$$

$$= \frac{x^{3/2}[5(x^2+3) - 4x^2]}{2(x^2+3)^2}$$

$$= \frac{x\sqrt{x}(5x^2 + 15 - 4x^2)}{2(x^2+3)^2} \qquad x^{3/2} = xx^{1/2} = x\sqrt{x}$$

$$= \frac{x\sqrt{x}(x^2+15)}{2(x^2+3)^2}.$$

23. $y = (x-1)(x-2)^{1/2}$. By the product rule,

$$y' = (x-1) \cdot \tfrac{1}{2}(x-2)^{-1/2} + (x-2)^{1/2} \cdot 1$$

$$= \frac{x-1}{2(x-2)^{1/2}} + \frac{(x-2)^{1/2}}{1} \cdot \frac{2(x-2)^{1/2}}{2(x-2)^{1/2}}$$

$$= \frac{x-1+2(x-2)}{2(x-2)^{1/2}} = \frac{3x-5}{2\sqrt{x-2}}.$$

25. $y = \dfrac{x\sqrt{x-1}}{2x+3} = \dfrac{x(x-1)^{1/2}}{2x+3}$. By the quotient rule:

$$
\begin{aligned}
y' &= \frac{(2x+3)\frac{d}{dx}[x(x-1)^{1/2}] - x(x-1)^{1/2}\frac{d}{dx}(2x+3)}{(2x+3)^2} \\[2mm]
&= \frac{(2x+3)[x(\frac{1}{2})(x-1)^{-1/2} + (x-1)^{1/2}] - x(x-1)^{1/2}(2)}{(2x+3)^2} \\[2mm]
&= \frac{(2x+3)[\frac{x}{2\sqrt{x-1}} + \sqrt{x-1}] - 2x\sqrt{x-1}}{(2x+3)^2} \cdot \frac{2\sqrt{x-1}}{2\sqrt{x-1}} \\[2mm]
&= \frac{(2x+3)[\frac{x}{2\sqrt{x-1}} + \sqrt{x-1}]\cdot 2\sqrt{x-1} - 2x\sqrt{x-1}\cdot 2\sqrt{x-1}}{(2x+3)^2 \cdot 2\sqrt{x-1}} \\[2mm]
&= \frac{(2x+3)[x + 2(x-1)] - 4x(\sqrt{x-1})^2}{2(2x+3)^2\sqrt{x-1}} \\[2mm]
&= \frac{(2x+3)(3x-2) - 4x(x-1)}{2(2x+3)^2\sqrt{x-1}} \\[2mm]
&= \frac{6x^2 + 5x - 6 - 4x^2 + 4x}{2(2x+3)^2\sqrt{x-1}} = \frac{2x^2 + 9x - 6}{2(2x+3)^2\sqrt{x-1}}.
\end{aligned}
$$

27. $y = \dfrac{x^2(x-5)^{1/2}}{(x+3)^{1/2}}$. By the quotient and product rules,

$$
\begin{aligned}
y' &= \frac{(x+3)^{1/2}\frac{d}{dx}[x^2(x-5)^{1/2}] - x^2(x-5)^{1/2}\frac{d}{dx}(x+3)^{1/2}}{[(x+3)^{1/2}]^2} \\[2mm]
&= \frac{(x+3)^{1/2}[x^2\cdot\frac{1}{2}(x-5)^{-1/2} + 2x(x-5)^{1/2}] - x^2(x-5)^{1/2}\cdot\frac{1}{2}(x+3)^{-1/2}}{x+3}.
\end{aligned}
$$

To start clearing fractions, multiply numerator and denominator by $2(x-5)^{1/2}$:

$$
y' = \frac{(x+3)^{1/2}[x^2 + 4x(x-5)] - x^2(x-5)(x+3)^{-1/2}}{2(x-5)^{1/2}(x+3)}.
$$

Now multiply numerator and denominator by $(x+3)^{1/2}$:

$$
y' = \frac{(x+3)(5x^2 - 20x) - x^2(x-5)}{2(x-5)^{1/2}(x+3)^{3/2}}.
$$

After collecting terms, we get

$$
y' = \frac{4x^3 - 60x}{2(x+3)^{3/2}(x-5)^{1/2}} = \frac{2x^3 - 30x}{(x+3)^{3/2}(x-5)^{1/2}}.
$$

29. By the product rule,

$$
\begin{aligned}
y' &= (x^3 - 1)(2x+3) + (3x^2)(x^2 + 3x + 2)\Big|_{x=1} \\[2mm]
&= 0 + (3)(1 + 3 + 2) = 18.
\end{aligned}
$$

31. $Z = \sqrt{16 + X^2}; \dfrac{dZ}{dX} = \dfrac{1}{2}(16 + X^2)^{-1/2}(2X) = \dfrac{X}{\sqrt{16 + X^2}}.$

33.
$$
\begin{aligned}
i &= \frac{dq}{dt} = \frac{d}{dt}\frac{t}{t^2 + 4} = \frac{(t^2 + 4)(1) - t(2t)}{(t^2 + 4)^2} \\[2mm]
&= \frac{t^2 + 4 - 2t^2}{(t^2 + 4)^2} = \frac{4 - t^2}{(t^2 + 4)^2} = 0
\end{aligned}
$$

Hence
$$
\begin{aligned}
4 - t^2 &= 0 \\
t &= \pm 2.
\end{aligned}
$$

Taking only the positive root, $t = 2\,\text{s}$.

2.8   Implicit Differentiation

1. $2x + 3y = 3$

$$2 + 3\frac{dy}{dx} = 0 \qquad \text{derivative of } y \text{ is } \frac{dy}{dx}$$

$$\frac{dy}{dx} = -\frac{2}{3}$$

3. $x^2 - y^2 = 2$

$$2x - 2y\frac{dy}{dx} = 0$$

$$-2y\frac{dy}{dx} = -2x$$

$$\frac{dy}{dx} = \frac{-2x}{-2y} = \frac{x}{y}$$

5. $2x^2 - 3y^2 = 1$

$$4x - 6y\frac{dy}{dx} = 0 \qquad \text{formula (2.11)}$$

$$-6y\frac{dy}{dx} = -4x$$

$$\frac{dy}{dx} = \frac{-4x}{-6y} = \frac{2x}{3y}$$

7. $2y^3 + x^2 + 1 = 0$

$$2 \cdot 3y^2\frac{dy}{dx} + 2x = 0$$

$$6y^2\frac{dy}{dx} = -2x$$

$$\frac{dy}{dx} = \frac{-2x}{6y^2} = -\frac{x}{3y^2}$$

9. $4y^4 - 3x^3 + 2 = 0$

$$16y^3\frac{dy}{dx} - 9x^2 = 0$$

$$16y^3\frac{dy}{dx} = 9x^2$$

$$\frac{dy}{dx} = \frac{9x^2}{16y^3}$$

11. $x - 5x^2 - 6y^3 = 0$

$$1 - 10x - 18y^2\frac{dy}{dx} = 0$$

$$-18y^2\frac{dy}{dx} = -1 + 10x$$

$$\frac{dy}{dx} = \frac{1 - 10x}{18y^2}$$

13.
$$\frac{x^2}{a^2} + \frac{y^2}{b^2} = 1$$

$$\frac{1}{a^2}x^2 + \frac{1}{b^2}y^2 = 1$$

$$\frac{1}{a^2}(2x) + \frac{1}{b^2}(2y)\frac{dy}{dx} = 0$$

$$\frac{2y}{b^2} \cdot \frac{dy}{dx} = -\frac{2x}{a^2}$$

$$\frac{dy}{dx} = -\frac{2x}{a^2} \cdot \frac{b^2}{2y}$$

$$= -\frac{b^2 x}{a^2 y}$$

15.
$$xy = 3$$

$$x\frac{dy}{dx} + y \cdot 1 = 0 \qquad \frac{d}{dx}(y) = \frac{dy}{dx}; \text{ product rule}$$

$$\frac{dy}{dx} = -\frac{y}{x}.$$

17.
$$x^2 y = 7$$

$$x^2 \frac{d}{dx}(y) + y\frac{d}{dx}(x^2) = 0 \qquad\qquad \text{product rule}$$

$$x^2 \frac{dy}{dx} + 2xy = 0$$

$$\frac{dy}{dx} = -\frac{2xy}{x^2} = -\frac{2y}{x}$$

19.
$$x^2 + x^2 y^2 + x = 0$$

$$2x + x^2 \frac{d}{dx}y^2 + y^2 \frac{d}{dx}x^2 + 1 = 0 \qquad\qquad \text{product rule}$$

$$2x + x^2 \cdot 2y\frac{dy}{dx} + y^2 \cdot 2x + 1 = 0$$

$$2x^2 y\frac{dy}{dx} = -(2x + 2xy^2 + 1)$$

$$\frac{dy}{dx} = -\frac{2x + 2xy^2 + 1}{2x^2 y}$$

21.
$$x^3 - 4x^2 y^2 + y^2 = 1$$

$$3x^2 - 4x^2 \frac{d}{dx}(y^2) + y^2 \frac{d}{dx}(-4x^2) + 2y\frac{dy}{dx} = 0 \qquad\qquad \text{product rule}$$

$$3x^2 - 4x^2(2y)\frac{dy}{dx} + y^2(-8x) + 2y\frac{dy}{dx} = 0$$

$$-8x^2 y\frac{dy}{dx} + 2y\frac{dy}{dx} = -3x^2 + 8xy^2$$

$$(-8x^2 y + 2y)\frac{dy}{dx} = 8xy^2 - 3x^2$$

$$\frac{dy}{dx} = \frac{8xy^2 - 3x^2}{2y - 8x^2 y}$$

23.
$$5x^2y^3 - y^4 = 2x^3$$

$$5\left(x^2 \cdot 3y^2\frac{dy}{dx} + y^3 \cdot 2x\right) - 4y^3\frac{dy}{dx} = 6x^2 \qquad \text{product rule}$$

$$15x^2y^2\frac{dy}{dx} - 4y^3\frac{dy}{dx} = 6x^2 - 10xy^3$$

$$(15x^2y^2 - 4y^3)\frac{dy}{dx} = 6x^2 - 10xy^3$$

$$\frac{dy}{dx} = \frac{6x^2 - 10xy^3}{15x^2y^2 - 4y^3}$$

25.
$$x^4y^4 - 3y^2 + 5x = 6$$

$$x^4\frac{d}{dx}(y^4) + y^4\frac{d}{dx}(x^4) - 6y\frac{dy}{dx} + 5 = 0$$

$$x^4(4y^3)\frac{dy}{dx} + y^4(4x^3) - 6y\frac{dy}{dx} + 5 = 0$$

$$4x^4y^3\frac{dy}{dx} - 6y\frac{dy}{dx} = -4x^3y^4 - 5$$

$$(4x^4y^3 - 6y)\frac{dy}{dx} = -(4x^3y^4 + 5)$$

$$\frac{dy}{dx} = -\frac{4x^3y^4 + 5}{4x^4y^3 - 6y}$$

27. $\quad x^2 + 4y^2 = 5$

$$2x + 8y\frac{dy}{dx} = 0$$

$$\frac{dy}{dx} = -\frac{x}{4y}\Big|_{(1,-1)} = \frac{1}{4}$$

Slope of tangent line: $\frac{1}{4}$; slope of normal line: -4; point: $(1,-1)$.

$$y - y_1 = m(x - x_1)$$

$$y + 1 = \tfrac{1}{4}(x - 1)$$

$$y = \tfrac{1}{4}(x - 5) \qquad \text{tangent line}$$

$$y + 1 = -4(x - 1)$$

$$y = -4x + 3 \qquad \text{normal line}$$

29. $\quad y^2 = -4x$

$$2y\frac{dy}{dx} = -4$$

$$\frac{dy}{dx} = -\frac{2}{y}\Big|_{(-1,-2)} = -\frac{2}{-2} = 1$$

Slope of tangent line: 1; slope of normal line: -1; point: $(-1,-2)$.

$$y - y_1 = m(x - x_1)$$

$$y + 2 = 1(x + 1)$$

$$y = x - 1 \qquad \text{tangent line}$$

$$y + 2 = -1(x + 1)$$

$$y = -x - 3 \qquad \text{normal line}$$

31.　　$x^2 - 2y^2 = 2$

$$2x - 4y\frac{dy}{dx} = 0$$

$$\frac{dy}{dx} = \frac{x}{2y}\Big|_{(-2,1)} = -1$$

Slope of tangent line: -1; slope of normal line: 1; point: $(-2, 1)$.

$$y - y_1 = m(x - x_1)$$

$$y - 1 = -1(x + 2)$$

$$y = -x - 1 \qquad \text{tangent line}$$

$$y - 1 = 1(x + 2)$$

$$y = x + 3 \qquad \text{normal line}$$

33.　　$2x^2 + y^2 = 17$

$$4x + 2y\frac{dy}{dx} = 0$$

$$\frac{dy}{dx} = -\frac{2x}{y}\Big|_{(-2,-3)} = -\frac{2(-2)}{-3} = -\frac{4}{3}$$

Slope of tangent line: $-\frac{4}{3}$; slope of normal line: $\frac{3}{4}$; point: $(-2, -3)$.

$$y - y_1 = m(x - x_1)$$

$$y + 3 = -\frac{4}{3}(x + 2) \qquad \text{tangent line}$$

$$y + 3 = \frac{3}{4}(x + 2) \qquad \text{normal line}$$

(Simplified forms and graphs are in the answer section in the back of the book.)

2.9　Higher Derivatives

1.　$y = 5x^4 + 5x^3 - 3x + 1$

$y' = 20x^3 + 15x^2 - 3$

$y'' = 60x^2 + 30x$

3.　$y = \sqrt{x - 1} = (x - 1)^{1/2}$

$y' = \frac{1}{2}(x - 1)^{-1/2}$

$y'' = -\frac{1}{4}(x - 1)^{-3/2} = \dfrac{-1}{4(x - 1)^{3/2}}$

5.　$y = x^6 - 2x^5 - x^4$

$\dfrac{dy}{dx} = 6x^5 - 10x^4 - 4x^3$

$\dfrac{d^2y}{dx^2} = 30x^4 - 40x^3 - 12x^2$

$\dfrac{d^3y}{dx^3} = 120x^3 - 120x^2 - 24x$

7.　$f(x) = (5 + x)^{1/2}$

$f'(x) = \frac{1}{2}(5 + x)^{-1/2}$

$f''(x) = \left(\frac{1}{2}\right)\left(-\frac{1}{2}\right)(5 + x)^{-3/2}$

$f'''(x) = \left(\frac{1}{2}\right)\left(-\frac{1}{2}\right)\left(-\frac{3}{2}\right)(5 + x)^{-5/2} = \dfrac{3}{8(5 + x)^{5/2}}$

9. $y = \dfrac{3+2x}{3-2x}$

$\dfrac{dy}{dx} = \dfrac{(3-2x)(2)-(3+2x)(-2)}{(3-2x)^2} = \dfrac{6-4x+6+4x}{(3-2x)^2} = \dfrac{12}{(3-2x)^2} = 12(3-2x)^{-2}$

$\dfrac{d^2y}{dx^2} = 12(-2)(3-2x)^{-3}(-2) = \dfrac{48}{(3-2x)^3}$

11. $x^2 + y^2 = 4$

$2x + 2y\dfrac{dy}{dx} = 0$

$\dfrac{dy}{dx} = -\dfrac{x}{y}$ or $y' = -\dfrac{x}{y}$

$\begin{aligned} y'' &= -\dfrac{y\cdot 1 - xy'}{y^2} & \text{quotient rule}\\[2mm] &= -\dfrac{y - x(-\frac{x}{y})}{y^2} & \text{since } y' = -\dfrac{x}{y}\\[2mm] &= -\dfrac{y^2 + x^2}{y^3}\\[2mm] &= -\dfrac{4}{y^3} & \text{from given equation, } x^2 + y^2 = 4 \end{aligned}$

13. $x^2 - y^2 = 4$

$2x - 2y\dfrac{dy}{dx} = 0$

$\dfrac{dy}{dx} = \dfrac{x}{y}$ or $y' = \dfrac{x}{y}$

$\begin{aligned} y'' &= \dfrac{y\cdot 1 - xy'}{y^2} & \text{quotient rule}\\[2mm] &= \dfrac{y - x\cdot\frac{x}{y}}{y^2} & \text{since } y' = \dfrac{x}{y}\\[2mm] &= \dfrac{y - \frac{x^2}{y}}{y^2}\cdot\dfrac{y}{y} & \text{clearing fractions}\\[2mm] &= \dfrac{y^2 - x^2}{y^3}\\[2mm] &= -\dfrac{x^2 - y^2}{y^3}\\[2mm] &= -\dfrac{4}{y^3} & \text{from given equation, } x^2 - y^2 = 4 \end{aligned}$

15. $y^2 = 4px$

$2y\dfrac{dy}{dx} = 4p$

$\dfrac{dy}{dx} = \dfrac{2p}{y}$ or $y' = 2py^{-1}$

$\begin{aligned} y'' &= -2py^{-2}\dfrac{dy}{dx}\\[2mm] &= -2py^{-2}y'\\[2mm] &= -2py^{-2}(2py^{-1}) & \text{since } y' = 2py^{-1}\\[2mm] &= -\dfrac{4p^2}{y^3} \end{aligned}$

17. $\sqrt{x} + \sqrt{y} = 1$ or $x^{1/2} + y^{1/2} = 1$

$$\frac{1}{2}x^{-1/2} + \frac{1}{2}y^{-1/2}y' = 0$$

$$y' = -\frac{x^{-1/2}}{y^{-1/2}} = -\frac{y^{1/2}}{x^{1/2}}$$

$$y'' = -\frac{x^{1/2} \cdot \frac{1}{2}y^{-1/2}y' - y^{1/2} \cdot \frac{1}{2}x^{-1/2}}{(x^{1/2})^2} \qquad \text{quotient rule}$$

$$= -\frac{1}{2} \cdot \frac{x^{1/2}y^{-1/2}y' - y^{1/2}x^{-1/2}}{x}$$

$$= -\frac{1}{2} \cdot \frac{x^{1/2}y^{-1/2}(-\frac{y^{1/2}}{x^{1/2}}) - y^{1/2}x^{-1/2}}{x} \qquad y' = -\frac{y^{1/2}}{x^{1/2}}$$

$$= \frac{1}{2} \cdot \frac{1 + y^{1/2}x^{-1/2}}{x} \cdot \frac{x^{1/2}}{x^{1/2}}$$

$$= \frac{x^{1/2} + y^{1/2}}{2x^{3/2}}$$

$$= \frac{1}{2x^{3/2}} \qquad\qquad\qquad \sqrt{x} + \sqrt{y} = 1$$

Chapter 2 Review

1. $\quad f(x) = x^2 - 1$

   $\quad f(0) = 0^2 - 1 = -1 \qquad x = 0$

   $\quad f(1) = 1^2 - 1 = 0 \qquad x = 1$

   $\quad f(\sqrt{2}) = (\sqrt{2})^2 - 1 = 1 \qquad x = \sqrt{2}$

3. $f(0) = 0$ and $f(\frac{1}{2}) = 0$ $\qquad\qquad$ since $0 \le x < 1$

   $f(\frac{5}{2}) = 2;$ $\qquad\qquad\qquad\qquad$ since $x > 2$

   $f(x)$ is not defined for $x = 1$.

5. (a) To avoid imaginary values, we must have $x \ge 1$. The range is $y \ge 0$, since y cannot be negative.

   (b) The cube root of any real number is a real number. So y is defined for all x.

7. $\displaystyle\lim_{x \to 4} \frac{16 - x^2}{4 - x} = \lim_{x \to 4} \frac{(4 - x)(4 + x)}{4 - x} = \lim_{x \to 4}(4 + x) = 8$

9. $\displaystyle\lim_{x \to 0} \frac{x^3 - x^2 + 3x}{x} = \lim_{x \to 0} \frac{x(x^2 - x + 3)}{x} \qquad$ factoring x

   $\qquad\qquad\qquad\quad = \lim_{x \to 0}(x^2 - x + 3) = 3$

11. (a) $f(x) = 1 - x^2$ is defined at $x = 2$. So we obtain the limit by inspection:

   $\displaystyle\lim_{x \to 2}(1 - x^2) = 1 - 2^2 = -3.$

   (b) $\displaystyle\lim_{x \to 1} \frac{x^2 - 5x + 4}{x - 1} = \lim_{x \to 1} \frac{(x - 1)(x - 4)}{x - 1} = \lim_{x \to 1}(x - 4) = -3$

13. Since the function is not defined at $x = 1$, we find the limit by rationalizing the numerator:

   $\displaystyle\lim_{x \to 1} \frac{\sqrt{x} - 1}{x - 1} = \lim_{x \to 1} \frac{\sqrt{x} - 1}{x - 1} \cdot \frac{\sqrt{x} + 1}{\sqrt{x} + 1} = \lim_{x \to 1} \frac{x - 1}{(x - 1)(\sqrt{x} + 1)}$

   $\qquad\qquad\qquad = \displaystyle\lim_{x \to 1} \frac{1}{\sqrt{x} + 1} = \frac{1}{2}.$

15. $\lim\limits_{x \to \infty} \dfrac{2x^2 - 3x + 2}{x^2 - 10x + 1} = \lim\limits_{x \to \infty} \dfrac{2 - 3/x + 2/x^2}{1 - 10/x + 1/x^2} = \dfrac{2 - 0 + 0}{1 - 0 + 0} = 2$

17. By inspection, $\sqrt{x - 4} \to 0$ as $x \to 4$ from the right. (The restriction 4+ is necessary to avoid imaginary values.)

19. $f(x) = x - 3x^2;\ f(\ ) = (\ ) - 3(\ )^2$

   Step 1. $y + \Delta y = (x + \Delta x) - 3(x + \Delta x)^2$

   Step 2. $\begin{aligned}\Delta y &= (x + \Delta x) - 3(x + \Delta x)^2 - (x - 3x^2) \\ &= x + \Delta x - 3[x^2 + 2x\Delta x + (\Delta x)^2] - x + 3x^2 \\ &= \Delta x - 6x\Delta x - 3(\Delta x)^2 \end{aligned}$

   Step 3. $\dfrac{\Delta y}{\Delta x} = \dfrac{\Delta x - 6x\Delta x - 3(\Delta x)^2}{\Delta x} = \dfrac{\Delta x(1 - 6x - 3\Delta x)}{\Delta x} = 1 - 6x - 3\Delta x$

   Step 4. $f'(x) = \lim\limits_{\Delta x \to 0} \dfrac{\Delta y}{\Delta x} = \lim\limits_{\Delta x \to 0}(1 - 6x - 3\Delta x) = 1 - 6x$

21. (a) $f(\ ) = \dfrac{1}{4 - (\ )}$

   Step 1. $y + \Delta y = f(x + \Delta x) = \dfrac{1}{4 - (x + \Delta x)}$

   Step 2. $\Delta y = \dfrac{1}{4 - (x + \Delta x)} - \dfrac{1}{4 - x}$

   Note that the common denominator is $[4 - (x + \Delta x)](4 - x)$. We now get

   $\begin{aligned}\Delta y &= \dfrac{4 - x}{[4 - (x + \Delta x)](4 - x)} - \dfrac{4 - (x + \Delta x)}{[4 - (x + \Delta x)](4 - x)} \\ &= \dfrac{4 - x - 4 + (x + \Delta x)}{[4 - (x + \Delta x)](4 - x)} = \dfrac{\Delta x}{[4 - (x + \Delta x)](4 - x)}. \end{aligned}$

   Step 3. $\begin{aligned}\dfrac{\Delta y}{\Delta x} &= \dfrac{\Delta x}{[4 - (x + \Delta x)](4 - x)} \cdot \dfrac{1}{\Delta x} \\ &= \dfrac{1}{[4 - (x + \Delta x)](4 - x)} \end{aligned}$

   Step 4. $\begin{aligned}f'(x) &= \lim\limits_{\Delta x \to 0} \dfrac{1}{[4 - (x + \Delta x)](4 - x)} \\ &= \dfrac{1}{(4 - x)^2} \end{aligned}$

   (b) $f(\ ) = \sqrt{\ }$

   Step 1. $y + \Delta y = f(x + \Delta x) = \sqrt{x + \Delta x}$

   Step 2. $\Delta y = \sqrt{x + \Delta x} - \sqrt{x}$

   Rationalizing the numerator,

   $\begin{aligned}\Delta y &= \dfrac{\sqrt{x + \Delta x} - \sqrt{x}}{1} \cdot \dfrac{\sqrt{x + \Delta x} + \sqrt{x}}{\sqrt{x + \Delta x} + \sqrt{x}} \\ &= \dfrac{(\sqrt{x + \Delta x})^2 - (\sqrt{x})^2}{\sqrt{x + \Delta x} + \sqrt{x}} = \dfrac{x + \Delta x - x}{\sqrt{x + \Delta x} + \sqrt{x}} = \dfrac{\Delta x}{\sqrt{x + \Delta x} + \sqrt{x}}. \end{aligned}$

   Step 3. $\dfrac{\Delta y}{\Delta x} = \dfrac{\Delta x}{\sqrt{x + \Delta x} + \sqrt{x}} \cdot \dfrac{1}{\Delta x} = \dfrac{1}{\sqrt{x + \Delta x} + \sqrt{x}}$

   Step 4. $f'(x) = \lim\limits_{\Delta x \to 0} \dfrac{1}{\sqrt{x + \Delta x} + \sqrt{x}} = \dfrac{1}{\sqrt{x} + \sqrt{x}} = \dfrac{1}{2\sqrt{x}}$

23. $y = (x^3 - 2)^4$. By the generalized power rule,

   $y' = 4(x^3 - 2)^3 \dfrac{d}{dx}(x^3 - 2) = 4(x^3 - 2)^3(3x^2) = 12x^2(x^3 - 2)^3.$

25. $y = \dfrac{x-4}{x+1}$. By the quotient rule, $y' = \dfrac{(x+1)-(x-4)}{(x+1)^2} = \dfrac{5}{(x+1)^2}$.

27. $y = \dfrac{x^2}{(4-x^2)^{1/2}}$. By the quotient rule,

$$y' = \frac{(4-x^2)^{1/2}(2x) - x^2 \cdot \frac{1}{2}(4-x^2)^{-1/2}(-2x)}{[(4-x^2)^{1/2}]^2}$$

$$= \frac{2x(4-x^2)^{1/2} + x^3(4-x^2)^{-1/2}}{4-x^2}.$$

Now multiply numerator and denominator by $(4-x^2)^{1/2}$:

$$y' = \frac{2x(4-x^2)+x^3}{(4-x^2)^{3/2}} = \frac{8x-x^3}{(4-x^2)^{3/2}}.$$

29. $y = x\sqrt{4-x^2} = x(4-x^2)^{1/2}$. By the product rule,

$$\begin{aligned}
y' &= x\frac{d}{dx}(4-x^2)^{1/2} + (4-x^2)^{1/2}\frac{d}{dx}(x) \\
&= x \cdot \tfrac{1}{2}(4-x^2)^{-1/2}(-2x) + (4-x^2)^{1/2} \\
&= \frac{-x^2}{(4-x^2)^{1/2}} + (4-x^2)^{1/2} \\
&= \frac{-x^2}{(4-x^2)^{1/2}} + \frac{(4-x^2)^{1/2}}{1} \cdot \frac{(4-x^2)^{1/2}}{(4-x^2)^{1/2}} \\
&= \frac{-x^2 + (4-x^2)}{(4-x^2)^{1/2}} \\
&= \frac{4-2x^2}{\sqrt{4-x^2}}.
\end{aligned}$$

31.
$$\begin{aligned}
x^2y + xy^2 + y^3 &= 1 \\
x^2\frac{d}{dx}(y) + y\frac{d}{dx}(x^2) + x\frac{d}{dx}(y^2) + y^2\frac{d}{dx}(x) + \frac{d}{dx}(y^3) &= 0 \\
x^2\frac{dy}{dx} + 2xy + 2xy\frac{dy}{dx} + y^2 + 3y^2\frac{dy}{dx} &= 0 \\
x^2\frac{dy}{dx} + 2xy\frac{dy}{dx} + 3y^2\frac{dy}{dx} &= -(2xy+y^2) \\
(x^2 + 2xy + 3y^2)\frac{dy}{dx} &= -(2xy+y^2) \\
\frac{dy}{dx} &= -\frac{2xy+y^2}{x^2+2xy+3y^2}
\end{aligned}$$

33.
$$\begin{aligned}
y^3 - xy &= 4 \\
3y^2y' - xy' - y &= 0 \\
y' &= \frac{y}{3y^2-x}
\end{aligned}$$

$$\begin{aligned}
y'' &= \frac{(3y^2-x)y' - y(6yy'-1)}{(3y^2-x)^2} \\
&= \frac{(3y^2-x)\frac{y}{3y^2-x} - 6y^2\frac{y}{3y^2-x} + y}{(3y^2-x)^2} \\
&= \frac{y - \frac{6y^3}{3y^2-x} + y}{(3y^2-x)^2} \cdot \frac{3y^2-x}{3y^2-x} \\
&= \frac{2y(3y^2-x) - 6y^3}{(3y^2-x)^3} = -\frac{2xy}{(3y^2-x)^3}
\end{aligned}$$

35.
$$8x^2 + 4xy + 5y^2 + 28x - 2y + 20 = 0$$

$$16x + (4x\frac{dy}{dx} + 4y) + 10y\frac{dy}{dx} + 28 - 2\frac{dy}{dx} = 0$$

$$(4x + 10y - 2)\frac{dy}{dx} = -16x - 4y - 28$$

$$\frac{dy}{dx} = \frac{-16x - 4y - 28}{4x + 10y - 2}\bigg|_{(-1,\,6/5)}$$

$$= \frac{-16(-1) - 4(6/5) - 28}{4(-1) + 10(6/5) - 2} \cdot \frac{5}{5}$$

$$= \frac{80 - 24 - 140}{-20 + 60 - 10} = \frac{-84}{30} = -\frac{14}{5}$$

37. (a)
$$f(x) = \frac{x}{\sqrt{x-1}} = \frac{x}{(x-1)^{1/2}}$$

$$f'(x) = \frac{(x-1)^{1/2} - x(\frac{1}{2})(x-1)^{-1/2}}{[(x-1)^{1/2}]^2}$$

$$= \frac{\sqrt{x-1} - \frac{x}{2\sqrt{x-1}}}{x-1} \cdot \frac{2\sqrt{x-1}}{2\sqrt{x-1}}$$

$$= \frac{2(x-1) - x}{2(x-1)\sqrt{x-1}} = \frac{x-2}{2(x-1)\sqrt{x-1}}$$

$$f'(2) = 0$$

Thus $f'(x) = 0$ when $x = 2$.

(b)
$$f(x) = 2x^3 - 6x^2 + 4$$

$$f'(x) = 6x^2 - 12x$$

$$f''(x) = 12x - 12$$

$$f''(1) = 0$$

Thus $f''(x) = 0$ when $x = 1$.

39. $\dfrac{dR}{dr} = k(-2r^{-3}) = -\dfrac{2k}{r^3}$

41. $V = IR = (4.12 + 0.020t)(0.010t^2)$. By the product rule,
$\dfrac{dV}{dt} = (4.12 + 0.020t)(0.020t) + (0.020)(0.010t^2)$. Letting $t = 2.5\,\text{s}$, we get

$$\frac{dV}{dt} = 0.21\,\frac{V}{s}.$$

43. $C'(y) = \dfrac{1}{2} - \dfrac{1}{250}y\bigg|_{y=50}$

$\qquad = \$0.30$ per widget $= 30$ cents per widget.

45. $F = \dfrac{1}{r^2} = r^{-2}$. Thus

$$\frac{dF}{dr} = -2r^{-3} = -\frac{2}{r^3}\bigg|_{r=5.02} = -\frac{2}{(5.02)^3} = -0.0158\,\text{dyne/cm}.$$

(The negative sign indicates that F decreases as r increases.)

Chapter 3

Applications of the Derivative

3.1 The First-Derivative Test

1. $y = -x^2 - 2x$; $y' = -2x - 2 = 0$; $x = -1$, the critical value.

   To the left of $x = -1$, y' is positive (for example, the test value $x = -2$ yields $y' = 2$). So the function is increasing. To the right of $x = -1$, y' is negative (for example, the test value $x = 0$ yields $y' = -2$). So the function is decreasing.

3. $y = \frac{1}{4}x^3 - 3x + 2$;

$$
\begin{aligned}
y' = \tfrac{3}{4}x^2 - 3 &= 0 \\
\tfrac{3}{4}x^2 &= 3 \\
x^2 &= 4 \\
x &= \pm 2, \text{ the critical values.}
\end{aligned}
$$

   To see where y' is negative and positive, we substitute certain test values.

	test values	$y' = \frac{3}{4}x^2 - 3$	
$x < -2$	-3	$+$	increasing
$-2 < x < 2$	0	$-$	decreasing
$x > 2$	3	$+$	increasing

   We conclude that the function is increasing on $(-\infty, -2]$, decreasing on $[-2, 2]$, and increasing on $[2, \infty)$.

5. $y = x^2 - 2x + 1$

$$
\begin{aligned}
y' &= 2x - 2 \\
2x - 2 &= 0 \\
x &= 1 \qquad \text{critical value}
\end{aligned}
$$

   Substituting $x = 1$ in $y = x^2 - 2x + 1$, we get $y = 0$. So $(1, 0)$ is the critical point.

   If $x < 1$, $y' < 0$; if $x > 1$, $y' > 0$. Since the function is decreasing to the left of $x = 1$ and increasing to the right, $(1, 0)$ is a minimum point.

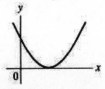

79

7. $y = 8 - 2x - x^2$;

$$y' = -2 - 2x = 0$$

$$x = -1 \text{ (critical value)}.$$

If $x = -1$, $y = 8 - 2(-1) - (-1)^2 = 9$; so $(-1, 9)$ is the critical point. To the left of $x = -1$, $y' > 0$. (For example, for the test value $x = -2$, $y' = 2$.) So the function is increasing. Similarly, to the right of $x = -1$, the function is decreasing. It follows that the critical point $(-1, 9)$ is a maximum.

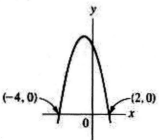

9. $y = 2x^3 + 3x^2 - 12x + 6$; $y' = 6x^2 + 6x - 12$

$$6x^2 + 6x - 12 = 0$$

$$6(x^2 + x - 2) = 0$$

$$6(x + 2)(x - 1) = 0 \qquad x = -2, 1$$

If $x = -2$, $y = 26$; if $x = 1$, $y = -1$. So $(-2, 26)$ and $(1, -1)$ are the critical points. To see where y' is positive or negative, we substitute certain convenient values between the critical values.

	test values	$x + 2$	$x - 1$	$y' = 6(x + 2)(x - 1)$
$x < -2$	-3	$-$	$-$	$+$
$-2 < x < 1$	0	$+$	$-$	$-$
$x > 1$	2	$+$	$+$	$+$

Summary:

If $x < -2$, $y' > 0$, so that $f(x)$ is increasing.

If $-2 < x < 1$, $y' < 0$ so that $f(x)$ is decreasing.

If $x > 1$, $y' > 0$, so that $f(x)$ is increasing.

We conclude that $(-2, 26)$ is a maximum point and $(1, -1)$ is a minimum point.

11. $y = -x^3 + 6x^2 - 9x - 5$; $y' = -3x^2 + 12x - 9$

$$-3x^2 + 12x - 9 = 0$$
$$-3(x^2 - 4x + 3) = 0$$
$$-3(x - 1)(x - 3) = 0$$
$$x = 1, 3 \text{ (critical values)}.$$

If $x = 1$, $y = -9$ and if $x = 3$, $y = -5$. So $(1, -9)$ and $(3, -5)$ are the critical points.

Now we use test values to determine the sign of y'.

	test values	$x - 1$	$x - 3$	$y' = -3(x-1)(x-3)$
$x < 1$	0	$-$	$-$	$-$
$1 < x < 3$	2	$+$	$-$	$+$
$x > 3$	4	$+$	$+$	$-$

Summary:

If $x < 1$, $y' < 0$, so that $f(x)$ is decreasing.

If $1 < x < 3$, $y' > 0$, so that $f(x)$ is increasing.

If $x > 3$, $y' < 0$, so that $f(x)$ is decreasing.

We conclude that $(1, -9)$ is a minimum and $(3, -5)$ is a maximum point.

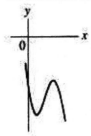

13. $y = x^4 - 2x^2 - 2$; $y' = 4x^3 - 4x$

$$4x^3 - 4x = 0$$
$$4x(x^2 - 1) = 0$$
$$4x(x + 1)(x - 1) = 0$$
$$x = -1, 0, 1 \qquad \text{(critical values)}.$$

To see where $y' < 0$ and where $y' > 0$, we substitute certain test values:

	test values	$4x$	$x + 1$	$x - 1$	$y' = 4x(x+1)(x-1)$
$x < -1$	-2	$-$	$-$	$-$	$-$
$-1 < x < 0$	$-1/2$	$-$	$+$	$-$	$+$
$0 < x < 1$	$1/2$	$+$	$+$	$-$	$-$
$x > 1$	2	$+$	$+$	$+$	$+$

Summary:

If $x < -1$, $y' < 0$: $f(x)$ is decreasing

If $-1 < x < 0$, $y' > 0$: $f(x)$ is increasing

If $0 < x < 1$, $y' < 0$: $f(x)$ is decreasing

If $x > 1$, $y' > 0$: $f(x)$ is increasing

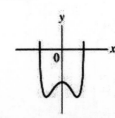

It follows that $(\pm 1, -3)$ are minimum points and $(0, -2)$ is a maximum point.

15. $y = x^4 + \frac{4}{3}x^3$; $y' = 4x^3 + 4x^2$

$$4x^3 + 4x^2 = 0$$
$$4x^2(x+1) = 0$$
$$x = 0, -1 \text{ (critical values)}.$$

Critical points: $(0, 0)$ and $(-1, -\frac{1}{3})$.

	test values	$4x^2$	$x+1$	$y' = 4x^2(x+1)$
$x < -1$	-2	$+$	$-$	$-$
$-1 < x < 0$	$-1/2$	$+$	$+$	$+$
$x > 0$	1	$+$	$+$	$+$

To the left of $x = -1$, $f(x)$ is decreasing. To the right of $x = -1$, $f(x)$ is increasing. So $(-1, -\frac{1}{3})$ is a minimum point. The function continues to increase to the right of $x = 0$. Consequently, $(0, 0)$ is neither a minimum nor a maximum. (The x-axis is nevertheless a horizontal tangent.)

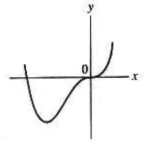

17. $y = 4 - 4x^3 - 3x^4$; $y' = -12x^2 - 12x^3$

$$-12x^2 - 12x^3 = 0$$
$$-12x^2(1+x) = 0$$
$$x = -1, 0$$

Critical points: $(-1, 5)$ and $(0, 4)$.

	test values	$-12x^2$	$1+x$	$y' = -12x^2(1+x)$
$x < -1$	-2	$-$	$-$	$+$
$-1 < x < 0$	$-1/2$	$-$	$+$	$-$
$x > 0$	1	$-$	$+$	$-$

Observe that to the left of $x = -1$, the function is increasing, and to the right of $x = -1$, the function is decreasing. It follows that $(-1, 5)$ is a maximum.

Since the function is decreasing to the right of $x = -1$, it is decreasing to the left and right of $x = 0$. So the point $(0, 4)$ is neither a minimum nor a maximum.

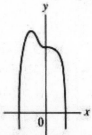

19. $y = \sqrt[3]{x} = x^{1/3}$; $y' = \frac{1}{3}x^{-2/3} = \dfrac{1}{3x^{2/3}} \neq 0$

While y' cannot be 0, y' is undefined at $x = 0$. In other words, there is a vertical tangent at $x = 0$, so that $(0,0)$ is a critical point. Now observe that

$$y' = \frac{1}{3(x^{1/3})^2} > 0 \text{ for all } x \neq 0,$$

so that $f(x)$ is an increasing function. We conclude that $(0,0)$ is neither a minimum nor a maximum.

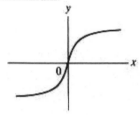

21. $s = 80t - 5t^2$
$$\frac{ds}{dt} = 80 - 10t = 0$$
$$t = 8$$

If $t < 8$, $\dfrac{ds}{dt} > 0$; if $t > 8$, $\dfrac{ds}{dt} < 0$. So s attains a maximum value at $t = 8$. The maximum altitude is

$$s = 80t - 5t^2\Big|_{t=8} = 320\,\text{m}.$$

23. Using the quotient rule, we find that $\dfrac{dP}{dR} = -\dfrac{16(R-2)}{(R+2)^3}$. $R = 2\,\Omega$ is the critical value. Since P is increasing to the left of $R = 2$ and decreasing to the right, $R = 2\,\Omega$ does correspond to a maximum.

3.2 The Second-Derivative Test

1. $y = 2x - x^2$; $y' = 2 - 2x$; $y'' = -2$. Since y'' is strictly negative, the graph is everywhere concave down.

3. $y = x^3 - 12x + 6$; $y' = 3x^2 - 12$; $y'' = 6x$. If $x < 0$, then $y'' < 0$ and if $x > 0$, then $y'' > 0$. We conclude that the graph is concave down on $(-\infty, 0]$ and concave up on $[0, \infty)$.

5. $y = x^4 - 4x^3 + 8x - 5$;
$y' = 4x^3 - 12x^2 + 8$;
$y'' = 12x^2 - 24x = 12x(x - 2) = 0$.
So $y'' = 0$ when $x = 0$ and $x = 2$. These are the only values for which y'' is zero; y'' is therefore different from zero everywhere else, and we may use arbitrary test values:

$x = -1$:	$y'' = 12(-1)(-3) > 0$	concave up for $x < 0$
$x = 1$:	$y'' = 12(1)(-1) < 0$	concave down on $[0, 2]$
$x = 3$:	$y'' = 12(3)(1) > 0$	concave up for $x > 2$

7. $y = x^3 - 9x^2 + 27x - 27;$

$y' = 3x^2 - 18x + 27;$

$y'' = 6x - 18 = 0$ when $x = 3$.

Observe that if $x < 3$, then $y'' < 0$ and if $x > 3$, then $y'' > 0$. It follows that the graph is concave down on $(-\infty, 3]$ and concave up on $[3, \infty)$.

9. $f(x) = 2x^2 - 4x;\ f'(x) = 4x - 4;\ f''(x) = 4.$

Step 1. Critical points: $f'(x) = 4x - 4 = 0$

$x = 1.$

Substituting $x = 1$ in the given equation $y = 2x^2 - 4x$ yields $y = -2$; thus $(1, -2)$ is the critical point.

Step 2. Test of critical point: Since $f''(x) = 4$ for all x, $f''(1) = 4 > 0$. Thus $(1, -2)$ is a minimum.

Step 3. Concavity: Since $f''(x) = 4 > 0$, the graph is concave up everywhere.

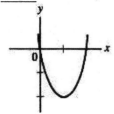

11. $f(x) = 6x - 6x^2;\ f'(x) = 6 - 12x;\ f''(x) = -12.$

Step 1. Critical points: $f'(x) = 6 - 12x = 0$

$x = \dfrac{1}{2}.$

Substituting $x = 1/2$ in the given equation yields $y = 3/2$. So $(\frac{1}{2}, \frac{3}{2})$ is the critical point.

Step 2. Test of critical point: Since $f''(x) = -12$ for all x, $f''(\frac{1}{2}) = -12 < 0$. So $(\frac{1}{2}, \frac{3}{2})$ is a maximum point.

Step 3. Concavity: Since $f''(x) < 0$ for all x, the graph is concave down everywhere.

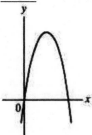

13. $f(x) = -4 - 3x - \frac{1}{2}x^2;\ f'(x) = -3 - x;\ f''(x) = -1.$

Step 1. Critical points: $f'(x) = -3 - x = 0$

$x = -3.$

Substituting $x = -3$ in $y = -4 - 3x - \frac{1}{2}x^2$, we get $y = \frac{1}{2}$. So $(-3, \frac{1}{2})$ is a critical point.

Step 2. Test of critical point: Since $f''(x) = -1$ for all x, $f''(-3) = -1 < 0$. So $(-3, \frac{1}{2})$ is a maximum.

Step 3. Concavity: Since $f''(x) = -1$, the graph is concave down everywhere.

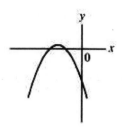

15. $f(x) = 2x^3 - 6x + 1;\ f'(x) = 6x^2 - 6;\ f''(x) = 12x.$

Step 1. Critical points:
$$f'(x) = 6x^2 - 6 = 0$$
$$x = \pm 1.$$

Substituting these values in $y = 2x^3 - 6x + 1$, we get $(1, -3)$ and $(-1, 5)$ for the critical points.

Step 2. Test of critical points:

$f''(-1) = -12 < 0;\ (-1, 5)$ is a maximum.

$f''(1) = 12 > 0;\ (1, -3)$ is a minimum.

Step 3. Concavity: We need to determine where $y'' < 0$ and where $y'' > 0$. To this end, we first determine where $y'' = 0$: $f''(x) = 12x = 0$ when $x = 0$; the point is $(0, 1)$.

	test values	$y'' = 12x$
$x < 0$	-1	$-$
$x > 0$	1	$+$

So $f(x)$ is concave down to the left of $(0, 1)$ and concave up to the right of $(0, 1)$.

Step 4. Inflection point: Since the concavity changes, $(0, 1)$ is an inflection point.

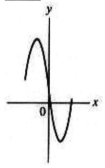

17. $f(x) = x^3 - 6x^2 + 9x - 3;\ f'(x) = 3x^2 - 12x + 9;\ f''(x) = 6x - 12.$

Step 1. Critical points:
$$f'(x) = 3x^2 - 12x + 9 = 0$$
$$3(x^2 - 4x + 3) = 0$$
$$3(x - 3)(x - 1) = 0$$
$$x = 1,\ 3.$$

Substituting these values in $y = x^3 - 6x^2 + 9x - 3$, we see that $(1, 1)$ and $(3, -3)$ are the critical points.

Step 2. Test of critical points:

$f''(1) = 6 - 12 = -6 < 0;$ thus $(1, 1)$ is a maximum.

$f''(3) = 18 - 12 = 6 > 0;$ thus $(3, -3)$ is a minimum.

Step 3. Concavity: We need to determine where $y'' < 0$ and where $y'' > 0$. To this end, we first determine where $y'' = 0$: $f''(x) = 6x - 12 = 0$ when $x = 2$. If $x = 2$, then $y = -1$; the point is $(2, -1)$.

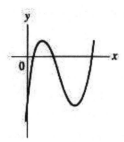

	test values	$y'' = 6x - 12$
$x < 2$	1	$-$
$x > 2$	3	$+$

So $f(x)$ is concave down to the left of $(2, -1)$ and $f(x)$ is concave up to the right of $(2, -1)$.

Step 4. Inflection point: Since the concavity changes, the point $(2, -1)$ is an inflection point.

19. $f(x) = x^3 - 3x^2 - 9x + 11$; $f'(x) = 3x^2 - 6x - 9$; $f''(x) = 6x - 6$.

Step 1. Critical points:
$$
\begin{aligned}
f'(x) = 3x^2 - 6x - 9 &= 0 \\
3(x^2 - 2x - 3) &= 0 \\
3(x + 1)(x - 3) &= 0 \\
x &= -1, 3.
\end{aligned}
$$

Substituting these values in $y = x^3 - 3x^2 - 9x + 11$, we get the following critical points: $(-1, 16)$ and $(3, -16)$.

Step 2. Test of critical points:

$f''(-1) = 6(-1) - 6 = -12 < 0$; $(-1, 16)$ is a maximum.

$f''(3) = 6(3) - 6 = 12 > 0$; $(3, -16)$ is a minimum.

Step 3. Concavity: $f''(x) = 6x - 6 = 0$ when $x = 1$. The point is $(1, 0)$.

	test values	$f''(x) = 6x - 6$	
$x < 1$	0	$-$	concave down
$x > 1$	2	$+$	concave up

The graph is concave down to the left of $x = 1$ and concave up to the right of $x = 1$.

Step 4. Since the concavity changes, $(1, 0)$ is an inflection point.

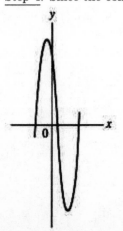

21. $f(x) = 2 + 3x - x^3$; $f'(x) = 3 - 3x^2$; $f''(x) = -6x$.

Step 1. Critical points:
$$\begin{aligned} f'(x) &= 3 - 3x^2 = 0 \\ 3(1 - x^2) &= 0 \\ x &= \pm 1. \end{aligned}$$

Substituting these values in $y = 2 + 3x - x^3$, we get the following critical points: $(-1, 0)$ and $(1, 4)$.

Step 2. Test of critical points:

$f''(-1) = -6(-1) = 6 > 0$; $(-1, 0)$ is a minimum.

$f''(1) = -6(1) = -6 < 0$; $(1, 4)$ is maximum.

Step 3. Concavity: $f''(x) = -6x$ is positive for x less than 0 (concave up) and negative for x greater than 0 (concave down). When $x = 0$, $y = 2$.

Step 4. Since the concavity changes at $x = 0$, the point $(0, 2)$ is an inflection point.

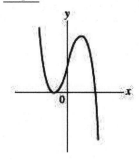

23. $f(x) = 3x^4 - 4x^3 + 2$; $f'(x) = 12x^3 - 12x^2$; $f''(x) = 36x^2 - 24x$.

Step 1. Critical points:
$$\begin{aligned} f'(x) = 12x^3 - 12x^2 &= 0 \\ 12x^2(x - 1) &= 0 \\ x &= 0, 1. \end{aligned}$$

Substituting these values in $y = 3x^4 - 4x^3 + 2$, we get $(0, 2)$ and $(1, 1)$ for the critical points.

Step 2. Test of critical points:

$f''(1) > 0$; $(1, 1)$ is a minimum.

$f''(0) = 0$; the test fails.

Using the first-derivative test, $f'(-\frac{1}{2}) < 0$ and $f'(\frac{1}{2}) < 0$. So $(0, 2)$ is neither a minimum nor a maximum.

Step 3. Concavity: We need to determine where $y'' < 0$ and where $y'' > 0$. To this end we first determine where $f''(x) = 0$:
$$\begin{aligned} f''(x) = 36x^2 - 24x &= 0 \\ 12x(3x - 2) &= 0 \\ x &= 0, \frac{2}{3}. \end{aligned}$$

Substituting in $y = 3x^4 - 4x^3 + 2$, we find that the points are $(0, 2)$ and $\left(\dfrac{2}{3}, \dfrac{38}{27}\right)$.

	test values	$12x$	$3x - 2$	$y'' = 12x(3x - 2)$
$x < 0$	-1	$-$	$-$	$+$
$0 < x < 2/3$	$1/2$	$+$	$-$	$-$
$x > 2/3$	1	$+$	$+$	$+$

Summary:

$f(x)$ is concave up for $x < 0$.

$f(x)$ is concave down for $0 < x < 2/3$.

$f(x)$ is concave up for $x > 2/3$.

Step 4. Inflection points: Since the concavity changes, $(0, 2)$ and $\left(\dfrac{2}{3}, \dfrac{38}{27}\right)$ are inflection points.

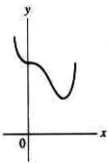

25. $f(x) = x^4 + x^3$; $f'(x) = 4x^3 + 3x^2$; $f''(x) = 12x^2 + 6x$.

Step 1. Critical points:
$$4x^3 + 3x^2 = 0$$
$$x^2(4x + 3) = 0$$
$$x = -\frac{3}{4}, 0.$$

Substituting in $y = x^4 + x^3$, we get $\left(-\dfrac{3}{4}, -\dfrac{27}{256}\right)$ and $(0, 0)$ for the critical points.

Step 2. Test of critical points:

$f''(-\frac{3}{4}) > 0$; the point is a minimum.

$f''(0) = 0$; the test fails.

Using the first derivative test, if $x = -\frac{1}{2}$, $y' > 0$; if $x = 1$, $y' > 0$. So $(0, 0)$ is neither a minimum nor a maximum.

Step 3. Concavity: We need to determine where $y'' < 0$ and where $y'' > 0$. To this end we find those values of x for which $f''(x) = 0$: $f''(x) = 12x^2 + 6x = 0$
$$6x(2x + 1) = 0$$
$$x = -\frac{1}{2}, 0.$$

Substituting in $y = x^4 + x^3$, we find that the points are $\left(-\dfrac{1}{2}, -\dfrac{1}{16}\right)$ and $(0, 0)$.

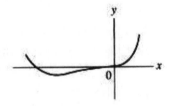

	test values	$6x$	$2x + 1$	$y'' = 6x(2x + 1)$
$x < -1/2$	-1	$-$	$-$	$+$
$-1/2 < x < 0$	$-1/4$	$-$	$+$	$-$
$x > 0$	1	$+$	$+$	$+$

Summary:

$f(x)$ is concave up for $x < -\dfrac{1}{2}$.

$f(x)$ is concave down for $-\dfrac{1}{2} < x < 0$.

$f(x)$ is concave up for $x > 0$.

Step 4. Inflection points: Since the concavity changes at each of the points, $\left(-\dfrac{1}{2}, -\dfrac{1}{16}\right)$ and $(0, 0)$ are points of inflection.

27. $f(x) = (x - 3)^4$; $f'(x) = 4(x - 3)^3$; $f''(x) = 12(x - 3)^2$.

Step 1. Critical points: $f'(x) = 4(x - 3)^3 = 0$
$$x = 3.$$

Substituting in $y = (x - 3)^4$, we get $(3, 0)$ for the critical point.

Step 2. Test of critical point:

$f''(3) = 0$; the test fails.

Using the first-derivative test, observe that $f'(2) < 0$ and $f'(4) > 0$. So the function decreases to the left of $x = 3$ and increases to the right of $x = 3$. The critical point is therefore a minimum.

Step 3. Concavity: $f''(x) = 12(x - 3)^2 > 0$ for all $x \neq 3$.

The graph is concave up everywhere.

Step 4. There are no inflection points.

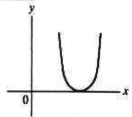

29. $f(x) = \dfrac{x}{x - 3}$; $f'(x) = -\dfrac{3}{(x - 3)^2}$; $f''(x) = \dfrac{6}{(x - 3)^3}$.

Step 1. Critical points: $f'(x) = -\dfrac{3}{(x - 3)^2} < 0$, $(x \neq 3)$.

Since $f'(x) \neq 0$, there are no critical points. In fact, the graph is strictly decreasing. [At $x = 3$, neither $f'(x)$ nor $f(x)$ is defined.]

Step 2. (Does not apply.)

Step 3. Concavity and points of inflection: Since $f''(x) \neq 0$, there are no inflections points. [Since $f''(3)$ does not exist, $x = 3$ is a possible point of inflection. However, since $f(x)$ is not defined at $x = 3$, this point cannot be an inflection point either.]

$\dfrac{6}{(x - 3)^3} > 0$ for $x > 3$ (concave up).

$\dfrac{6}{(x - 3)^3} < 0$ for $x < 3$ (concave down).

Step 4. Other: Since the denominator of $f(x) = \dfrac{x}{x - 3}$ is 0 when $x = 3$, the line $x = 3$ is a vertical asymptote. Also, since $\lim\limits_{x \to \infty} \dfrac{x}{x - 3} = \lim\limits_{x \to \infty} \dfrac{1}{1 - 3/x} = 1$, $y = 1$ is a horizontal asymptote. (Note that the concavity changes at the vertical asymptote.)

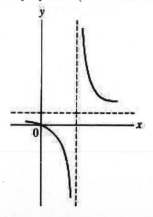

31. $f(x) = x^2 + \dfrac{8}{x}$; $f'(x) = 2x - \dfrac{8}{x^2}$; $f''(x) = 2 + \dfrac{16}{x^3}$.

    Step 1. Critical points:

$$f'(x) = 2x - \frac{8}{x^2} = \frac{2x^3 - 8}{x^2} = 0, \text{ or } x = \sqrt[3]{4}.$$

Step 2. Test of critical points: $f''(\sqrt[3]{4}) > 0$ (minimum).

Step 3. Concavity: We need to determine where $y'' < 0$ and $y'' > 0$. To this end we first determine those values of x for which $f''(x) = 0$:

$$\begin{aligned} f''(x) = 2 + \frac{16}{x^3} = \frac{2x^3 + 16}{x^3} &= 0 \\ 2x^3 + 16 &= 0 \\ x^3 &= -8 \\ x &= -2. \end{aligned}$$

From $y = x^2 + \dfrac{8}{x}$, the point is $(-2, 0)$.

If $x = -3$, $y'' > 0$; so $f(x)$ is concave up for $x < -2$.

If $x = -1$, $y'' < 0$; so $f(x)$ is concave down for $-2 < x < 0$.

We see that the concavity changes at $(-2, 0)$. Since $y'' > 0$ for $x > 0$, the graph is also concave up for $x > 0$.

Step 4. Inflection points: Since the concavity changes, the point $(-2, 0)$ is an inflection point.

Step 5. Other: If x gets large, $y = x^2 + \dfrac{8}{x}$ approaches $y = x^2$, which is therefore an asymptotic curve. Since the denominator of $f(x)$ is 0 when $x = 0$, the y-axis is a vertical asymptote. (Note that the concavity changes at the asymptote.)

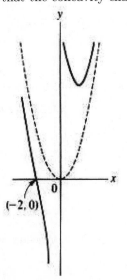

33. $f(x) = \dfrac{x+1}{x-2}$. Vertical asymptote: $x = 2$. Horizontal asymptote: $y = 1$ since

$$\lim_{x \to \infty} \frac{x+1}{x-2} = \lim_{x \to \infty} \frac{1 + 1/x}{1 - 2/x} = 1.$$

$$f'(x) = \frac{-3}{(x-2)^2}; \quad f''(x) = \frac{6}{(x-2)^3}.$$

Step 1. Critical points: Since $f'(x) = \dfrac{-3}{(x-2)^2} < 0$ for all $x \neq 2$, there are no critical points. Moreover, $f(x)$ is always decreasing.

Step 2. (Does not apply.)

Step 3. Concavity: $f''(x) = \dfrac{6}{(x-2)^3}$.

$f''(x) > 0$ for $x > 2$ (concave up).

$f''(x) < 0$ for $x < 2$ (concave down).

Step 4. Inflection points: None. (Note that the concavity changes at $x = 2$, the vertical asymptote.)

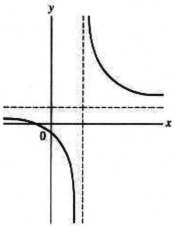

35. $f(x) = x^{3/4}$; $f'(x) = \frac{3}{4}x^{-1/4} = \dfrac{3}{4x^{1/4}}$; $f''(x) = -\frac{3}{16}x^{-5/4} = -\dfrac{3}{16x^{5/4}}$.

Step 1. $f'(x)$ is undefined (infinitely large) at $x = 0$, so that $(0,0)$ is a critical point. The graph has a vertical tangent at $x = 0$.

Step 2. $f(x) = x^{3/4} = (\sqrt[4]{x})^3$; its domain is $[0, \infty)$. As a result, the origin is an endpoint and hence a minimum.

Step 3. For $x > 0$, $f''(x) < 0$. The graph is concave down.

Step 4. There are no inflection points.

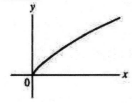

37. $f(x) = (x-3)^{1/3}$; $f'(x) = \dfrac{1}{3(x-3)^{2/3}}$; $f''(x) = -\dfrac{2}{9(x-3)^{5/3}}$.

Neither $f'(x)$ nor $f''(x)$ are 0 for any value of x, but $f'(3)$ and $f''(3)$ do not exist. Thus $(3,0)$ is a critical point. Since $f'(x) > 0$ for all $x \neq 3$, $f(x)$ is strictly increasing. Also, $f''(x) > 0$ for $x < 3$ and $f''(x) < 0$ for $x > 3$. So $(3,0)$ is an inflection point. Finally, since $f'(x)$ approaches infinity when x approaches 3, there is a vertical tangent line at $(3,0)$.

39. $y = \dfrac{2x}{(x+1)^2}$. There is a vertical asymptote at $x = -1$.

$f'(x) = -\dfrac{2(x-1)}{(x+1)^3}$; $f''(x) = \dfrac{4(x-2)}{(x+1)^4}$.

<u>Step 1</u>. Critical points: $f'(x) = 0$ when $x = 1$. The critical point is $\left(1, \dfrac{1}{2}\right)$.

<u>Step 2</u>. Test of critical point: $f''(1) < 0$; $\left(1, \dfrac{1}{2}\right)$ is a maximum.

<u>Step 3</u>. Concavity: $y'' = 0$ when $x = 2$. Now observe that for $x < 2$, $y'' < 0$. So the graph is concave down to the left of the asymptote and also concave down on $(-1, 2]$. For $x > 2$, $y'' > 0$. The graph is concave up on $[2, \infty)$.

<u>Step 4</u>. Because the concavity changes, $\left(2, \dfrac{4}{9}\right)$ is an inflection point.

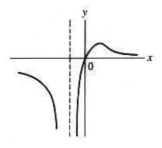

41. $f(x) = \dfrac{6x}{x^2+3}$; $f'(x) = \dfrac{6(3-x^2)}{(x^2+3)^2}$; $f''(x) = \dfrac{12x(x^2-9)}{(x^2+3)^3}$.

<u>Step 1</u>. Critical points: $6(3 - x^2) = 0$ when $x = -\sqrt{3}, \sqrt{3}$.

<u>Step 2</u>. Test of critical points:

$f''(-\sqrt{3}) > 0$ (minimum).

$f''(\sqrt{3}) < 0$ (maximum).

<u>Step 3</u>. Concavity: We need to determine where $y'' < 0$ and where $y'' > 0$. To this end we first determine where $y'' = 0$:

$$12x(x^2 - 9) = 0$$
$$x = 0, \pm 3.$$

test values	$12x$	$x^2 - 9$	$y'' = \dfrac{12x(x^2-9)}{(x^2+3)^3}$
$x < -3$ -4	$-$	$+$	$-$
$-3 < x < 0$ -1	$-$	$-$	$+$
$0 < x < 3$ 1	$+$	$-$	$-$
$x > 3$ 4	$+$	$+$	$+$

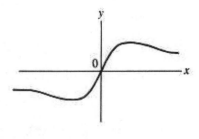

Summary:

$f(x)$ is concave down for $x < -3$.

$f(x)$ is concave up for $-3 < x < 0$.

$f(x)$ is concave down for $0 < x < 3$.

$f(x)$ is concave up for $x > 3$.

<u>Step 4</u>. Inflection points: Since the concavity changes, we get inflection points at $x = 0$ and $x = \pm 3$.

3.3 Exploring with Graphing Utilities

1. $y = x^5 + x^3 + 1$

$$y' = 5x^4 + 3x^2 \;=\; 0$$
$$x^2(5x^2 + 3) \;=\; 0$$
$$x = 0, \; 5x^2 + 3 \;>\; 0$$

Critical point: $(0, 1)$.

$$y'' = 20x^3 + 6x \;=\; 0$$
$$x(20x^2 + 6) \;=\; 0$$
$$x = 0, \; 20x^2 + 6 \;>\; 0$$

Inflection point: $(0, 1)$.

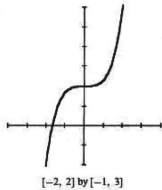

$[-2, \, 2]$ by $[-1, \, 3]$

3. $y = 4x^2 - \frac{4}{5}x^5 + 2$

$$y' = 8x - 4x^4 \;=\; 0$$
$$4x(2 - x^3) \;=\; 0$$
$$x \;=\; 0, \sqrt[3]{2}$$

Critical points: $(0, 2)$ and $(\sqrt[3]{2}, 5.8)$.

$$y'' = 8 - 16x^3 \;=\; 0$$
$$1 - 2x^3 \;=\; 0$$
$$x \;=\; \frac{1}{\sqrt[3]{2}}$$

Inflection point: $\left(\dfrac{1}{\sqrt[3]{2}}, 4.27 \right)$.

$[-2, \, 2]$ by $[-1, \, 6]$

5. $y = 3.0x^2 + 1.2x^5 - 1.0$

$$
\begin{aligned}
y' = 6x + 1.2(5x^4) &= 0 \\
6x + 6x^4 &= 0 \\
6x(1 + x^3) &= 0 \\
x &= 0, -1
\end{aligned}
$$

$$
\begin{aligned}
y'' = 6 + 24x^3 &= 0 \\
6(1 + 4x^3) &= 0 \\
4x^3 &= -1 \\
x^3 &= -\frac{1}{4} \\
x &= -\frac{1}{\sqrt[3]{4}} \approx -0.63
\end{aligned}
$$

If $x = 0$, $y = -1.0$.

If $x = -1$, $y = 3.0(-1)^2 + 1.2(-1)^5 - 1.0 = 0.80$.

If $x = -\dfrac{1}{\sqrt[3]{4}}$, $y = 3.0\left(-\dfrac{1}{\sqrt[3]{4}}\right)^2 + 1.2\left(-\dfrac{1}{\sqrt[3]{4}}\right)^5 - 1.0 = 0.071$.

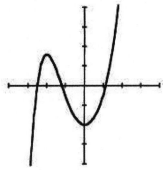

[−2, 2] by [−2, 2]

7. $y = 3.7x^6 + 2.4x^4 + 1.5$

$$
\begin{aligned}
y' = 22.2x^5 + 9.6x^3 &= 0 \\
x^3(22.2x^2 + 9.6) &= 0
\end{aligned}
$$

$x = 0$ is the only real solution. By the graph, $(0, 1.5)$ is a minimum.

$$
\begin{aligned}
y'' = 111x^4 + 28.8x^2 &= 0 \\
x^2(111x^2 + 28.8) &= 0
\end{aligned}
$$

$x = 0$ is the only real solution. Even though $y'' = 0$ at $x = 0$, $(0, 1.5)$ is not an inflection point.

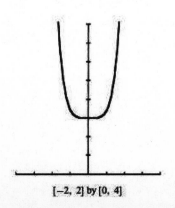

[−2, 2] by [0, 4]

9. $y = 1.5x^4 - 0.50x^6 + 0.20$

$$\begin{aligned} y' = 1.5(4x^3) - 0.50(6x^5) &= 0 \\ 6x^3 - 3x^5 &= 0 \\ 3x^3(2 - x^2) &= 0 \\ x &= 0, \pm\sqrt{2} \end{aligned}$$

$$\begin{aligned} y'' = 18x^2 - 15x^4 &= 0 \\ x^2(18 - 15x^2) &= 0 \\ x = 0, \; x^2 &= \frac{18}{15} \\ x &= \pm\sqrt{\frac{18}{15}} \approx \pm 1.1 \end{aligned}$$

If $x = 0$, $y = 0.20$.

If $x = \pm\sqrt{2}$, $y = 1.5(\pm\sqrt{2})^4 - 0.50(\pm\sqrt{2})^6 + 0.20 = 2.2$.

If $x = \pm\sqrt{\frac{18}{15}}$, $y = 1.5\left(\pm\sqrt{\frac{18}{15}}\right)^4 - 0.50\left(\pm\sqrt{\frac{18}{15}}\right)^6 + 0.20 = 1.5$.

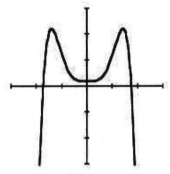

$[-3, 3]$ by $[-3, 3]$

11. $y = \frac{3}{5}x^5 - 2x^3$

$$\begin{aligned} y' = 3x^4 - 6x^2 &= 0 \\ 3x^2(x^2 - 2) &= 0 \\ x &= 0, \pm\sqrt{2} \end{aligned}$$

(Critical points.)

$$\begin{aligned} y'' = 12x^3 - 12x &= 0 \\ 12x(x^2 - 1) &= 0 \\ x &= 0, \pm 1 \end{aligned}$$

Inflection points at $(0, 0)$,

$(-1, 1.4)$, $(1, -1.4)$.

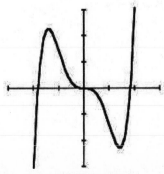

$[-3, 3]$ by $[-3, 3]$

13. $y = \frac{1}{7}x^7 - \frac{1}{5}x^5$

$$\begin{aligned} y' = x^6 - x^4 &= 0 \\ x^4(x^2 - 1) &= 0 \\ x &= 0, \pm 1 \end{aligned}$$

$$\begin{aligned} y'' = 6x^5 - 4x^3 &= 0 \\ x^3(6x^2 - 4) &= 0 \\ x = 0, \; x^2 &= \frac{2}{3} \\ x &= \pm\frac{\sqrt{6}}{3} \end{aligned}$$

If $x = 0$, $y = 0$.

If $x = 1$, $y = \dfrac{1}{7} - \dfrac{1}{5} = -\dfrac{2}{35}$.

If $x = -1$, $y = -\dfrac{1}{7} + \dfrac{1}{5} = \dfrac{2}{35}$.

If $x = \dfrac{\sqrt{6}}{3}$, $y = \dfrac{1}{7}\left(\dfrac{\sqrt{6}}{3}\right)^7 - \dfrac{1}{5}\left(\dfrac{\sqrt{6}}{3}\right)^5 \approx -0.038$.

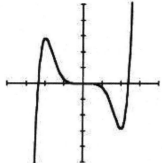

[-2, 2] by [-0.1, 0.1]

15. $y = 1.2x + \dfrac{4.0}{\sqrt{x}}$

Vertical asymptote: y-axis.

$$\begin{aligned} y' = 1.2 - \frac{2}{x^{3/2}} &= 0 \\ 1.2x^{3/2} - 2 &= 0 \\ x^{3/2} &= \frac{2}{1.2} \\ x &= 1.41 \end{aligned}$$

Thus $(1.41, 5.06)$ is a minimum. (See graph in answer section of book.)

17. By the quotient rule,

$$\begin{aligned} \frac{dy}{dx} &= \frac{(x^2 - 1) \cdot 1 - x(2x)}{(x^2 - 1)^2} = \frac{x^2 - 1 - 2x^2}{(x^2 - 1)^2} \\ &= \frac{-x^2 - 1}{(x^2 - 1)^2} = -\frac{x^2 + 1}{(x^2 - 1)^2} \neq 0. \end{aligned}$$

In fact, $y' < 0$ for all x, so that the slope of the tangent line is negative everywhere. (See graph in answer section of book.)

19. $y = \dfrac{x^3}{x^2 - 3}$. By the quotient rule,

$$y' = \frac{x^2(x^2 - 9)}{(x^2 - 3)^2} = 0$$

$$x = 0, \pm 3 \text{ (critical points)}.$$

Vertical asymptotes: $x = \pm\sqrt{3}$. (See graph in answer section of book.)

21. $y = 0.50x^3 + 2.34x^{-1}$

$y' = 0.50(3x^2) - (2.34)x^{-2}$

$= 1.50x^2 - \dfrac{2.34}{x^2} = 0$

Multiplying both sides of the equation by x^2, we get

$1.50x^4 - 2.34 = 0$

$$x = \pm\sqrt[4]{\dfrac{2.34}{1.50}} \approx \pm 1.12$$

If $x = 1.12$, $y = 0.50(1.12)^3 + \dfrac{2.34}{1.12} = 2.79$.

If $x = -1.12$, $y = -2.79$.

(See graph in answer section of book.)

23. $y = 2x - \dfrac{1}{x}$

Vertical asymptote: y-axis.

$y' = 2 + \dfrac{1}{x^2} = 0$ has no real solutions. (See graph in answer section of book.)

3.4   Applications of Minima and Maxima

1. Find the critical point:

$\dfrac{dP}{di} = 0 + 12.8 - 6.40i = 0$

$i = \dfrac{12.8}{6.40} = 2.0\,\text{A}.$

Since $\dfrac{d^2P}{di^2} = -6.40 < 0$, $i = 2.0$ leads to a maximum. Thus

$P = 4.50 + 12.8(2.0) - 3.20(2.0)^2 = 17.3\,\text{W}.$

3. $d(x) = 1.5 \times 10^{-6}x^2(40 - x)^2$; by the product rule,

$d'(x) = 1.5 \times 10^{-6}[x^2 \cdot 2(40 - x)(-1) + (40 - x)^2 \cdot 2x]$

$= 1.5 \times 10^{-6}(2x)(40 - x)[-x + (40 - x)]$

$= 1.5 \times 10^{-6}(2x)(40 - x)(40 - 2x) = 0$

when $x = 0$, 40, and 20. At $x = 0$ and $x = 40$, $d(x) = 0$ (no deflection). At $x = 20$, $d(x)$ is positive. The value $x = 20\,\text{in.}$ leads to a maximum by the first-derivative test: $d'(19) > 0$ and $d'(21) < 0$.

5. $C = k_1A + k_2A^{-1}.$

$\dfrac{dC}{dA} = k_1 + k_2(-1A^{-2})$ since k_1 and k_2 are constants

$\dfrac{dC}{dA} = 0$ when

$k_1 + k_2(-A^{-2}) = 0$

$k_1 - \dfrac{k_2}{A^2} = 0$

$k_1A^2 - k_2 = 0$ multiplying by A^2

$A^2 = \dfrac{k_2}{k_1}$

$A = \sqrt{\dfrac{k_2}{k_1}}$

Since $\dfrac{d^2C}{dA^2} = k_2(2A^{-3}) = \dfrac{2k_2}{A^3} > 0$, C is a minimum at the critical value.

7. $q(t) = \dfrac{t}{t^2 + 1}$; by the quotient rule, $q'(t) = \dfrac{1 - t^2}{(t^2 + 1)^2} = 0$, whence $t = \pm 1$.

   Consider the critical value $t = 1$: for t slightly less than 1, $q' > 0$; for t slightly more than 1, $q' < 0$. So t does correspond to a maximum. When $t = 1\,\mathrm{s}$,

   $$q(1) = \frac{1}{1^2 + 1} = \frac{1}{2}\,\mathrm{C}.$$

9. Let $M(x) =$ the marginal cost, that is,
$$\begin{aligned}
M(x) = C'(x) &= 3(2.0 \times 10^{-6})x^2 - 0.0030x + 2.5 \\
&= 6.0 \times 10^{-6}x^2 - 0.0030x + 2.5.
\end{aligned}$$
   Then
$$\begin{aligned}
M'(x) &= 2(6.0 \times 10^{-6})x - 0.0030 = 0 \\
x &= \frac{0.0030}{2(6.0 \times 10^{-6})} = 250\,\text{units}.
\end{aligned}$$
   Since $M''(x) = 2(6.0 \times 10^{-6}) > 0$, $x = 250$ corresponds to the minimum.

11. Let $x =$ the first number; then $60 - x =$ the second number. For the product P we have
$$P(x) = x(60 - x) = 60x - x^2.$$
$$\begin{aligned}
\frac{dP}{dx} = 60 - 2x &= 0 & \frac{d^2P}{dx^2} &= -2 \\
x &= 30 & &\text{maximum} \\
60 - x &= 30 &
\end{aligned}$$

13. Referring to Figure 3.35, the quantity to be minimized is $L = x + 2y$.

   To eliminate one of the variables, we use the fact that the area is $200\,\mathrm{m}^2$, that is, length times width is 200:

   $$xy = 200 \quad \text{or} \quad x = \frac{200}{y}.$$

   Thus
$$\begin{aligned}
L &= \frac{200}{y} + 2y = 200y^{-1} + 2y \\
L' &= -200y^{-2} + 2 = 0.
\end{aligned}$$
   Multiplying by y^2, we get
$$\begin{aligned}
-200 + 2y^2 &= 0 \\
2y^2 &= 200 \text{ and } y = 10, \\
y &= 10\,\mathrm{m} \\
x &= \frac{200}{10} = 20\,\mathrm{m}.
\end{aligned}$$
   From $L'' = 400y^{-3} = \dfrac{400}{y^3}$, we get $L'' = \dfrac{400}{10^3} > 0$, so that $y = 10\,\mathrm{m}$ leads to the minimum value for L.

15.

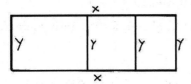

Quantity to be maximized: $A = xy$. To eliminate one of the variables, we make use of the 600 m, the amount of fencing available:

$$2x + 4y = 600$$
$$y = \tfrac{1}{4}(600 - 2x) = \tfrac{1}{2}(300 - x).$$

It follows that

$$A(x) = xy = x \cdot \tfrac{1}{2}(300 - x) = \tfrac{1}{2}(300x - x^2)$$
$$A'(x) = \tfrac{1}{2}(300 - 2x) = 0 \qquad\qquad A''(x) = -1$$
$$x = 150\,\text{m} \qquad\qquad\qquad\qquad \text{maximum}$$
$$y = \tfrac{1}{2}(300 - 150) = 75\,\text{m}.$$

17.

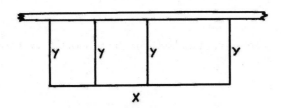

Quantity to be minimized: $L = x + 4y$.

Eliminate x: $xy = 1600$ or $x = \dfrac{1600}{y}$.   lenght times width=1600

$$L = \frac{1600}{y} + 4y = 1600y^{-1} + 4y$$
$$L' = -1600y^{-2} + 4 = 0$$
$$-400y^{-2} + 1 = 0$$

After multiplying by y^2, we get $-400 + y^2 = 0$, or $y = 20\,\text{ft}$.

19. From Figure 3.37,

$$V(x) = (12 - 2x)(12 - 2x)x \qquad\qquad \text{length} \times \text{width} \times \text{height}$$
$$V(x) = x(12 - 2x)^2$$
$$V'(x) = x \cdot 2(12 - 2x)(-2) + (12 - 2x)^2 \cdot 1$$
$$= (12 - 2x)[-4x + (12 - 2x)]$$
$$= (12 - 2x)(12 - 6x) = 0$$
$$x = 6, 2.$$

The only meaningful solution is $x = 2\,\text{cm}$.

$$V''(x) = -96 + 24x \Big|_{x=2} < 0 \text{ (maximum)}.$$

21. Let $x =$ width and $y =$ depth. Then S, the strength of the beam, is $S = kxy^2$, k a constant. This is the quantity to be maximized. To eliminate y, note that from the Pythagorean theorem,

$$x^2 + y^2 = d^2$$
$$y^2 = 9 - x^2. \qquad \text{since } d = 3\,\text{ft}$$

So
$$S = kx(9 - x^2)$$
$$= k(9x - x^3)$$
$$S' = k(9 - 3x^2) = 0$$
$$9 - 3x^2 = 0$$
$$x^2 = 3,$$
$$y^2 = 9 - 3 = 6.$$

So $x = \sqrt{3}\,\text{ft}$ and $y = \sqrt{6}\,\text{ft}$.

23. From Figure 3.39, the quantity to be minimized is $S = x^2 + 4xy$. To eliminate y:

$$32 = x^2 y \quad \text{or} \quad y = \frac{32}{x^2}.$$

It follows that $S(x) = x^2 + 4x\left(\dfrac{32}{x^2}\right) = x^2 + 128x^{-1}$.

$$S'(x) = 2x - 128x^{-2} = 0$$
$$x^3 - 64 = 0$$
$$x = 4\,\text{in.}, y = 2\,\text{in.}$$

$$S''(x) = 2 + \left.\frac{256}{x^3}\right|_{x=4} > 0 \; (\text{minimum}).$$

Dimensions: $4\,\text{in.} \times 4\,\text{in.} \times 2\,\text{in.}$

25. From Figure 3.40: Area of rectangle: $2rx$; Area of semicircle: $\frac{1}{2}\pi r^2$.

Quantity to be maximized: $A = 2rx + \frac{1}{2}\pi r^2$. To eliminate x, note that the perimeter is

$$5 = 2x + 2r + \pi r$$
$$x = \tfrac{1}{2}(5 - 2r - \pi r)$$

Substituting in A, we get

$$A = 2r \cdot \tfrac{1}{2}(5 - 2r - \pi r) + \tfrac{1}{2}\pi r^2$$
$$= 5r - 2r^2 - \pi r^2 + \tfrac{1}{2}\pi r^2$$
$$= 5r - 2r^2 - \tfrac{1}{2}\pi r^2$$
$$\frac{dA}{dr} = 5 - 4r - \pi r = 0$$
$$(4 + \pi)r = 5$$
$$r = \frac{5}{4 + \pi}\,\text{m}.$$

27. Following the hint in Exercise 26, if the cylinder has an open top, then the quantity to be minimized must be $A = 2\pi rh + \pi r^2$. (The area of the side, $2\pi rh$, can be obtained by unrolling the side after cutting it vertically.) To eliminate h:

$$V = \pi r^2 h \quad \text{or} \quad h = \frac{V}{\pi r^2}.$$

So $A(r) = 2\pi r \dfrac{V}{\pi r^2} + \pi r^2 = 2Vr^{-1} + \pi r^2$.

$$\begin{aligned} A'(r) = -2Vr^{-2} + 2\pi r &= 0 \\ -V + \pi r^3 &= 0 \\ r^3 &= \frac{V}{\pi}. \end{aligned}$$

Now calculate h/r:

$$\frac{h}{r} = \frac{hr^2}{r^3} = \frac{\frac{V}{\pi r^2} \cdot r^2}{\frac{V}{\pi}} = 1.$$

So $h = r$.

29. Let (x, y) be a point on the curve. Then

$$d^2 = (x - 1)^2 + (y - 2)^2.$$

Since $y = \frac{1}{4}x^2$,

$$\begin{aligned} d^2 &= (x - 1)^2 + (\tfrac{1}{4}x^2 - 2)^2 \\ &= x^2 - 2x + 1 + \tfrac{1}{16}x^4 - x^2 + 4 \\ &= \tfrac{1}{16}x^4 - 2x + 5. \end{aligned}$$

Critical value:

$$\begin{aligned} \tfrac{1}{16}(4x^3) - 2 &= 0 \\ 4x^3 &= 32 \\ x^3 &= 8 \quad \text{and} \quad x = 2, \\ y &= \tfrac{1}{4}(2)^2 = 1. \end{aligned}$$

31. From Figure 3.41, $A = xy$, the quantity to be maximized. Since $y = 9 - x^2$, the equation of the curve, $A = x(9 - x^2) = 9x - x^3$.

$$\begin{aligned} A' = 9 - 3x^2 &= 0 \\ x &= \sqrt{3} \text{ and } y = 6. \end{aligned}$$

33. Let $x =$ the number of passengers <u>above</u> 30. Then $400 - 10x =$ price per ticket.

Intake: $I =$ price per ticket times the number of passengers, which is $30 + x$.

$$\begin{aligned} I &= (400 - 10x)(30 + x) \\ &= -10x^2 + 100x + 12,000 \\ I' &= -20x + 100 = 0 \quad \text{or} \quad x = 5 \end{aligned}$$

So $30 + x = 35$ passengers.

35.

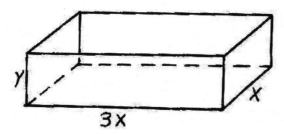

Quantity to be minimized:

S = area of base + area of sides

    = $(3x)(x) + 2(3xy + xy)$

    = $3x^2 + 8xy$.

To eliminate y:

$$486 = 3x \cdot x \cdot y \ \text{ or } \ y = \frac{162}{x^2}.$$

It follows that $S = 3x^2 + 8x \cdot \dfrac{162}{x^2} = 3x^2 + 1296x^{-1}$.

$$S' = 6x - 1296x^{-2} = 0$$

$$x - 216x^{-2} = 0$$

$$x^3 = 216$$

$$x = 6\,\text{in.}$$

$$3x = 18\,\text{in.}$$

$$y = \frac{162}{6^2} = 4.5\,\text{in.}$$

37. From Figure 3.42, the quantity to be maximized is the volume

$$V = x^2y.$$

The combined length and perimeter of a cross-section is $4x + y = 108$, which can be used to eliminate one of the variables: $y = 108 - 4x$ and

$$V = x^2(108 - 4x)$$

   = $108x^2 - 4x^3$.

$$\frac{dV}{dx} = 216x - 12x^2$$

    = $12x(18 - x) = 0$

$$x = 0, \ 18,$$

$$y = 108 - 4(18) = 36.$$

39. Quantity to be minimized: $A = (x+2)(y+4)$. To eliminate x or y:

$$xy = 72 \ \text{ or } \ y = \frac{72}{x}.$$

So $A = (x+2)\left(\dfrac{72}{x} + 4\right) = 4x + 144x^{-1} + 80$.

$$A' = 4 - 144x^{-2} = 0$$

$$x^2 - 36 = 0$$

$$x = 6\,\text{in. and } y = 12\,\text{in.}$$

So the width of the poster is $6 + 2 = 8$ in. and the height is $12 + 4 = 16$ in.

41.

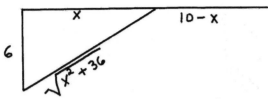

Since distance = rate × time, time = $\dfrac{\text{distance}}{\text{rate}}$. From the figure, total time T is

$$T = \frac{\sqrt{x^2 + 36}}{4} + \frac{10 - x}{5}.$$

$$
\begin{aligned}
\frac{dT}{dx} = \frac{1}{4} \cdot \frac{1}{2}(x^2 + 36)^{-1/2}(2x) - \frac{1}{5} &= 0 \\
\frac{1}{4} \cdot \frac{x}{\sqrt{x^2 + 36}} &= \frac{1}{5} \\
\frac{x}{\sqrt{x^2 + 36}} &= \frac{4}{5} \\
\frac{x^2}{x^2 + 36} &= \frac{16}{25} \\
25x^2 &= 16x^2 + 16 \cdot 36 \\
9x^2 &= 16 \cdot 36 \\
3x &= 4 \cdot 6 \\
x &= 8\,\text{km}
\end{aligned}
$$

43. Let h and r be the height and radius of the cylinder and h_1 and r_1 the height and radius of the cone, so that h_1 and r_1 are fixed quantities.

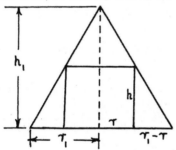

Quantity to be maximized: $V = \pi r^2 h$. To eliminate h, note that by similar triangles we get

$$\frac{h}{h_1} = \frac{r_1 - r}{r_1} \quad \text{or} \quad h = \frac{h_1(r_1 - r)}{r_1}. \qquad (1)$$

Substituting in V:

$$
\begin{aligned}
V &= \pi r^2 \frac{h_1(r_1 - r)}{r_1} \\
&= \frac{\pi h_1}{r_1}(r_1 r^2 - r^3). \qquad h_1 \text{ and } r_1 \text{ constants} \\
\frac{dV}{dr} = \frac{\pi h_1}{r_1}(2r_1 r - 3r^2) &= 0 \\
r &= \frac{2}{3}r_1
\end{aligned}
$$

By (1), $h = \dfrac{h_1}{r_1}\left(r_1 - \dfrac{2}{3}r_1\right) = \dfrac{1}{3}h_1$.

So the height of the cylinder is one-third the height of the cone.

3.5   Related Rates

1. $\dfrac{dy}{dt} = 2x\dfrac{dx}{dt} = 2(2)(1) = 4$

3. $x^2 + y^2 = 25$

$$\begin{aligned} 2x\frac{dx}{dt} + 2y\frac{dy}{dt} &= 0 \\ x\frac{dx}{dt} + y\frac{dy}{dt} &= 0 \\ x\frac{dx}{dt} + y(3) &= 0 \qquad\qquad \frac{dy}{dt} = 3 \\ \frac{dx}{dt} &= -\frac{3y}{x} \end{aligned}$$

If $y = 3$, then $x = \sqrt{25 - 3^2} = 4$. Substituting $y = 3$ and $x = 4$, we get

$$\frac{dx}{dt} = -\frac{3(3)}{4} = -\frac{9}{4}.$$

5. $I = \dfrac{E}{R} = \dfrac{100}{R} = 100R^{-1}$. Given $\dfrac{dR}{dt} = 2$, find $\dfrac{dI}{dt}$ when $R = 10$.

$$\frac{dI}{dt} = 100(-1)R^{-2}\frac{dR}{dt} = -\frac{100}{R^2}\frac{dR}{dt} = -\frac{(100)(2)}{R^2}\Big|_{R=10} = -2\,\text{A/s}$$

7. Given: $\dfrac{di}{dt} = 0.20\,\text{A/s}$, find: $\dfrac{dP}{dt}$ when $i = 2.0\,\text{A}$. $P = 100i^2$ (since $R = 100\,\Omega$).

$\dfrac{dP}{dt} = 100\left(2i\dfrac{di}{dt}\right) = 200i(0.20) = 40i$. At the instant when $i = 2.0$, we have

$$\frac{dP}{di} = 40i\Big|_{i=2.0} = 80\,\text{W/s}.$$

9. In the diagram we label all quantities that change and all quantities that remain fixed.

Given: $\dfrac{dx}{dt} = 2$,

find: $\dfrac{dy}{dt}$ when $x = 4$.

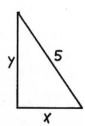

$$\begin{aligned} x^2 + y^2 &= 25 \\ 2x\frac{dx}{dt} + 2y\frac{dy}{dt} &= 0 \\ x\frac{dx}{dt} + y\frac{dy}{dt} &= 0 \\ 2x + y\frac{dy}{dt} &= 0 \qquad \text{since } \frac{dx}{dt} = 2 \\ \frac{dy}{dt} &= -\frac{2x}{y} \end{aligned}$$

At the instant when $x = 4$, we have $y = 3$. Thus $\dfrac{dy}{dt} = -\dfrac{(2)(4)}{3} = -\dfrac{8}{3}\,\text{m/min}.$

11. $V = \frac{4}{3}\pi r^3$. Given: $\dfrac{dV}{dt} = 20.0\,\text{cm}^3/\text{min}$, find: $\dfrac{dr}{dt}$ when $r = 10.0\,\text{cm}$.

$$
\begin{aligned}
\frac{dV}{dt} &= 4\pi r^2 \frac{dr}{dt}\\
20.0 &= 4\pi r^2 \frac{dr}{dt}\\
\frac{dr}{dt} &= \frac{20.0}{4\pi r^2}.
\end{aligned}
$$

At the instant when $r = 10.0$,

$$\frac{dr}{dt} = \frac{20.0}{4\pi r^2}\bigg|_{r=10.0} = \frac{1}{20\pi} \approx 0.0159\,\text{cm}/\text{min}.$$

13. First we need to compute k from the formula $PV^{1.4} = k$. At the instant in question, $V = 2.0$ and $P = 76$. Thus $k = 76(2.0)^{1.4}$.

Given $\dfrac{dV}{dt} = -1.0$, find $\dfrac{dP}{dt}$ when $V = 2.0$. From $PV^{1.4} = k$, we have

$$
\begin{aligned}
P &= kV^{-1.4}.\\
\frac{dP}{dt} &= k(-1.4)V^{-2.4}\frac{dV}{dt}\\
&= k(-1.4)V^{-2.4}(-1.0) = \frac{1.4k}{V^{2.4}}\bigg|_{V=2.0}\\
&= \frac{(1.4)(76)(2.0)^{1.4}}{(2.0)^{2.4}} \qquad\qquad k = 76(2.0)^{1.4}\\
&= 53\,\text{Pa}/\text{min}.
\end{aligned}
$$

15.

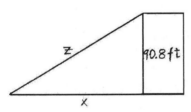

Given: $\dfrac{dx}{dt} = 11.8\,\text{ft/s}$, find: $\dfrac{dz}{dt}$ when $x = 151\,\text{ft}$.

$$
\begin{aligned}
z^2 &= x^2 + (90.8)^2\\
2z\frac{dz}{dt} &= 2x\frac{dx}{dt} + 0\\
z\frac{dz}{dt} &= x\frac{dx}{dt} = x(11.8)\\
\frac{dz}{dt} &= \frac{11.8x}{z}.
\end{aligned}
$$

At the instant when $x = 151$, $z = \sqrt{151^2 + (90.8)^2} = 176.2\,\text{ft}$. Substituting these values,

$$\frac{dz}{dt} = \frac{11.8(151)}{176.2} = 10.1\,\text{ft/s}.$$

17. In the diagram, we label all quantities that change and all quantities that remain fixed.

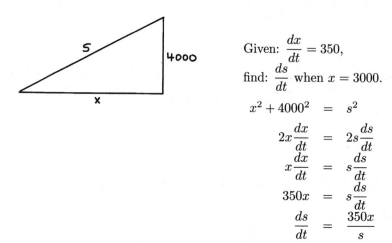

Given: $\dfrac{dx}{dt} = 350$,

find: $\dfrac{ds}{dt}$ when $x = 3000$.

$$x^2 + 4000^2 \;=\; s^2$$
$$2x\frac{dx}{dt} \;=\; 2s\frac{ds}{dt}$$
$$x\frac{dx}{dt} \;=\; s\frac{ds}{dt}$$
$$350x \;=\; s\frac{ds}{dt}$$
$$\frac{ds}{dt} \;=\; \frac{350x}{s}$$

At the instant when $x = 3000$, we have $s = 5000$. Thus

$$\frac{ds}{dt} = \frac{(350)(3000)}{5000} = \frac{(350)(3)}{5} = (70)(3) = 210 \,\text{km/h}.$$

19.

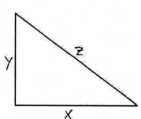

Given: $\dfrac{dx}{dt} = 20\,\text{km/h}$ and $\dfrac{dy}{dt} = 15\,\text{km/h}$, find: $\dfrac{dz}{dt}$ when $x = 40\,\text{km}$ and $y = 30\,\text{km}$.

$$z^2 \;=\; x^2 + y^2$$
$$2z\frac{dz}{dt} \;=\; 2x\frac{dx}{dt} + 2y\frac{dy}{dt}$$
$$z\frac{dz}{dt} \;=\; x\frac{dx}{dt} + y\frac{dy}{dt} = x(20) + y(15)$$
$$\frac{dz}{dt} \;=\; \frac{20x + 15y}{z}.$$

When $x = 40$ and $y = 30$, $z = \sqrt{40^2 + 30^2} = 50$. So

$$\frac{dz}{dt} = \frac{20(40) + 15(30)}{50} = 25 \,\text{km/h}.$$

21. Given $\dfrac{dx}{dt} = -2$, find $\dfrac{dy}{dt}$ when $x = 16$ and $y = 4$. From $y^2 = x$, we have

$$2y\frac{dy}{dt} = \frac{dx}{dt} \;=\; -2$$
$$\frac{dy}{dt} \;=\; -\frac{1}{y}\bigg|_{y=4} = -\frac{1}{4}\,\frac{\text{unit}}{\text{min}}.$$

23.

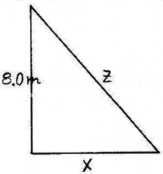

Given: $\dfrac{dz}{dt} = -2.0\,\text{m/min}$ (decreasing quantity), find: $\dfrac{dx}{dt}$ when $x = 12.0\,\text{m}$.

$$z^2 = x^2 + 8.0^2$$

$$2z\frac{dz}{dt} = 2x\frac{dx}{dt} + 0$$

$$z\frac{dz}{dt} = x\frac{dx}{dt}$$

$$z(-2.0) = x\frac{dx}{dt}$$

$$\frac{dx}{dt} = -\frac{2.0z}{x}.$$

At the instant when $x = 12.0$, $z = \sqrt{(12.0)^2 + (8.0)^2} = \sqrt{208} = \sqrt{16 \cdot 13} = 4\sqrt{13}$.

$$\frac{dx}{dt} = -\frac{2.0(4\sqrt{13})}{12.0} = -\frac{2\sqrt{13}}{3} \approx -2.4\,\text{m/min.} \quad \text{(decreasing)}$$

25.

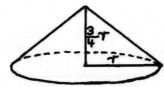

Since $h = \dfrac{3}{4}r$, $r = \dfrac{4}{3}h$.

$$V = \frac{1}{3}\pi r^2 h = \frac{1}{3}\pi \left(\frac{4}{3}h\right)^2 h$$

$$V = \frac{16\pi}{27}h^3$$

Given $\dfrac{dV}{dt} = 12$, find $\dfrac{dh}{dt}$ when $h = 6.0$.

$$\frac{dV}{dt} = \frac{16\pi}{27}(3h^2)\frac{dh}{dt}$$

$$12 = \frac{16\pi}{9}h^2\frac{dh}{dt} \qquad\qquad \text{since } \frac{dV}{dt} = 12$$

$$\frac{dh}{dt} = \frac{108}{16\pi h^2}\Big|_{h=6.0} = \frac{108}{16\pi(36)} = \frac{3}{16\pi} = 0.060\,\text{ft/s.}$$

27.

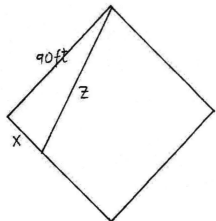

Given: $\dfrac{dx}{dt} = 24\,\text{ft/s}$, find: $\dfrac{dz}{dt}$ when $x = 90 - 25 = 65\,\text{ft}$.

$$z^2 = x^2 + 90^2$$

$$2z\frac{dz}{dt} = 2x\frac{dx}{dt} + 0$$

$$z\frac{dz}{dt} = x\frac{dx}{dt} = x(24)$$

$$\frac{dz}{dt} = \frac{24x}{z}.$$

When $x = 65$, then $z = \sqrt{65^2 + 90^2} = 111.0\,\text{ft}$. So

$$\frac{dz}{dt} = \frac{24(65)}{111.0} = 14\,\text{ft/s}.$$

29.

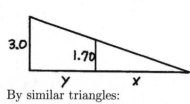

Given: $\dfrac{dy}{dt} = 2.0$,

find: $\dfrac{dx}{dt}$.

By similar triangles:

$$\frac{x+y}{3.0} = \frac{x}{1.70}$$

$$x+y = \frac{3.0}{1.70}x$$

$$y = \frac{1.3}{1.70}x$$

$$\frac{dy}{dt} = \frac{1.3}{1.70} \cdot \frac{dx}{dt}$$

$$2.0 = \frac{1.3}{1.70} \cdot \frac{dx}{dt}$$

$$\frac{dx}{dt} = 2.6\,\text{m/s}.$$

31. By similar triangles, $\dfrac{x}{h} = \dfrac{6.0}{8.0}$

$\qquad\qquad\qquad x = \dfrac{3}{4}h.$

$V = \frac{1}{3}\pi r^2 h$

$\quad = \frac{1}{3}\pi(\frac{3}{4}h)^2 h$

$\quad = \dfrac{3\pi}{16}h^3.$

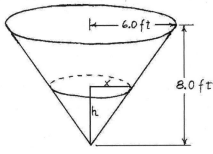

Given: $\dfrac{dV}{dt} = 9.0\,\text{ft}^3/\text{min}$,

find: $\dfrac{dh}{dt}$ when $h = 2.0\,\text{ft}$.

$\dfrac{dV}{dt} = \dfrac{3\pi}{16} \cdot 3h^2 \dfrac{dh}{dt}$

$9.0 = \dfrac{9\pi}{16}h^2 \dfrac{dh}{dt}.$

$\dfrac{dh}{dt} = \dfrac{16}{\pi h^2}\Big|_{h=2.0} = \dfrac{4}{\pi} \approx 1.3\,\text{ft/min}.$

33.

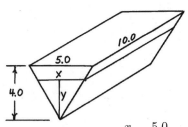

Given: $\dfrac{dV}{dt} = 2.0$,

find: $\dfrac{dy}{dt}$ when $y = 3.0$.

By similar triangles: $\dfrac{x}{y} = \dfrac{5.0}{4.0}$ or $x = \dfrac{5}{4}y.$

Volume $= $ Base $\times$ height $= \dfrac{1}{2}xy \times 10.0$

$V = 5.0xy$

$V = 5.0\left(\dfrac{5}{4}y\right)y \qquad\qquad\qquad$ since $x = \dfrac{5}{4}y$

$V = \dfrac{25}{4}y^2$

$\dfrac{dV}{dt} = \dfrac{25}{4}(2y)\dfrac{dy}{dt}$

$\dfrac{dV}{dt} = \dfrac{25y}{2}\dfrac{dy}{dt}$

$2.0 = \dfrac{25y}{2}\dfrac{dy}{dt} \qquad\qquad\qquad$ since $\dfrac{dV}{dt} = 2.0$

$\dfrac{dy}{dt} = \dfrac{4.0}{25y}\Big|_{y=3.0} = \dfrac{4}{75} = 0.053\,\text{m/min}.$

35.

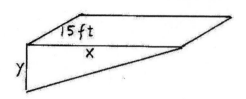

Referring to Figure 3.57 and the diagram, the volume is equal to the area of the triangle times 15 ft, the width of the pool. Using similar triangles,

$$\frac{x}{y} = \frac{50}{12} \quad \text{or} \quad x = \frac{25}{6}y.$$

$$
\begin{aligned}
V &= \frac{1}{2}y\left(\frac{25}{6}y\right) \cdot 15 = \frac{125}{4}y^2 \\
\frac{dV}{dt} &= \frac{125}{2}y\frac{dy}{dt} \\
120 &= \frac{125}{2}y\frac{dy}{dt} \\
\frac{dy}{dt} &= \frac{240}{125y}\Big|_{y=8} = \frac{240}{(125)(8)} = \frac{6}{25}\text{ ft/min.}
\end{aligned}
$$

$$\frac{dV}{dt} = 120\,\text{ft}^3/\text{min}$$

3.6   Differentials

1. Since $\dfrac{dy}{dx} = 3x^2 - 1$, $dy = (3x^2 - 1)dx$.

3. By the quotient rule, $\dfrac{dy}{dx} = -\dfrac{1}{(x-1)^2}$; hence $dy = -\dfrac{dx}{(x-1)^2}$.

5. If $x = 2$, $y = 2^2 - 2 = 2$. If $x = 2.1$, $y = (2.1)^2 - 2.1 = 2.31$. Hence $\Delta y = 2.31 - 2 = 0.31$. From $y = x^2 - x$, we get the differential $dy = (2x - 1)dx$. So if $x = 2$ and $dx = 0.1$, $dy = (4 - 1)(0.1) = 0.3$.

7. $A = x^2$, where x is the length of a side. Given: $x = 6.00$ cm, $dx = 0.02$ cm.
   $\Delta A \approx dA = 2x\,dx = 2(6.00)(0.02) = 0.24\,\text{cm}^2.$
   Percentage error: $\dfrac{dA}{A} \times 100 = \dfrac{0.24}{(6.00)^2} \times 100 = 0.67\%.$

9. $V = s^3$, where s is the length of a side. $s = 5.00$ in. and $ds = \pm 0.01$ in.
   $\Delta V \approx dV = 3s^2\,ds = 3(5.00)^2(\pm 0.01) = \pm 0.75\,\text{in.}^3$
   Percentage error: $\dfrac{dV}{V} \times 100 = \dfrac{0.75}{(5.00)^3} \times 100 = 0.6\%.$

11. $V = \frac{4}{3}\pi r^3$ and $S = 4\pi r^2$. Given: $r = 10.00$ cm, $dr = \pm 1$ mm$= \pm 0.1$ cm.
    $\Delta V \approx dV = 4\pi r^2\,dr = 4\pi(10.00)^2(\pm 0.1) = \pm 125.66\,\text{cm}^3.$
    Percentage error: $\dfrac{dV}{V} \times 100 = \dfrac{125.66}{(4/3)\pi(10.00)^3} \times 100 = 3\%.$
    $\Delta S \approx dS = 8\pi r\,dr = 8\pi(10.00)(\pm 0.1) = \pm 25.13\,\text{cm}^2.$
    Percentage error: $\dfrac{dS}{S} \times 100 = \dfrac{25.13}{4\pi(10.00)^2} \times 100 = 2\%.$

13. $T = 2\pi\sqrt{\dfrac{L}{10}} = \dfrac{2\pi}{\sqrt{10}}L^{1/2}$. $L = 2.0\,\text{m}$ and $dL = \pm 0.1\,\text{m}$.

$$\Delta T \approx dT \quad = \quad \dfrac{2\pi}{\sqrt{10}} \cdot \dfrac{1}{2}L^{-1/2}dL$$

$$= \quad \dfrac{\pi}{\sqrt{10}}\dfrac{1}{\sqrt{L}}dL$$

$$= \quad \dfrac{\pi}{\sqrt{10}}\dfrac{1}{\sqrt{2.0}}(\pm 0.1) = \pm 0.07\,\text{s}.$$

Percentage error: $\dfrac{dT}{T} \times 100 = \dfrac{0.07}{2\pi\sqrt{\dfrac{2.0}{10}}} \times 100 = 2.5\%.$

15. $P = 10.0i^2$. Given: $i = 2.1\,\text{A}$, $di = (2.2 - 2.1) = 0.1\,\text{A}$.

$\Delta P \approx dP = 20.0i\,di = 20.0(2.1)(0.1) = 4.2\,\text{W}.$

17. $A = \pi r^2$; $\Delta A \approx dA = 2\pi r\,dr$. The geometric interpretation of this formula can be seen from the following diagram:

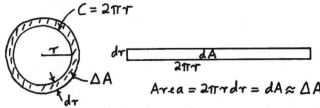

Chapter 3 Review

1. $y = (x-2)^{1/2}$, $y' = \frac{1}{2}(x-2)^{-1/2} = \dfrac{1}{2\sqrt{x-2}}\Big|_{x=3} = \dfrac{1}{2}$

Tangent line:
$$y - 1 \;=\; \tfrac{1}{2}(x-3) \qquad m = \tfrac{1}{2}$$
$$2y - 2 \;=\; x - 3$$
$$x - 2y - 1 \;=\; 0$$

Normal line:
$$y - 1 \;=\; -2(x-3) \qquad m = -\dfrac{1}{1/2} = -2$$
$$y - 1 \;=\; -2x + 6$$
$$2x + y - 7 \;=\; 0$$

3. $f(x) = x^2 - 4x + 3$; $f'(x) = 2x - 4$; $f''(x) = 2.$

Step 1. Critical points: $f'(x) = 2x - 4 \;=\; 0$
$$x \;=\; 2.$$

Substituting in $y = x^2 - 4x + 3$, we get $(2, -1)$ for the critical point.

Step 2. Test of critical point: $f''(2) = 2 > 0$. The critical point is a minimum.

Step 3. Concavity: $y'' = 2 > 0$; the graph is concave up everywhere.

Step 4. There are no inflection points.

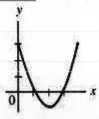

5. $f(x) = -x^3 + 12x + 2$; $f'(x) = -3x^2 + 12$; $f''(x) = -6x$.

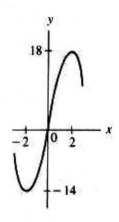

Step 1. Critical points:

$$f'(x) = -3x^2 + 12 = 0$$
$$x = \pm 2.$$

The points are $(2, 18)$ and $(-2, -14)$.

Step 2. Test of critical points:

$f''(2) = -12 < 0$ (maximum).

$f''(-2) = 12 > 0$ (minimum).

Step 3. Concavity: $f''(x) = -6x = 0$ when $x = 0$.

If $x < 0$, $y'' > 0$ (concave up).

If $x > 0$, $y'' < 0$ (concave down).

Step 4. Inflection points: Since the concavity changes, the point $(0, 2)$ is a point of inflection.

7. $f(x) = 3x^4 - 4x^3 + 1$; $f'(x) = 12x^3 - 12x^2$; $f''(x) = 36x^2 - 24x$.

Step 1. Critical points:
$$f'(x) = 12x^3 - 12x^2 = 0$$
$$12x^2(x - 1) = 0$$
$$x = 0, 1.$$

Substituting in the given function, we find that $(0, 1)$ and $(1, 0)$ are the critical points.

Step 2. Test of critical points:

$f''(1) = 36 - 24 > 0$; $(1, 0)$ is a minimum.

$f''(0) = 0$; the test fails.

By the first-derivative test, using $x = -1/2$ and $x = 1/2$ for the test values, we see that $f(x)$ is a decreasing function in the neighborhood of $x = 0$. So $(0, 1)$ is neither a minimum nor a maximum.

Step 3. Concavity:
$$f''(x) = 36x^2 - 24x = 0$$
$$12x(3x - 2) = 0$$
$$x = 0, \frac{2}{3}.$$

	test values	$12x$	$3x - 2$	$y'' = 12x(3x - 2)$
$x < 0$	-1	$-$	$-$	$+$
$0 < x < 2/3$	$1/2$	$+$	$-$	$-$
$x > 2/3$	1	$+$	$+$	$+$

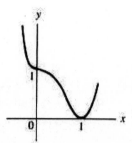

We conclude that the graph is concave up on $(-\infty, 0]$, concave down on $[0, 2/3]$, and concave up on $[2/3, \infty)$.

Step 4. Since the concavity changes, the points $(0, 1)$ and $\left(\frac{2}{3}, \frac{11}{27}\right)$ are inflection points.

9. $f(x) = x^2 - \dfrac{1}{x}; \quad f'(x) = 2x + \dfrac{1}{x^2}; \quad f''(x) = 2 - \dfrac{2}{x^3}.$

<u>Step 1</u>. Critical points:
$$\begin{aligned} f'(x) &= 2x + \frac{1}{x^2} = 0 \\ 2x^3 + 1 &= 0 \\ 2x^3 &= -1 \\ x &= -\frac{1}{\sqrt[3]{2}}. \end{aligned}$$

<u>Step 2</u>. Test of critical points: $f''\left(-\dfrac{1}{\sqrt[3]{2}}\right) > 0$ (minimum).

<u>Step 3</u>. Concavity: We need to determine where $y'' < 0$ and where $y'' > 0$. To this end we first determine those values of x for which $f''(x) = 0$:
$$\begin{aligned} f''(x) = 2 - \frac{2}{x^3} &= 0 \\ 2x^3 - 2 &= 0 \\ x &= 1. \end{aligned}$$
(The concavity may also change at the vertical asymptote $x = 0$.)

	test values	$y'' = 2 - \dfrac{2}{x^3}$
$x > 1$	2	+
$0 < x < 1$	1/2	−
$x < 0$	−1	+

<u>Summary:</u>

If $x > 1$, $f(x)$ is concave up.

If $0 < x < 1$, $f(x)$ is concave down.

If $x < 0$, $f(x)$ is concave up.

<u>Step 4</u>. Inflection points: Because of the change in concavity, $(1, 0)$ is a point of inflection.

(Note that the concavity also changes at $x = 0$, the vertical asymptote.)

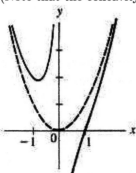

11. $v = kr^2(a - r) = k(ar^2 - r^3).$
$$\begin{aligned} \frac{dv}{dr} = k(2ar - 3r^2) &= 0 \\ kr(2a - 3r) &= 0 \\ r &= \frac{2a}{3}. \qquad \text{positive root} \end{aligned}$$
$$\frac{d^2v}{dr^2} = k(2a - 6r)\Big|_{r=2a/3} < 0.$$
So the critical value $r = 2a/3$ leads to a maximum.

13.

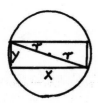

Denote the radius of the log by r,

the width of the beam by x,

and the depth by y.

Quantity to be maximized: $S = kxy^3$. To eliminate x, note that

$$x^2 + y^2 = 4r^2$$
$$x = \sqrt{4r^2 - y^2}.$$

Substituting in the expression for S, we get

$$S = ky^3(4r^2 - y^2)^{1/2}$$
$$S' = ky^3 \cdot \tfrac{1}{2}(4r^2 - y^2)^{-1/2}(-2y) + k(3y^2)(4r^2 - y^2)^{1/2}$$
$$= k\left[\frac{-y^4}{\sqrt{4r^2 - y^2}} + 3y^2\sqrt{4r^2 - y^2}\right] = 0.$$

Multiplying by $\dfrac{1}{k}\sqrt{4r^2 - y^2}$, we get

$$-y^4 + 3y^2(4r^2 - y^2) = 0$$
$$-y^4 + 12r^2y^2 - 3y^4 = 0$$
$$12r^2y^2 - 4y^4 = 0$$
$$4y^2(3r^2 - y^2) = 0$$
$$y^2 = 3r^2$$
$$y = \sqrt{3}r.$$

Hence $x = \sqrt{4r^2 - 3r^2} = r$. So the desired ratio is $\dfrac{y}{x} = \dfrac{\sqrt{3}r}{r} = \sqrt{3}$.

15. $$P(x) = R(x) - C(x)$$
$$P'(x) = R'(x) - C'(x) = 0 \text{ and}$$
$$R'(x) = C'(x).$$

17.

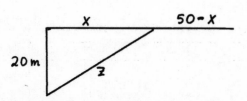

$$z = \sqrt{400 + x^2}$$

Quantity to be minimized:

$$C = 45\sqrt{400 + x^2} + 30(50 - x)$$
$$C' = 45 \cdot \tfrac{1}{2}(400 + x^2)^{-1/2}(2x) - 30 = 0$$

$$\frac{45x}{\sqrt{400 + x^2}} = 30$$

$$\frac{3x}{\sqrt{400 + x^2}} = 2$$

$$3x = 2\sqrt{400 + x^2}$$

$$9x^2 = 4(400 + x^2)$$

$$5x^2 = 1600$$

$$x^2 = 320 = (64)(5)$$

$$x = 8\sqrt{5}$$

So the distance on land is $(50 - 8\sqrt{5})$ m.

19. The quantity to be maximized is the cross-sectional area $A = \frac{1}{2}(x + x)h = xh$. From Figure 3.59, $x = \sqrt{8^2 - h^2}$, and $A = h(64 - h^2)^{1/2}$.

$$\frac{dA}{dh} = h \cdot \frac{1}{2}(64 - h^2)^{-1/2}(-2h) + (64 - h^2)^{1/2}$$

$$= \frac{-h^2}{\sqrt{64 - h^2}} + \sqrt{64 - h^2} = 0.$$

Multiplying by $\sqrt{64 - h^2}$, we get

$$-h^2 + 64 - h^2 = 0$$

$$h^2 = 32$$

$$h = \sqrt{32} = \sqrt{16 \cdot 2} = 4\sqrt{2} \text{ in.}$$

21. $R = 25 + 0.020T^2$

Given: $\dfrac{dT}{dt} = 1.6 \dfrac{°C}{\min}$, find: $\dfrac{dR}{dt}$ when $T = 50.0\,°C$.

$$\frac{dR}{dt} = 0 + 0.020(2T)\frac{dT}{dt}$$

$$= 0.020(2T)(1.6)$$

$$= 0.020(2T)(1.6)\Big|_{T=50.0} = 3.2 \frac{\Omega}{\min}$$

23. Given: $\dfrac{dE}{dt} = 3.00\,\text{V/s}$, find: $\dfrac{dI}{dt}$. $E = \dfrac{0.1138I}{(0.060)^2}$.

$$\frac{dE}{dt} = \frac{0.1138}{(0.060)^2}\frac{dI}{dt}$$

$$3.00 = \frac{0.1138}{(0.060)^2}\frac{dI}{dt}$$

$$\frac{dI}{dt} = \frac{3.00(0.060)^2}{0.1138} = 0.095\,\text{A/s.}$$

25.

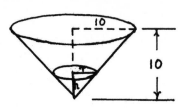

$$\frac{r}{h} = \frac{10}{10} = 1, \text{ or } r = h$$

$$V = \tfrac{1}{3}\pi r^2 h = \tfrac{1}{3}\pi h^3$$

Given $\dfrac{dV}{dt} = 9$, find $\dfrac{dh}{dt}$ when $h = 3$.

$$\begin{aligned}
\frac{dV}{dt} &= \pi h^2 \frac{dh}{dt} \\
9 &= \pi h^2 \frac{dh}{dt} \\
\frac{dh}{dt} &= \frac{9}{\pi h^2}\bigg|_{h=3} = \frac{1}{\pi}\frac{\text{m}}{\text{min}}
\end{aligned}$$

27. First we need to determine the constant k:

$$k = PV^{1.4} = 300(40)^{1.4} = 52481.38.$$

$$P = 52481.38 V^{-1.4}.$$

Given: $\dfrac{dV}{dt} = -80 \text{ in.}^3/\text{s}$, find: $\dfrac{dP}{dt}$ when $V = 40 \text{ in.}^3$

$$\begin{aligned}
\frac{dP}{dt} &= 52481.38(-1.4)V^{-2.4}\frac{dV}{dt} \\
&= 52481.38(-1.4)V^{-2.4}(-80).
\end{aligned}$$

When $V = 40$, $\dfrac{dP}{dt} = 840 \text{ (lb/in.}^2)$ per second.

29. $V = s^3$, $s = 10.00 \text{ cm}$, $ds = \pm 0.02 \text{ cm}$.

$$\Delta V \approx dV = 3s^2 ds = 3(10.00)^2(\pm 0.02) = \pm 6 \text{ cm}^3.$$

Percentage error: $\dfrac{dV}{V} \times 100 = \dfrac{6}{(10.00)^3} \times 100 = 0.6\%.$

31. $A = \pi r^2$.

$$\Delta A \approx dA = 2\pi r\, dr = 2\pi(5.00)(\pm 0.02) = \pm 0.63 \text{ in.}^2$$

Percentage error: $\dfrac{dA}{A} \times 100 = \dfrac{0.63}{\pi(5.00)^2} \times 100 = 0.8\%.$

Chapter 4

The Integral

4.1 Antiderivatives

1. By (4.3), $F(x) = 3x + C$.

3. $f(x) = 1 - 3x^2$. By (4.3) and (4.2), $F(x) = x - 3\dfrac{x^3}{3} + C = x - x^3 + C$.

5. $f(x) = 2x^3 - 3x^2 + x$. By (4.2), $F(x) = 2\dfrac{x^4}{4} - 3\dfrac{x^3}{3} + \dfrac{x^2}{2} + C = \dfrac{1}{2}x^4 - x^3 + \dfrac{1}{2}x^2 + C$.

7. $f(x) = x^3 - 3x^2$. By (4.1) and (4.2), $F(x) = \dfrac{x^4}{4} - 3\dfrac{x^3}{3} + C = \dfrac{1}{4}x^4 - x^3 + C$.

9. $f(x) = x^5 - 6x^4 + 2x^3 + 3$.
   $F(x) = \dfrac{x^6}{6} - 6\dfrac{x^5}{5} + 2\dfrac{x^4}{4} + 3x + C = \dfrac{1}{6}x^6 - \dfrac{6}{5}x^5 + \dfrac{1}{2}x^4 + 3x + C$.

11. $f(x) = \dfrac{1}{x^2} - 2 = x^{-2} - 2$. By (4.1) and (4.3), $F(x) = \dfrac{x^{-1}}{-1} - 2x + C = -\dfrac{1}{x} - 2x + C$.

13. $f(x) = \dfrac{3}{x^2} + \dfrac{2}{\sqrt[3]{x}} = 3x^{-2} + 2x^{-1/3}$. By (4.2), $F(x) = 3\dfrac{x^{-1}}{-1} + 2\dfrac{x^{2/3}}{2/3} + C = -\dfrac{3}{x} + 3x^{2/3} + C$.

15. $f(x) = \dfrac{2}{3x^2} + \dfrac{5}{4x^3} + \sqrt{x} = \dfrac{2}{3}x^{-2} + \dfrac{5}{4}x^{-3} + x^{1/2}$.
    $F(x) = \dfrac{2}{3}\dfrac{x^{-1}}{-1} + \dfrac{5}{4}\dfrac{x^{-2}}{-2} + \dfrac{x^{3/2}}{3/2} + C = -\dfrac{2}{3x} - \dfrac{5}{8x^2} + \dfrac{2}{3}x^{3/2} + C$.

117

4.2   The Area Problem

1. Subdivide the interval $[0, 1]$ into n equal parts, each of length $1/n$. Thus $\Delta x_i = 1/n$. Choosing the right endpoint of each subinterval for x_i, we get

$$x_1 = 1 \cdot \frac{1}{n}, \ x_2 = 2 \cdot \frac{1}{n}, \ x_3 = 3 \cdot \frac{1}{n}, \dots, \ x_i = i \cdot \frac{1}{n}, \dots, \ x_n = n \cdot \frac{1}{n} = 1.$$

Since $f(x) = x$, we get, for the corresponding altitudes,

$$f(x_1) = 1 \cdot \frac{1}{n}, \ f(x_2) = 2 \cdot \frac{1}{n}, \dots, \ f(x_i) = i \cdot \frac{1}{n}, \dots, \ f(x_n) = n \cdot \frac{1}{n}.$$

Since $\Delta x_i = 1/n$, the sum of the areas is given by

$$\sum_{i=1}^{n} f(x_i) \Delta x_i = \sum_{i=1}^{n} i \cdot \frac{1}{n} \cdot \frac{1}{n} = \frac{1}{n^2} \sum_{i=1}^{n} i = \frac{1}{n^2} \cdot \frac{n(n+1)}{2}$$

by formula A. So the exact area is

$$\int_0^1 x \, dx = \lim_{n \to \infty} \frac{n(n+1)}{2n^2} = \lim_{n \to \infty} \frac{n^2 + n}{2n^2} = \lim_{n \to \infty} \frac{1 + 1/n}{2} = \frac{1}{2}.$$

3. Subdivide the interval $[0, 1]$ into n equal parts, each of length $1/n$. Thus $\Delta x_i = 1/n$. Choosing the right endpoint of each subinterval for x_i, we get

$$x_1 = 1 \cdot \frac{1}{n}, \ x_2 = 2 \cdot \frac{1}{n}, \ x_3 = 3 \cdot \frac{1}{n}, \dots, \ x_i = i \cdot \frac{1}{n}, \dots, \ x_n = n \cdot \frac{1}{n} = 1.$$

Since $f(x) = x^2$, we get, for the corresponding altitudes,

$$f(x_1) = \left(1 \cdot \frac{1}{n}\right)^2, \ f(x_2) = \left(2 \cdot \frac{1}{n}\right)^2, \dots, \ f(x_i) = \left(i \cdot \frac{1}{n}\right)^2, \dots, \ f(x_n) = \left(n \cdot \frac{1}{n}\right)^2 = 1.$$

Since $\Delta x_i = 1/n$, the sum of the areas is given by

$$\sum_{i=1}^{n} f(x_i) \Delta x_i = \sum_{i=1}^{n} \left(i \cdot \frac{1}{n}\right)^2 \frac{1}{n} = \frac{1}{n^3} \sum_{i=1}^{n} i^2 = \frac{1}{n^3} \frac{n(n+1)(2n+1)}{6}$$

by Formula B. So the exact area is

$$\int_0^1 x^2 \, dx = \lim_{n \to \infty} \frac{n(n+1)(2n+1)}{6n^3} = \lim_{n \to \infty} \frac{2n^3 + 3n^2 + n}{6n^3} = \lim_{n \to \infty} \frac{2 + 3/n + 1/n^2}{6} = \frac{2}{6} = \frac{1}{3}.$$

5. Subdividing the interval $[0, 2]$ into n equal parts, we get $\Delta x_i = 2/n$. Choosing the right endpoint in each subinterval for x_i again, we have

$$x_1 = 1 \cdot \frac{2}{n}, \ x_2 = 2 \cdot \frac{2}{n}, \dots, \ x_i = i \cdot \frac{2}{n}, \dots, \ x_n = n \cdot \frac{2}{n} = 2.$$

For the altitudes, we get from $f(x) = 3x^2$

$$f(x_1) = 3(1)^2 \frac{4}{n^2}, \ f(x_2) = 3(2)^2 \frac{4}{n^2}, \dots, \ f(x_i) = 3(i)^2 \frac{4}{n^2}, \dots, \ f(x_n) = 3(n)^2 \frac{4}{n^2}.$$

Since $\Delta x_i = 2/n$, the sum of the areas is

$$\sum_{i=1}^{n} f(x_i) \Delta x_i = \sum_{i=1}^{n} 3i^2 \frac{4}{n^2} \cdot \frac{2}{n} = \frac{24}{n^3} \sum_{i=1}^{n} i^2 = \frac{24}{n^3} \cdot \frac{n(n+1)(2n+1)}{6} = \frac{4n(n+1)(2n+1)}{n^3}$$

by Formula B. Hence

$$\int_0^2 3x^2 \, dx = \lim_{n \to \infty} \frac{4n(n+1)(2n+1)}{n^3} = \lim_{n \to \infty} \frac{8n^3 + 12n^2 + 4n}{n^3} = \lim_{n \to \infty} \frac{8 + 12/n + 4/n^2}{1} = 8.$$

4.3   The Fundamental Theorem of Calculus

1. Partial solution:

   (5) $\displaystyle\int_0^2 3x^2\,dx = 3\dfrac{x^3}{3}\bigg|_0^2 = x^3\bigg|_0^2 = 2^3 - 0^3 = 8.$

   (6) $\displaystyle\int_0^2 (1+2x)\,dx = x + 2\dfrac{x^2}{2}\bigg|_0^2 = x + x^2\bigg|_0^2 = (2 + 2^2) - 0 = 6.$

3. $\displaystyle\int_0^2 \frac{1}{2}x\,dx = \frac{1}{2}\frac{x^2}{2}\bigg|_0^2 = \frac{2^2}{4} - \frac{0^2}{4} = 1.$

5. $\displaystyle\int_0^1 (x^3+1)\,dx = \frac{1}{4}x^4 + x\bigg|_0^1 = \left(\frac{1}{4}+1\right) - (0) = \frac{5}{4}.$

7. For the function $y = 4 - x^2$, the x-intercepts are $x = \pm 2$.

$$\begin{aligned}
A &= \int_{-2}^2 (4-x^2)\,dx = 4x - \frac{x^3}{3}\bigg|_{-2}^2 = \left(8 - \frac{8}{3}\right) - \left(-8 + \frac{8}{3}\right) \\
&= 2\left(8 - \frac{8}{3}\right) = 2\left(\frac{24}{3} - \frac{8}{3}\right) = 2\cdot\frac{16}{3} = \frac{32}{3}.
\end{aligned}$$

9. $\displaystyle\int_1^3 \frac{1}{x^3}\,dx = \int_1^3 x^{-3}\,dx = \frac{x^{-2}}{-2}\bigg|_1^3 = -\frac{1}{2x^2}\bigg|_1^3 = -\frac{1}{2\cdot 3^2} + \frac{1}{2\cdot 1^2} = -\frac{1}{18} + \frac{1}{2} = \frac{8}{18} = \frac{4}{9}.$

4.5   Basic Integration Formulas

1. $\displaystyle\int \sqrt{x}\,dx = \int x^{1/2}\,dx = \frac{x^{3/2}}{3/2} + C = \frac{2}{3}x^{3/2} + C.$

3. $\displaystyle\int \left(\frac{1}{x^3} - \frac{3}{x^2}\right)dx = \int (x^{-3} - 3x^{-2})\,dx = \frac{x^{-2}}{-2} - 3\frac{x^{-1}}{-1} + C = -\frac{1}{2x^2} + \frac{3}{x} + C.$

5. $\displaystyle\int (2\sqrt{x} - 3x^2 + 1)\,dx = 2\frac{x^{3/2}}{3/2} - 3\frac{x^3}{3} + x + C = \frac{4}{3}x^{3/2} - x^3 + x + C.$

7. $\displaystyle\int \left(\frac{1}{x^4} + \frac{1}{\sqrt{x}} - 4\right)dx = \int (x^{-4} + x^{-1/2} - 4)\,dx = \frac{x^{-3}}{-3} + \frac{x^{1/2}}{1/2} - 4x + C$

   $\displaystyle\phantom{\int \left(\frac{1}{x^4} + \frac{1}{\sqrt{x}} - 4\right)dx} = -\frac{1}{3x^3} + 2\sqrt{x} - 4x + C.$

9. $\displaystyle\int (2x^2 - 3)^3(4x)\,dx.$ Let $u = 2x^2 - 3$; then $du = 4x\,dx$.

   $\displaystyle\int u^3\,du = \frac{1}{4}u^4 + C = \frac{1}{4}(2x^2 - 3)^4 + C.$

11. $\displaystyle\int (2 - x^2)^4 x\,dx.$ Let $u = 2 - x^2$; then $du = -2x\,dx$.

$$\begin{aligned}
\int (2 - x^2)^4 x\,dx &= -\frac{1}{2}\int (2 - x^2)^4(-2x)\,dx = -\frac{1}{2}\int u^4\,du \\
&= -\frac{1}{2}\frac{u^5}{5} + C = -\frac{1}{10}(2 - x^2)^5 + C.
\end{aligned}$$

13. (a) $\int (1-x)\,dx = x - \dfrac{1}{2}x^2 + C.$

(b) $\int (1-x)^4\,dx.$ Let $u = 1-x$; then $du = -dx.$

$$\int (1-x)^4\,dx = -\int (1-x)^4\,(-dx) = -\int u^4\,du = -\frac{1}{5}u^5 + C = -\frac{1}{5}(1-x)^5 + C.$$

15. $\int \dfrac{x\,dx}{(x^2-1)^2} = \int (x^2-1)^{-2}x\,dx.$ Let $u = x^2-1$; then $du = 2x\,dx.$

$$\frac{1}{2}\int (x^2-1)^{-2}(2x)\,dx = \frac{1}{2}\int u^{-2}\,du = \frac{1}{2}\frac{u^{-1}}{-1} + C = -\frac{1}{2(x^2-1)} + C.$$

17. $\int (2x^2+x)^3(4x+1)\,dx.$ Let $u = 2x^2+x$; then $du = (4x+1)\,dx.$

$$\int u^3\,du = \frac{1}{4}u^4 + C = \frac{1}{4}(2x^2+x)^4 + C.$$

19. $\int \dfrac{dt}{\sqrt{1-t}} = \int (1-t)^{-1/2}\,dt.$ Let $u = 1-t$; then $du = -dt.$

$$-\int (1-t)^{-1/2}(-dt) = -\int u^{-1/2}\,du = -\frac{u^{1/2}}{1/2} + C = -2\sqrt{1-t} + C.$$

21. $\int \dfrac{x\,dx}{\sqrt{1-x^2}} = \int (1-x^2)^{-1/2}x\,dx.$ Let $u = 1-x^2$; then $du = -2x\,dx.$

$$\begin{aligned}
\int (1-x^2)^{-1/2}x\,dx &= -\frac{1}{2}\int (1-x^2)^{-1/2}(-2x\,dx) = -\frac{1}{2}\int u^{-1/2}\,du \\
&= -\frac{1}{2}\frac{u^{1/2}}{1/2} + C = -u^{1/2} + C = -\sqrt{1-x^2} + C.
\end{aligned}$$

23. $\int (2x^3-1)\sqrt[5]{x^4-2x}\,dx = \int (x^4-2x)^{1/5}(2x^3-1)\,dx.$ Let $u = x^4-2x$;
then $du = (4x^3-2)\,dx = 2(2x^3-1)\,dx.$

$$\begin{aligned}
\int (x^4-2x)^{1/5}(2x^3-1)\,dx &= \frac{1}{2}\int (x^4-2x)^{1/5}2(2x^3-1)\,dx = \frac{1}{2}\int u^{1/5}\,du \\
&= \frac{1}{2}\frac{u^{6/5}}{6/5} + C = \frac{5}{12}(x^4-2x)^{6/5} + C.
\end{aligned}$$

25. $\int (x^2+1)^2\,dx.$ If we try letting $u = x^2+1$, then $du = 2x\,dx$ and no substitution can be made. As noted in Example 4, the integral is not of the proper form. Consequently, we must multiply out the binomial and integrate term by term using (4.8):

$$\int (x^2+1)^2\,dx = \int (x^4+2x^2+1)\,dx = \frac{1}{5}x^5 + \frac{2}{3}x^3 + x + C.$$

27. $\int (1-x^2)^2x\,dx.$ Let $u = 1-x^2$; then $du = -2x\,dx.$

$$\int (1-x^2)^2x\,dx = -\frac{1}{2}\int (1-x^2)^2(-2x)\,dx = -\frac{1}{2}\int u^2\,du = -\frac{1}{2}\frac{u^3}{3} + C = -\frac{1}{6}(1-x^2)^3 + C.$$

29. As in Exercise 25, we must multiply out the integrand:

$$\int (1+\sqrt{x})^2\,dx = \int (1+2x^{1/2}+x)\,dx = x + 2\frac{x^{3/2}}{3/2} + \frac{x^2}{2} + C = x + \frac{4}{3}x^{3/2} + \frac{1}{2}x^2 + C.$$

31. $\int (1-5s)^{4/3} \, ds$. Let $u = 1 - 5s$; then $du = -5 \, ds$.

$$\int (1-5s)^{4/3} \, ds = -\frac{1}{5} \int (1-5s)^{4/3}(-5) \, ds = -\frac{1}{5} \int u^{4/3} \, du$$

$$= -\frac{1}{5} \frac{u^{7/3}}{7/3} + C = -\frac{3}{35}(1-5s)^{7/3} + C.$$

33. $\int \left(x^{-1/4} + \frac{1}{x\sqrt{x}} - x \right) dx = \int (x^{-1/4} + x^{-3/2} - x) \, dx = \frac{x^{3/4}}{3/4} + \frac{x^{-1/2}}{-1/2} - \frac{x^2}{2} + C$

$$= \frac{4}{3} x^{3/4} - \frac{2}{\sqrt{x}} - \frac{1}{2} x^2 + C.$$

35. $\int (x^3 + 1)^2 (3x) \, dx$. If we let $u = x^3 + 1$, then $du = 3x^2 \, dx$, which does not match $3x \, dx$. So we need to multiply out the integrand and integrate term by term.

$$\int (x^6 + 2x^3 + 1)(3x) \, dx = 3 \int (x^7 + 2x^4 + x) \, dx$$

$$= 3 \left(\frac{x^8}{8} + 2\frac{x^5}{5} + \frac{x^2}{2} \right) + C$$

$$= \frac{3}{8} x^8 + \frac{6}{5} x^5 + \frac{3}{2} x^2 + C.$$

37. $\int (x^3 + 1)^3 (5x^2) \, dx$. Let $u = x^3 + 1$; then $du = 3x^2 \, dx$.

$$\int (x^3 + 1)^3 (5x^2) \, dx = 5 \int (x^3 + 1)^3 x^2 \, dx = \frac{5}{3} \int (x^3 + 1)(3x^2) \, dx$$

$$= \frac{5}{3} \int u^3 \, du = \frac{5}{3} \frac{u^4}{4} + C = \frac{5}{12}(x^3 + 1)^4 + C.$$

39. $\int (4x^3 - 1)^2 (12x) \, dx$. If we let $u = 4x^3 - 1$, then $du = 12x^2 \, dx$, which does not match $12x \, dx$. So we need to multiply out the binomial and integrate term by term:

$$\int (16x^6 - 8x^3 + 1)(12x) \, dx = 12 \int (16x^7 - 8x^4 + x) \, dx$$

$$= 12 \left(16\frac{x^8}{8} - 8\frac{x^5}{5} + \frac{x^2}{2} \right) + C$$

$$= 24x^8 - \frac{96}{5} x^5 + 6x^2 + C.$$

41. $\int (1 + x^3)^4 (3x^2) \, dx$. Let $u = 1 + x^3$; then $du = 3x^2 \, dx$.

$$\int u^4 \, du = \frac{1}{5} u^5 + C = \frac{1}{5}(1 + x^3)^5 + C.$$

43. $\int (1 + x^3)^4 (x^2) \, dx$. Let $u = 1 + x^3$; then $du = 3x^2 \, dx$.

$$\frac{1}{3} \int (1 + x^3)^4 (3x^2) \, dx = \frac{1}{3} \int u^4 \, du = \frac{1}{3} \frac{u^5}{5} + C = \frac{1}{15}(1 + x^3)^5 + C.$$

45. $\int (2x^3 + 1)^2 (6x) \, dx$. If we let $u = 2x^3 + 1$, then $du = 6x^2 \, dx$, which is not the same as $6x \, dx$. So the substitution cannot be made and the expression in the integrand must be multiplied out instead: $\int (2x^3 + 1)^2 (6x) \, dx = \int (4x^6 + 4x^3 + 1)6x \, dx = \int (24x^7 + 24x^4 + 6x) \, dx$

$$= 24\frac{x^8}{8} + 24\frac{x^5}{5} + 6\frac{x^2}{2} + C = 3x^8 + \frac{24}{5} x^5 + 3x^2 + C.$$

47. $\int (x^4 + 2)^2 (4x^3) \, dx = \frac{1}{3}(x^4 + 2)^3 + C.$ $\qquad u = x^4 + 2$

$$du = 4x^3 \, dx$$

49. As in Exercise 45, if $u = x^3 + 1$, then $du = 3x^2\,dx$, which cannot be made to match $x\,dx$ in the integral. Instead, we multiply out the integrand and integrate term by term:

$$\int (x^3+1)^2 x\,dx = \int (x^6 + 2x^3 + 1)x\,dx = \int (x^7 + 2x^4 + x)\,dx = \frac{1}{8}x^8 + \frac{2}{5}x^5 + \frac{1}{2}x^2 + C.$$

51. $\int (3-t^4)^2 t^3\,dt$. Let $u = 3 - t^4$; then $du = -4t^3\,dt$.

$$-\frac{1}{4}\int (3-t^4)^2(-4t^3)\,dt = -\frac{1}{4}\int u^2\,du = -\frac{1}{4}\frac{u^3}{3} + C = -\frac{1}{12}(3-t^4)^3 + C.$$

53. $\int_0^1 (1-x)\,dx = x - \frac{1}{2}x^2\Big|_0^1 = \left[1 - \frac{1}{2}(1)^2\right] - \left[0 - \frac{1}{2}(0)^2\right] = \frac{1}{2}.$

55. $\int_1^8 \sqrt[3]{x}\,dx = \int_1^8 x^{1/3}\,dx = \frac{x^{4/3}}{4/3}\Big|_1^8 = \frac{3}{4}x^{4/3}\Big|_1^8 = \frac{3}{4}(8^{4/3} - 1^{4/3}) = \frac{3}{4}(16-1) = \frac{45}{4}.$

57. $\int_0^1 \sqrt{1-x}\,dx;\qquad u = 1-x,\ du = -dx.$
Lower limit: if $x = 0$, then $u = 1 - x = 1 - 0 = 1$.
Upper limit: if $x = 1$, then $u = 1 - x = 1 - 1 = 0$.

$$\int_0^1 (1-x)^{1/2}\,dx = -\int_0^1 (1-x)^{1/2}(-dx) = -\int_1^0 u^{1/2}\,du = -\frac{2}{3}u^{3/2}\Big|_1^0 = 0 - \left(-\frac{2}{3}\right) = \frac{2}{3}.$$

Alternatively, we can find the indefinite integral first and substitute $x = 0$ and $x = 1$:

$$\int_0^1 (1-x)^{1/2}\,dx = -\int_0^1 (1-x)^{1/2}(-dx) \qquad u = 1-x;\ du = -dx$$
$$= -\frac{2}{3}(1-x)^{3/2}\Big|_0^1 = 0 - \left(-\frac{2}{3}\cdot 1\right) = \frac{2}{3}.$$

59. $\int_0^1 (x^2-1)^2\,dx = \int_0^1 (x^4 - 2x^2 + 1)\,dx = \frac{1}{5}x^5 - \frac{2}{3}x^3 + x\Big|_0^1$
$$= \left(\frac{1}{5} - \frac{2}{3} + 1\right) - 0 = \frac{3}{15} - \frac{10}{15} + \frac{15}{15} = \frac{8}{15}.$$

61. $\int_2^7 \frac{dx}{\sqrt{x+2}} = \int_2^7 (x+2)^{-1/2}\,dx;\qquad u = x+2,\ du = dx.$
Lower limit: if $x = 2$, then $u = x + 2 = 2 + 2 = 4$.
Upper limit: if $x = 7$, then $u = x + 2 = 7 + 2 = 9$.

$$\int_2^7 (x+2)^{-1/2}\,dx = \int_4^9 u^{-1/2}\,du = 2u^{1/2}\Big|_4^9 = 2\sqrt{u}\Big|_4^9 = 2\sqrt{9} - 2\sqrt{4} = 6 - 4 = 2.$$

Alternatively, we can find the indefinite integral first and substitute $x = 2$ and $x = 7$:

$$\int_2^7 (x+2)^{-1/2}\,dx = 2(x+2)^{1/2}\Big|_2^7 \qquad u = x+2,\ du = dx$$
$$= 2(9)^{1/2} - 2(4)^{1/2} = 2.$$

63. $\int_{-4}^0 \sqrt{1-2x}\,dx = \int_{-4}^0 (1-2x)^{1/2}\,dx$. Let $u = 1 - 2x$; then $du = -2\,dx$.
Lower limit: if $x = -4$, then $u = 9$.
Upper limit: if $x = 0$, then $u = 1$.

$$-\frac{1}{2}\int_{-4}^0 (1-2x)^{1/2}(-2)\,dx = -\frac{1}{2}\int_9^1 u^{1/2}\,du = -\frac{1}{2}\frac{2}{3}u^{3/2}\Big|_9^1 = -\frac{1}{3}(1 - 9^{3/2}) = -\frac{1}{3}(1-27) = \frac{26}{3}.$$

65. $\displaystyle\int_4^9 \frac{1+\sqrt{r}}{\sqrt{r}}\,dr = \int_4^9 \frac{1+r^{1/2}}{r^{1/2}}\,dr = \int_4^9 \left(\frac{1}{r^{1/2}} + \frac{r^{1/2}}{r^{1/2}}\right) dr = \int_4^9 (r^{-1/2}+1)\,dr$

$\displaystyle = 2r^{1/2} + r\Big|_4^9 = [2(9)^{1/2}+9] - [2(4)^{1/2}+4] = (6+9)-(4+4) = 15-8 = 7.$

67. $\displaystyle\int_1^2 \theta\sqrt{4-\theta^2}\,d\theta = \int_1^2 (4-\theta^2)^{1/2}\theta\,d\theta.$ Let $u = 4-\theta^2$; then $du = -2\theta\,d\theta.$
Lower limit: if $\theta = 1$, then $u = 3.$
Upper limit: if $\theta = 2$, then $u = 0.$

$\displaystyle -\frac{1}{2}\int_1^2 (4-\theta^2)^{1/2}(-2\theta)\,d\theta = -\frac{1}{2}\int_3^0 u^{1/2}\,du = -\frac{1}{2}\cdot\frac{2}{3}u^{3/2}\Big|_3^0 = -\frac{1}{3}(0-3^{3/2}) = 3^{1/2} = \sqrt{3}.$

4.6 Area Between Curves

1.

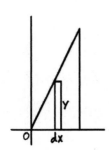

$\displaystyle\int_0^1 y\,dx = \int_0^1 2x\,dx = x^2\Big|_0^1 = 1.$

3.

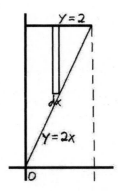

$\displaystyle\int_0^1 (2-2x)\,dx = 2x - x^2\Big|_0^1 = 2-1 = 1.$

5.

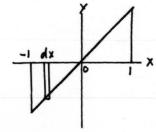

For the region on the left,
$\displaystyle\int_{-1}^0 (0-x)\,dx = -\frac{x^2}{2}\Big|_{-1}^0$
$\displaystyle = 0 - \left(-\frac{1}{2}\right) = \frac{1}{2}.$

For the region on the right: $\displaystyle\int_0^1 x\,dx = \frac{x^2}{2}\Big|_0^1 = \frac{1}{2}$ for a total of $\dfrac{1}{2}+\dfrac{1}{2}=1.$

7.

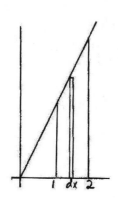

$$\int_1^2 3x\, dx = \frac{3}{2}x^2\Big|_1^2 = \frac{3}{2}(4-1) = \frac{9}{2}.$$

9.

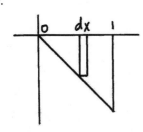

$$\int_0^1 [0 - (-x)\, dx] = \int_0^1 x\, dx = \frac{1}{2}.$$

11.

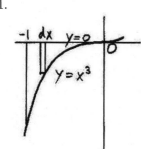

$$\int_{-1}^0 (0 - x^3)\, dx = -\frac{x^4}{4}\Big|_{-1}^0 = 0 + \frac{1}{4} = \frac{1}{4}.$$

13.

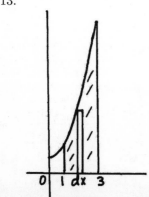

$$\int_1^3 (x^2 + 1)\, dx = \frac{1}{3}x^3 + x\Big|_1^3$$
$$= (9+3) - \left(\frac{1}{3} + 1\right) = 12 - \frac{4}{3} = \frac{32}{3}.$$

15.

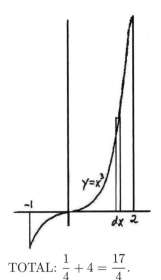

$Y = x^3$

-1

dx 2

TOTAL: $\dfrac{1}{4} + 4 = \dfrac{17}{4}$.

Left region: $\dfrac{1}{4}$ by Exercise 11.

Right region: $\displaystyle\int_0^2 x^3\, dx = \dfrac{1}{4}x^4 \Big|_0^2 = 4.$

17.

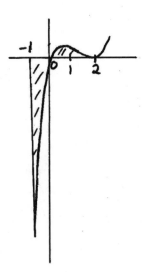

-1

0 1 2

TOTAL: $\dfrac{43}{12} + \dfrac{4}{3} = \dfrac{43}{12} + \dfrac{16}{12} = \dfrac{59}{12}$.

Left region: $\displaystyle\int_{-1}^0 [0 - x(x-2)^2]\, dx$

$= \displaystyle\int_{-1}^0 (-x^3 + 4x^2 - 4x)\, dx$

$= -\dfrac{1}{4}x^4 + \dfrac{4}{3}x^3 - 2x^2 \Big|_{-1}^0$

$= 0 - \left(-\dfrac{1}{4} - \dfrac{4}{3} - 2\right) = \dfrac{43}{12}.$

Right region: $\displaystyle\int_0^2 x(x-2)^2\, dx$

$= \displaystyle\int_0^2 (x^3 - 4x^2 + 4x)\, dx = \dfrac{1}{4}x^4 - \dfrac{4}{3}x^3 + 2x^2 \Big|_0^2$

$= 4 - \dfrac{32}{3} + 8 = \dfrac{36}{3} - \dfrac{32}{3} = \dfrac{4}{3}.$

19.

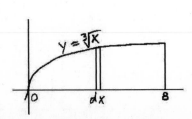

$y = \sqrt[3]{x}$

0 dx 8

$\displaystyle\int_0^8 x^{1/3}\, dx = \dfrac{3}{4}x^{4/3} \Big|_0^8 = \dfrac{3}{4}(8^{4/3} - 0) = \dfrac{3}{4}(16) = 12.$

21. Here the typical element is drawn horizontally.

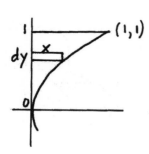

$$\int_0^1 x\,dy = \int_0^1 y^2\,dy = \frac{1}{3}y^3\Big|_0^1 = \frac{1}{3}.$$

23.

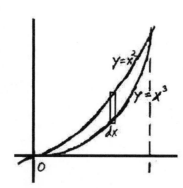

$$\int_0^1 (x^2 - x^3)\,dx = \frac{x^3}{3} - \frac{x^4}{4}\Big|_0^1$$
$$= \frac{1}{3} - \frac{1}{4} = \frac{1}{12}.$$

25.

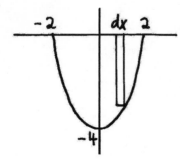

$$\int_{-2}^2 [0 - (x^2 - 4)]\,dx = \int_{-2}^2 (-x^2 + 4)\,dx$$
$$= -\frac{1}{3}x^3 + 4x\Big|_{-2}^2 = \left(-\frac{8}{3} + 8\right) - \left(\frac{8}{3} - 8\right)$$
$$= -\frac{8}{3} + 8 - \frac{8}{3} + 8 = 16 - \frac{16}{3} = \frac{32}{3}.$$

27.

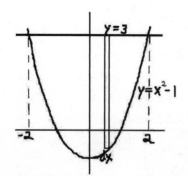

$$y = x^2 - 1$$
$$\underline{\;\;\;y = 3\;\;\;}$$
$$3 = x^2 - 1 \quad \text{(substitution)}$$
$$x^2 = 4$$
$$x = \pm 2.$$

$$\int_{-2}^2 [3 - (x^2 - 1)]\,dx = \int_{-2}^2 (4 - x^2)\,dx = 4x - \frac{1}{3}x^3\Big|_{-2}^2$$
$$= \left(8 - \frac{8}{3}\right) - \left(-8 + \frac{8}{3}\right) = 2\left(8 - \frac{8}{3}\right) = 2\left(\frac{24}{3} - \frac{8}{3}\right)$$
$$= 2\left(\frac{16}{3}\right) = \frac{32}{3}.$$

29. To see where the curves intersect, we need to solve the equations simultaneously:

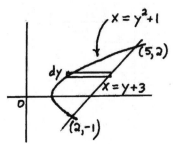

$$
\begin{aligned}
y^2 &= x - 1 \\
y &= x - 3 \\
\hline
y^2 - y &= 2 \quad \text{(subtracting)} \\
y^2 - y - 2 &= 0 \\
(y - 2)(y + 1) &= 0 \\
y &= -1, 2.
\end{aligned}
$$

The points are $(2, -1)$ and $(5, 2)$.

Note that in the resulting region the typical element should be drawn sideways. Solving the given equations for x in terms of y, we get

$$x = y^2 + 1 \quad \text{and} \quad x = y + 3,$$

respectively. The length of the typical element is now seen to be: $(y + 3) - (y^2 + 1)$. Thus

$$
\begin{aligned}
\int_{-1}^{2} [(y + 3) - (y^2 + 1)]\, dy &= \int_{-1}^{2} (y + 3 - y^2 - 1)\, dy \\
&= \int_{-1}^{2} (-y^2 + y + 2)\, dy \\
&= -\frac{1}{3}y^3 + \frac{1}{2}y^2 + 2y \Big|_{-1}^{2} \\
&= \left(-\frac{8}{3} + 2 + 4\right) - \left(\frac{1}{3} + \frac{1}{2} - 2\right) \\
&= -\frac{8}{3} + 2 + 4 - \frac{1}{3} - \frac{1}{2} + 2 \\
&= -\frac{8}{3} - \frac{1}{3} - \frac{1}{2} + 8 = \frac{9}{2}.
\end{aligned}
$$

31.
$$
\begin{aligned}
x^2 + 4y &= 0 \\
x^2 - 4y - 8 &= 0 \\
\hline
8y + 8 &= 0 \\
y &= -1; \\
x^2 &= 4, \ x = \pm 2.
\end{aligned}
\qquad
\begin{aligned}
y &= -\tfrac{1}{4}x^2 \\
y &= \tfrac{1}{4}(x^2 - 8)
\end{aligned}
$$

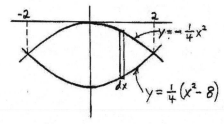

$$
\begin{aligned}
\int_{-2}^{2} \left[-\frac{1}{4}x^2 - \frac{1}{4}(x^2 - 8)\right] dx &= \frac{1}{4}\int_{-2}^{2} (-x^2 - x^2 + 8)\, dx \\
&= \frac{1}{4}\int_{-2}^{2} (8 - 2x^2)\, dx = \frac{1}{4}\left(8x - \frac{2}{3}x^3\right)\Big|_{-2}^{2} \\
&= \frac{1}{4}\left(16 - \frac{16}{3}\right) - \frac{1}{4}\left(-16 + \frac{16}{3}\right) = \frac{1}{2}\left(16 - \frac{16}{3}\right) \\
&= \frac{1}{2} \cdot \frac{48 - 16}{3} = \frac{16}{3}.
\end{aligned}
$$

33.

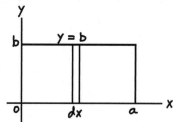

(b) Since the function is $y = b$, $A = \displaystyle\int_0^a b\,dx = bx\Big|_0^a = ab.$

35. We can see from the graph in the Answer Section that the limits of integration are -2 and 4.
Alternatively, we can solve the equations simultaneously:

$$
\begin{aligned}
y &= \tfrac{1}{2}x^2 \\
y &= x + 4 \\
\hline
0 &= \tfrac{1}{2}x^2 - x - 4 \qquad \text{(subtraction)} \\
x^2 - 2x - 8 &= 0 \\
(x + 2)(x - 4) &= 0 \\
x &= -2, 4.
\end{aligned}
$$

Using the integration capability of a graphing utility, we get

$$
\int_{-2}^4 \left(x + 4 - \frac{1}{2}x^2 \right) dx = 18.
$$

37. We can see from the graph in the Answer Section that the limits of integration are -1 and 2.
Alternatively, solving the equations simultaneously,

$$
\begin{aligned}
y &= 2 - x^2 \\
y &= -x \\
\hline
-x &= 2 - x^2 \qquad \text{(substitution)} \\
x^2 - x - 2 &= 0 \\
(x + 1)(x - 2) &= 0 \\
x &= -1, 2.
\end{aligned}
$$

Using the integration capability of a graphing utility, we get $A = \displaystyle\int_{-1}^2 (2 - x^2 + x)\,dx = \frac{9}{2}.$

39. We can see from the graph in the Answer Section that the limits of integration are -2 and 2.
As a check:

If $x = -2$, then $y = 0$ for both equations; if $x = 2$, then $y = 8$ for both equations.

$$
\int_{-2}^2 [(2x + 4) - (x^2 + 2x)]\,dx = \frac{32}{3}.
$$

41. From the graph in the Answer Section, the limits of integration are 0 and 4. Alternatively,

$$y = x^2 - 4x + 2$$
$$y = 2 + 4x - x^2$$
$$\overline{x^2 - 4x + 2 = 2 + 4x - x^2} \qquad \text{(substitution)}$$
$$2x^2 - 8x = 0$$
$$2x(x - 4) = 0$$
$$x = 0, 4.$$

Using the integration capability of a graphing utility, we get

$$A = \int_0^4 [(2 + 4x - x^2) - (x^2 - 4x + 2)]\, dx = \int_0^4 (8x - 2x^2)\, dx = \frac{64}{3}.$$

43. According to the graphs in the Answer Section, the graphs intersect at $(\pm 1, 1)$ and $(\pm 2, 4)$. Using the integration capability of a graphing utility, we can find the sum of the areas of the regions on the right and double the results:

$$2\left(\int_0^1 [(x^4 - 4x^2 + 4) - x^2]\, dx + \int_1^2 [x^2 - (x^4 - 4x^2 + 4)]\, dx \right) = 2\left(\frac{38}{15} + \frac{22}{15} \right) = 2 \cdot \frac{60}{15} = 8.$$

4.7   Improper Integrals

1. $\displaystyle\int_1^\infty \frac{2}{x^4}\, dx = \lim_{b \to \infty} \int_1^b 2x^{-4}\, dx = \lim_{b \to \infty} \left(2\frac{x^{-3}}{-3} \right) \Big|_1^b$

$\displaystyle\phantom{\int_1^\infty \frac{2}{x^4}\, dx} = \lim_{b \to \infty} \left(-\frac{2}{3x^3} \right) \Big|_1^b = \lim_{b \to \infty} \left(-\frac{2}{3b^3} + \frac{2}{3} \right) = \frac{2}{3}$

3. $\displaystyle\int_{-\infty}^0 \frac{4}{(x - 3)^2}\, dx = \lim_{b \to -\infty} \int_b^0 4(x - 3)^{-2}\, dx \qquad u = x - 3$

$\displaystyle\phantom{\int_{-\infty}^0 \frac{4}{(x-3)^2}\, dx} = \lim_{b \to -\infty} 4\frac{(x - 3)^{-1}}{-1} \Big|_b^0 \qquad\qquad du = dx$

$\displaystyle\phantom{\int_{-\infty}^0 \frac{4}{(x-3)^2}\, dx} = \lim_{b \to -\infty} \frac{-4}{x - 3} \Big|_b^0$

$\displaystyle\phantom{\int_{-\infty}^0 \frac{4}{(x-3)^2}\, dx} = \lim_{b \to -\infty} \left[\frac{4}{3} + \frac{4}{b - 3} \right] = \frac{4}{3} + 0 = \frac{4}{3}$

5. $\displaystyle\int_0^\infty \frac{x}{(x^2 + 4)^2}\, dx = \lim_{b \to \infty} \int_0^b (x^2 + 4)^{-2} x\, dx$

$\displaystyle\phantom{\int_0^\infty \frac{x}{(x^2+4)^2}\, dx} = \lim_{b \to \infty} \frac{1}{2} \int_0^b (x^2 + 4)^{-2}(2x\, dx) \qquad u = x^2 + 4$

$\displaystyle\phantom{\int_0^\infty \frac{x}{(x^2+4)^2}\, dx} = \lim_{b \to \infty} \frac{1}{2} \frac{(x^2 + 4)^{-1}}{-1} \Big|_0^b \qquad\qquad du = 2x\, dx$

$\displaystyle\phantom{\int_0^\infty \frac{x}{(x^2+4)^2}\, dx} = \lim_{b \to \infty} \left(-\frac{1}{2(x^2 + 4)} \right) \Big|_0^b$

$\displaystyle\phantom{\int_0^\infty \frac{x}{(x^2+4)^2}\, dx} = \lim_{b \to \infty} \left(-\frac{1}{2(b^2 + 4)} + \frac{1}{8} \right) = \frac{1}{8}$

7.
$$\int_{-\infty}^{0} \frac{dz}{(2z-3)^3} = \lim_{b \to -\infty} \int_{b}^{0} (2z-3)^{-3}\, dz$$

$u = 2z - 3$

$$= \lim_{b \to -\infty} \frac{1}{2} \int_{b}^{0} (2z-3)^{-3}(2)\, dz$$

$du = 2\, dz$

$$= \lim_{b \to -\infty} \frac{1}{2} \left. \frac{(2z-3)^{-2}}{-2} \right|_{b}^{0}$$

$$= \lim_{b \to -\infty} \left(-\frac{1}{4(2z-3)^2} \right) \bigg|_{b}^{0}$$

$$= \lim_{b \to -\infty} \left(-\frac{1}{36} + \frac{1}{4(2b-3)} \right) = -\frac{1}{36}$$

9.

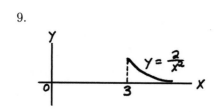

$$\int_{3}^{\infty} \frac{2}{x^2}\, dx = \lim_{b \to \infty} \int_{3}^{b} 2x^{-2}\, dx$$

$$= \lim_{b \to \infty} 2 \left. \frac{x^{-1}}{-1} \right|_{3}^{b} = \lim_{b \to \infty} \left(-\frac{2}{x} \right) \bigg|_{3}^{b}$$

$$= \lim_{b \to \infty} \left(-\frac{2}{b} + \frac{2}{3} \right) = \frac{2}{3}$$

11.

$$\int_{2}^{\infty} \frac{1}{(x-1)^2}\, dx = \lim_{b \to \infty} \int_{2}^{b} (x-1)^{-2}\, dx$$

$u = x - 1$

$$= \lim_{b \to \infty} \left. \frac{-1}{x-1} \right|_{2}^{b}$$

$du = dx$

$$= \lim_{b \to \infty} \left(\frac{-1}{b-1} + 1 \right)$$

$$= 0 + 1 = 1$$

13.

$$\int_{1}^{\infty} \frac{2}{x^3}\, dx = \lim_{b \to \infty} \int_{1}^{b} 2x^{-3}\, dx$$

$$= \lim_{b \to \infty} \left(2 \frac{x^{-2}}{-2} \right) \bigg|_{1}^{b}$$

$$= \lim_{b \to \infty} \left(-\frac{1}{x^2} \right) \bigg|_{1}^{b}$$

$$= \lim_{b \to \infty} \left(-\frac{1}{b^2} + 1 \right) = 1$$

15.
$$\int_{0}^{\infty} \frac{1}{(x+1)^{3/2}}\, dx = \lim_{b \to \infty} \int_{0}^{b} (x+1)^{-3/2}\, dx$$

$u = x + 1$

$$= \lim_{b \to \infty} \left. \frac{(x+1)^{-1/2}}{-1/2} \right|_{0}^{b}$$

$du = dx$

$$= \lim_{b \to \infty} \left(\frac{-2}{(x+1)^{1/2}} \right) \bigg|_{0}^{b}$$

$$= \lim_{b \to \infty} \left(\frac{-2}{(b+1)^{1/2}} + \frac{2}{1} \right)$$

$$= 0 + 2 = 2$$

17. $A = \int_{-\infty}^{0} \left[0 - \frac{1}{(2x-3)^3} \right] dx$

$= \lim_{b \to -\infty} \int_{b}^{0} [-(2x-3)^{-3}] \, dx$

$= \lim_{b \to -\infty} \frac{1}{2} \int_{b}^{0} [-(2x-3)^{-3}] 2 \, dx$

$= \lim_{b \to -\infty} \frac{1}{2} \left(-\frac{(2x-3)^{-2}}{-2} \right) \Big|_{b}^{0}$

$= \lim_{b \to -\infty} \frac{1}{4} \frac{1}{(2x-3)^2} \Big|_{b}^{0}$

$= \lim_{b \to -\infty} \left(\frac{1}{4} \cdot \frac{1}{9} - \frac{1}{4} \cdot \frac{1}{(2b-3)^2} \right) = \frac{1}{36}$

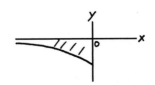

$u = 2x - 3$

$du = 2 \, dx$

19. $\int_{-\infty}^{-2} \frac{3x^2 \, dx}{(x^3+1)^2} = \lim_{b \to -\infty} \int_{b}^{-2} (x^3+1)^{-2}(3x^2) \, dx$

$= \lim_{b \to -\infty} \frac{-1}{x^3+1} \Big|_{b}^{-2}$

$= \lim_{b \to -\infty} \left(\frac{-1}{-8+1} + \frac{1}{b^3+1} \right)$

$= \frac{1}{7} + 0 = \frac{1}{7}$

$u = x^3 + 1$

$du = 3x^2 \, dx$

21.

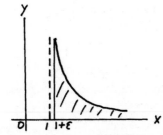

vertical asymptote: $x = 1$

We avoid the vertical asymptote by integrating from $1 + \epsilon$ to 5:

$\int_{1}^{5} \frac{dx}{\sqrt{x-1}} = \lim_{\epsilon \to 0} \int_{1+\epsilon}^{5} (x-1)^{-1/2} \, dx$ $u = x - 1; \; du = dx$

$= \lim_{\epsilon \to 0} 2(x-1)^{1/2} \Big|_{1+\epsilon}^{5}$

$= \lim_{\epsilon \to 0} [2(4)^{1/2} - 2(\epsilon)^{1/2}] = 2\sqrt{4} = 4$

23.

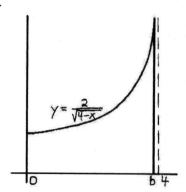

$$y = \frac{2}{\sqrt{4-x}}$$

$$
\begin{aligned}
A &= \int_0^4 \frac{2}{\sqrt{4-x}}\, dx = \lim_{b\to 4-} \int_0^b 2(4-x)^{-1/2}\, dx \qquad\qquad u = 4 - x \\
&= \lim_{b\to 4-} (-2)\frac{(4-x)^{1/2}}{1/2}\Big|_0^b \qquad\qquad\qquad\qquad\qquad du = -dx \\
&= \lim_{b\to 4-} (-4\sqrt{4-x})\Big|_0^b \\
&= \lim_{b\to 4-} (-4\sqrt{4-b} + 4\sqrt{4}) \\
&= 0 + 4\sqrt{4} = 8
\end{aligned}
$$

25. $\displaystyle \int_1^\infty \frac{1}{\sqrt{x}}\, dx = \lim_{b\to\infty}\int_1^b x^{-1/2}\, dx = \lim_{b\to\infty} 2x^{1/2}\Big|_1^b = \lim_{b\to\infty}(2\sqrt{b} - 2) = \infty$

27. Since the y-axis is a vertical asymptote, we need to split the integral as follows:

$$
\begin{aligned}
\int_{-1}^2 \frac{dx}{\sqrt[3]{x}} &= \lim_{\eta\to 0-}\int_{-1}^\eta x^{-1/3}\, dx + \lim_{\epsilon\to 0+}\int_\epsilon^2 x^{-1/3}\, dx \\
&= \lim_{\eta\to 0-} \frac{3}{2}x^{2/3}\Big|_{-1}^\eta + \lim_{\epsilon\to 0+} \frac{3}{2}x^{2/3}\Big|_\epsilon^2 \\
&= \lim_{\eta\to 0-} \frac{3}{2}[\eta^{2/3} - (-1)^{2/3}] + \lim_{\epsilon\to 0+} \frac{3}{2}(2^{2/3} - \epsilon^{2/3}) \\
&= \frac{3}{2}(0 - 1) + \frac{3}{2}(2^{2/3} - 0) \\
&= \frac{3}{2}(2^{2/3} - 1) \approx 0.881
\end{aligned}
$$

29. Since $x = 3$ is a vertical asymptote, we proceed as in Example 4:

$$
\begin{aligned}
\int_0^4 \frac{x}{(9-x^2)^2}\, dx &= \int_0^3 \frac{x}{(9-x^2)^2}\, dx + \int_3^4 \frac{x}{(9-x^2)^2}\, dx \\
&= \lim_{\epsilon\to 0}\int_0^{3-\epsilon} (9-x^2)^{-2}x\, dx + \lim_{\eta\to 0}\int_{3+\eta}^4 (9-x^2)^{-2}x\, dx.
\end{aligned}
$$

Let $u = 9 - x^2$; then $du = -2x\, dx$:

$$
\begin{aligned}
&= \lim_{\epsilon\to 0}\left(-\frac{1}{2}\frac{(9-x^2)^{-1}}{-1}\right)\Big|_0^{3-\epsilon} + \lim_{\eta\to 0}\left(-\frac{1}{2}\frac{(9-x^2)^{-1}}{-1}\right)\Big|_{3+\eta}^4 \\
&= \lim_{\epsilon\to 0}\frac{1}{2(9-x^2)}\Big|_0^{3-\epsilon} + \lim_{\eta\to 0}\frac{1}{2(9-x^2)}\Big|_{3+\eta}^4 \\
&= \lim_{\epsilon\to 0}\left\{\frac{1}{2[9-(3-\epsilon)^2]} - \frac{1}{18}\right\} + \lim_{\eta\to 0}\left\{\frac{1}{2(-7)} - \frac{1}{2[9-(3+\eta)^2]}\right\}.
\end{aligned}
$$

Neither limit exists.

31. $\displaystyle\int_1^4 \frac{3\,dx}{(x-3)^{2/5}}$ $=$ $\displaystyle\lim_{\eta\to 0}\int_1^{3-\eta} 3(x-3)^{-2/5}\,dx + \lim_{\epsilon\to 0}\int_{3+\epsilon}^4 3(x-3)^{-2/5}\,dx$ vertical

   $=$ $\displaystyle\lim_{\eta\to 0} 5(x-3)^{3/5}\Big|_1^{3-\eta} + \lim_{\epsilon\to 0} 5(x-3)^{3/5}\Big|_{3+\epsilon}^4$ asymptote $x = 3$

   $=$ $\displaystyle\lim_{\eta\to 0} 5[(3-\eta-3)^{3/5} - (-2)^{3/5}] + \lim_{\epsilon\to 0} 5[1 - (3+\epsilon-3)^{3/5}]$

   $=$ $5(0 + 2^{3/5}) + 5(1-0) = 5(1 + 2^{3/5}) \approx 12.58$

33. Since $x = -2$ is a vertical asymptote, we proceed as in Example 4:

   $\displaystyle\int_{-3}^{-1} \frac{1}{(x+2)^4}\,dx$ $=$ $\displaystyle\lim_{\epsilon\to 0}\int_{-3}^{-2-\epsilon}(x+2)^{-4}\,dx + \lim_{\eta\to 0}\int_{-2+\eta}^{-1}(x+2)^{-4}\,dx$

   $=$ $\displaystyle\lim_{\epsilon\to 0}\left[-\frac{1}{3(x+2)^3}\right]\Big|_{-3}^{-2-\epsilon} + \lim_{\eta\to 0}\left[-\frac{1}{3(x+2)^3}\right]\Big|_{-2+\eta}^{-1}$

   $=$ $\displaystyle\lim_{\epsilon\to 0}\left(-\frac{1}{3(-2-\epsilon+2)^3} - \frac{1}{3}\right) + \lim_{\eta\to 0}\left(-\frac{1}{3} + \frac{1}{3(-2+\eta+2)}\right)$

   $=$ $\displaystyle\lim_{\epsilon\to 0}\left(\frac{1}{3\epsilon^3} - \frac{1}{3}\right) + \lim_{\eta\to 0}\left(-\frac{1}{3} + \frac{1}{3\eta^3}\right).$

Neither limit exists.

4.8   The Constant of Integration

1. From $\dfrac{dy}{dx} = 3x$, we get $y = \dfrac{3}{2}x^2 + C$. Substituting $(0,1)$ in the equation, we get $1 = 0 + C$ or $C = 1$. The resulting function is

$$y = \frac{3}{2}x^2 + 1.$$

3. From $\dfrac{dy}{dx} = 6x^2 + 1$, we get $y = 2x^3 + x + C$. Substituting $(-1,1)$ in the equation, we get $1 = 2(-1)^3 + (-1) + C$ or $C = 4$. The resulting function is

$$y = 2x^3 + x + 4.$$

5. From $\dfrac{dy}{dx} = 3x^2 + 2$, we get $y = x^3 + 2x + C$. Substituting $(1,0)$ in the equation, we get $0 = 3 + C$, or $C = -3$. It follows that

$$y = x^3 + 2x - 3.$$

7.

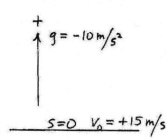

Taking the upward direction as positive, we get $g = -10 \,\text{m/s}^2$ and $v_0 = +15 \,\text{m/s}$, since the stone is moving in the upward direction initially. Integrating, we have

$$v = -10t + C.$$

If $t = 0$, $v = +15$, so that $15 = 0 + C$.

Thus

$$v = -10t + 15.$$

To see how long the stone takes to reach its highest point, we set v equal to 0:

$$0 = -10t + 15 \quad \text{and} \quad t = 1.5 \,\text{s}.$$

So the entire trip takes $2(1.5) = 3 \,\text{s}$.

9.

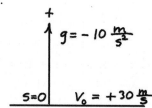

Taking the upward direction as positive (as was done in the examples), we get $g = -10 \,\text{m/s}^2$ and $v_0 = +30 \,\text{m/s}$, since the ball is moving in the positive direction initially. Integrating, we have

$$v = -10t + C.$$

If $t = 0$, $v = +30$, so that $30 = 0 + C$.

Thus

$$v = -10t + 30.$$

To find s, we integrate v:

$$s = -5t^2 + 30t + k.$$

If the origin is on the ground, as in the figure, $s = 0$ when $t = 0$. Thus $0 = 0 + 0 + k$ and

$$s = -5t^2 + 30t.$$

To find how high the ball rises, we first determine how long it takes to reach the highest point by setting v to 0:

$$v = -10t + 30 = 0, \quad \text{whence } t = 3.$$

Thus

$$s = -5t^2 + 30t \big|_{t=3} = -45 + 90 = 45.$$

In the coordinate system chosen, 45 corresponds to 45 m above the ground.

11.

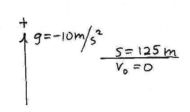

Taking the upward direction to be positive, $g = -10 \, \text{m/s}^2$. Since the initial velocity is 0,

$$v = -10t. \qquad v = 0 \text{ when } t = 0$$

Integrating v, we have

$$s = -5t^2 + k.$$

Since the origin ($s = 0$) is on the ground, $s = 125 \, \text{m}$ initially (when $t = 0$). So

$$125 = 0 + k$$

and

$$s = -5t^2 + 125. \qquad\qquad k = 125$$

To see how long the object takes to reach the ground, we let $s = 0$:

$$0 = -5t^2 + 125$$
$$t^2 = 25 \quad \text{and} \quad t = 5 \, \text{s}.$$

13.

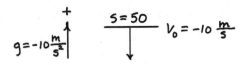

Taking the upward direction as positive, $g = -10 \, \text{m/s}^2$, as usual, and $v_0 = -10 \, \text{m/s}$ (since the object is moving in the downward direction). Thus $v = -10t + C$ and

$$v = -10t - 10. \qquad \text{if } t = 0, \, v = -10 \, \text{m/s}$$

Hence

$$s = -5t^2 - 10t + k. \qquad\qquad \text{integrating } v$$

Now observe that when $t = 0$, $s = 50 \, \text{m}$. So

$$50 = 0 + 0 + k$$

and

$$s = -5t^2 - 10t + 50.$$

To see how long it takes the object to reach the ground, we let $s = 0$ and solve for t:

$$0 = -5t^2 - 10t + 50$$
$$t^2 + 2t - 10 = 0 \qquad\qquad \text{dividing by } -5$$
$$t = \frac{-2 \pm \sqrt{4 + 40}}{2} = \frac{-2 \pm 2\sqrt{11}}{2} = -1 \pm \sqrt{11}.$$

Taking the positive root, $t = -1 + \sqrt{11} \approx 2.3 \, \text{s}$. Finally, when $t = 2.3$,

$$v = -10(2.3) - 10 = -33.$$

We conclude that the velocity is $33 \, \text{m/s}$ in the downward direction.

15.

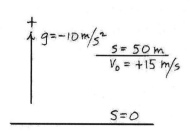

Taking the upward direction to be positive, $g = -10\,\text{m/s}^2$ and $v_0 = +15\,\text{m/s}$ (upward direction).

$$
\begin{aligned}
v &= -10t + C \\
+15 &= 0 + C \qquad \text{when } t = 0,\, v = +15\,\text{m/s} \\
v &= -10t + 15.
\end{aligned}
$$

So

$$s = -5t^2 + 15t + k.$$

When $t = 0$, $s = 50\,\text{m}$:

$$50 = 0 + k$$

and

$$s = -5t^2 + 15t + 50.$$

To see how long it takes for the rock to reach the ground, we let $s = 0$ and solve for t:

$$
\begin{aligned}
0 &= -5t^2 + 15t + 50 \\
t^2 - 3t - 10 &= 0 \\
(t - 5)(t + 2) &= 0 \\
t &= 5\,\text{s}. \qquad \text{(positive root)}
\end{aligned}
$$

Finally, when $t = 5$, $v = -10(5) + 15 = -35\,\text{m/s}$. We conclude that the velocity is $35\,\text{m/s}$ in the downward direction.

17. Taking the upward direction as positive, $g = -10\,\text{m/s}^2$ and $v_0 = +10\,\text{m/s}$ (since the object is moving in the upward direction). Thus $v = -10t + C$, and

$$v = -10t + 10. \hspace{3cm} \text{if } t = 0,\, v = +10\,\text{m/s}$$

So

$$s = -5t^2 + 10t + k \hspace{3cm} \text{integrating } v$$

and

$$s = -5t^2 + 10t + 40. \hspace{3cm} \text{if } t = 0,\, s = 40\,\text{m}$$

To see how long the object travels before striking the ground, we let $s = 0$:

$$
\begin{aligned}
0 &= -5t^2 + 10t + 40 \\
t^2 - 2t - 8 &= 0 \hspace{3cm} \text{dividing by -5} \\
(t - 4)(t + 2) &= 0 \\
t &= +4. \hspace{3cm} \text{positive root}
\end{aligned}
$$

From $v = -10t + 10$, we get $v = -30\,\text{m/s}$, that is, the object is moving in the downward direction.

19. The simplest way to solve this problem is to take the bottom of the pit to be the origin ($s = 0$). However, to be consistent with our general method, let $s = 0$ at ground level, so that $s = -20$ m initially, with $v_0 = +25$ m/s.

$$
\begin{aligned}
v &= -10t + 25 \\
s &= -5t^2 + 25t + k \\
-20 &= 0 + k
\end{aligned}
$$

and

$$s = -5t^2 + 25t - 20.$$

To find how long it takes for the stone to reach the top of the pit (ground level), we let $s = 0$ and solve for t:

$$
\begin{aligned}
0 &= -5t^2 + 25t - 20 \\
t^2 - 5t + 4 &= 0 \\
(t - 1)(t - 4) &= 0 \\
t &= 1, 4.
\end{aligned}
$$

Answer: $t = 1$ s to reach the top. (The stone actually overshoots and reaches the top of the pit again after 4 s on the way down.)

21. If we let the direction of motion be the positive direction, then $a = -7$ and $v_0 = 28$. For convenience, we let $s = 0$ be the point at which the car starts to decelerate. We now get

$$
\begin{aligned}
v &= -7t + 28, \\
s &= -\frac{7}{2}t^2 + 28t + 0. \qquad\qquad k = 0 \text{ (since } s = 0 \text{ when } t = 0)
\end{aligned}
$$

If we let $v = 0$, we find that the car stops in four seconds. From

$$s = -\frac{7}{2}t^2 + 28t \Big|_{t=4} = 56$$

we see that the car stops in 56 m.

23. The easiest way to solve this problem is to assume that the keys are dropped from rest at
    $s = 15\,\text{m}$ and determine the speed when it hits the ground. Alternatively, let v_0 be the un-
    known initial velocity. Then

$$
\begin{aligned}
v &= -10t + v_0 \\
s &= -5t^2 + v_0 t + k
\end{aligned}
$$

and

$\qquad$ (1) $\; s = -5t^2 + v_0 t.$ $\qquad\qquad\qquad\qquad\qquad$ $s = 0$ when $t = 0$

By assumption, the object reaches the dorm window when $v = 0$ and $s = 15$. If $v = 0$,

$$
0 = -10t + v_0 \;\text{ and }\; t = \frac{v_0}{10}.
$$

Substituting in Equation (1),

$$
\begin{aligned}
15 &= -5\left(\frac{v_0}{10}\right)^2 + v_0\left(\frac{v_0}{10}\right) \\
15 &= \left(-\frac{5}{10^2} + \frac{1}{10}\right)v_0^2 \\
v_0^2 &= 3\cdot 10^2 \\
v_0 &= 10\sqrt{3}\,\text{m/s}.
\end{aligned}
$$

25. $a = \dfrac{t}{(t^2+1)^2}$

$\quad v = \displaystyle\int \frac{t}{(t^2+1)^2}\,dt = \int (t^2+1)^{-2}t\,dt$ $\qquad$ $u = t^2+1;\; du = 2t\,dt$

$\qquad = \dfrac{1}{2}\displaystyle\int (t^2+1)^{-2}(2t\,dt)$

$\qquad = \dfrac{1}{2}\dfrac{(t^2+1)^{-1}}{-1} + C = -\dfrac{1}{2(t^2+1)} + C$

Since $v = 10$ when $t = 0$, we get from

$\quad v = -\dfrac{1}{2(t^2+1)} + C,$

$\quad 10 = -\dfrac{1}{2(0+1)} + C \quad$ or $C = \dfrac{21}{2}.$

So

$\quad v = -\dfrac{1}{2(t^2+1)} + \dfrac{21}{2}$

$\qquad = \dfrac{1}{2}\left(21 - \dfrac{1}{t^2+1}\right).$

27. $q = \displaystyle\int i\,dt = \int (t+1)^{1/2}\,dt = \dfrac{2}{3}(t+1)^{3/2} + C.$

$\quad 0 = \dfrac{2}{3} + C$ and $C = -\dfrac{2}{3}.$

$\quad q = \dfrac{2}{3}(t+1)^{3/2} - \dfrac{2}{3}\Big|_{t=3.0} = 4.7\,\text{C}.$

29. Given: $i = 0.010t + 0.10$ and $q = 0.030\,\text{C}$ when $t = 0$.
$$q = \int i\,dt = \int (0.010t + 0.10)\,dt = 0.010\left(\frac{t^2}{2}\right) + 0.10t + C$$
Substituting $t = 0$ and $q = 0.030$, we get
$$0.030 = 0 + C \quad \text{or} \quad C = 0.030,$$

so that
$$q = 0.010\left(\frac{t^2}{2}\right) + 0.10t + 0.030\big|_{t=3.0} = 0.38\,\text{C}.$$

31. $v = \dfrac{1}{C}\displaystyle\int i\,dt = \dfrac{1}{0.00300}\displaystyle\int 0.0100t\,dt = \dfrac{1}{0.00300}(0.0100)\dfrac{t^2}{2} + C$. Since $v = 30$ when $t = 0$, we find that $C = 30$. So
$$v = \frac{1}{0.00300}(0.0100)\frac{t^2}{2} + 30.$$
When $t = 4.00\,\text{s}$, $v = 56.7\,\text{V}$.

33. Integrate the given expression:
$$\begin{aligned} R &= 0.00060\frac{T^3}{3} + 0.0080\frac{T^2}{2} + 0.14T + C \\ R &= 0.00020T^3 + 0.0040T^2 + 0.14T + C. \end{aligned}$$
Let $R = 50$ and $T = 0$:
$$50 = 0 + C \quad \text{or} \quad C = 50.$$
$$R = 0.00020T^3 + 0.0040T^2 + 0.14T + 50\big|_{T=20} = 56\,\Omega$$

35. $\begin{aligned} P &= 3(t+1)^{1/2} \\ W &= \int P\,dt = \int 3(t+1)^{1/2}\,dt = 2(t+1)^{3/2} + C \\ 5 &= 2 + C \text{ and } C = 3 \hspace{3cm} W = 5 \text{ when } t = 0 \\ W &= 2(t+1)^{3/2} + 3. \end{aligned}$
When $t = 8\,\text{s}$, $W = 2(8+1)^{3/2} + 3 = 57\,\text{J}$.

4.9   Numerical Integration

1. If $n = 6$, we have $h = \dfrac{b-a}{6} = \dfrac{1}{3}$. Thus $x_0 = 1$, $x_1 = \dfrac{4}{3}$, $x_2 = \dfrac{5}{3}$, $x_3 = \dfrac{6}{3}$, $x_4 = \dfrac{7}{3}$, $x_5 = \dfrac{8}{3}$, $x_6 = \dfrac{9}{3} = 3$.
By the trapezoidal rule:

$$\int_1^3 x^2\,dx \approx \frac{1}{3}\left[\frac{1}{2}(1)^2 + \left(\frac{4}{3}\right)^2 + \left(\frac{5}{3}\right)^2 + \left(\frac{6}{3}\right)^2 + \left(\frac{7}{3}\right)^2 + \left(\frac{8}{3}\right)^2 + \frac{1}{2}\left(\frac{9}{3}\right)^2\right] = 8.704.$$

By Simpson's rule:

$$\int_1^3 x^2\,dx \approx \frac{1/3}{3}\left[1^2 + 4\left(\frac{4}{3}\right)^2 + 2\left(\frac{5}{3}\right)^2 + 4\left(\frac{6}{3}\right)^2 + 2\left(\frac{7}{3}\right)^2 + 4\left(\frac{8}{3}\right)^2 + \left(\frac{9}{3}\right)^2\right] = 8.667.$$

By direct integration:

$$\int_1^3 x^2\,dx = \frac{1}{3}x^3\Big|_1^3 = \frac{26}{3}.$$

3. Since $n = 4$, we have $h = \dfrac{b - a}{4} = \dfrac{1}{4}$. Thus $x_0 = 0$, $x_1 = \dfrac{1}{4}$, $x_2 = \dfrac{1}{2}$, $x_3 = \dfrac{3}{4}$, $x_4 = 1$.

By the trapezoidal rule:

$$\int_0^1 \frac{1}{1 + x^2}\, dx \approx \frac{1}{4}\left[\frac{1}{2}\frac{1}{1 + 0^2} + \frac{1}{1 + (1/4)^2} + \frac{1}{1 + (1/2)^2} + \frac{1}{1 + (3/4)^2} + \frac{1}{2}\frac{1}{1 + 1^2}\right] = 0.783.$$

By Simpson's rule:

$$\int_0^1 \frac{1}{1 + x^2}\, dx \approx \frac{1/4}{3}\left[\frac{1}{1 + 0^2} + 4\frac{1}{1 + (1/4)^2} + 2\frac{1}{1 + (1/2)^2} + 4\frac{1}{1 + (3/4)^2} + \frac{1}{1 + 1^2}\right] = 0.785.$$

5. If $n = 4$, we have $h = \dfrac{b - a}{4} = \dfrac{1}{2}$. Thus $x_0 = 0$, $x_1 = \dfrac{1}{2}$, $x_2 = 1$, $x_3 = \dfrac{3}{2}$, $x_4 = 2$.

By the trapezoidal rule:

$$\int_0^2 \sqrt{1 + x}\, dx \approx \frac{1}{2}\left[\frac{1}{2}\sqrt{1} + \sqrt{\frac{3}{2}} + \sqrt{2} + \sqrt{\frac{5}{2}} + \frac{1}{2}\sqrt{3}\right] = 2.793.$$

By Simpson's rule:

$$\int_0^2 \sqrt{1 + x}\, dx \approx \frac{1/2}{3}\left[\sqrt{1} + 4\sqrt{\frac{3}{2}} + 2\sqrt{2} + 4\sqrt{\frac{5}{2}} + \sqrt{3}\right]$$

$$= \frac{1}{6}\left[1 + 4\sqrt{1.5} + 2\sqrt{2} + 4\sqrt{2.5} + \sqrt{3}\right] = 2.797$$

7. Since $n = 5$, $h = \dfrac{3 - 0}{5} = \dfrac{3}{5}$.

Thus $x_0 = 0$, $x_1 = \dfrac{3}{5}$, $x_2 = \dfrac{6}{5}$, $x_3 = \dfrac{9}{5}$, $x_4 = \dfrac{12}{5}$, $x_5 = \dfrac{15}{5} = 3$.

$$\int_0^3 \sqrt{x}\, dx \approx \frac{3}{5}\left(0 + \sqrt{\frac{3}{5}} + \sqrt{\frac{6}{5}} + \sqrt{\frac{9}{5}} + \sqrt{\frac{12}{5}} + \frac{1}{2}\sqrt{3}\right) = 3.376.$$

9. $\displaystyle\int_{-1}^2 \frac{dx}{x^3 + 2}$ $\qquad h = \dfrac{2 - (-1)}{12} = 0.25$

$x_0 = -1$, $x_1 = -1 + 0.25 = -0.75$, $x_2 = -0.75 + 0.25 = -0.5$, $x_3 = -0.5 + 0.25 = -0.25$, etc.

The function values are listed next:

$\begin{aligned}
f(x_0) &= 1/(-1 + 2) = 1 \\
f(x_1) &= 1/[(-0.75)^3 + 2] = 0.63366 \\
f(x_2) &= 1/[(-0.5)^3 + 2] = 0.53333 \\
f(x_3) &= 1/[(-0.25)^3 + 2] = 0.50394 \\
f(x_4) &= 1/(0 + 2) = 0.5 \\
f(x_5) &= 1/(0.25^3 + 2) = 0.49612 \\
f(x_6) &= 1/(0.5^3 + 2) = 0.47059 \\
f(x_7) &= 1/(0.75^3 + 2) = 0.41290 \\
f(x_8) &= 1/(1 + 2) = 0.33333 \\
f(x_9) &= 1/(1.25^3 + 2) = 0.25296 \\
f(x_{10}) &= 1/(1.5^3 + 2) = 0.18605 \\
f(x_{11}) &= 1/(1.75^3 + 2) = 0.13588 \\
f(x_{12}) &= 1/(2^3 + 2) = 0.1
\end{aligned}$

By the trapezoidal rule:

$$\int_{-1}^2 \frac{dx}{x^3 + 2} \approx 0.25\left[\frac{1}{2}(1) + 0.63366 + 0.53333 + \ldots + \frac{1}{2}(0.1)\right] = 1.252.$$

11. Since $n = 8$, $h = \dfrac{6-0}{8} = \dfrac{3}{4}$. Thus $x_0 = 0$, $x_1 = \dfrac{3}{4}$, $x_2 = \dfrac{6}{4} = \dfrac{3}{2}$, $x_3 = \dfrac{9}{4}$, $x_4 = \dfrac{12}{4} = 3$,

$x_5 = \dfrac{15}{4}$, $x_6 = \dfrac{18}{4} = \dfrac{9}{2}$, $x_7 = \dfrac{21}{4}$, $x_8 = \dfrac{24}{4} = 6$.

The function values are listed next:

$$f(x_0) = \frac{3}{7} = 0.4286 \qquad\qquad f(x) = \frac{\sqrt{x}+3}{x+7}$$

$$f(x_1) = \frac{\sqrt{3/4}+3}{3/4+7} = 0.4988$$

$$f(x_2) = \frac{\sqrt{3/2}+3}{3/2+7} = 0.4970$$

$$f(x_3) = \frac{\sqrt{9/4}+3}{9/4+7} = 0.4865$$

$$f(x_4) = \frac{\sqrt{3}+3}{3+7} = 0.4732$$

$$f(x_5) = \frac{\sqrt{15/4}+3}{15/4+7} = 0.4592$$

$$f(x_6) = \frac{\sqrt{9/2}+3}{9/2+7} = 0.4453$$

$$f(x_7) = \frac{\sqrt{21/4}+3}{21/4+7} = 0.4319$$

$$f(x_8) = \frac{\sqrt{6}+3}{6+7} = 0.4192.$$

$$\int_0^6 \frac{\sqrt{x}+3}{x+7}\,dx \approx \frac{3}{4}\left[\frac{1}{2}(0.4286) + 0.4988 + 0.4970 + \cdots + \frac{1}{2}(0.4192)\right] = 2.787.$$

13. Since $n = 6$, we get $h = \dfrac{b-a}{6} = \dfrac{1}{2}$. Thus $x_0 = 1$, $x_1 = \dfrac{3}{2}$, $x_2 = 2$, $x_3 = \dfrac{5}{2}$,

$x_4 = 3$, $x_5 = \dfrac{7}{2}$, $x_6 = 4$.

By Simpson's rule:

$$\int_1^4 \sqrt{1+x^2}\,dx \approx \frac{1/2}{3}\left[\sqrt{1+1^2} + 4\sqrt{1+(3/2)^2} + 2\sqrt{1+2^2} + 4\sqrt{1+(5/2)^2}\right.$$
$$\left. + 2\sqrt{1+3^2} + 4\sqrt{1+(7/2)^2} + \sqrt{1+4^2}\right] = 8.146.$$

15. $h = \dfrac{2-0}{6} = \dfrac{1}{3}$; $x_0 = 0$, $x_1 = \dfrac{1}{3}$, $x_2 = \dfrac{2}{3}$, $x_3 = 1$, $x_4 = \dfrac{4}{3}$, $x_5 = \dfrac{5}{3}$, $x_6 = 2$.

$$\int_0^2 \sqrt{1+x^4}\,dx \approx \frac{1/3}{3}\left[\sqrt{1} + 4\sqrt{1+\left(\frac{1}{3}\right)^4} + 2\sqrt{1+\left(\frac{2}{3}\right)^4}\right.$$
$$+ \quad 4\sqrt{1+1^4} + 2\sqrt{1+\left(\frac{4}{3}\right)^4} + 4\sqrt{1+\left(\frac{5}{3}\right)^4} + \sqrt{1+2^4}\bigg]$$
$$= \quad 3.6535.$$

17. From the table, $h = 1$. Using the given y-values, we get, by Simpson's rule:

$$\frac{1}{3}\left[1.3 + 4(1.9) + 2(3.2) + 4(3.8) + 2(4.7) + 4(6.8) + 2(10.2) + 4(15.6) + 20.3\right] = 56.7.$$

19. From the table, $h = 0.2$. By the trapezoidal rule, the approximate value of the integral is

$$0.2\left(\frac{1}{2}\cdot 0.10 + 0.40 + 0.46 + 0.57 + 0.64 + 0.54 + 0.43 + \frac{1}{2}\cdot 0.31\right) = 0.65\,\text{C}.$$

Chapter 4 Review

1. Subdivide the interval $[0, 3]$ into n equal parts, each of length $3/n$. Thus $\Delta x_i = 3/n$. Choosing the right endpoint of each subinterval for x, we get

$$x_1 = 1 \cdot \frac{3}{n}, \; x_2 = 2 \cdot \frac{3}{n}, \; \ldots, \; x_i = i \cdot \frac{3}{n}, \; \ldots, \; x_n = n \cdot \frac{3}{n} = 3.$$

For the altitudes, we get from $f(x) = 3x^2$

$$f(x_1) = 3(1)^2 \frac{9}{n^2}, \; f(x_2) = 3(2)^2 \frac{9}{n^2}, \; \ldots, \; f(x_i) = 3(i)^2 \frac{9}{n^2}, \; \ldots, \; f(x_n) = 3(n)^2 \frac{9}{n^2}.$$

Since $\Delta x_i = 3/n$, the sum of the areas is

$$\sum_{i=1}^{n} f(x_i)\Delta x_i = \sum_{i=1}^{n} 3i^2 \frac{9}{n^2} \frac{3}{n} = \frac{81}{n^3} \frac{n(n+1)(2n+1)}{6} = \frac{27n(n+1)(2n+1)}{2n^3},$$

by Formula B. Hence

$$\begin{aligned}
\int_0^3 3x^2 \, dx &= \lim_{n\to\infty} \frac{27n(n+1)(2n+1)}{2n^3} = \lim_{n\to\infty} \frac{54n^3 + 81n^2 + 27n}{2n^3} \\
&= \lim_{n\to\infty} \frac{54 + 81/n + 27/n^2}{2} = 27.
\end{aligned}$$

3. $$\begin{aligned}
\int (3\sqrt{x} - x^{-4} + 1) \, dx &= \int (3x^{1/2} - x^{-4} + 1) \, dx \\
&= 3\frac{x^{3/2}}{3/2} - \frac{x^{-3}}{-3} + x + C \\
&= 2x^{3/2} + \frac{1}{3}x^{-3} + x + C
\end{aligned}$$

Since $x^{3/2} = x^{1+1/2} = x^1 x^{1/2} = x\sqrt{x}$, the answer can also be written

$$2x\sqrt{x} + \frac{1}{3}x^{-3} + x + C.$$

5. $\int (1 - x^2)^5 x \, dx$. Let $u = 1 - x^2$; then $du = -2x \, dx$.

$$\begin{aligned}
\int (1 - x^2)^5 x \, dx &= -\frac{1}{2} \int (1 - x^2)^5 (-2x) \, dx = -\frac{1}{2} \int u^5 \, du \\
&= -\frac{1}{2}\frac{u^6}{6} + C = -\frac{1}{12}(1 - x^2)^6 + C.
\end{aligned}$$

7. $\int \frac{3x \, dx}{\sqrt{x^2 - 2}} = 3 \int (x^2 - 2)^{-1/2} x \, dx$. Let $u = x^2 - 2$; then $du = 2x \, dx$. Thus

$$\begin{aligned}
3 \int (x^2 - 2)^{-1/2} x \, dx &= \frac{3}{2} \int (x^2 - 2)^{-1/2}(2x) \, dx = \frac{3}{2} \int u^{-1/2} \, du \\
&= \frac{3}{2}\frac{u^{1/2}}{1/2} + C = 3\sqrt{x^2 - 2} + C.
\end{aligned}$$

9. $\int (x^3 + 1)^2 (3x) \, dx$. If $u = x^3 + 1$, then $du = 3x^2 \, dx$, which is different from $3x \, dx$. Since the substitution cannot be made, we multiply out the integrand and integrate term by term:

$$\int (x^6 + 2x^3 + 1)(3x) \, dx = \int (3x^7 + 6x^4 + 3x) \, dx = \frac{3}{8}x^8 + \frac{6}{5}x^5 + \frac{3}{2}x^2 + C.$$

11. $$\begin{aligned}
\int (\sqrt{x} - 1)^2 \, dx &= \int (x - 2x^{1/2} + 1) \, dx = \frac{1}{2}x^2 - 2\frac{x^{3/2}}{3/2} + x + C \\
&= \frac{1}{2}x^2 - \frac{4}{3}x^{1+1/2} + x + C = \frac{1}{2}x^2 - \frac{4}{3}x\sqrt{x} + x + C.
\end{aligned}$$

13. $\int (x-2)\sqrt{x^2-4x}\,dx = \int (x^2-4x)^{1/2}(x-2)\,dx.$

Let $u = x^2 - 4x$; then $du = (2x-4)\,dx = 2(x-2)\,dx.$

$$\int (x^2-4x)^{1/2}(x-2)\,dx \;=\; \frac{1}{2}\int (x^2-4x)^{1/2}2(x-2)\,dx$$

$$=\; \frac{1}{2}\int u^{1/2}\,du = \frac{1}{2}\frac{u^{3/2}}{3/2} + C$$

$$=\; \frac{1}{3}(x^2-4x)^{3/2} + C.$$

15.

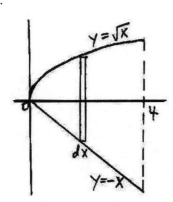

$$A \;=\; \int_0^4 [\sqrt{x} - (-x)]\,dx = \int_0^4 (x^{1/2}+x)\,dx = \frac{2}{3}x^{3/2} + \frac{1}{2}x^2 \Big|_0^4$$

$$=\; \frac{2}{3}(4)^{3/2} + \frac{1}{2}(4)^2 - 0 = \frac{16}{3} + 8 = \frac{16}{3} + \frac{24}{3} = \frac{40}{3}.$$

17.

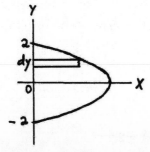

$$\int_{-2}^{2} x\,dy \;=\; \int_{-2}^{2} (4-y^2)\,dy$$

$$=\; 4y - \frac{y^3}{3}\Big|_{-2}^{2}$$

$$=\; \left(8 - \frac{8}{3}\right) - \left(-8 + \frac{8}{3}\right)$$

$$=\; 16 - \frac{16}{3} = \frac{32}{3}.$$

19. $x \;\;=\;\; y^2 - 4y + 4$

$\;\;\;\;\underline{x \;\;=\;\; 4 - y}$

$0 \;\;=\;\; y^2 - 3y$ $\qquad$ (subtracting)

$y(y-3) \;\;=\;\; 0$

$y \;\;=\;\; 0, 3$

The points of intersection are $(4, 0)$ and $(1, 3)$.

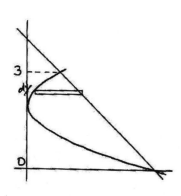

The typical element is drawn horizontally.

$$
\begin{aligned}
A &= \int_0^3 [(4-y) - (y^2 - 4y + 4)]\,dy = \int_0^3 (-y^2 + 3y)\,dy = -\frac{1}{3}y^3 + \frac{3}{2}y^2 \Big|_0^3 \\
&= -\frac{1}{3}\cdot 27 + \frac{3}{2}\cdot 9 = -9 + \frac{27}{2} = \frac{-18}{2} + \frac{27}{2} = \frac{9}{2}.
\end{aligned}
$$

21. $\displaystyle\int_{-\infty}^0 \frac{2x}{(x^2+4)^2}\,dx = \lim_{b\to-\infty} \int_b^0 (x^2+4)^{-2}(2x)\,dx$

$\qquad\qquad\qquad\qquad\qquad = \displaystyle\lim_{b\to-\infty} \frac{(x^2+4)^{-1}}{-1}\Big|_b^0$ $\qquad$ $u = x^2 + 4$

$\qquad\qquad\qquad\qquad\qquad\qquad\qquad\qquad\qquad\qquad\qquad\qquad\qquad\qquad du = 2x\,dx$

$\qquad\qquad\qquad\qquad\qquad = \displaystyle\lim_{b\to-\infty} \left(-\frac{1}{x^2+4}\right)\Big|_b^0$

$\qquad\qquad\qquad\qquad\qquad = \displaystyle\lim_{b\to-\infty} \left(-\frac{1}{4} + \frac{1}{b^2+4}\right) = -\frac{1}{4}.$

23. $\displaystyle\int_{-1}^1 \frac{3}{4x^2}\,dx.$ Since the y-axis is a vertical asymptote, we need to split the integral as follows:

$$
\int_{-1}^0 \frac{3}{4x^2}\,dx + \int_0^1 \frac{3}{4x^2}\,dx.
$$

Consider the second integral:

$$
\int_0^1 \frac{3}{4}x^{-2}\,dx = \lim_{\epsilon\to 0+} \int_\epsilon^1 \frac{3}{4}x^{-2}\,dx = \lim_{\epsilon\to 0+}\left(-\frac{3}{4x}\right)\Big|_\epsilon^1 = \lim_{\epsilon\to 0+}\left(-\frac{3}{4} + \frac{3}{4\epsilon}\right).
$$

This limit does not exist $(+\infty)$.

25. $\displaystyle\int_2^\infty \frac{1}{x^3}\,dx = \lim_{b\to\infty}\int_2^b x^{-3}\,dx = \lim_{b\to\infty} \frac{x^{-2}}{-2}\Big|_2^b = \lim_{b\to\infty}\left(-\frac{1}{2x^2}\right)\Big|_2^b$

$\qquad\qquad\qquad = \displaystyle\lim_{b\to\infty}\left(-\frac{1}{2b^2} + \frac{1}{8}\right) = \frac{1}{8}.$

27. $\displaystyle\int_0^1 \frac{1}{\sqrt{x}}\,dx = \lim_{\epsilon\to 0+}\int_\epsilon^1 x^{-1/2}\,dx = \lim_{\epsilon\to 0+} 2x^{1/2}\Big|_\epsilon^1 = \lim_{\epsilon\to 0+}(2 - 2\sqrt{\epsilon}) = 2.$

29. $q = \int i\,dt = \int 3.08 t^{1/2}\,dt = 3.08\left(\dfrac{2}{3}\right)t^{3/2} + C$. If $t = 0$, $q = 0$, so that $C = 0$. Thus

$q = 3.08\left(\dfrac{2}{3}\right)t^{3/2}\Big|_{t=1.75} = 4.75\,\text{C}.$

31.

Taking the upward direction to be positive, $g = -32\,\text{ft/s}^2$ with $v_0 = +29\,\text{ft/s}$. We get

$$
\begin{aligned}
v &= -32t + C \\
+29 &= 0 + C && \text{when } t = 0,\ v = 29\,\text{ft/s} \\
v &= -32t + 29 \\
s &= -16t^2 + 29t + k \\
121 &= 0 + k && \text{when } t = 0,\ s = 121\,\text{ft} \\
s &= -16t^2 + 29t + 121.
\end{aligned}
$$

To see how long the stone takes to reach the ground, we let $s = 0$ and solve for t:

$$
\begin{aligned}
0 &= -16t^2 + 29t + 121 \\
t &= \frac{-29 \pm \sqrt{29^2 + 4(16)(121)}}{2(-16)} = 3.8\,\text{s}.
\end{aligned}
$$

33. If the direction of motion is the positive direction, then $a = -20\,\text{ft/s}^2$. Let $s = 0$ be the point at which the brakes are first applied. Then

$$
\begin{aligned}
v &= -20t + v_0, && \text{unknown initial velocity} \\
s &= -10t^2 + v_0 t + k, && \text{integrating } v \\
s &= -10t^2 + v_0 t. && \text{since } k = 0
\end{aligned}
$$

At the instant when the minivan comes to a stop, we have $v = 0$ and $s = 180\,\text{ft}$. From the first equation, $0 = -20t + v_0$ and $t = \dfrac{v_0}{20}$.

Substituting in the third equation,

$$
\begin{aligned}
180 &= -10\left(\frac{v_0}{20}\right)^2 + v_0\frac{v_0}{20} \\
180 &= -10\frac{v_0^2}{20^2} + \frac{v_0^2}{20}\cdot\frac{20}{20} \\
180 &= \frac{-10v_0^2 + 20v_0^2}{20^2} = \frac{10v_0^2}{20^2} \\
v_0^2 &= \frac{180(20)^2}{10} \\
v_0 &\approx 85\,\text{ft/s}.
\end{aligned}
$$

35. $h = \dfrac{2 - (-1)}{6} = \dfrac{1}{2}$; $x_0 = -1$, $x_1 = -\dfrac{1}{2}$, $x_2 = 0$, $x_3 = \dfrac{1}{2}$, $x_4 = 1$, $x_5 = \dfrac{3}{2}$, $x_6 = 2$.

$$
\int_{-1}^{2}\frac{1}{x+3}\,dx = \frac{1/2}{3}\left[\frac{1}{-1+3} + 4\frac{1}{-1/2+3} + 2\frac{1}{0+3} + 4\frac{1}{1/2+3} + 2\frac{1}{1+3} + 4\frac{1}{3/2+3} + \frac{1}{2+3}\right]
$$

$$
= 0.916.
$$

Chapter 5

Applications of the Integral

5.1 Means and Root Mean Squares

1. $f_{av} = \dfrac{1}{16-1} \displaystyle\int_1^{16} x^{1/2}\, dx = \dfrac{1}{15} \cdot \dfrac{2}{3} x^{3/2} \Big|_1^{16} = \dfrac{2}{45}(16^{3/2} - 1) = \dfrac{2}{45}(64 - 1) = \dfrac{2(63)}{45} = \dfrac{14}{5}.$

3. $f_{av} = \dfrac{1}{2 - (-2)} \displaystyle\int_{-2}^2 x\sqrt{x^2 + 1}\, dx.$ Let $u = x^2 + 1$, then $du = 2x\, dx$.

$$f_{av} = \frac{1}{4} \int_{-2}^2 (x^2 + 1)^{1/2}\, dx = \frac{1}{4} \cdot \frac{1}{2} \int_{-2}^2 (x^2 + 1)^{1/2}(2x)\, dx = \frac{1}{8} \frac{(x^2 + 1)^{3/2}}{3/2} \Big|_{-2}^2 = 0.$$

5. $f_{rms}^2 = \dfrac{1}{2 - 1} \displaystyle\int_1^2 \left(\dfrac{1}{x}\right)^2 dx = \int_1^2 x^{-2}\, dx = \dfrac{x^{-1}}{-1} \Big|_1^2 = -\dfrac{1}{x} \Big|_1^2 = -\dfrac{1}{2} + 1 = \dfrac{1}{2}.$

Thus $f_{rms} = \sqrt{\dfrac{1}{2}} = \dfrac{\sqrt{2}}{2}.$

7. $f_{rms}^2 = \dfrac{1}{1 - 0} \displaystyle\int_0^1 [\sqrt{x}(x^2 + 1)]^2\, dx = \int_0^1 (x^5 + 2x^3 + x)\, dx = \dfrac{1}{6}x^6 + \dfrac{1}{2}x^4 + \dfrac{1}{2}x^2 \Big|_0^1$

$\qquad = \dfrac{1}{6} + \dfrac{1}{2} + \dfrac{1}{2} = \dfrac{7}{6}.$

$f_{rms} = \sqrt{\dfrac{7}{6}} = \dfrac{\sqrt{42}}{6}.$

9. Taking the upward direction as positive, we have $g = -10\,\text{m/s}^2$. Also, when $t = 0$, we have $v = 0$ and $s = 180$. Thus

$\quad v \;=\; -10t \qquad\qquad$ since $v_0 = 0$

$\quad s \;=\; -5t^2 + k$

$\quad s \;=\; -5t^2 + 180 \qquad$ since $s = 180$ when $t = 0$

Now let $s = 0$ and solve for t: $0 = -5t^2 + 180$ or $t = 6\,\text{s}$.

So it takes the object 6 s to reach the ground. We now get

$$v_{av} = \frac{1}{6 - 0} \int_0^6 (-10t)\, dt = -\frac{1}{6}(5t^2) \Big|_0^6 = -30\,\text{m/s},$$

so the average velocity is 30 m/s in the downward direction. (This agrees with the usual notion of average speed: 180 m in 6 s results in an average speed of $180\,\text{m}/6\,\text{s} = 30\,\text{m/s}$.)

11. Mean current $= \dfrac{1}{4.0 - 0.0} \displaystyle\int_{0.0}^{4.0} (1.0t + 1.0\sqrt{t})\, dt = 3.3\,\text{A}.$

13.
$$
\begin{aligned}
i_{\text{rms}}^2 &= \frac{1}{3-0} \int_0^3 (1-t^2)^2\, dt \qquad\qquad \text{(leaving out final zeros)}\\
&= \frac{1}{3} \int_0^3 (1 - 2t^2 + t^4)\, dt = \frac{1}{3}\left(t - \frac{2}{3}t^3 + \frac{1}{5}t^5 \right)\Bigg|_0^3 \\
&= \frac{1}{3}\left(3 - 18 + \frac{243}{5} \right) = \frac{1}{3}\frac{-75+243}{5} = \frac{168}{15}.
\end{aligned}
$$

Since $R = 5$, we get $P = i_{\text{rms}}^2 R = \dfrac{168}{15}(5) = 56\,\text{W}$ to two significant digits.

5.2   Volumes of Revolution: Disk and Washer Methods

1.

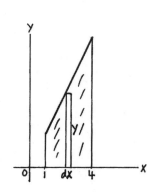

Volume of disk: $\pi(\text{radius})^2 \cdot \text{thickness} = \pi y^2\, dx.$
$$
\begin{aligned}
V &= \int_1^4 \pi(2x)^2\, dx = \pi \int_1^4 4x^2\, dx \\
&= \pi\left(\frac{4}{3}\right) x^3 \Bigg|_1^4 = \frac{4\pi}{3}(64 - 1) \\
&= \frac{252\pi}{3} = 84\pi.
\end{aligned}
$$

3.

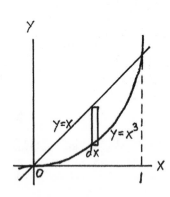

Volume of typical washer: $\pi(y_2^2 - y_1^2)\, dx.$
$$
\begin{aligned}
V &= \int_0^1 \pi[x^2 - (x^3)^2]\, dx \\
&= \pi\left(\frac{1}{3}x^3 - \frac{1}{7}x^7\right)\Bigg|_0^1 \\
&= \pi\left(\frac{1}{3} - \frac{1}{7}\right) = \frac{4\pi}{21}.
\end{aligned}
$$

5.

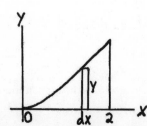

$$
\begin{aligned}
V &= \pi \int_0^2 (x^{3/2})^2\, dx \\
&= \pi \int_0^2 x^3\, dx \\
&= \pi\left(\frac{1}{4}x^4\right)\Bigg|_0^2 = 4\pi.
\end{aligned}
$$

7.

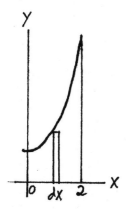

Volume of disk: $\pi(\text{radius})^2 \cdot (\text{thickness}) = \pi y^2 \, dx$.

Integrating (summing) from $x = 0$ to $x = 2$, we get

$$
\begin{aligned}
V &= \int_0^2 \pi(x^2 + 1)^2 \, dx = \pi \int_0^2 (x^4 + 2x^2 + 1) \, dx \\
&= \pi \left(\frac{1}{5}x^5 + \frac{2}{3}x^3 + x \right) \Big|_0^2 = \pi \left(\frac{32}{5} + \frac{16}{3} + 2 \right) = \frac{206\pi}{15}.
\end{aligned}
$$

9. $V = \int_1^3 \pi(\sqrt{x^2 + 1})^2 \, dx = \pi \int_1^3 (x^2 + 1) \, dx = \frac{32\pi}{3}.$

11.

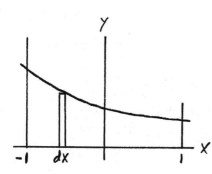

Volume of disk: $\pi(\text{radius})^2 \cdot (\text{thickness}) = \pi y^2 \, dx$.

$$
\begin{aligned}
V &= \int_{-1}^1 \pi \left(\frac{1}{x + 2} \right)^2 \, dx \\
&= \pi \int_{-1}^1 (x + 2)^{-2} \, dx = -\frac{\pi}{x + 2} \Big|_{-1}^1 \\
&= \pi \left(-\frac{1}{3} + 1 \right) = \frac{2\pi}{3}.
\end{aligned}
$$

13.

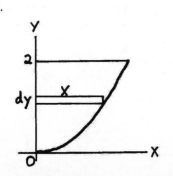

Volume of disk: $\pi(\text{radius})^2 \cdot \text{thickness} = \pi x^2 \, dy$.

From $y = \frac{1}{2}x^2$, we get $x^2 = 2y$.

$$
V = \int_0^2 \pi(2y) \, dy = \pi \int_0^2 2y \, dy = \pi y^2 \Big|_0^2 = 4\pi.
$$

15.

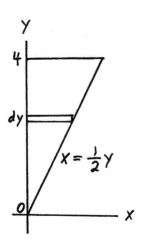

Volume of disk: $\pi(\text{radius})^2 \cdot (\text{thickness}) = \pi x^2 \, dy$.

Integrating (summing) from $y = 0$ to $y = 4$, we get

$$V = \int_0^4 \pi \left(\frac{1}{2}y\right)^2 \, dy = \frac{\pi}{4}\frac{y^3}{3}\bigg|_0^4 = \frac{16\pi}{3}.$$

17.

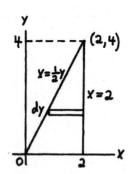

Note that the typical element has to be drawn horizontally to generate washers.

Volume of typical washer: $\pi(x_2^2 - x_1^2)\, dy = \pi\left[2^2 - \left(\frac{1}{2}y\right)^2\right].$ formula(5.4)

Integrating (summing) from $y = 0$ to $y = 4$, we get

$$
\begin{aligned}
V &= \int_0^4 \pi\left[2^2 - \left(\frac{1}{2}y\right)^2\right]\, dy = \pi\int_0^4 \left(4 - \frac{1}{4}y^2\right)\, dy = \pi\left(4y - \frac{1}{12}y^3\right)\bigg|_0^4 \\
&= \pi\left(4^2 - \frac{1}{12}4^3\right) = \pi\left(4^2 - \frac{1}{3}\cdot 4^2\right) = 4^2\pi\left(1 - \frac{1}{3}\right) = 16\pi \cdot \frac{2}{3} = \frac{32\pi}{3}.
\end{aligned}
$$

19.

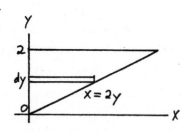

Volume of disk: $\pi(\text{radius})^2 \cdot (\text{thickness}) = \pi x^2 \, dy$.

$$V = \int_0^2 \pi(2y)^2 \, dy = 4\pi\frac{y^3}{3}\bigg|_0^2 = \frac{32\pi}{3}.$$

21. Since $y = \frac{1}{2}x$, we have $x = 2y$.

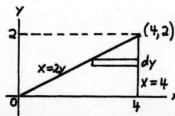

Volume of typical washer:

$$\pi\left(x_2^2 - x_1^2\right)\, dy = \pi[4^2 - (2y)^2]\, dy.$$

Summing from $y = 0$ to $y = 2$, we get

$$V = \int_0^2 \pi[4^2 - (2y)^2]\,dy = \pi \int_0^2 (16 - 4y^2)\,dy = \pi\left(16y - \frac{4}{3}y^3\right)\bigg|_0^2$$

$$= \pi\left(16 \cdot 2 - \frac{4}{3} \cdot 2^3\right) = \pi\left(32 - \frac{1}{3} \cdot 32\right) = 32\pi\left(1 - \frac{1}{3}\right) = 32\pi\left(\frac{2}{3}\right) = \frac{64\pi}{3}.$$

23.
$$y = x$$
$$y = 2 - x$$
$$\overline{\rule{6em}{0.4pt}}$$
$$0 = 2x - 2$$
$$x = 1, y = 1$$

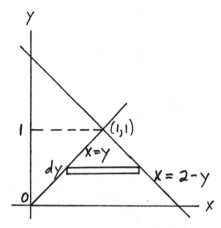

The lines intersect at $(1, 1)$. The typical element has to be drawn horizontally to generate washers. Volume of typical washer: $\pi(x_2^2 - x_1^2)\,dy$.

$$V = \int_0^1 \pi[(2-y)^2 - y^2]\,dy = \pi \int_0^1 (4 - 4y)\,dy = 2\pi.$$

25. To find the points of intersection, we need to solve the equations simultaneously:

$$x = y^2$$
$$x = y + 2$$
$$\overline{\rule{8em}{0.4pt}}$$
$$0 = y^2 - y - 2 \qquad \text{(subtracting)}$$
$$(y + 1)(y - 2) = 0$$
$$y = -1, 2.$$

The points of intersection are $(1, -1)$ and $(4, 2)$.

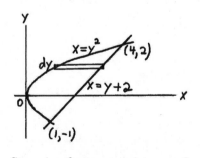

Note that the typical element has to be drawn horizontally. Now recall that the volume of a typical washer is $\pi(x_2^2 - x_1^2)\,dy = \pi[(y + 2)^2 - (y^2)^2]\,dy$.

Summing from $y = -1$ to $y = 2$, we get

$$V = \int_{-1}^2 \pi[(y+2)^2 - (y^2)^2]\,dy = \pi \int_{-1}^2 (y^2 + 4y + 4 - y^4)\,dy$$

$$= \pi\left(\frac{1}{3}y^3 + 2y^2 + 4y - \frac{1}{5}y^5\right)\bigg|_{-1}^2 = \pi\left[\left(\frac{8}{3} + 8 + 8 - \frac{32}{5}\right) - \left(-\frac{1}{3} + 2 - 4 + \frac{1}{5}\right)\right]$$

$$= \pi\left(\frac{8}{3} + 8 + 8 - \frac{32}{5} + \frac{1}{3} - 2 + 4 - \frac{1}{5}\right) = \pi\left(\frac{9}{3} - \frac{33}{5} + 18\right)$$

$$= \pi\left(21 - \frac{33}{5}\right) = \pi\frac{105 - 33}{5} = \frac{72\pi}{5}.$$

27.

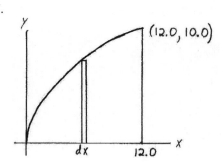

Assume that the equation of the parabola has the form $y^2 = 4px$ (axis horizontal). From the given information, the point $(12.0, 10.0)$ lies on the curve:

$$10.0^2 = 4p(12.0) \quad \text{or} \quad 4p = \frac{10.0^2}{12.0} = \frac{25}{3}.$$

Equation: $y^2 = \dfrac{25}{3}x$. Volume of disk: $\pi(\text{radius})^2 \cdot (\text{thickness}) = \pi y^2\, dx = \pi\left(\dfrac{25}{3}x\right)\, dx$.

$$V = \int_0^{12.0} \pi\left(\frac{25}{3}x\right) dx = \frac{25\pi}{3}\frac{x^2}{2}\Big|_0^{12.0} = 600\pi \approx 1880\,\text{cm}^3.$$

29. By turning the tank on the side, we can find the volume of the water by rotating the region in the figure about the x-axis. (Since the radius is $12\,\text{ft}$, note that the equation of the circle is $x^2 + y^2 = 12^2$.)

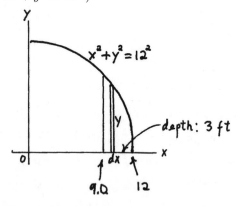

From $x^2 + y^2 = 144$, we get $y^2 = 144 - x^2$. Volume of typical disk: $\pi y^2\, dx = \pi(144 - x^2)\, dx$. Integrating from $x = 9$ to $x = 12$,

$$\begin{aligned}
V &= \pi\int_9^{12}(144 - x^2)\, dx = \pi\left(144x - \frac{1}{3}x^3\right)\Big|_9^{12}\\
&= \pi\left[\left(144\cdot 12 - \frac{1}{3}(12)^3\right) - \left(144\cdot 9 - \frac{1}{3}\cdot 9^3\right)\right]\\
&= \pi[(1728 - 576) - (1296 - 243)] = 99\pi \approx 310\,\text{ft}^3.
\end{aligned}$$

31. $V = \displaystyle\int_1^{\infty} \pi\left(\frac{1}{x}\right)^2 dx = \pi \lim_{b\to\infty}\int_1^b x^{-2}\, dx = \pi \lim_{b\to\infty}\left(-\frac{1}{x}\right)\Big|_1^b = \pi \lim_{b\to\infty}\left(-\frac{1}{b} + 1\right) = \pi.$

33. $y = \dfrac{1}{x^{3/4}} = x^{-3/4}$. Volume of typical disk: $\pi y^2 \, dx = \pi (x^{-3/4})^2 \, dx = \pi x^{-3/2} \, dx$.

$$
\begin{aligned}
V & = \pi \int_4^\infty x^{-3/2} \, dx = \lim_{b \to \infty} \pi \int_4^b x^{-3/2} \, dx = \lim_{b \to \infty} \pi (-2) x^{-1/2} \Big|_4^b \\
& = \lim_{b \to \infty} \frac{-2\pi}{\sqrt{x}} \Big|_4^b = \lim_{b \to \infty} \left[\frac{-2\pi}{\sqrt{b}} + \frac{2\pi}{\sqrt{4}} \right] = \frac{2\pi}{2} = \pi. \\
A & = \int_4^\infty x^{-3/4} \, dx = \lim_{b \to \infty} \int_4^b x^{-3/4} \, dx = \lim_{b \to \infty} 4 x^{1/4} \Big|_4^b \\
& = \lim_{b \to \infty} (4b^{1/4} - 4(4)^{1/4}) = \infty \ (\text{does not exist}).
\end{aligned}
$$

5.3 Volumes of Revolution: Shell Method

1. Volume of shell: $2\pi (\text{radius}) \cdot (\text{height}) \cdot (\text{thickness}) = 2\pi \cdot x \cdot y \cdot dx = 2\pi x \cdot x \, dx$.

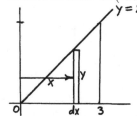

$$
V = \int_0^3 2\pi x(x) \, dx = 2\pi \int_0^3 x^2 \, dx = 2\pi \frac{x^3}{3} \Big|_0^3 = 18\pi.
$$

3.

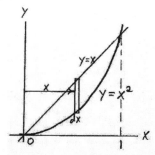

Volume of shell: $2\pi (\text{radius}) \cdot (\text{height}) \cdot (\text{thickness})$
$= 2\pi x (x - x^2) \, dx.$

Integrating (summing) from $x = 0$ to $x = 1$, we get

$$
V = 2\pi \int_0^1 x(x - x^2) \, dx = 2\pi \left(\frac{1}{3} x^3 - \frac{1}{4} x^4 \right) \Big|_0^1 = 2\pi \left(\frac{1}{3} - \frac{1}{4} \right) = 2\pi \left(\frac{1}{12} \right) = \frac{\pi}{6}.
$$

5. Substituting $y = x^2$ in $y^2 = 8x$, we get

$$x^4 = 8x$$
$$x^4 - 8x = 0$$
$$x(x^3 - 8) = 0$$
$$x = 0, 2.$$

The points of intersection are $(0, 0)$ and $(2, 4)$.

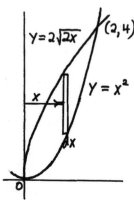

The functions are $y = x^2$ and $y = \sqrt{8x} = 2\sqrt{2x}$.

Note that the height of the typical element is $(2\sqrt{2x} - x^2)\,dx$.

Volume of shell: $2\pi(\text{radius}) \cdot (\text{height}) \cdot (\text{thickness}) = 2\pi \cdot x \cdot (2\sqrt{2x} - x^2)\,dx$.

$$
\begin{aligned}
V &= \int_0^2 2\pi x(2\sqrt{2x} - x^2)\,dx = 2\pi \int_0^2 (2\sqrt{2}x^{3/2} - x^3)\,dx = 2\pi\left[2\sqrt{2}\left(\frac{2}{5}\right)x^{5/2} - \frac{1}{4}x^4\right]\Big|_0^2 \\
&= 2\pi\left[2\sqrt{2}\left(\frac{2}{5}\right)2^{5/2} - 4\right] = 2\pi\left[\frac{4}{5}(2)^{6/2} - 4\right] = 2\pi\left(\frac{4}{5}\cdot 8 - 4\right) = 2\pi\frac{32 - 20}{5} = \frac{24\pi}{5}.
\end{aligned}
$$

7.

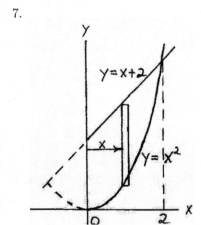

$$
\begin{aligned}
y &= x^2 \\
y &= x + 2 \\
\hline
0 &= x^2 - x - 2 \qquad \text{(subtracting)} \\
(x+1)(x-2) &= 0 \\
x &= -1, 2
\end{aligned}
$$

Volume of shell: $2\pi(\text{radius}) \cdot (\text{height}) \cdot (\text{thickness}) = 2\pi x[(x + 2) - x^2]\,dx$.

$$V = 2\pi \int_0^2 x(x + 2 - x^2)\,dx = 2\pi\left(\frac{1}{3}x^3 + x^2 - \frac{1}{4}x^4\right)\Big|_0^2 = 2\pi\left(\frac{8}{3} + 4 - 4\right) = \frac{16\pi}{3}.$$

9.

Volume of typical shell:

$2\pi(\text{radius}) \cdot (\text{height}) \cdot (\text{thickness})$

$= 2\pi x[0 - (x^2 - 2x)]\,dx$.

Summing from $x = 0$ to $x = 2$, we get

$$V = \int_0^2 2\pi x[0 - (x^2 - 2x)]\,dx = 2\pi \int_0^2 (-x^3 + 2x^2)\,dx = 2\pi \left(-\frac{1}{4}x^4 + \frac{2}{3}x^3 \right) \Big|_0^2$$

$$= 2\pi \left(-4 + \frac{16}{3} \right) = 2\pi \left(-\frac{12}{3} + \frac{16}{3} \right) = 2\pi \left(\frac{4}{3} \right) = \frac{8\pi}{3}.$$

11. $y = x^2 - 4x + 3 = (x-1)(x-3) = 0$, gives $x = 1, 3$ as the intercepts.

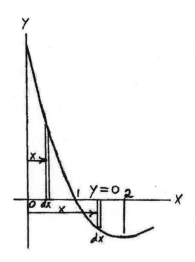

Volume of shell: $2\pi(\text{radius}) \cdot (\text{height}) \cdot (\text{thickness})$.

Left region: $2\pi \int_0^1 x(x^2 - 4x + 3)\,dx = \frac{5\pi}{6}$.

Right region: $2\pi \int_1^2 x[0 - (x^2 - 4x + 3)]\,dx = \frac{13\pi}{6}$.

TOTAL: $\frac{18\pi}{6} = 3\pi$.

13.

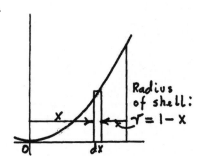

Volume of typical shell: $2\pi(\text{radius}) \cdot (\text{height}) \cdot (\text{thickness}) = 2\pi(1-x) \cdot x^2\,dx$.

$$V = \int_0^1 2\pi(1-x)x^2\,dx = 2\pi \int_0^1 (x^2 - x^3)\,dx = 2\pi \left(\frac{1}{3}x^3 - \frac{1}{4}x^4 \right) \Big|_0^1 = 2\pi \left(\frac{1}{3} - \frac{1}{4} \right) = \frac{2\pi}{12} = \frac{\pi}{6}.$$

15.

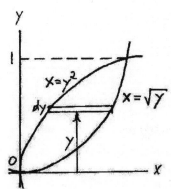

To generate shells, the typical element has to be drawn horizontally.

Volume of shell: $2\pi(\text{radius}) \cdot (\text{height}) \cdot (\text{thickness})$.

$$V = 2\pi \int_0^1 y(\sqrt{y} - y^2)\,dy = 2\pi \int_0^1 (y^{3/2} - y^3)\,dy = 2\pi \left(\frac{2}{5}y^{5/2} - \frac{1}{4}y^4 \right) \Big|_0^1$$

$$= 2\pi \left(\frac{2}{5} - \frac{1}{4} \right) = 2\pi \left(\frac{3}{20} \right) = \frac{3\pi}{10}.$$

17.

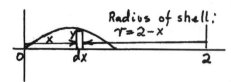

Note that the intercepts of $y = x - x^2 = x(1 - x)$ are $x = 0$ and $x = 1$. So the limits of integration are $x = 0$ and $x = 1$, regardless of the position of the (vertical) axis of rotation. (In other words, if we were using Riemann sums instead of the shortcut, we would subdivide the region under the graph.)

Volume of shell: $2\pi(\text{radius}) \cdot (\text{height}) \cdot (\text{thickness}) = 2\pi(2 - x) \cdot (x - x^2)\, dx$.

$$
\begin{aligned}
V &= \int_0^1 2\pi(2 - x)(x - x^2)\, dx = 2\pi \int_0^1 (2x - 3x^2 + x^3)\, dx \\
&= 2\pi \left(x^2 - x^3 + \frac{1}{4}x^4 \right) \Big|_0^1 = 2\pi \left(1 - 1 + \frac{1}{4} \right) = \frac{\pi}{2}.
\end{aligned}
$$

19.

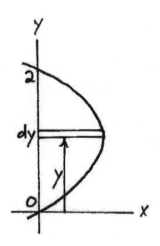

$$
\begin{aligned}
x = 2y - y^2 &= y(2 - y) = 0 \\
y &= 0, 2
\end{aligned}
$$

$$
V = 2\pi \int_0^2 y(2y - y^2)\, dy = \frac{8\pi}{3}.
$$

21.

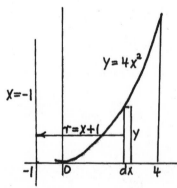

Radius of shell: $x - (-1) = x + 1$.

Volume of shell: $2\pi(\text{radius}) \cdot (\text{height}) \cdot (\text{thickness})$

$\qquad = 2\pi(x + 1) \cdot 4x^2\, dx.$

The region we are summing over extends from $x = 0$ to $x = 4$:

$$
\begin{aligned}
V &= \int_0^4 2\pi(x + 1)(4x^2)\, dx = 8\pi \int_0^4 (x^3 + x^2)\, dx = 8\pi \left(\frac{1}{4}x^4 + \frac{1}{3}x^3 \right) \Big|_0^4 \\
&= 8\pi \left(\frac{1}{4} \cdot 4^4 + \frac{1}{3} \cdot 4^3 \right) = 8\pi \left(4^3 + \frac{1}{3} \cdot 4^3 \right) = 8\pi(4^3) \left(1 + \frac{1}{3} \right) \\
&= 8\pi(64) \left(\frac{4}{3} \right) = \frac{2048\pi}{3}.
\end{aligned}
$$

23.

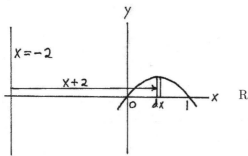

Radius of shell: $x - (-2) = x + 2$.

Summing from $x = 0$ to $x = 1$, the vertical lines that bound the region to be rotated:

$$
\begin{aligned}
V &= 2\pi \int_0^1 (x+2)(2x - 2x^2)\, dx = 2\pi \int_0^1 (-2x^3 - 2x^2 + 4x)\, dx \\
&= 2\pi \left(-\frac{1}{2}x^4 - \frac{2}{3}x^3 + 2x^2 \right) \Big|_0^1 = 2\pi \left(-\frac{1}{2} - \frac{2}{3} + 2 \right) = 2\pi \left(\frac{5}{6} \right) = \frac{5\pi}{3}.
\end{aligned}
$$

25.

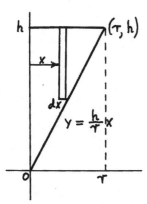

We can obtain a cone by rotating the region in the figure about the y-axis. Note that the slope of the line is $m = \dfrac{h}{r}$, so that the equation is $y = \dfrac{h}{r}x$.

Height of typical element: $h - \dfrac{h}{r}x$.

Volume of shell: $2\pi(\text{radius}) \cdot (\text{height}) \cdot (\text{thickness}) = 2\pi \cdot x \cdot \left(h - \dfrac{h}{r}x \right)\, dx$.

$$
\begin{aligned}
V &= \int_0^r 2\pi x \left(h - \frac{h}{r}x \right) dx = 2\pi \int_0^r x \cdot h \left(1 - \frac{1}{r}x \right) dx = 2\pi h \int_0^r \left(x - \frac{1}{r}x^2 \right) dx \\
&= 2\pi h \left(\frac{1}{2}x^2 - \frac{1}{r} \cdot \frac{1}{3}x^3 \right) \Big|_0^r = 2\pi h \left(\frac{1}{2}r^2 - \frac{1}{3}r^2 \right) = 2\pi h r^2 \left(\frac{1}{2} - \frac{1}{3} \right) \\
&= 2\pi h r^2 \frac{1}{6} = \frac{1}{3}\pi r^2 h.
\end{aligned}
$$

27. (a)

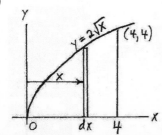

Volume of disk: $\pi(\text{radius})^2 \cdot (\text{thickness})$.

$$
V = \pi \int_0^4 (2\sqrt{x})^2\, dx = \pi \int_0^4 4x\, dx = 32\pi.
$$

(b)

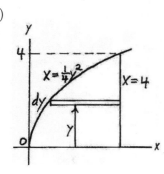

Volume of shell: $2\pi(\text{radius}) \cdot (\text{height}) \cdot (\text{thickness})$.

$$V = 2\pi \int_0^4 y \left(4 - \frac{1}{4}y^2\right) dy = 32\pi.$$

29. (a)

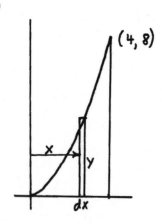

The simplest way to find this volume is by the disk method: $\pi \cdot (\text{radius})^2 \cdot \text{thickness} = \pi y^2 \, dx$. Since $y^2 = x^3$, we get

$$V = \pi \int_0^4 x^3 \, dx = \pi \frac{x^4}{4}\bigg|_0^4 = 64\pi.$$

(b) The simplest way to find this volume is by the shell method. Since $y = x^{3/2}$, we have
$2\pi(\text{radius}) \cdot (\text{height}) \cdot (\text{thickness}) = 2\pi x \cdot x^{3/2} \, dx$.

$$V = 2\pi \int_0^4 x \cdot x^{3/2} \, dx = 2\pi \int_0^4 x^{5/2} \, dx = 2\pi \left(\frac{2}{7}\right) x^{7/2}\bigg|_0^4 = \frac{4\pi}{7}(4)^{7/2} = \frac{4\pi}{7}(128) = \frac{512\pi}{7}.$$

(b) alternate

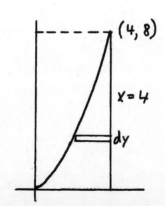

By the washer method, we get, for the typical washer, $\pi(x_2^2 - x_1^2)\, dy$ and $x = y^{2/3}$.

$$V = \pi \int_0^8 [4^2 - (y^{2/3})^2]\, dy,$$

which is harder to evaluate.

(c) Shell: $2\pi(\text{radius}) \cdot (\text{height}) \cdot (\text{thickness})$ with $r = 4 - x$:

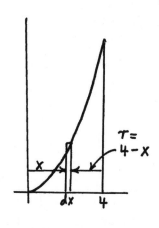

$$
\begin{aligned}
V &= 2\pi \int_0^4 (4 - x)x^{3/2}\, dx \\
&= 2\pi \int_0^4 (4x^{3/2} - x^{5/2})\, dx \\
&= 2\pi \left[4\left(\frac{2}{5}\right)x^{5/2} - \frac{2}{7}x^{7/2} \right]\Bigg|_0^4 \\
&= 2\pi \left[\frac{8}{5}(4)^{5/2} - \frac{2}{7}(4)^{7/2} \right] \\
&= 2\pi \left[\frac{8}{5}(32) - \frac{2}{7}(128) \right] \\
&= 2\pi(256)\left(\frac{1}{5} - \frac{1}{7}\right) \\
&= 512\pi\left(\frac{2}{35}\right) = \frac{1024\pi}{35}.
\end{aligned}
$$

(d)

Volume of shell:

$2\pi(\text{radius}) \cdot (\text{height}) \cdot (\text{thickness})$.

Since $r = 8 - y$ and the height of the typical element is $(4 - y^{2/3})\, dy$, we get:

$$
\begin{aligned}
V &= 2\pi \int_0^8 (8 - y)(4 - y^{2/3})\, dy = 2\pi \int_0^8 (32 - 4y - 8y^{2/3} + y^{5/3})\, dy \\
&= 2\pi \left[32y - 2y^2 - 8\left(\frac{3}{5}\right)y^{5/3} + \left(\frac{3}{8}\right)y^{8/3} \right]\Bigg|_0^8 \\
&= 2\pi \left[256 - 128 - \frac{24}{5}(8)^{5/3} + \frac{3}{8}(8)^{8/3} \right] = 2\pi \left[128 - \frac{24}{5}(32) + \frac{3}{8}(256) \right] \\
&= 2\pi \left(128 - \frac{768}{5} + \frac{768}{8} \right) = 2\pi(128)\left(1 - \frac{6}{5} + \frac{6}{8} \right) \\
&= 256\pi \frac{20 - 24 + 15}{20} = 256\pi\left(\frac{11}{20}\right) = \frac{64\pi(11)}{5} = \frac{704\pi}{5}.
\end{aligned}
$$

31. (a)

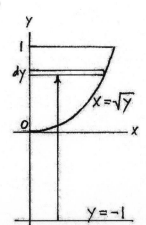

Radius of shell: $y - (-1) = y + 1$.

$$
V = 2\pi \int_0^1 (y + 1)\sqrt{y}\, dy = \frac{32\pi}{15}.
$$

(b)

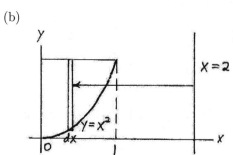

Radius of shell: $2 - x$.

$$V = 2\pi \int_0^1 (2 - x)(1 - x^2)\, dx = \frac{13\pi}{6}.$$

33.

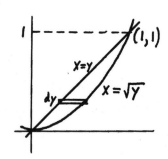

Volume of typical washer: $\pi[(\sqrt{y})^2 - y^2]\, dy$.

$$
\begin{aligned}
V &= \int_0^1 \pi[(\sqrt{y})^2 - y^2]\, dy = \pi \int_0^1 (y - y^2)\, dy \\
&= \pi \left(\frac{1}{2}y^2 - \frac{1}{3}y^3 \right) \Big|_0^1 = \pi \left(\frac{1}{2} - \frac{1}{3} \right) = \frac{\pi}{6}.
\end{aligned}
$$

35. (a)

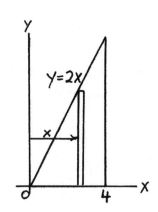

Volume of disk: $\pi(\text{radius})^2 \cdot (\text{thickness})$.

$$V = \int_0^4 \pi(2x)^2\, dx = \frac{256\pi}{3}.$$

(b)

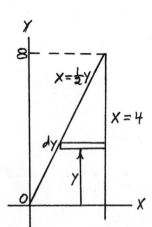

Volume of shell: $2\pi(\text{radius}) \cdot (\text{height}) \cdot (\text{thickness})$.

$$V = 2\pi \int_0^8 y \left(4 - \frac{1}{2}y \right) dy = \frac{256\pi}{3}.$$

37. Volume of shell: $2\pi(\text{radius}) \cdot (\text{height}) \cdot (\text{thickness})$.

$$\int_0^2 2\pi x \sqrt{1 + \sqrt{x}}\, dx$$

Using the integration capability of a graphing utility, we get 18.317006.

39.

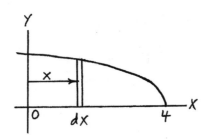

$$V = 2\pi \int_0^4 x \sqrt[3]{4-x}\, dx = 51.294523.$$

41. Placing the vertex at the origin, the form of the equation is $x^2 = 4py$. One point on the curve is $(3, 6)$. Substituting these values,

$$3^2 = 4p(6) \quad \text{or} \quad 4p = \frac{3}{2}.$$

The equation is $x^2 = \frac{3}{2}y$ or $y = \frac{2}{3}x^2$. The "upper" curve is $y = 6$. By the shell method,

$$
\begin{aligned}
V &= 2\pi \int_0^3 x \left(6 - \frac{2}{3}x^2 \right) dx = 2\pi \left(3x^2 - \frac{2}{3}\frac{x^4}{4} \right) \Big|_0^3 \\
&= 2\pi \left(3^3 - \frac{1}{2} \cdot 3^3 \right) = 2\pi \cdot 3^3 \left(1 - \frac{1}{2} \right) = 27\pi\,\text{m}^3.
\end{aligned}
$$

5.4   Centroids

1.

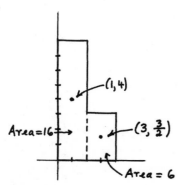

$$\overline{x} = \frac{1 \cdot 16 + 3 \cdot 6}{16 + 6} = \frac{34}{22} = \frac{17}{11}$$

$$\overline{y} = \frac{4 \cdot 16 + \frac{3}{2} \cdot 6}{16 + 6} = \frac{73}{22}$$

3.

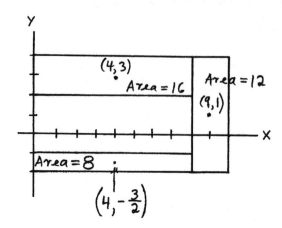

$$\overline{x} = \frac{4 \cdot 16 + 9 \cdot 12 + 4 \cdot 8}{16 + 12 + 8} = \frac{204}{36} = \frac{17}{3}$$

$$\overline{y} = \frac{3 \cdot 16 + 1 \cdot 12 + \left(-\frac{3}{2}\right) \cdot 8}{36} = \frac{48}{36} = \frac{4}{3}$$

5.

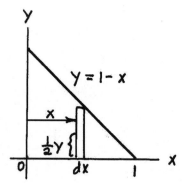

$\overline{x}$: moment of typical element: $xy\,dx = x(1-x)\,dx$.

$$\overline{x} = \frac{M_y}{A} = \frac{\int_0^1 xy\,dx}{\int_0^1 y\,dx} = \frac{\int_0^1 x(1-x)\,dx}{\int_0^1 (1-x)\,dx} = \frac{\frac{1}{2}x^2 - \frac{1}{3}x^3\Big|_0^1}{x - \frac{1}{2}x^2\Big|_0^1} = \frac{\frac{1}{2} - \frac{1}{3}}{1 - \frac{1}{2}} = \frac{\frac{1}{6}}{\frac{1}{2}} = \frac{1}{3}.$$

$\overline{y}$: moment of typical element: $\left(\frac{1}{2}y\right)(y\,dx)$.

$$\overline{y} = \frac{M_x}{A} = \frac{\int_0^1 \left(\frac{1}{2}y\right)(y\,dx)}{A} = \frac{\frac{1}{2}\int_0^1 (1-x)(1-x)\,dx}{1/2} = 2 \cdot \frac{1}{2} \int_0^1 (1 - 2x + x^2)\,dx$$

$$= x - x^2 + \frac{1}{3}x^3\Big|_0^1 = \frac{1}{3}.$$

$\overline{y}$ (alternate):

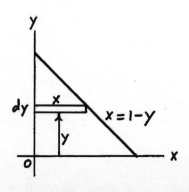

Interchanging the roles of x and y.

Moment of typical element with respect to x-axis: $y \cdot x \, dy = y(1-y)\,dy$.

$$\overline{y} = \frac{M_x}{A} = \frac{\int_0^1 yx\,dy}{A} = \frac{\int_0^1 y(1-y)\,dy}{1/2} = \frac{1}{3}.$$

7.

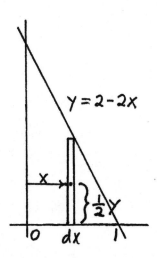

$\overline{x}$: moment of typical element: $xy\,dx = x(2-2x)\,dx$.

$$\overline{x} = \frac{M_y}{A} = \frac{\int_0^1 xy\,dx}{\int_0^1 y\,dx} = \frac{\int_0^1 x(2-2x)\,dx}{\int_0^1 (2-2x)\,dx} = \frac{x^2 - \frac{2}{3}x^3 \Big|_0^1}{2x - x^2 \Big|_0^1} = \frac{1 - \frac{2}{3}}{1} = \frac{1}{3}.$$

$\overline{y}$: moment of typical element: $\left(\dfrac{1}{2}y\right)y\,dx$.

$$\overline{y} = \frac{\int_0^1 \frac{1}{2}(2-2x)(2-2x)\,dx}{1} = \frac{1}{2}\int_0^1 (4 - 8x + 4x^2)\,dx = \frac{1}{2}\left(4x - 4x^2 + \frac{4}{3}x^3\right)\Big|_0^1 = \frac{2}{3}.$$

$\overline{y}$ (alternate):

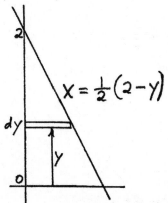

Moment of typical element with respect to the x-axis: $y \cdot x\,dy = y \cdot \dfrac{1}{2}(2-y)\,dy$.

$$\overline{y} = \frac{M_x}{A} = \frac{\int_0^2 y \cdot \frac{1}{2}(2-y)\,dy}{1} = \frac{2}{3}.$$

9.

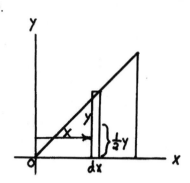

$\bar{x}$: moment of typical element: $xy\,dx$. $\bar{x} = \dfrac{M_y}{A} = \dfrac{\int_0^1 xy\,dx}{\int_0^1 y\,dx} = \dfrac{\int_0^1 x(x)\,dx}{\int_0^1 x\,dx} = \dfrac{\frac{1}{3}x^3\big|_0^1}{\frac{1}{2}x^2\big|_0^1} = \dfrac{\frac{1}{3}}{\frac{1}{2}} = \dfrac{2}{3}.$

$\bar{y}$: moment of typical element: $\left(\frac{1}{2}\right)(y\,dx)$.

$\bar{y} = \dfrac{M_x}{A} = \dfrac{\int_0^1 \left(\frac{1}{2}y\right)(y\,dx)}{A} = \dfrac{\frac{1}{2}\int_0^1 x\cdot x\,dx}{1/2} = \int_0^1 x^2\,dx = \dfrac{1}{3}.$

$\bar{y}$ (alternate): moment (with respect to x-axis) of typical element: $y\cdot(\text{height})\cdot dy = y(1-y)\,dy$.

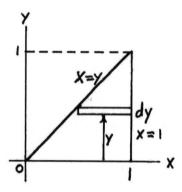

$$\bar{y} = \dfrac{\int_0^1 y(1-y)\,dy}{1/2}$$
$$= 2\int_0^1 (y - y^2)\,dy$$
$$= \dfrac{1}{3}.$$

11.

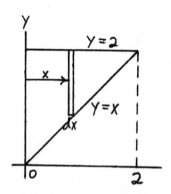

$\bar{x}$: moment of typical element: $x(2-x)\,dx$.

$\bar{x} = \dfrac{\int_0^2 x(2-x)\,dx}{\int_0^2 (2-x)\,dx} = \dfrac{x^2 - \frac{1}{3}x^3\big|_0^2}{2x - \frac{1}{2}x^2\big|_0^2} = \dfrac{4 - \frac{8}{3}}{4 - 2} = \dfrac{\frac{12}{3} - \frac{8}{3}}{2} = \dfrac{2}{3}.$

To obtain $\bar{y}$, observe that

(a) distance from x-axis to center of typical element: $\dfrac{1}{2}(2+x)$;

(b) height of typical element: $2 - x$.

$$\overline{y} = \frac{M_x}{A} = \frac{\int_0^2 \frac{1}{2}(2+x)(2-x)\,dx}{2} = \frac{4}{3}.$$

However, the alternate method is simpler in this case:

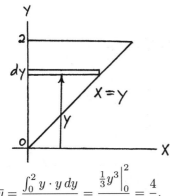

$$\overline{y} = \frac{\int_0^2 y \cdot y\,dy}{2} = \frac{\frac{1}{3}y^3\Big|_0^2}{2} = \frac{4}{3}.$$

13. Intercepts: $4 - x^2 = 0$ when $x = \pm 2$.

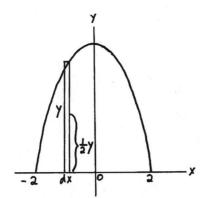

$$A = \int_{-2}^2 (4 - x^2)\,dx = \frac{32}{3}.$$

$\overline{y}$: moment of typical element: $\left(\dfrac{1}{2}y\right)(y\,dx).$

$$
\begin{aligned}
\overline{y} &= \frac{M_x}{A} = \frac{\int_{-2}^2 \left(\frac{1}{2}y\right)(y\,dx)}{A} = \frac{\frac{1}{2}\int_{-2}^2 (4-x^2)(4-x^2)\,dx}{32/3} = \frac{3}{32}\cdot\frac{1}{2}\int_{-2}^2 (16 - 8x^2 + x^4)\,dx \\
&= \frac{3}{64}\left(16x - \frac{8x^3}{3} + \frac{x^5}{5}\right)\Big|_{-2}^2 = \frac{3}{64}\left(32 - \frac{64}{3} + \frac{32}{5}\right) - \frac{3}{64}\left(-32 + \frac{64}{3} - \frac{32}{5}\right) \\
&= 2\cdot\frac{3}{64}\left(32 - \frac{64}{3} + \frac{32}{5}\right) = \frac{3}{32}\frac{480 - 320 + 96}{15} = \frac{3}{32}\cdot\frac{256}{15} = \frac{8}{5}.
\end{aligned}
$$

$\overline{x} = 0$ by symmetry.

15.

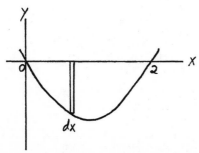

$\overline{x} = 1$ by symmetry.

$\overline{y}$: moment with respect to the x-axis: $\left(\dfrac{1}{2}y\right)y\,dx = \dfrac{1}{2}y^2\,dx.$

$$\bar{y} = \frac{\int_0^2 \frac{1}{2}y^2\,dx}{\int_0^2 y\,dx} = \frac{\int_0^2 \frac{1}{2}(x^2 - 2x)^2\,dx}{\int_0^2 (x^2 - 2x)\,dx} = \frac{\frac{8}{15}}{-\frac{4}{3}} = -\frac{2}{5}.$$

That the denominator is negative can be seen from the Riemann sum:

$$\bar{y} = \frac{\displaystyle\int_0^2 \frac{1}{2}y^2\,dx}{\displaystyle\lim_{n\to\infty}\sum_{i=1}^n f(x_i)\Delta x_i}\bigg/ n. \quad \text{Each } f(x_i) < 0.$$

17.

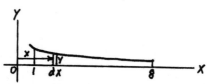

$$y = \frac{1}{\sqrt[3]{x}} = x^{-1/3}$$

$\bar{x}$: moment of typical element: $xy\,dx$.

$$
\begin{aligned}
\bar{x} &= \frac{M_y}{A} = \frac{\int_1^8 x \cdot x^{-1/3}\,dx}{\int_1^8 x^{-1/3}\,dx} = \frac{\int_1^8 x^{2/3}\,dx}{\int_1^8 x^{-1/3}\,dx} = \frac{\left(\frac{3}{5}\right)x^{5/3}\Big|_1^8}{\left(\frac{3}{2}\right)x^{2/3}\Big|_1^8} = \frac{\left(\frac{3}{5}\right)(8^{5/3}-1)}{\left(\frac{3}{2}\right)(8^{2/3}-1)} \\[2mm]
&= \frac{(3/5)(32-1)}{(3/2)(4-1)} = \frac{(3/5)(31)}{(3/2)(3)} = \frac{93/5}{9/2} = \frac{93}{5}\frac{2}{9} = \frac{31}{5}\frac{2}{3} = \frac{62}{15}.
\end{aligned}
$$

$$
\begin{aligned}
\bar{y} &= \frac{M_x}{A} = \frac{\int_1^8 \left(\frac{1}{2}y\right)y\,dx}{9/2} = \frac{2}{9}\int_1^8 \frac{1}{2}y^2\,dx = \frac{1}{9}\int_1^8 (x^{-1/3})^2\,dx = \frac{1}{9}\int_1^8 x^{-2/3}\,dx \\[2mm]
&= \frac{1}{9}(3x^{1/3})\Big|_1^8 = \frac{1}{3}(2-1) = \frac{1}{3}.
\end{aligned}
$$

19.

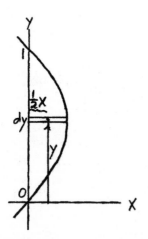

$\bar{y}$: moment of typical element with respect to the x-axis: $yx\,dy = y(y - y^2)\,dy$.

$\bar{x}$: moment of typical element with respect to the y-axis: $\frac{1}{2}x \cdot x\,dy = \frac{1}{2}(y - y^2)(y - y^2)\,dy$.

$$\bar{y} = \frac{\int_0^1 y(y - y^2)\,dy}{\int_0^1 (y - y^2)\,dy} = \frac{\frac{1}{3}y^3 - \frac{1}{4}y^4\Big|_0^1}{\frac{1}{2}y^2 - \frac{1}{3}y^3\Big|_0^1} = \frac{\frac{1}{3} - \frac{1}{4}}{\frac{1}{2} - \frac{1}{3}} = \frac{\frac{1}{12}}{\frac{1}{6}} = \frac{1}{2}.$$

$$\bar{x} = \frac{\int_0^1 \frac{1}{2}(y - y^2)(y - y^2)\,dy}{1/6} = \frac{1/60}{1/6} = \frac{1}{10}.$$

21.

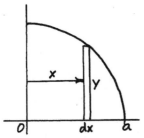

Since the region is a quarter of a circle, the area is:
$\frac{1}{4}\pi a^2$.

$$\bar{x} = \frac{M_y}{A} = \frac{\int_0^a x\sqrt{a^2 - x^2}\,dx}{\frac{1}{4}\pi a^2} = \frac{4}{\pi a^2}\int_0^a (a^2 - x^2)^{1/2}x\,dx.$$

Let $u = a^2 - x^2$; then $du = -2x\,dx$. We now get

$$\frac{4}{\pi a^2}\left(-\frac{1}{2}\right)\int_0^a (a^2 - x^2)^{1/2}(-2x)\,dx \quad = \quad -\frac{2}{\pi a^2}\left.\frac{(a^2 - x^2)^{3/2}}{3/2}\right|_0^a = -\frac{2}{\pi a^2}\left(\frac{2}{3}\right)(a^2 - x^2)^{3/2}\Big|_0^a$$

$$= \quad 0 + \frac{4}{3\pi a^2}(a^2)^{3/2} = \frac{4}{3\pi a^2}(a^3) = \frac{4a}{3\pi}.$$

By symmetry, $\bar{y} = \dfrac{4a}{3\pi}$.

23.

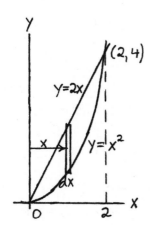

$\bar{x}$: moment with respect to the y-axis: $x(2x - x^2)\,dx$.

$$\bar{x} = \frac{M_y}{A} = \frac{\int_0^2 x(2x - x^2)\,dx}{\int_0^2 (2x - x^2)\,dx} = \frac{\frac{4}{3}}{\frac{4}{3}} = 1.$$

To find $\bar{y}$, observe that

(a) distance from x-axis to center of typical element: $\dfrac{1}{2}(2x + x^2)$;

(b) height of typical element: $2x - x^2$.

$$\bar{y} = \frac{M_x}{A} = \frac{\int_0^2 \frac{1}{2}(2x + x^2)(2x - x^2)\,dx}{4/3} = \frac{3}{4}\cdot\frac{1}{2}\int_0^2 (4x^2 - x^4)\,dx = \frac{3}{8}\left(\frac{4}{3}x^3 - \frac{1}{5}x^5\right)\Big|_0^2$$

$$= \frac{3}{8}\left(\frac{32}{3} - \frac{32}{5}\right) = \frac{3}{8}\cdot 32\left(\frac{1}{3} - \frac{1}{5}\right) = 12\left(\frac{2}{15}\right) = \frac{8}{5}.$$

$\bar{y}$ (alternate):

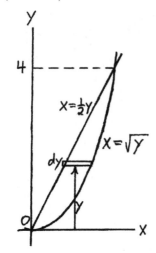

$$\bar{y} = \frac{M_x}{A} = \frac{\int_0^4 y\left(\sqrt{y} - \frac{1}{2}y\right) dy}{4/3} = \frac{32/15}{4/3} = \frac{8}{5}.$$

25. Substituting $y = 2x$ in $y^2 = 4x$, we obtain:

$$\begin{aligned} 4x^2 &= 4x \\ x^2 - x &= 0 \\ x(x-1) &= 0 \\ x &= 0,\ 1. \end{aligned}$$

The points of intersection are $(0,0)$ and $(1,2)$.

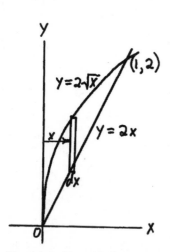

$\bar{x}$: moment (with respect to y-axis) of typical element: $x \cdot (\text{height}) \cdot dx = x(2\sqrt{x} - 2x)\,dx$.

$$\begin{aligned} \bar{x} &= \frac{M_y}{A} = \frac{\int_0^1 x(2\sqrt{x} - 2x)\,dx}{\int_0^1 (2\sqrt{x} - 2x)\,dx} \\ &= \frac{\int_0^1 (2x^{3/2} - 2x^2)\,dx}{\int_0^1 (2x^{1/2} - 2x)\,dx} \\ &= \frac{\frac{4}{5}x^{5/2} - \frac{2}{3}x^3 \Big|_0^1}{\frac{4}{3}x^{3/2} - x^2 \Big|_0^1} = \frac{\frac{4}{5} - \frac{2}{3}}{\frac{4}{3} - 1} \\ &= \frac{\frac{12-10}{15}}{\frac{1}{3}} = \frac{2}{15} \cdot \frac{3}{1} = \frac{2}{5}. \end{aligned}$$

To obtain $\bar{y}$, note that

(a) distance from x-axis to center of element $= \frac{1}{2}(2\sqrt{x} + 2x)$;

(b) height of typical element $= (2\sqrt{x} - 2x)\,dx$.

$$\begin{aligned} \bar{y} &= \frac{M_x}{A} = \frac{\int_0^1 \frac{1}{2}(2\sqrt{x} + 2x)(2\sqrt{x} - 2x)\,dx}{1/3} = \frac{3}{2}\int_0^1 (4x - 4x^2)\,dx \\ &= \frac{3}{2}\left(2x^2 - \frac{4}{3}x^3\right)\Big|_0^1 = \frac{3}{2}\left(2 - \frac{4}{3}\right) = \frac{3}{2} \cdot \frac{2}{3} = 1. \end{aligned}$$

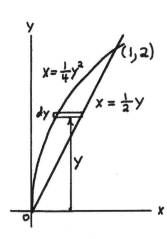

$\overline{y}$ (alternate): moment (with respect to x-axis) of typical element: $y \cdot (\text{height}) \cdot dy = y \left(\frac{1}{2}y - \frac{1}{4}y^2 \right) dy$. Thus

$$\overline{y} = \frac{\int_0^2 y \left(\frac{1}{2}y - \frac{1}{4}y^2 \right) dy}{1/3}$$

$$= 3 \int_0^2 \left(\frac{1}{2}y^2 - \frac{1}{4}y^3 \right) dy$$

$$= 3 \left(\frac{1}{6}y^3 - \frac{1}{16}y^4 \right) \Big|_0^2$$

$$= 3 \left(\frac{4}{3} - 1 \right) = 3 \left(\frac{1}{3} \right) = 1.$$

27.

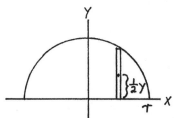

$\overline{x} = 0$ by symmetry.

$\overline{y}$: Area $= \frac{1}{2}\pi r^2$.

$$\overline{y} = \frac{\int_{-r}^r \frac{1}{2}\sqrt{r^2 - x^2} \cdot \sqrt{r^2 - x^2} \, dx}{A} = \frac{\int_{-r}^r \frac{1}{2}(r^2 - x^2) \, dx}{\frac{1}{2}\pi r^2} = \frac{\frac{1}{2} \left(r^2 x - \frac{1}{3}x^3 \right) \Big|_{-r}^r}{\frac{1}{2}\pi r^2}$$

$$= \frac{\frac{1}{2} \left(r^3 - \frac{1}{3}r^3 \right) - \frac{1}{2} \left(-r^3 + \frac{1}{3}r^3 \right)}{\frac{1}{2}\pi r^2} = \frac{r^3 - \frac{1}{3}r^3}{\frac{1}{2}\pi r^2} = \frac{\frac{2}{3}r^3}{\frac{1}{2}\pi r^2} = \frac{4r}{3\pi}.$$

Answer: along the axis of the semicircle, $\dfrac{4r}{3\pi}$ units from the center.

29.

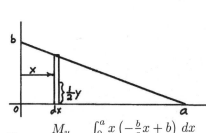

Slope of line: $-\dfrac{b}{a}$; y-intercept: $(0, b)$.

By the slope-intercept form, $y = -\dfrac{b}{a}x + b$.

$$\overline{x} = \frac{M_y}{A} = \frac{\int_0^a x \left(-\frac{b}{a}x + b \right) dx}{\frac{1}{2}ab} = \frac{2}{ab} \int_0^a \left(-\frac{b}{a}x^2 + bx \right) dx = \frac{2}{ab} \left(-\frac{b}{a} \cdot \frac{x^3}{3} + b\frac{x^2}{2} \right) \Big|_0^a$$

$$= \frac{2}{ab} \left(-\frac{b}{a} \cdot \frac{a^3}{3} + b\frac{a^2}{2} \right) = \frac{2}{ab} \left(-\frac{a^2 b}{3} + \frac{a^2 b}{2} \right) = \frac{2}{ab}(a^2 b) \left(\frac{1}{2} - \frac{1}{3} \right) = 2a \left(\frac{1}{6} \right) = \frac{a}{3}.$$

$\overline{y}$: moment (with respect to x-axis) of typical element: $\left(\frac{1}{2}y \right) y \, dx = \frac{1}{2}y^2 \, dx$.

$$\overline{y} = \frac{M_x}{A} = \frac{\frac{1}{2} \int_0^a \left(-\frac{b}{a}x + b \right)^2 dx}{\frac{1}{2}ab} = \frac{1}{ab} \int_0^a \left(\frac{b^2}{a^2}x^2 - \frac{2b^2}{a}x + b^2 \right) dx$$

$$= \frac{1}{ab} \left(\frac{b^2}{a^2} \cdot \frac{x^3}{3} - \frac{2b^2}{a} \cdot \frac{x^2}{2} + b^2 x \right) \Big|_0^a = \frac{1}{ab} \left(\frac{b^2}{a^2} \cdot \frac{a^3}{3} - \frac{b^2 a^2}{a} + ab^2 \right)$$

$$= \frac{1}{ab} \left(\frac{1}{3}ab^2 - ab^2 + ab^2 \right) = \frac{1}{ab} \left(\frac{1}{3}ab^2 \right) = \frac{b}{3}.$$

31. (a)

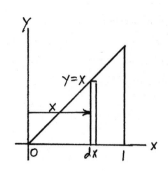

Volume of typical disk:
$$\pi(\text{radius})^2 \cdot (\text{thickness}) = \pi y^2 \, dx.$$

Moment of typical disk: $x \cdot \pi y^2 \, dx$.

$$\bar{x} = \frac{M_y}{V} = \frac{\int_0^1 x \cdot \pi x^2 \, dx}{\int_0^1 \pi x^2 \, dx} = \frac{\pi \frac{x^4}{4}\big|_0^1}{\pi \frac{x^3}{3}\big|_0^1} = \frac{3}{4}.$$

$$\bar{y} = 0.$$

(b)

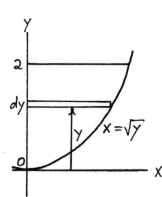

Volume of typical disk:
$$\pi(\text{radius})^2 \cdot (\text{thickness}) = \pi \left(\sqrt{y}\right)^2 \, dy.$$

Moment of typical disk: $y \cdot \pi \left(\sqrt{y}\right)^2 \, dy$.

$$\bar{y} = \frac{M_x}{V} = \frac{\int_0^2 y \cdot \pi \left(\sqrt{y}\right)^2 \, dy}{\int_0^2 \pi \left(\sqrt{y}\right)^2 \, dy} = \frac{\pi \frac{y^3}{3}\big|_0^2}{\pi \frac{y^2}{2}\big|_0^2} = \frac{4}{3}.$$

$$\bar{x} = 0.$$

33. Volume of typical disk: $\pi(\text{radius})^2 \cdot (\text{thickness}) = \pi y^2 \, dx$.

Moment of typical disk: $x \cdot \pi y^2 \, dx = x \cdot \pi(2x^2)^2 \, dx$.

$$\bar{x} = \frac{M_y}{V} = \frac{\int_0^1 x \cdot \pi(2x^2)^2 \, dx}{\int_0^1 \pi(2x^2)^2 \, dx} = \frac{4\pi \int_0^1 x^5 \, dx}{4\pi \int_0^1 x^4 \, dx} = \frac{1/6}{1/5} = \frac{5}{6}.$$

$$\bar{y} = 0.$$

35.

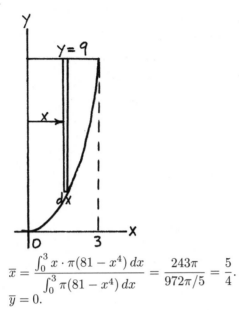

Volume of typical washer:

$\pi \left[9^2 - (x^2)^2 \right] \, dx$.

Moment of typical element:

$x \cdot \pi \left[9^2 - (x^2)^2 \right] \, dx$.

$\bar{x} = \dfrac{\int_0^3 x \cdot \pi (81 - x^4) \, dx}{\int_0^3 \pi (81 - x^4) \, dx} = \dfrac{243\pi}{972\pi/5} = \dfrac{5}{4}$.

$\bar{y} = 0$.

37.

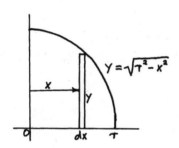

Volume of typical disk: $\pi y^2 \, dx$.

Moment of typical disk:

$x(\pi y^2 \, dx) = \pi x \left(\sqrt{r^2 - x^2} \right)^2 \, dx$.

$$
\begin{aligned}
\bar{x} &= \frac{M_y}{V} = \frac{\pi \int_0^r x \left(\sqrt{r^2 - x^2} \right)^2 \, dx}{\pi \int_0^r \left(\sqrt{r^2 - x^2} \right)^2 \, dx} = \frac{\int_0^r x(r^2 - x^2) \, dx}{\int_0^r (r^2 - x^2) \, dx} = \frac{\int_0^r (r^2 x - x^3) \, dx}{\int_0^r (r^2 - x^2) \, dx} \\
&= \frac{\left. r^2 \frac{x^2}{2} - \frac{x^4}{4} \right|_0^r}{\left. r^2 x - \frac{x^3}{3} \right|_0^r} = \frac{\frac{r^4}{2} - \frac{r^4}{4}}{r^3 - \frac{r^3}{3}} = \frac{r^4 \left(\frac{1}{2} - \frac{1}{4} \right)}{r^3 \left(1 - \frac{1}{3} \right)} = \frac{r \left(\frac{1}{4} \right)}{\frac{2}{3}} = \frac{3}{8} r.
\end{aligned}
$$

$\bar{y} = 0$.

39. (a)

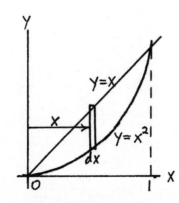

Volume of typical washer:

$\pi \left[x^2 - (x^2)^2 \right] \, dx$.

$\bar{x} = \dfrac{\int_0^1 x \cdot \pi \left[x^2 - (x^2)^2 \right] \, dx}{\int_0^1 \pi \left[x^2 - (x^2)^2 \right] \, dx} = \dfrac{\pi/12}{2\pi/15} = \dfrac{5}{8}$.

(b)

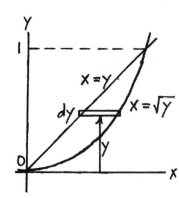

Volume of typical washer:

$$\pi \left[(\sqrt{y})^2 - y^2 \right] dy.$$

$$\bar{y} = \frac{\int_0^1 y \cdot \pi \left[(\sqrt{y})^2 - y^2 \right] dy}{\int_0^1 \pi \left[(\sqrt{y})^2 - y^2 \right] dy} = \frac{\pi/12}{\pi/6} = \frac{1}{2}.$$

41.

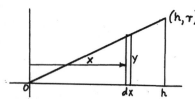

Slope of line: $\dfrac{r}{h}$.

Equation: $y = \dfrac{r}{h} x$.

Moment of typical disk: $x(\pi y^2 \, dx) = \pi x \left(\dfrac{r}{h} x \right)^2 dx.$

$$\bar{x} = \frac{\pi \int_0^h x \left(\frac{r}{h} x \right)^2 dx}{\pi \int_0^h \left(\frac{r}{h} x \right)^2 dx} = \frac{(r^2/h^2) \int_0^h x^3 \, dx}{(r^2/h^2) \int_0^h x^2 \, dx} = \frac{\int_0^h x^3 \, dx}{\int_0^h x^2 \, dx} = \frac{\left. \frac{x^4}{4} \right|_0^h}{\left. \frac{x^3}{3} \right|_0^h} = \frac{h^4}{4} \cdot \frac{3}{h^3} = \frac{3}{4} h.$$

Since $\bar{y} = 0$, we conclude that the centroid lies on the axis, one-fourth of the way from the base.

43.

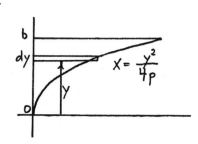

Volume of typical disk:

$$\pi \left(\frac{y^2}{4p} \right)^2 dy.$$

$$\bar{y} = \frac{\int_0^b y \cdot \pi \left(\frac{y^4}{16p^2} \right) dy}{\int_0^b \pi \left(\frac{y^4}{16p^2} \right) dy} = \frac{\pi/(16p^2) \int_0^b y^5 \, dy}{\pi/(16p^2) \int_0^b y^4 \, dy} = \frac{\left. \frac{1}{6} y^6 \right|_0^b}{\left. \frac{1}{5} y^5 \right|_0^b} = \frac{\frac{1}{6} b^6}{\frac{1}{5} b^5} = \frac{5}{6} b.$$

$$\bar{x} = 0.$$

45. Form of upper parabola: $x^2 = 4p_1(y - 4).$

Form of lower parabola: $x^2 = 4p_2 y.$

Both parabolas pass through the point $(5, 10)$:

$$25 = 4p_1(10 - 4) \quad \text{and} \quad 4p_1 = \frac{25}{6},$$

$$25 = 4p_2 \cdot 10 \quad \text{and} \quad 4p_2 = \frac{5}{2}.$$

Equations:

$$x^2 = \frac{25}{6}(y-4) \quad \text{or} \quad y = 0.24(x^2 + 50/3),$$

$$x^2 = \frac{5}{2}y \quad \text{or} \quad y = 0.4x^2.$$

$\overline{y}$: moment of typical element:

$$\frac{1}{2}\left[0.24(x^2 + 50/3) + 0.4x^2\right]\left[0.24(x^2 + 50/3) - 0.4x^2\right]\,dx$$

$$= \frac{1}{2}\left\{\left[0.24(x^2 + 50/3)\right]^2 - \left[0.4x^2\right]^2\right\}\,dx.$$

Using the integration capability of a graphing utility,

$$\overline{y} = \frac{\frac{1}{2}\int_0^5 \left\{\left[0.24(x^2 + 50/3)\right]^2 - \left[0.4x^2\right]^2\right\}\,dx}{\int_0^5 \left[0.24(x^2 + 50/3) - 0.4x^2\right]\,dx} = 3.6\,\text{ft}.$$

Finally, $\overline{x} = 0$.

5.5 Moments of Inertia

1.

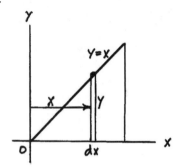

Moment of inertia of typical element:

$$x^2 \cdot \rho y\,dx = \rho x^2 \cdot x\,dx.$$

$$I_y = \int_0^1 \rho x^2 x\,dx = \rho\left.\frac{x^4}{4}\right|_0^1 = \frac{\rho}{4}.$$

Mass: $\rho\displaystyle\int_0^1 x\,dx = \frac{\rho}{2}.$

$$R_y = \sqrt{\frac{\rho}{4}\frac{2}{\rho}} = \sqrt{\frac{1}{2}} = \frac{\sqrt{2}}{2}.$$

3.

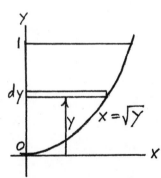

Moment of inertia of typical element:

$$y^2 \cdot \rho x\,dy = y^2 \rho\sqrt{y}\,dy.$$

Summing from $y = 0$ to $y = 1$, we get

$$I_x = \int_0^1 y^2 \cdot \rho\sqrt{y}\,dy = \rho\int_0^1 y^{5/2}\,dy = \rho\left.\frac{y^{7/2}}{7/2}\right|_0^1 = \frac{2\rho}{7}.$$

Mass: $\rho\displaystyle\int_0^1 \sqrt{y}\,dy = \rho\left(\frac{2}{3}\right)\left.y^{3/2}\right|_0^1 = \frac{2\rho}{3}.$

$$R_x = \sqrt{\frac{2\rho}{7}\cdot\frac{3}{2\rho}} = \sqrt{\frac{3}{7}} = \frac{\sqrt{21}}{7}.$$

5.

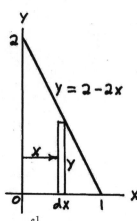

I_y: moment of inertia of typical element:

$$x^2 \cdot \rho y \, dx = \rho x^2 \cdot (2 - 2x) \, dx.$$

$$I_y = \int_0^1 \rho x^2 (2 - 2x) \, dx = \rho \int_0^1 (2x^2 - 2x^3) \, dx = \rho \left(\frac{2}{3} x^3 - \frac{1}{2} x^4 \right) \bigg|_0^1 = \rho \left(\frac{2}{3} - \frac{1}{2} \right) = \frac{\rho}{6}.$$

Since $A = \frac{1}{2} bh = \frac{1}{2}(1)(2) = 1$, the mass $= \rho$. Thus $R_y = \sqrt{\frac{\rho}{6} \frac{1}{\rho}} = \sqrt{\frac{1}{6}} = \frac{\sqrt{6}}{6}$.

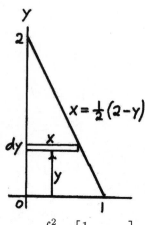

I_x: moment of inertia of typical element:

$$y^2 \cdot \rho x \, dy = \rho y^2 \left[\frac{1}{2}(2 - y) \right] dy.$$

$$I_x = \rho \int_0^2 y^2 \left[\frac{1}{2}(2 - y) \right] dy = \frac{\rho}{2} \int_0^2 (2y^2 - y^3) \, dy = \frac{2\rho}{3}.$$

$$R_x = \sqrt{\frac{2\rho}{3} \frac{1}{\rho}} = \sqrt{\frac{2}{3}} = \frac{\sqrt{6}}{3}.$$

7.

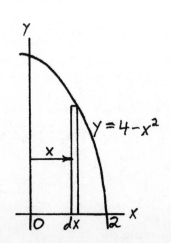

Moment of inertia of typical element:

$$x^2 \cdot \rho y \, dx = x^2 \cdot \rho(4 - x^2) \, dx.$$

$$I_y = \int_0^2 x^2 \cdot \rho(4 - x^2)\,dx = \rho\left(\frac{4}{3}x^3 - \frac{1}{5}x^5\right)\Big|_0^2$$

$$= \rho\left(\frac{32}{3} - \frac{32}{5}\right) = 32\rho\left(\frac{1}{3} - \frac{1}{5}\right) = 32\rho \cdot \frac{5-3}{15} = \frac{64\rho}{15}.$$

Mass: $\rho \int_0^2 (4 - x^2)\,dx = \rho\left(4x - \frac{1}{3}x^3\right)\Big|_0^2 = \rho\left(8 - \frac{8}{3}\right) = \frac{16\rho}{3}.$

$$R_y = \sqrt{\frac{64\rho}{15} \cdot \frac{3}{16\rho}} = \sqrt{\frac{4}{5}} = \frac{2}{\sqrt{5}} = \frac{2\sqrt{5}}{5}.$$

9. Substituting $y = \frac{1}{2}x$ in $y^2 = x$, we get

$$\frac{1}{4}x^2 = x$$
$$x^2 - 4x = 0$$
$$x(x - 4) = 0$$
$$x = 0, 4.$$

The points of intersection are $(0,0)$ and $(4,2)$.

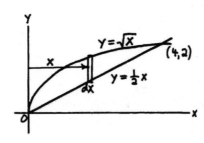

Moment of inertia (with respect to y-axis) of typical element:

$$x^2 \cdot \rho \cdot (\text{height}) \cdot dx = \rho x^2 \left(\sqrt{x} - \frac{1}{2}x\right)dx.$$

$$I_y = \rho \int_0^4 x^2\left(\sqrt{x} - \frac{1}{2}x\right)dx = \rho \int_0^4 \left(x^{5/2} - \frac{1}{2}x^3\right)dx = \rho\left(\frac{2}{7}x^{7/2} - \frac{1}{8}x^4\right)\Big|_0^4$$

$$= \rho\left[\frac{2}{7}(4^{7/2}) - \frac{1}{8}(4^4)\right] = \rho\left(\frac{2}{7} \cdot 128 - 32\right) = 32\rho\left(\frac{8}{7} - 1\right) = 32\rho\left(\frac{8-7}{7}\right) = \frac{32\rho}{7}.$$

Mass: $\rho \int_0^4 \left(\sqrt{x} - \frac{1}{2}x\right)dx = \rho\left(\frac{2}{3}x^{3/2} - \frac{1}{4}x^2\right)\Big|_0^4 = \rho\left(\frac{2}{3} \cdot 8 - 4\right) = \frac{4\rho}{3}.$

$$R_y = \sqrt{\frac{32\rho}{7} \frac{3}{4\rho}} = \sqrt{\frac{8 \cdot 3}{7}} = \frac{2\sqrt{2 \cdot 3}}{\sqrt{7}} = \frac{2\sqrt{42}}{7}.$$

11. $y = 9 - 3x$

$\underline{y = 9 - x^2}$

$0 = -3x + x^2$ (subtracting)

$x(x-3) = 0;\ x = 0, 3.$

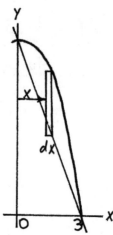

Moment of typical element: $x^2 \cdot \rho(\text{height})\, dx = x^2 \cdot \rho\left[(9-x^2)-(9-3x)\right]\, dx.$

$$I_y = \rho \int_0^3 x^2(9-x^2-9+3x)\, dx = \rho\left(-\frac{1}{5}x^5+\frac{3}{4}x^4\right)\Big|_0^3$$

$$= \rho\left(-\frac{1}{5}\cdot 3^5+\frac{3}{4}\cdot 3^4\right) = \rho(3^5)\left(-\frac{1}{5}+\frac{1}{4}\right) = \rho\frac{3^5}{20} = \frac{243\rho}{20}.$$

Mass: $\rho \displaystyle\int_0^3 \left[(9-x^2)-(9-3x)\right]\, dx = \frac{9\rho}{2}.$

$$R_y = \sqrt{\frac{3^5\rho}{20}\cdot\frac{2}{3^2\rho}} = \sqrt{\frac{3^3}{10}} = \frac{3\sqrt{3}}{\sqrt{10}} = \frac{3\sqrt{30}}{10}.$$

13. Solving the equations simultaneously, we get

$x = y^2 + 2$

$\underline{x = y + 2}$

$0 = y^2 - y$ (subtracting)

$y(y-1) = 0$

$\qquad y = 0, 1.$

The points of intersection are $(2, 0)$ and $(3, 1)$. Note that the typical element has to be drawn horizontally.

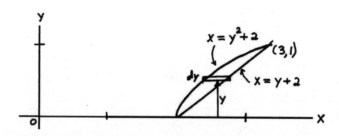

Moment of inertia (with respect to x-axis) of typical element:

$\rho y^2 \cdot (\text{height}) \cdot dy = \rho y^2 \left[(y+2)-(y^2+2)\right]\, dy.$

$$I_x = \rho \int_0^1 y^2\left[(y+2)-(y^2+2)\right]\, dy = \rho \int_0^1 y^2(y-y^2)\, dy$$

$$= \rho \int_0^1 (y^3-y^4)\, dy = \rho\left(\frac{1}{4}y^4-\frac{1}{5}y^5\right)\Big|_0^1 = \rho\left(\frac{1}{4}-\frac{1}{5}\right) = \frac{\rho}{20}.$$

Mass: $\rho \displaystyle\int_0^1 \left[(y+2) - (y^2+2)\right] dy = \rho \displaystyle\int_0^1 (y - y^2)\, dy = \dfrac{\rho}{6}.$

$R_x = \sqrt{\dfrac{\rho}{20}\dfrac{6}{\rho}} = \sqrt{\dfrac{3}{10}} = \dfrac{\sqrt{30}}{10}.$

15. $y \;=\; 2x^2$

$\dfrac{y \;=\; 4x + 6}{}$

$0 \;=\; 2x^2 - 4x - 6$ (subtracting)

$x^2 - 2x - 3 \;=\; 0$

$(x+1)(x-3) \;=\; 0; \; x = -1, 3.$

Moment of inertia of typical element:

$x^2 \cdot \rho(\text{height})\, dx.$

$I_y = \displaystyle\int_{-1}^{3} x^2 \cdot \rho(4x + 6 - 2x^2)\, dx = \dfrac{192\rho}{5}.$

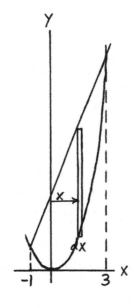

17. Volume of shell: $2\pi(\text{radius}) \cdot (\text{height}) \cdot (\text{thickness}) = 2\pi xy\, dx.$

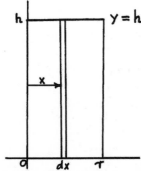

Mass of shell: $\rho(2\pi xy\, dx).$

Moment of inertia of typical shell (since $y = h$):

$x^2 \cdot \rho(2\pi xy\, dx) = 2\pi\rho x^3 h\, dx.$

$I_y = 2\pi\rho \displaystyle\int_0^r x^3 h\, dx = 2\pi\rho h \dfrac{x^4}{4}\Big|_0^r = \dfrac{1}{2}\pi r^4 h\rho.$

Since the mass of the cylinder is $\rho\pi r^2 h$, I_y can also be written as follows:

$I_y = \dfrac{1}{2}(\rho\pi r^2 h)r^2 = \dfrac{1}{2}mr^2.$

$R_y = \sqrt{\dfrac{\pi r^4 h\rho}{2}\dfrac{1}{\rho\pi r^2 h}} = \sqrt{\dfrac{r^2}{2}} = \dfrac{r}{\sqrt{2}} = \dfrac{r\sqrt{2}}{2}.$

19.

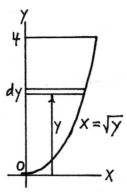

Volume of shell: $2\pi(\text{radius}) \cdot (\text{height}) \cdot (\text{thickness})$

$= 2\pi y \cdot x \, dy = 2\pi y \cdot \sqrt{y} \, dy.$

Mass of shell: $\rho \cdot 2\pi y \sqrt{y} \, dy.$

Moment of inertia of typical shell:

$y^2 \cdot \rho \cdot 2\pi y \sqrt{y} \, dy.$

$$I_x = \int_0^4 y^2 \cdot \rho \cdot 2\pi y \sqrt{y} \, dy = 2\pi\rho \int_0^4 y^{7/2} \, dy = 2\pi\rho \cdot \frac{2}{9} y^{9/2} \Big|_0^4$$

$$= 2\pi\rho \cdot \frac{2}{9}(4)^{9/2} = 2\pi\rho \left(\frac{2}{9}\right) 2^9 = \frac{2^{11}\pi\rho}{9}.$$

Mass: $\displaystyle\int_0^4 \rho \cdot 2\pi y \sqrt{y} \, dy = 2\pi\rho \left(\frac{2}{5}\right) y^{5/2} \Big|_0^4 = 2\pi\rho \left(\frac{2}{5}\right) 2^5 = \frac{2^7\pi\rho}{5}.$

$$R_x = \sqrt{\frac{2^{11}\pi\rho}{9} \cdot \frac{5}{2^7\pi\rho}} = \frac{4\sqrt{5}}{3}.$$

21.

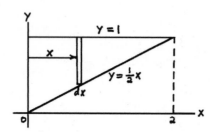

Mass of shell:

$\rho \cdot 2\pi(\text{radius}) \cdot (\text{height}) \cdot (\text{thickness}) = \rho \cdot 2\pi \cdot x \cdot \left(1 - \frac{1}{2}x\right) \, dx = 2\pi\rho x \left(1 - \frac{1}{2}x\right) \, dx.$

Moment of inertia of typical shell: $x^2 \cdot 2\pi\rho x \left(1 - \frac{1}{2}x\right) \, dx.$

$$I_y = \int_0^2 x^2 \cdot 2\pi\rho x \left(1 - \frac{1}{2}x\right) \, dx = 2\pi\rho \int_0^2 \left(x^3 - \frac{1}{2}x^4\right) \, dx = 2\pi\rho \left(\frac{x^4}{4} - \frac{x^5}{10}\right) \Big|_0^2$$

$$= 2\pi\rho \left(4 - \frac{16}{5}\right) = 2\pi\rho \left(\frac{4}{5}\right) = \frac{8\pi\rho}{5}.$$

23.

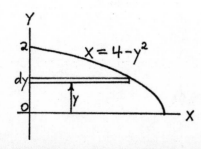

Volume of shell:

$2\pi(\text{radius}) \cdot (\text{height}) \cdot (\text{thickness})$

$= 2\pi y(4 - y^2) \, dy.$

Mass of shell: $\rho \cdot 2\pi y(4 - y^2) \, dy.$

$$I_x = \int_0^2 y^2 \cdot \rho \cdot 2\pi y(4 - y^2) \, dy = \frac{32\pi\rho}{3}.$$

Mass: $\int_0^2 \rho \cdot 2\pi y (4 - y^2)\, dy = 8\pi\rho.$

$R_x = \sqrt{\dfrac{32\pi\rho}{3} \cdot \dfrac{1}{8\pi\rho}} = \sqrt{\dfrac{4}{3}} = \dfrac{2}{\sqrt{3}} = \dfrac{2\sqrt{3}}{3}.$

25.

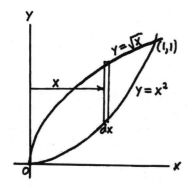

Mass of shell: $\rho \cdot 2\pi x \cdot (\text{height}) \cdot dx$
$= \rho \cdot 2\pi x \left(\sqrt{x} - x^2\right) dx.$

Moment of inertia of shell:

$x^2 \cdot \rho \cdot 2\pi x \left(\sqrt{x} - x^2\right) dx = 2\pi\rho x^3 \left(\sqrt{x} - x^2\right) dx.$

$$
\begin{aligned}
I_y &= 2\pi\rho \int_0^1 x^3 \left(\sqrt{x} - x^2\right) dx = 2\pi\rho \int_0^1 \left(x^{7/2} - x^5\right) dx = 2\pi\rho \left(\frac{2}{9}x^{9/2} - \frac{1}{6}x^6\right)\Big|_0^1 \\
&= 2\pi\rho \left(\frac{2}{9} - \frac{1}{6}\right) = 2\pi\rho \left(\frac{4}{18} - \frac{3}{18}\right) = \frac{\pi\rho}{9}.
\end{aligned}
$$

$$
\begin{aligned}
\text{Mass: } 2\pi\rho \int_0^1 x \left(\sqrt{x} - x^2\right) dx &= 2\pi\rho \int_0^1 \left(x^{3/2} - x^3\right) dx \\
&= 2\pi\rho \left(\frac{2}{5}x^{5/2} - \frac{1}{4}x^4\right)\Big|_0^1 \\
&= 2\pi\rho \left(\frac{2}{5} - \frac{1}{4}\right) = \frac{3\pi\rho}{10}.
\end{aligned}
$$

$R_y = \sqrt{\dfrac{\pi\rho}{9}\dfrac{10}{3\pi\rho}} = \sqrt{\dfrac{10}{9 \cdot 3}} = \dfrac{\sqrt{10}}{3\sqrt{3}} = \dfrac{\sqrt{30}}{9}.$

27. By Exercise 17, $I_y = \dfrac{1}{2}mr^2.$

$\dfrac{360 \text{ rev}}{1 \text{ min}} \times \dfrac{2\pi \text{ rad}}{1 \text{ rev}} \times \dfrac{1 \text{ min}}{60 \text{ s}} = 12\pi \text{ rad/s.}$

$K = \dfrac{1}{2}I\omega^2 = \dfrac{1}{2}\left[\dfrac{1}{2}(2.0)(0.10)^2\right](12\pi)^2 = 7.1 \text{ J.}$

$L = I\omega = \left[\dfrac{1}{2}(2.0)(0.10)^2\right](12\pi) = 0.38 \text{ kg} \cdot \text{m}^2/\text{s.}$

5.6 Work and Fluid Pressure

1. From Hooke's law,

$$
\begin{aligned}
F &= kx \\
6 &= k \cdot \frac{1}{8} \text{ or } k = 48.
\end{aligned}
$$

So $F = 48x$. Since the spring is stretched to 2 ft, $W = \displaystyle\int_0^2 48x\, dx = 24x^2\Big|_0^2 = 24 \cdot 4 = 96 \text{ ft-lb.}$

3. From Hooke's law,

$$F = kx$$
$$12 = k \cdot 2 \quad \text{or} \quad k = 6.$$

So $F = 6x$.

(a) Since the spring is being compressed from its natural length of 8 ft to 6 ft, a total of 2 ft, we obtain

$$W = \int_0^2 6x \, dx = 3x^2 \Big|_0^2 = 12 \, \text{ft-lb}.$$

(b) The spring is stretched from 2 ft (beyond its natural length) to 5 ft:

$$W = \int_2^5 6x \, dx = 3x^2 \Big|_2^5 = 3(25 - 4) = 63 \, \text{ft-lb}.$$

5.

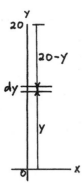

Weight of typical element:

$3 \, \text{N/m} \times \, dy \, \text{m} = 3 \, dy \, \text{N}.$

Work done in moving the typical element to the top:

$3 \, dy \cdot (20 - y) = 3(20 - y) \, dy.$

Summing from $y = 0$ to $y = 20$, we get $W = \int_0^{20} 3(20 - y) \, dy = 3\left(20y - \frac{1}{2}y^2\right)\Big|_0^{20} = 600 \, \text{J}.$

7.

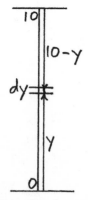

Weight of typical element:

$10 \, \text{lb/ft} \times \, dy \, \text{ft} = 10 \, dy \, \text{lb}.$

Work done in moving the typical element to the top:

$10 \, dy \cdot (10 - y).$

Summing from $y = 0$ to $y = 10$, we get

$$W = 10 \int_0^{10} (10 - y) \, dy = 10\left(10y - \frac{1}{2}y^2\right)\Big|_0^{10} = 10(100 - 50) = 500 \, \text{ft-lb}.$$

The 20 lb-weight attached to the end is moved 10 ft: $20 \, \text{lb} \times 10 \, \text{ft} = 200 \, \text{ft-lb}.$

Total: $500 + 200 = 700 \, \text{ft-lb}.$

9.

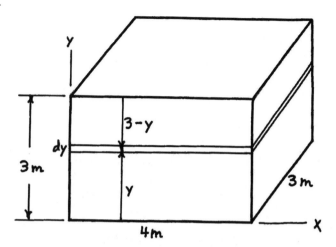

Volume of typical element: $4 \cdot 3 \cdot dy = 12\,dy$.

Weight of typical element: $(12\,dy)w = 12w\,dy$.

Work done in moving the typical element to the top: $(3 - y) \cdot 12w\,dy$.

Summing from $y = 0$ to $y = 3$, we get

$$W = \int_0^3 (3 - y) \cdot 12w\,dy = 12w \left(3y - \frac{1}{2}y^2 \right) \Big|_0^3 = 12w \left(9 - \frac{9}{2} \right) = 54w \text{ J.}$$

11.

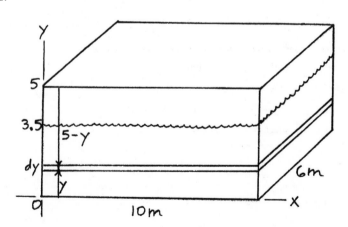

Volume of typical element: $10 \cdot 6 \cdot dy = 60\,dy$.

Weight of typical element: $w \cdot 60\,dy = 60w\,dy$.

Work done in moving the typical element to the top: $(5 - y)(60w\,dy)$.

We sum from $y = 0$ to $y = 3.5$, the level to which the tank is filled:

$$W = \int_0^{3.5} (5 - y)(60w\,dy) = 60w \left(5y - \frac{1}{2}y^2 \right) \Big|_0^{3.5} = 682.5w \text{ J.}$$

13.

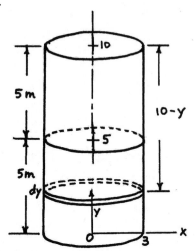

Volume of typical element: $\pi(3^2)\, dy = 9\pi\, dy$.

Weight of typical element: $9\pi\, dy \cdot w = 9\pi w\, dy$.

Work done in moving the typical element to the top: $(10 - y) \cdot 9\pi w\, dy$.

Summing from $y = 0$ to $y = 5$, we get

$$
\begin{aligned}
W &= \int_0^5 (10 - y) \cdot 9\pi w\, dy = 9\pi w \int_0^5 (10 - y)\, dy = 9\pi w \left(10y - \frac{1}{2}y^2 \right) \Big|_0^5 \\
&= 9\pi w \left(50 - \frac{25}{2} \right) = \frac{675\pi w}{2}\ \text{J}.
\end{aligned}
$$

15.

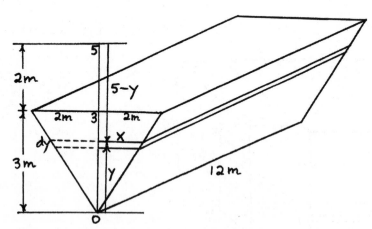

By similar triangles, $\dfrac{x}{y} = \dfrac{2}{3}$ or $x = \dfrac{2}{3}y$.

Volume of typical element: $2x\, dy \cdot 12 = 24x\, dy = 24 \left(\dfrac{2}{3}y \right) dy = 16y\, dy$.

Weight of typical element: $w \cdot 16y\, dy = 16wy\, dy$.

Work done in moving the typical element to the top: $(5 - y)(16wy\, dy)$.

Summing from $y = 0$ to $y = 3$,

$$
W = \int_0^3 (5 - y)(16wy\, dy) = 16w \int_0^3 (5y - y^2)\, dy = 16w \left(\frac{5}{2}y^2 - \frac{1}{3}y^3 \right) \Big|_0^3 = 216w\ \text{J}.
$$

17.

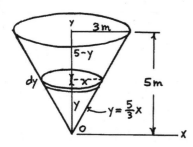

To find the volume and weight of the typical element, we need to determine the radius x. Since the line on the right has slope $m = \dfrac{5}{3}$, its equation is $y = \dfrac{5}{3}x$, or $x = \dfrac{3}{5}y$. Thus:

Weight of typical element: $\pi \left(\dfrac{3}{5}y\right)^2 dy \cdot w$.

Work done in moving the typical element to the top:

$$(5 - y) \cdot \pi \left(\frac{3}{5}y\right)^2 dy \cdot w = \frac{9\pi w}{25}(5y^2 - y^3)\, dy.$$

Summing from $y = 0$ to $y = 5$, we get

$$
\begin{aligned}
W &= \int_0^5 \frac{9\pi w}{25}(5y^2 - y^3)\, dy = \frac{9\pi w}{25}\left(\frac{5y^3}{3} - \frac{y^4}{4}\right)\Bigg|_0^5 \\
&= \frac{9\pi w}{25}\left(\frac{5^4}{3} - \frac{5^4}{4}\right) = \frac{9\pi w}{25}\cdot 5^4\left(\frac{1}{3} - \frac{1}{4}\right) = 9\pi w(25)\frac{1}{12} = \frac{75\pi w}{4}\ \text{J.}
\end{aligned}
$$

19.

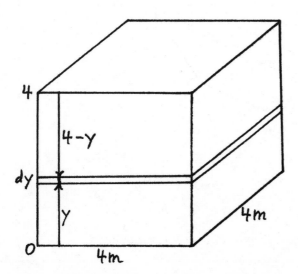

Weight of typical element: $w \cdot 16y\, dy = 16w\, dy$.

Work done in moving the typical element from the bottom: $y \cdot 16w\, dy$.

$$W = \int_0^4 y \cdot 16w\, dy = w(8y^2)\Big|_0^4 = 128w\ \text{J.}$$

21.

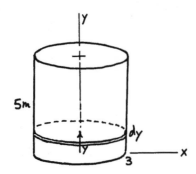

Volume of typical element: $\pi(3^2)\,dy$.

Weight of typical element: $\pi(3^2)\,dy \cdot w = 9\pi w\,dy$.

Work done in moving the typical element from the bottom: $y \cdot 9\pi w\,dy$.

Summing from $y = 0$ to $y = 2$, $W = \displaystyle\int_0^2 y \cdot 9\pi w\,dy = 9\pi w \left.\dfrac{y^2}{2}\right|_0^2 = 18\pi w$ J.

23. The easiest way to solve this problem is to retain the coordinate system in Exercise 19 and to change the y to $y + 2$:

$$W = \int_0^4 (y + 2) \cdot 16w\,dy = 256w \text{ J.}$$

Alternatively, we can use the coordinate system in the diagram.

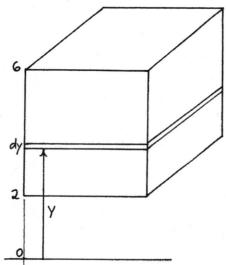

$$W = \int_2^6 16wy\,dy = \left. w(8y^2) \right|_2^6 = 8w(36 - 4) = 256w \text{ J.}$$

25. Since $k = 1600$, $P = 1600V^{-1.4}$, and

$$W = \int_{0.059}^{0.417} 1600 V^{-1.4}\,dV = 1600 \left. \dfrac{V^{-0.4}}{-0.4} \right|_{0.059}^{0.417} = -\dfrac{1600}{0.4} \left. \dfrac{1}{V^{0.4}} \right|_{0.059}^{0.417} = 6733 \approx 6700 \text{ ft-lb.}$$

27. $F = \dfrac{k}{s^2}$

   $20.0 = \dfrac{k}{(1.0)^2}$ $F = 20.0\,\text{dynes}, \ s = 1.0\,\text{cm}.$

   $k = 20.0$ and $F = \dfrac{20.0}{s^2}$.

   (a) $W = \displaystyle\int_{1.0}^{10.0} \dfrac{20.0}{s^2}\,ds = 18\,\text{ergs}.$

(b) $\quad W = \int_{1.0}^{\infty} \frac{20.0}{s^2}\,ds = \lim_{b\to\infty} \int_{1.0}^{b} 20.0 s^{-2}\,ds = \lim_{b\to\infty} \frac{-20.0}{s}\Big|_{1.0}^{b}$

$\qquad = \lim_{b\to\infty} \left(\frac{-20.0}{b} + \frac{20.0}{1.0} \right) = 20\,\text{ergs}.$

29.

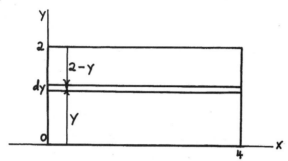

Pressure on strip: $(2-y)w$.

Area of strip: $4\,dy$.

Force against strip: $(2-y)w \cdot 4\,dy = 4w(2-y)\,dy$.

Summing from $y=0$ to $y=2$:

$$F = \int_0^2 4w(2-y)\,dy = 4w\left(2y - \frac{1}{2}y^2 \right)\Big|_0^2 = 8w\,\text{N}.$$

31.

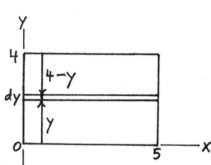

Pressure on strip: $(4-y)w$.

Area of strip: $5\,dy$.

Force against strip: $(4-y)w \cdot 5\,dy$.

Summing from $y=0$ to $y=4$:

$$F = \int_0^4 (4-y)w \cdot 5\,dy = \int_0^4 5w(4-y)\,dy = 5w\left(4y - \frac{1}{2}y^2 \right)\Big|_0^4 = 40w\,\text{N}.$$

33.

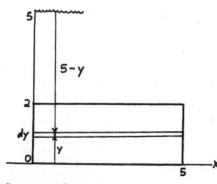

Pressure on strip: $(5-y)w$.

Area of strip: $5\,dy$.

Force against strip:

$\qquad (5-y)w \cdot 5\,dy = 5w(5-y)\,dy$.

Summing from $y=0$ to $y=2$:

$$F = \int_0^2 5w(5-y)\,dy = 5w\left(5y - \frac{1}{2}y^2 \right)\Big|_0^2 = 40w\,\text{N}.$$

35.

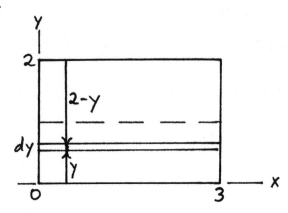

Pressure on strip: $(2 - y)w$.

Area of strip: $3\,dy$.

Force against strip: $(2 - y)w \cdot 3\,dy = 3w(2 - y)\,dy$.

Summing from $y = 0$ to $y = 1$ (lower half):

$$F = \int_0^1 3w(2 - y)\,dy = 3w\left(2y - \frac{1}{2}y^2\right)\Big|_0^1 = \frac{9}{2}w\,\text{N}.$$

37.

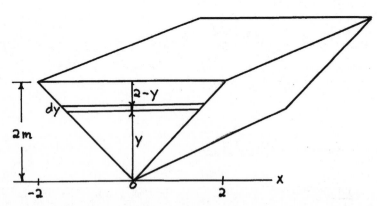

Since the slope of the line on the right is 1, the equation is $y = x$.

Area of strip: $2x\,dy = 2y\,dy$.

Pressure on strip: $(2 - y)w$.

Force against strip: $(2 - y)w \cdot 2y\,dy = 2w(2y - y^2)\,dy$.

Summing from $y = 0$ to $y = 2$, we get

$$F = \int_0^2 2w(2y - y^2)\,dy = 2w\left(y^2 - \frac{1}{3}y^3\right)\Big|_0^2 = 2w\left(4 - \frac{8}{3}\right) = \frac{8w}{3}\,\text{N}.$$

39. We find the force against one side and double the result.

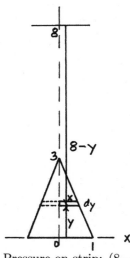

The slope of the line in the first quadrant: $m = -3$.
Thus
$$y = mx + b$$
$$y = -3x + 3$$
and
$$x = 1 - \frac{1}{3}y.$$

Pressure on strip: $(8 - y)w$.

Area of strip: $2x\,dy = 2\left(1 - \frac{1}{3}y\right)\,dy$.

Force against strip: $(8 - y)w \cdot 2\left(1 - \frac{1}{3}y\right)\,dy$.

Summing from $y = 0$ to $y = 3$,
$$F = \int_0^3 (8 - y)w \cdot 2\left(1 - \frac{1}{3}y\right)\,dy = 2w \int_0^3 (8 - y)\left(1 - \frac{1}{3}y\right)\,dy = 21w\,\text{N}.$$
So the force against both sides is $42w\,\text{N}$.

41.

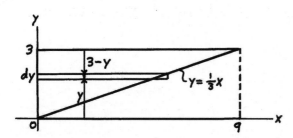

Equation of line: $y = \frac{1}{3}x$, or $x = 3y$.

Area of strip: $x\,dy = 3y\,dy$.

Pressure on strip: $(3 - y)w$.

Force against strip: $(3 - y)w \cdot 3y\,dy$.

Summing from $y = 0$ to $y = 3$:
$$\begin{aligned}
F &= \int_0^3 (3 - y)w \cdot 3y\,dy = 3w \int_0^3 (3y - y^2)\,dy = 3w\left(3\frac{y^2}{2} - \frac{y^3}{3}\right)\Bigg|_0^3 \\
&= 3w\left(\frac{3^3}{2} - \frac{3^3}{3}\right) = 3^4 w\left(\frac{1}{2} - \frac{1}{3}\right) = 81w\left(\frac{1}{6}\right) = \frac{27w}{2}\,\text{N}.
\end{aligned}$$

43.

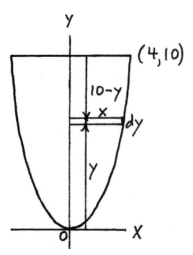

Form of the parabola: $x^2 = 4py$.

From the given information, the point $(4, 10)$ lies on the curve:

$$16 = 4p(10) \quad \text{or} \quad 4p = \frac{16}{10} = \frac{8}{5}.$$

Equation: $x^2 = \frac{8}{5}y$ or $x = \sqrt{\frac{8}{5}y}$.

Pressure on strip: $(10 - y)w$.

Area of strip: $2x\,dy = 2\sqrt{\frac{8}{5}y}\,dy$.

Force against strip: $(10 - y)w \cdot 2\sqrt{\frac{8}{5}y}\,dy$.

$$F = \int_0^{10} (10 - y)w \cdot 2\sqrt{\frac{8}{5}y}\,dy = \frac{640w}{3} \text{ N}.$$

45.

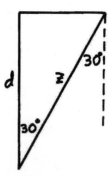

If the end of the trough is tilted, then a point which is z units from the top of the trough is at depth d, as shown in the diagram.

It follows that

$$d = z \cos 30° = \frac{\sqrt{3}}{2}z.$$

The other calculations follow Exercise 37:

Pressure on strip: $\frac{\sqrt{3}}{2}(2 - y)w$.

Force against strip: $\frac{\sqrt{3}}{2}(2 - y)w \cdot 2y\,dy$.

$$F = \int_0^2 \frac{\sqrt{3}}{2}(2 - y)w \cdot 2y\,dy = \frac{\sqrt{3}}{2}\int_0^2 2w(2y - y^2)\,dy = \frac{\sqrt{3}}{2}\left(\frac{8w}{3}\right) \text{ N by Exercise 37.}$$

Chapter 5 Review

1. Taking the upward direction to be positive, $g = -32$. So

$$v = -32t \qquad v = 0 \text{ when } t = 0$$

and

$$s = -16t^2 + k.$$

Since $s = 256$ when $t = 0$, we have

$$256 = 0 + k$$

so that

$$s = -16t^2 + 256.$$

To see how long it takes for the object to reach the ground, we let $s = 0$ and solve for t:

$$0 = -16t^2 + 256 \quad \text{or} \quad t = 4\,\text{s}.$$

Taking v to be $32t$, we get

$$v_{\text{av}} = \frac{1}{4-0} \int_0^4 32t\,dt = \frac{1}{4}(16t^2)\Big|_0^4 = 64\,\text{ft/s}.$$

3. $Ri_{\text{rms}}^2 = 10i_{\text{rms}}^2 = \dfrac{10}{4.0 - 0.0} \displaystyle\int_{0.0}^{4.0} (2.1 - 0.18t^{5/2})^2\,dt = 30.28 \approx 30\,\text{W}.$

5. See Exercise 29, Section 5.3.

7.

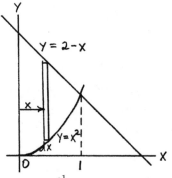

Volume of shell: $2\pi x(\text{height})\,dx$.

$$\begin{aligned}
V &= \int_0^1 2\pi x(2 - x - x^2)\,dx = 2\pi \int_0^1 (2x - x^2 - x^3)\,dx \\
&= 2\pi\left(x^2 - \frac{1}{3}x^3 - \frac{1}{4}x^4\right)\Big|_0^1 = 2\pi\left(1 - \frac{1}{3} - \frac{1}{4}\right) = \frac{5\pi}{6}.
\end{aligned}$$

9.

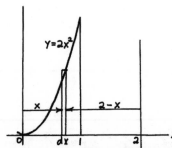

Radius of shell $= 2 - x$.

Volume of typical shell: $2\pi(\text{radius}) \cdot (\text{height}) \cdot (\text{thickness}) = 2\pi(2 - x)(2x^2)\,dx$.

$$V = \int_0^1 2\pi(2-x)(2x^2)\,dx = 4\pi\int_0^1(2-x)(x^2)\,dx = 4\pi\int_0^1(2x^2-x^3)\,dx$$

$$= 4\pi\left(\frac{2}{3}x^3 - \frac{1}{4}x^4\right)\bigg|_0^1 = 4\pi\left(\frac{2}{3} - \frac{1}{4}\right) = 4\pi\frac{8-3}{12} = \frac{5\pi}{3}.$$

11.

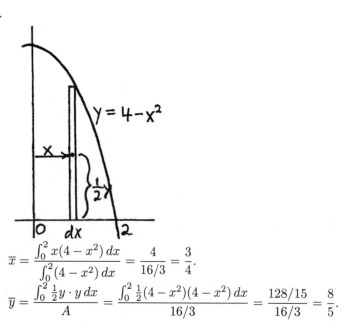

$$\bar{x} = \frac{\int_0^2 x(4-x^2)\,dx}{\int_0^2(4-x^2)\,dx} = \frac{4}{16/3} = \frac{3}{4}.$$

$$\bar{y} = \frac{\int_0^2 \frac{1}{2}y\cdot y\,dx}{A} = \frac{\int_0^2 \frac{1}{2}(4-x^2)(4-x^2)\,dx}{16/3} = \frac{128/15}{16/3} = \frac{8}{5}.$$

13.

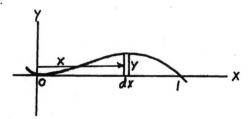

From $y = x^2 - x^3 = x^2(1-x) = 0$, the x-intercepts are 0 and 1.

$\bar{x}$: moment (with respect to y-axis) of typical element: $xy\,dx$.

$$\bar{x} = \frac{M_y}{A} = \frac{\int_0^1 x(x^2-x^3)\,dx}{\int_0^1(x^2-x^3)\,dx} = \frac{\frac{1}{4}x^4 - \frac{1}{5}x^5\big|_0^1}{\frac{1}{3}x^3 - \frac{1}{4}x^4\big|_0^1} = \frac{1/20}{1/12} = \frac{3}{5}.$$

$\bar{y}$: moment (with respect to x-axis) of typical element $\left(\frac{1}{2}y\right)(y\,dx)$.

$$\bar{y} = \frac{M_x}{A} = \frac{\int_0^1\left(\frac{1}{2}y\right)(y\,dx)}{A} = \frac{\frac{1}{2}\int_0^1(x^2-x^3)(x^2-x^3)\,dx}{1/12}$$

$$= 12\left(\frac{1}{2}\right)\int_0^1(x^4-2x^5+x^6)\,dx = 6\left(\frac{1}{5}x^5 - \frac{1}{3}x^6 + \frac{1}{7}x^7\right)\bigg|_0^1$$

$$= 6\left(\frac{1}{5} - \frac{1}{3} + \frac{1}{7}\right) = 6\left(\frac{21-35+15}{105}\right) = \frac{6}{105} = \frac{2}{35}.$$

15.

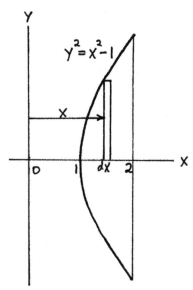

Volume of disk: $\pi(\text{radius})^2 \cdot (\text{thickness})$

$= \pi y^2 \, dx = \pi(x^2 - 1) \, dx.$

$$\overline{x} = \frac{\int_1^2 x \cdot \pi(x^2 - 1) \, dx}{\int_1^2 \pi(x^2 - 1) \, dx} = \frac{\frac{9}{4}\pi}{\frac{4}{3}\pi} = \frac{27}{16}.$$

$$\overline{y} = 0.$$

17.

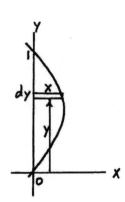

Moment of inertia (with respect to x-axis) of typical element:

$$y^2 \cdot \rho x \, dy = y^2 \cdot \rho(y - y^2) \, dy = \rho y^2 (y - y^2) \, dy.$$

$$\begin{aligned}
I_x &= \rho \int_0^1 y^2(y - y^2) \, dy = \rho \left(\frac{1}{4}y^4 - \frac{1}{5}y^5 \right) \Bigg|_0^1 \\
&= \rho \left(\frac{1}{4} - \frac{1}{5} \right) = \frac{\rho}{20}.
\end{aligned}$$

Mass: $\rho \int_0^1 (y - y^2) \, dy = \frac{1}{2}y^2 - \frac{1}{3}y^3 \Big|_0^1 \rho = \frac{\rho}{6}.$

$$R_x = \sqrt{\frac{\rho}{20} \frac{6}{\rho}} = \sqrt{\frac{3}{10}} = \frac{\sqrt{30}}{10}.$$

19.

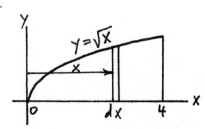

Volume of shell: $2\pi(\text{radius}) \cdot (\text{height}) \cdot (\text{thickness})$

$= 2\pi x \sqrt{x} \, dx.$

Mass of typical shell:

$\rho \cdot 2\pi x \sqrt{x} \, dx.$

Moment of inertia of typical shell:

$x^2 \cdot 2\pi \rho x \sqrt{x} \, dx.$

$$I_y = \int_0^4 x^2 \cdot 2\pi \rho x \sqrt{x} \, dx = \frac{2048\pi\rho}{9}.$$

Mass: $\int_0^4 2\pi \rho x \sqrt{x} \, dx = \frac{128\pi\rho}{5}.$

$$R_y = \sqrt{\frac{2048\pi\rho}{9} \cdot \frac{5}{128\pi\rho}} = \sqrt{\frac{80}{9}} = \sqrt{\frac{16 \cdot 5}{9}} = \frac{4\sqrt{5}}{3}.$$

21. By Hooke's law, $F = kx$, $2 = k \cdot \frac{1}{2}$, or $k = 4$. Thus $F = 4x$.

   (a) The spring is stretched $2\,\mathrm{ft}$ beyond its natural length. Thus
   $$W = \int_0^2 4x\,dx = 2x^2 \Big|_0^2 = 8\,\text{ft-lb.}$$

   (b) $W = \int_1^3 4x\,dx = 2x^2 \Big|_1^3 = 16\,\text{ft-lb.}$

23.

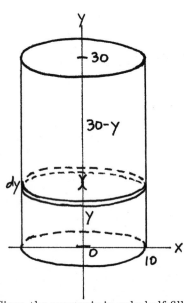

Volume of typical element:

$\pi(10^2)\,dy = 100\pi\,dy.$

Weight of typical element:

$w \cdot 100\pi\,dy.$

Work done in moving the typical element to the top:

$(30 - y)(w \cdot 100\pi\,dy).$

Since the reservoir is only half-filled, we sum from $y = 0$ to $y = 15$:

$$W = \int_0^{15} (30 - y)(w \cdot 100\pi\,dy) = 100\pi w \int_0^{15} (30 - y)\,dy = 100\pi w(337.5) = 1.06 \times 10^9 \,\text{J}$$
$(w = 10,000\,\text{N/m}^3).$

25.

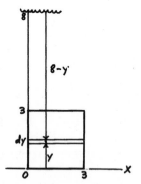

Pressure on strip: $(8 - y)w.$

Area of strip: $3\,dy.$

Force against strip: $(8 - y)w \cdot 3\,dy.$

Summing from $y = 0$ to $y = 3$, we get

$$\begin{aligned}
F &= \int_0^3 (8 - y)w \cdot 3\,dy = 3w \int_0^3 (8 - y)\,dy = 3w \left(8y - \frac{1}{2}y^2\right) \Big|_0^3 \\
&= 3w \left(24 - \frac{9}{2}\right) = \frac{117w}{2}\,\text{N} = \frac{117}{2}(10,000) = 585,000\,\text{N.}
\end{aligned}$$

Chapter 6

Derivatives of Transcendental Functions

6.1   Review of Trigonometry

3.

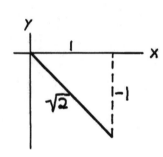

$$\tan(-45°) = -1$$

5.

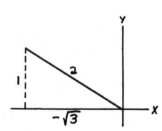

$$\sec 150° = \frac{2}{-\sqrt{3}} = -\frac{2\sqrt{3}}{3}$$

9.

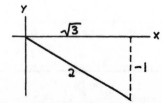

$$\csc(-30°) = -2$$

13.

$$\tan 90° = \frac{1}{0} \text{ (undefined)}$$

25. $60° = 60° \cdot \dfrac{\pi}{180°} = \dfrac{\pi}{3}$

29. $135° = 135° \cdot \dfrac{\pi}{180°} = \dfrac{3\pi}{4}$

33. $20° = 20° \cdot \dfrac{\pi}{180°} = \dfrac{\pi}{9}$

37. $\dfrac{\pi}{6} = \dfrac{\pi}{6} \cdot \dfrac{180°}{\pi} = 30°$

41. $\dfrac{11\pi}{10} = \dfrac{11\pi}{10} \cdot \dfrac{180°}{\pi} = 198°$

45. amplitude $= \dfrac{1}{3}$ (coefficient of $\sin 2x$)

   period $= \dfrac{2\pi}{2} = \pi$

   (see drawing in answer section)

49. $\cos\theta \tan\theta = \cos\theta \dfrac{\sin\theta}{\cos\theta} = \sin\theta$

51. $\tan\theta + \sec\theta = \dfrac{\sin\theta}{\cos\theta} + \dfrac{1}{\cos\theta} = \dfrac{\sin\theta + 1}{\cos\theta}$

53. Since $1 + \tan^2\theta = \sec^2\theta$, we have $\tan^2\theta - \sec^2\theta = -1$. So
   $$\dfrac{\tan^2\theta - \sec^2\theta}{\sec\theta} = \dfrac{-1}{\sec\theta} = -\cos\theta.$$

55. $\csc^2\theta - \cot^2\theta = \dfrac{1}{\sin^2\theta} - \dfrac{\cos^2\theta}{\sin^2\theta} = \dfrac{1 - \cos^2\theta}{\sin^2\theta} = \dfrac{\sin^2\theta}{\sin^2\theta} = 1$
   (This also follows from the identity $1 + \cot^2\theta = \csc^2\theta$.)

57.
$$\dfrac{1}{\sec^2\theta + \tan^2\theta + \cos^2\theta} = \dfrac{1}{(\sin^2\theta + \cos^2\theta) + \tan^2\theta} \qquad \text{rearranging}$$

$$= \dfrac{1}{1 + \tan^2\theta} \qquad \sin^2\theta + \cos^2\theta = 1$$

$$= \dfrac{1}{\sec^2\theta} \qquad 1 + \tan^2\theta = \sec^2\theta$$

$$= \cos^2\theta \qquad \cos\theta = \dfrac{1}{\sec\theta}$$

59. $\sec^2 \theta + \csc^2 \theta = \dfrac{1}{\cos^2 \theta} + \dfrac{1}{\sin^2 \theta} = \dfrac{1}{\cos^2 \theta}\dfrac{\sin^2 \theta}{\sin^2 \theta} + \dfrac{1}{\sin^2 \theta}\dfrac{\cos^2 \theta}{\cos^2 \theta}$

$\qquad\qquad\quad = \dfrac{\sin^2 \theta + \cos^2 \theta}{\sin^2 \theta \cos^2 \theta} = \dfrac{1}{\cos^2 \theta \sin^2 \theta}$

61. $\cos \theta \cot \theta + \sin \theta = \cos \theta \dfrac{\cos \theta}{\sin \theta} + \sin \theta \qquad \cot \theta = \dfrac{\cos \theta}{\sin \theta}$

$\qquad\qquad\qquad\qquad = \dfrac{\cos^2 \theta}{\sin \theta} + \dfrac{\sin^2 \theta}{\sin \theta} \qquad\quad \text{common denominator} = \sin \theta$

$\qquad\qquad\qquad\qquad = \dfrac{\cos^2 \theta + \sin^2 \theta}{\sin \theta}$

$\qquad\qquad\qquad\qquad = \dfrac{1}{\sin \theta} \qquad\qquad\qquad \cos^2 \theta + \sin^2 \theta = 1$

81. $\cos \left(x - \dfrac{\pi}{2}\right) = \cos x \cos \dfrac{\pi}{2} + \sin x \sin \dfrac{\pi}{2}$

$\qquad\qquad\qquad = \cos x \cdot 0 + \sin x \cdot 1 = \sin x$

85. Since $\sin 2\theta = 2 \sin \theta \cos \theta$, $2 \sin 5x \cos 5x = \sin(2 \cdot 5x) = \sin 10x$.

89. Since $\cos 2\theta = \cos^2 \theta - \sin^2 \theta$, $\cos^2 3x - \sin^2 3x = \cos(2 \cdot 3x) = \cos 6x$.

93. Since $\sin^2 \theta = \dfrac{1}{2}(1 - \cos 2\theta)$, $\sin^2 3x = \dfrac{1}{2}[1 - \cos(2 \cdot 3x)] = \dfrac{1}{2}(1 - \cos 6x)$.

97. $\sin^2 \dfrac{1}{2}x = \dfrac{1}{2}\left[1 - \cos\left(2 \cdot \dfrac{1}{2}x\right)\right] = \dfrac{1}{2}(1 - \cos x)$

6.2 Derivatives of Sine and Cosine Functions

1. $y = \cos 5x$

$y' = (-\sin 5x)\dfrac{d}{dx}(5x) = (-\sin 5x)(5) = -5 \sin 5x$.

3. $y = 2 \cos 4x$

$y' = 2(-\sin 4x)\dfrac{d}{dx}(4x) = 2(-\sin 4x)(4) = -8 \sin 4x$.

5. $y = \sin x^2$

$y' = (\cos x^2)\dfrac{d}{dx}(x^2) = (\cos x^2)(2x) = 2x \cos x^2$.

7. $s = 3 \cos t^3$

$\dfrac{ds}{dt} = 3(-\sin t^3)\dfrac{d}{dt}(t^3) = 3(-\sin t^3)(3t^2) = -9t^2 \sin t^3$.

9. $y = \sin 3x$

$y' = (\cos 3x)\dfrac{d}{dx}(3x) = 3 \cos 3x$.

11. $y = x \sin x$. By the product rule,
$$y' = x\frac{d}{dx}\sin x + (\sin x)\frac{d}{dx}(x) = x\cos x + \sin x.$$

13. $y = (\sin x)^2$. By the power rule,
$$y' = 2\sin x\frac{d}{dx}\sin x = 2\sin x\cos x \text{ and } 2\sin x\cos x = \sin 2x \text{ by the double-angle identity.}$$

15. $w = \cos^2 4v = (\cos 4v)^2$. By the power rule,
$$\begin{aligned}
\frac{dw}{dv} &= 2(\cos 4v)\frac{d}{dv}\cos 4v \\
&= 2(\cos 4v)(-\sin 4v)(4) \\
&= -8\cos 4v \sin 4v \\
&= -4(2\sin 4v \cos 4v) \\
&= -4\sin(2 \cdot 4v) \qquad \text{double-angle formula} \\
&= -4\sin 8v.
\end{aligned}$$

17. $y = \dfrac{\sin x}{x}$. By the quotient rule,
$$y' = \frac{x\left(\frac{d}{dx}\right)\sin x - \sin x\left(\frac{dx}{dx}\right)}{x^2} = \frac{x\cos x - \sin x}{x^2}.$$

19. $w = \cos(v^2 + 3)$
$$\frac{dw}{dv} = -\sin(v^2 + 3)\frac{d}{dv}(v^2 + 3) = -2v\sin(v^2 + 3).$$

21. $y = x\cos 2x$. By the product rule,
$$\begin{aligned}
y' &= x\frac{d}{dx}\cos 2x + \cos 2x\frac{d}{dx}(x) \\
&= x(-\sin 2x)\frac{d}{dx}(2x) + \cos 2x \cdot 1 \\
&= x(-\sin 2x)(2) + \cos 2x \\
&= -2x\sin 2x + \cos 2x \\
&= \cos 2x - 2x\sin 2x.
\end{aligned}$$

23. $y = 2x\sin(2x + 2)$. By the product rule,
$$\begin{aligned}
y' &= 2x\frac{d}{dx}\sin(2x + 2) + \sin(2x + 2)\frac{d}{dx}(2x) \\
&= 2x\cos(2x + 2) \cdot 2 + \sin(2x + 2) \cdot 2 \\
&= 4x\cos(2x + 2) + 2\sin(2x + 2).
\end{aligned}$$

25. $y = \sin\dfrac{1}{x}$
$$y' = \left(\cos\frac{1}{x}\right)\frac{d}{dx}\left(\frac{1}{x}\right) = \left(\cos\frac{1}{x}\right)\frac{d}{dx}(x^{-1}) = \left(\cos\frac{1}{x}\right)(-1x^{-2}) = -\frac{\cos(1/x)}{x^2}.$$

27. $y = \dfrac{x}{\cos x}$. By the quotient rule,
$$y' = \frac{\cos x\left(\frac{d}{dx}\right)(x) - x\left(\frac{d}{dx}\right)(\cos x)}{(\cos x)^2} = \frac{\cos x - x(-\sin x)}{\cos^2 x} = \frac{\cos x + x\sin x}{\cos^2 x}.$$

29. $y = \dfrac{x}{\sin 4x}$. By the quotient rule,

$$
\begin{aligned}
y' &= \frac{(\sin 4x)\left(\frac{d}{dx}\right)x - x\left(\frac{d}{dx}\right)\sin 4x}{(\sin 4x)^2} \\[2mm]
&= \frac{(\sin 4x)\cdot 1 - x\cos 4x\cdot 4}{\sin^2 4x} \\[2mm]
&= \frac{\sin 4x - 4x\cos 4x}{\sin^2 4x}.
\end{aligned}
$$

31. $N = \dfrac{\cos 2\theta}{3\theta}$. By the quotient rule,

$$
\begin{aligned}
\frac{dN}{d\theta} &= \frac{3\theta\left(\frac{d}{d\theta}\right)\cos 2\theta - \cos 2\theta\left(\frac{d}{d\theta}\right)(3\theta)}{(3\theta)^2} \\[2mm]
&= \frac{3\theta(-\sin 2\theta)(2) - \cos 2\theta(3)}{9\theta^2} \\[2mm]
&= -\frac{3(2\theta\sin 2\theta + \cos 2\theta)}{9\theta^2} \\[2mm]
&= -\frac{2\theta\sin 2\theta + \cos 2\theta}{3\theta^2}.
\end{aligned}
$$

33. $y = \sqrt{x}\sin x$. By the product rule,

$$
\begin{aligned}
y' &= x^{1/2}\frac{d}{dx}\sin x + \sin x\frac{d}{dx}(x^{1/2}) = x^{1/2}\cos x + (\sin x)\frac{1}{2}x^{-1/2} \\[2mm]
&= \sqrt{x}\cos x + \frac{1}{2\sqrt{x}}\sin x.
\end{aligned}
$$

35. $y = \cos^2 x^3 = (\cos x^3)^2$. By the power rule,

$$
y' = 2(\cos x^3)\frac{d}{dx}(\cos x^3) = 2(\cos x^3)(-\sin x^3)(3x^2) = -6x^2\cos x^3\sin x^3.
$$

37. $y = \sin x\cos x$. By the product rule,

$$
\begin{aligned}
y' &= (\sin x)\frac{d}{dx}\cos x + (\cos x)\frac{d}{dx}\sin x = \sin x(-\sin x) + \cos x(\cos x) \\[2mm]
&= -\sin^2 x + \cos^2 x = \cos^2 x - \sin^2 x = \cos 2x
\end{aligned}
$$

by the double-angle identity.

39. $y = \dfrac{\sin^3 x}{x} = \dfrac{(\sin x)^3}{x}$. By the quotient rule,

$$
\begin{aligned}
y' &= \frac{x\left(\frac{d}{dx}\right)(\sin x)^3 - (\sin x)^3\left(\frac{d}{dx}\right)x}{x^2} \\[2mm]
&= \frac{x\cdot 3(\sin x)^2\cos x - (\sin x)^3\cdot 1}{x^2} \\[2mm]
&= \frac{\sin^2 x(3x\cos x - \sin x)}{x^2}.
\end{aligned}
$$

41. $y = x\cos^2 3x = x(\cos 3x)^2$. By the product rule,

$$
y' = x\frac{d}{dx}(\cos 3x)^2 + (\cos 3x)^2\frac{d}{dx}(x) = x\frac{d}{dx}(\cos 3x)^2 + \cos^2 3x\cdot 1.
$$

Next, by the generalized power rule,

$$
\begin{aligned}
y' &= x\cdot 2(\cos 3x)\frac{d}{dx}\cos 3x + \cos^2 3x \\[2mm]
&= x\cdot 2(\cos 3x)(-\sin 3x\cdot 3) + \cos^2 3x \\[2mm]
&= -6x\cos 3x\sin 3x + \cos^2 3x.
\end{aligned}
$$

After factoring $\cos 3x$, we get $y' = \cos 3x(\cos 3x - 6x\sin 3x)$.

43. $y = \sin x$, $y' = \cos x$, $y'' = -\sin x$.

45. $y = \sin x$, $y' = \cos x$, $y'' = -\sin x$, $y^{(3)} = -\cos x$, and $y^{(4)} = \sin x$. Thus $\dfrac{d^4}{dx^4}\sin x = \sin x$.

47. $y = x\sin 2x$. By the product rule,

$$
\begin{aligned}
y' &= x\cos 2x \cdot 2 + \sin 2x \cdot 1 \\
&= 2x\cos 2x + \sin 2x \Big|_{x=\pi/4} \\
&= 2\left(\frac{\pi}{4}\right)\cos\frac{\pi}{2} + \sin\frac{\pi}{2} = 0 + 1 = 1.
\end{aligned}
$$

49. $i = 20.0\sin 4.0t$.

$$
\begin{aligned}
v = L\frac{di}{dt} &= 0.0050(20.0\cos 4.0t)\frac{d}{dt}(4.0t) \\
&= 0.0050(20.0\cos 4.0t)(4.0)\Big|_{t=0.20} \\
&= 0.28\,\text{V}.
\end{aligned}
$$

51. $s = \dfrac{1}{8}\sin(20\pi t)$.

$$
v = \frac{ds}{dt} = \frac{1}{8}\cos(20\pi t)(20\pi)\Big|_{t=0.1} = \frac{20\pi}{8} = \frac{5\pi}{2}\,\text{cm/s}.
$$

6.3   Other Trigonometric Functions

1. $y = \sec 5x$.

$y' = (\sec 5x\tan 5x)\dfrac{d}{dx}(5x) = (\sec 5x\tan 5x)(5) = 5\sec 5x\tan 5x$.

3. $y = 2\csc 3t$.

$\dfrac{dy}{dt} = 2\left(-\csc 3t\cot 3t\right)\dfrac{d}{dt}(3t) = 2\left(-\csc 3t\cot 3t\right)(3) = -6\csc 3t\cot 3t$.

5. $y = 3\cot 4x$.

$y' = 3\dfrac{d}{dx}\cot 4x = 3(-\csc^2 4x)\dfrac{d}{dx}(4x) = 3(-\csc^2 4x)(4) = -12\csc^2 4x$.

7. $z = 2\csc w^2$.

$\dfrac{dz}{dw} = 2\left(-\csc w^2\cot w^2\right)\dfrac{d}{dw}w^2 = -4w\csc w^2\cot w^2$.

9. $s = \tan 2t$.

$\dfrac{ds}{dt} = (\sec^2 2t)\dfrac{d}{dt}(2t) = (\sec^2 2t)(2) = 2\sec^2 2t$.

11. $y = x\cot 2x$. By the product rule,

$y' = x\dfrac{d}{dx}(\cot 2x) + \cot 2x\dfrac{d}{dx}(x) = x(-\csc^2 2x)(2) + \cot 2x \cdot 1 = \cot 2x - 2x\csc^2 2x$.

13. $y = \sec(x^3 + 1)$.

$\quad y' = [\sec(x^3 + 1)\tan(x^3 + 1)]\dfrac{d}{dx}(x^3 + 1) = 3x^2\sec(x^3 + 1)\tan(x^3 + 1)$.

15. $r = \dfrac{\sec\theta}{\theta}$. By the quotient rule,

$\quad \dfrac{dr}{d\theta} = \dfrac{\theta\left(\frac{d}{d\theta}\right)\sec\theta - \sec\theta\left(\frac{d}{d\theta}\right)\theta}{\theta^2} = \dfrac{\theta\sec\theta\tan\theta - \sec\theta\cdot 1}{\theta^2} = \dfrac{\sec\theta(\theta\tan\theta - 1)}{\theta^2}$.

17. $y = \cot\sqrt{3x} = \cot(3x)^{1/2}$.

$\quad y' = [-\csc^2(3x)^{1/2}]\dfrac{d}{dx}(3x)^{1/2}$. By the power rule,

$\quad \dfrac{d}{dx}(3x)^{1/2} = \dfrac{1}{2}(3x)^{-1/2}\dfrac{d}{dx}(3x) = \dfrac{1}{2}(3x)^{-1/2}\cdot 3$.

$\quad$ So $y' = [-\csc^2(3x)^{1/2}]\dfrac{1}{2}(3x)^{-1/2}(3) = \dfrac{-3\csc^2(3x)^{1/2}}{2(3x)^{1/2}} = -\dfrac{3\csc^2\sqrt{3x}}{2\sqrt{3x}}$.

19. $y = \sqrt{\tan 2x} = (\tan 2x)^{1/2}$. By the power rule,

$\quad y' = \dfrac{1}{2}(\tan 2x)^{-1/2}\dfrac{d}{dx}\tan 2x = \dfrac{1}{2}(\tan 2x)^{-1/2}(\sec^2 2x)(2) = \dfrac{\sec^2 2x}{\sqrt{\tan 2x}}$.

21. $y = 2\tan^4 4x = 2(\tan 4x)^4$. By the power rule,

$\quad y' = 2\cdot 4(\tan 4x)^3\dfrac{d}{dx}\tan 4x = 8(\tan 4x)^3\sec^2 4x\cdot 4 = 32\tan^3 4x\sec^2 4x$.

23. $r = \sqrt{\csc\omega^2} = (\csc\omega^2)^{1/2}$. By the power rule,

$\quad \dfrac{dr}{d\omega} = \dfrac{1}{2}(\csc\omega^2)^{-1/2}(-\csc\omega^2\cot\omega^2)(2\omega) = -\omega(\csc\omega^2)^{1/2}\cot\omega^2 = -\omega\sqrt{\csc\omega^2}\cot\omega^2$.

25. $T_1 = T_2^2\csc T_2$. By the product rule,

$\quad \begin{aligned}\dfrac{dT_1}{dT_2} &= T_2^2\dfrac{d}{dT_2}\csc T_2 + \csc T_2\dfrac{d}{dT_2}(T_2^2)\\ &= T_2^2(-\csc T_2\cot T_2) + \csc T_2(2T_2)\\ &= -T_2^2\csc T_2\cot T_2 + 2T_2\csc T_2\\ &= T_2\csc T_2(2 - T_2\cot T_2).\end{aligned}$

27. $y = \cos^2 x\cot x = (\cos x)^2\cot x$. By the product and power rules,

$\quad \begin{aligned}y' &= (\cos x)^2\dfrac{d}{dx}\cot x + \cot x\dfrac{d}{dx}(\cos x)^2 = \cos^2 x(-\csc^2 x) + \cot x\cdot 2(\cos x)(-\sin x)\\ &= -\dfrac{\cos^2 x}{\sin^2 x} - 2\left(\dfrac{\cos x}{\sin x}\right)\cos x\sin x = -\cot^2 x - 2\cos^2 x.\end{aligned}$

29. $y = \dfrac{1 + \tan x}{\sin x}$. By the quotient rule,

$\quad \begin{aligned}y' &= \dfrac{(\sin x)\left(\frac{d}{dx}\right)(1 + \tan x) - (1 + \tan x)\left(\frac{d}{dx}\right)(\sin x)}{(\sin x)^2}\\ &= \dfrac{(\sin x)(\sec^2 x) - (1 + \tan x)(\cos x)}{\sin^2 x}\\ &= \dfrac{\sin x\sec^2 x - \cos x - \tan x\cos x}{\sin^2 x}.\end{aligned}$

$\quad$ Since $\tan x\cos x = \dfrac{\sin x}{\cos x}\cos x = \sin x$, we get $y' = \dfrac{\sin x\sec^2 x - \cos x - \sin x}{\sin^2 x}$.

31. $y = x \sin(1 - x)^2$.

$$
\begin{aligned}
y' &= x\frac{d}{dx}\sin(1-x)^2 + \sin(1-x)^2\frac{d}{dx}(x) \\
&= x\cos(1-x)^2\frac{d}{dx}(1-x)^2 + \sin(1-x)^2 \cdot 1 \\
&= x\cos(1-x)^2 \cdot 2(1-x)(-1) + \sin(1-x)^2 \\
&= 2x(x-1)\cos(1-x)^2 + \sin(1-x)^2.
\end{aligned}
$$

33. $y = \dfrac{x^3}{\tan 3x}$. By the quotient rule,

$$
\begin{aligned}
y' &= \frac{\tan 3x\left(\frac{d}{dx}\right)x^3 - x^3\left(\frac{d}{dx}\right)\tan 3x}{(\tan 3x)^2} = \frac{\tan 3x(3x^2) - x^3(\sec^2 3x \cdot 3)}{\tan^2 3x} \\
&= \frac{3x^2\tan 3x - 3x^3\sec^2 3x}{\tan^2 3x} = \frac{3x^2(\tan 3x - x\sec^2 3x)}{\tan^2 3x}.
\end{aligned}
$$

35. $y = \dfrac{x}{(\csc 5x)^2}$.

$$
\begin{aligned}
y' &= \frac{(\csc 5x)^2\left(\frac{d}{dx}\right)(x) - x\left(\frac{d}{dx}\right)(\csc 5x)^2}{(\csc 5x)^4} \\
&= \frac{(\csc 5x)^2 \cdot 1 - x \cdot 2(\csc 5x)\left(\frac{d}{dx}\right)\csc 5x}{(\csc 5x)^4} \\
&= \frac{\csc^2 5x - 2x(\csc 5x)(-\csc 5x\cot 5x)(5)}{(\csc 5x)^4} \\
&= \frac{\csc^2 5x(1 + 10x\cot 5x)}{(\csc 5x)^4} \\
&= \frac{1 + 10x\cot 5x}{\csc^2 5x}.
\end{aligned}
$$

37. $y = \dfrac{\csc 2x^2}{4x}$. By the quotient rule,

$$
\begin{aligned}
y' &= \frac{4x\left(\frac{d}{dx}\right)\csc 2x^2 - \csc 2x^2\left(\frac{d}{dx}\right)(4x)}{(4x)^2} = \frac{4x(-\csc 2x^2\cot 2x^2)(4x) - \csc 2x^2 \cdot 4}{4^2x^2} \\
&= -\frac{4x^2\csc 2x^2\cot 2x^2 + \csc 2x^2}{4x^2}.
\end{aligned}
$$

39. $y = \dfrac{\cos 3x}{1 - \cot x^2}$.

$$
\begin{aligned}
y' &= \frac{(1 - \cot x^2)\left(\frac{d}{dx}\right)(\cos 3x) - \cos 3x\left(\frac{d}{dx}\right)(1 - \cot x^2)}{(1 - \cot x^2)^2} \\
&= \frac{(1 - \cot x^2)(-\sin 3x)(3) - \cos 3x(\csc^2 x^2 \cdot 2x)}{(1 - \cot x^2)^2} \\
&= \frac{-3\sin 3x + 3\sin 3x\cot x^2 - 2x\cos 3x\csc^2 x^2}{(1 - \cot x^2)^2}.
\end{aligned}
$$

41. By the power rule, $\dfrac{d}{dx}\tan^3 x = (3\tan^2 x)\dfrac{d}{dx}\tan x = 3\tan^2 x\sec^2 x$.

Thus

$$
\begin{aligned}
\frac{d}{dx}\left(\frac{1}{3}\tan^3 x + \tan x\right) &= \frac{1}{3} \cdot 3\tan^2 x\sec^2 x + \sec^2 x \\
&= \sec^2 x(\tan^2 x + 1) \qquad\qquad \text{facoring} \\
&= \sec^2 x\sec^2 x \qquad\qquad\qquad 1 + \tan^2 x = \sec^2 x \\
&= \sec^4 x.
\end{aligned}
$$

43. $y = \sec 4x$.

$y' = 4\sec 4x\tan 4x$.

$y'' = 4\left[\sec 4x(\sec^2 4x)(4) + \tan 4x(\sec 4x\tan 4x)(4)\right] = 16\sec 4x(\sec^2 4x + \tan^2 4x)$.

45. $y = x \tan x$. By the product rule, $y' = x \sec^2 x + \tan x$. By the product and power rules,

$$
\begin{aligned}
y'' &= x\frac{d}{dx}(\sec x)^2 + (\sec^2 x)\cdot 1 + \sec^2 x = x(2\sec x)(\sec x \tan x) + \sec^2 x + \sec^2 x \\
&= 2x\sec^2 x \tan x + 2\sec^2 x = 2\sec^2 x(x\tan x + 1).
\end{aligned}
$$

47. $y^2 = \tan x$.

$$
2y\frac{dy}{dx} = \sec^2 x \quad\text{and}\quad \frac{dy}{dx} = \frac{1}{2y}\sec^2 x.
$$

49. Treating y as a function of x, we get $\dfrac{d}{dx}(y) = \dfrac{dy}{dx}$ and $\dfrac{d}{dx}(y^2) = 2y\dfrac{dy}{dx}$:

$$
\begin{aligned}
y^2 &= x\sec x \\
2y\frac{dy}{dx} &= x\sec x \tan x + \sec x \qquad\text{product rule} \\
\frac{dy}{dx} &= \frac{x\sec x \tan x + \sec x}{2y} \\
&= \frac{\sec x(x\tan x + 1)}{2y}.
\end{aligned}
$$

51. $y^2 = \sin(x + y^2)$.

$$
\begin{aligned}
2y\frac{dy}{dx} &= \cos(x + y^2)\frac{d}{dx}(x + y^2) \\
&= \cos(x + y^2)\left(1 + 2y\frac{dy}{dx}\right) \\
&= \cos(x + y^2) + 2y\cos(x + y^2)\frac{dy}{dx} \\
\frac{dy}{dx}\left[2y - 2y\cos(x + y^2)\right] &= \cos(x + y^2) \\
\frac{dy}{dx} &= \frac{\cos(x + y^2)}{2y - 2y\cos(x + y^2)}.
\end{aligned}
$$

53. $y = x\cot y^2$.

$$
\begin{aligned}
\frac{dy}{dx} &= x\frac{d}{dx}\cot y^2 + \cot y^2 \cdot 1 && \text{product rule} \\
\frac{dy}{dx} &= x(-\csc^2 y^2)\left(2y\frac{dy}{dx}\right) + \cot y^2 && \frac{d}{dx}y^2 = 2y\frac{dy}{dx} \\
\frac{dy}{dx} + (2xy\csc^2 y^2)\frac{dy}{dx} &= \cot y^2 \\
\frac{dy}{dx}(1 + 2xy\csc^2 y^2) &= \cot y^2 && \text{factoring } \frac{dy}{dx} \\
\frac{dy}{dx} &= \frac{\cot y^2}{1 + 2xy\csc^2 y^2}.
\end{aligned}
$$

55.

$$
\begin{aligned}
\cos y &= x^2 y - 2x \\
-\sin y\frac{dy}{dx} &= x^2\frac{dy}{dx} + 2xy - 2 \\
2 - 2xy &= x^2\frac{dy}{dx} + \sin y\frac{dy}{dx} \\
\frac{dy}{dx} &= \frac{2 - 2xy}{x^2 + \sin y}.
\end{aligned}
$$

57. Slope of tangent line:

$$\frac{dy}{dx} = 2(-\csc^2 2x) \cdot 2 = -4\csc^2 2x \Big|_{x=\pi/8} = -4\left(\csc \frac{\pi}{4}\right)^2 = -4(\sqrt{2})^2 = -8.$$

Slope of normal line: $-\dfrac{1}{-8} = \dfrac{1}{8}$.   negative reciprocal

6.4   Inverse Trigonometric Functions

19. Let $\theta = \text{Arctan}\, 2 = \text{Arctan}\, \dfrac{2}{1}$. We place 2 on the side opposite θ and 1 on the side adjacent.
    Since $\sqrt{2^2 + 1^2} = \sqrt{5}$, the length of the hypotenuse, we get

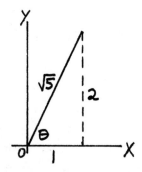

$$\sin(\text{Arctan}\, 2) = \sin \theta = \frac{2}{\sqrt{5}} = \frac{2\sqrt{5}}{5}.$$

21. Let $\theta = \text{Arctan}\, 6 = \text{Arctan}\, \dfrac{6}{1}$. We place 6 on the side opposite θ and 1 on the side adjacent.
    Since $\sqrt{6^2 + 1^2} = \sqrt{37}$, the length of the hypotenuse, we get

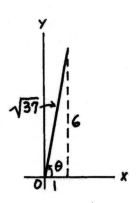

$$\cos(\text{Arctan}\, 6) = \cos \theta = \frac{1}{\sqrt{37}} = \frac{\sqrt{37}}{37}.$$

23. Let $\theta = \text{Arcsin}\, \dfrac{2}{3}$. So we place 2 on the side opposite θ and 3 on the hypotenuse.
    From $x^2 + 2^2 = 3^2$, we get $x = \sqrt{5}$.

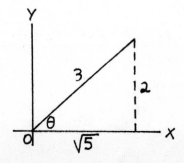

$$\tan\left(\text{Arcsin}\, \frac{2}{3}\right) = \tan \theta = \frac{2}{\sqrt{5}} = \frac{2\sqrt{5}}{5}.$$

25. Let $\theta = \text{Arcsin}\left(-\dfrac{1}{3}\right)$, where θ is between 0 and $-\pi/2$. So we place -1 on the side opposite (see figure) and 3 on the hypotenuse. From $x^2 + (-1)^2 = 3^2$, we get $x = 2\sqrt{2}$.

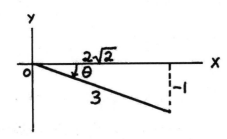

$$\sec\left[\text{Arcsin}\left(-\frac{1}{3}\right)\right] = \sec\theta = \frac{3}{2\sqrt{2}} = \frac{3\sqrt{2}}{4}.$$

27.

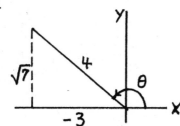

$$\tan\left[\text{Arccos}\left(-\frac{3}{4}\right)\right] = \tan\theta = -\frac{\sqrt{7}}{3}.$$

29.

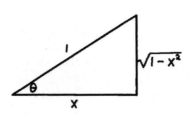

Let $\theta = \text{Arccos}\dfrac{x}{1}$. By the Pythagorean theorem, the length of the remaining side is $\sqrt{1-x^2}$. Thus $\sin(\text{Arccos}\,x) = \sin\theta = \sqrt{1-x^2}$.

31.

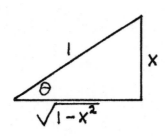

Let $\theta = \text{Arcsin}\dfrac{x}{1}$. By the Pythagorean theorem, the length of the remaining side is $\sqrt{1-x^2}$. $\csc(\text{Arcsin}\,x) = \csc\theta = \dfrac{1}{x}$.

33.

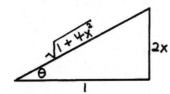

Let $\theta = \text{Arctan}\dfrac{2x}{1}$.
Length of hypotenuse:
$\sqrt{1^2 + (2x)^2} = \sqrt{1+4x^2}$.
$\cos(\text{Arctan}\,2x) = \cos\theta = \dfrac{1}{\sqrt{1+4x^2}}$.

35.

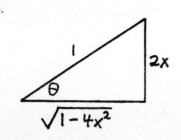

Let $\theta = \text{Arcsin}\dfrac{2x}{1}$. By the Pythagorean theorem, the length of the remaining side is $\sqrt{1-4x^2}$.
$\cot\left(\text{Arcsin}\dfrac{2x}{1}\right) = \cot\theta = \dfrac{\sqrt{1-4x^2}}{2x}$.

37.

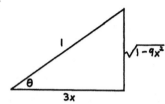

Let $\theta = \text{Arccos}\,\dfrac{3x}{1}$. By the Pythagorean theorem, the length of the remaining side is $\sqrt{1 - 9x^2}$. Thus

$$\sin(\text{Arccos}\,3x) = \sin\theta = \frac{\sqrt{1 - 9x^2}}{1}.$$

39. $\begin{aligned} y &= 2\text{Arctan}\,\frac{x}{2} \\[1mm] \frac{y}{2} &= \text{Arctan}\,\frac{x}{2} \\[1mm] \frac{x}{2} &= \tan\frac{y}{2} \\[1mm] x &= 2\tan\frac{y}{2}. \end{aligned}$

41. $\begin{aligned} y &= 1 + \sin x \\[1mm] \sin x &= y - 1 \\[1mm] x &= \text{Arcsin}\,(y - 1). \end{aligned}$

43. $\begin{aligned} y &= 1 + \cos 3x \\[1mm] \cos 3x &= y - 1 \\[1mm] 3x &= \text{Arccos}\,(y - 1) \\[1mm] x &= \frac{1}{3}\text{Arccos}\,(y - 1). \end{aligned}$

45. $\begin{aligned} y &= 3\tan 4x + 1 \\[1mm] y - 1 &= 3\tan 4x \\[1mm] \frac{y - 1}{3} &= \tan 4x \\[1mm] \tan 4x &= \frac{y - 1}{3} \\[1mm] 4x &= \text{Arctan}\,\frac{y - 1}{3} \\[1mm] x &= \frac{1}{4}\text{Arctan}\,\frac{y - 1}{3}. \end{aligned}$

6.5   Derivatives of Inverse Trigonometric Functions

1. $y = \text{Arctan}\,3x$.
$$y' = \frac{1}{1 + (3x)^2}\frac{d}{dx}(3x) = \frac{3}{1 + 9x^2}.$$

3. $y = \text{Arccos}\,5x$.
$$y' = -\frac{1}{\sqrt{1 - (5x)^2}}\frac{d}{dx}(5x) = \frac{-5}{\sqrt{1 - 25x^2}}.$$

5. $s = \text{Arctan}\,2t^2$.
$$\frac{ds}{dt} = \frac{1}{1 + (2t^2)^2}\frac{d}{dt}(2t^2) = \frac{4t}{1 + 4t^4}.$$

7. $u = \text{Arcsin } 3v^2$.
$$\frac{du}{dv} = \frac{1}{\sqrt{1 - (3v^2)^2}} \frac{d}{dv}(3v^2) = \frac{6v}{\sqrt{1 - 9v^4}}.$$

9. $y = \text{Arcsin } 2w$.
$$\frac{dy}{dw} = \frac{1}{\sqrt{1 - (2w)^2}} \frac{d}{dw}(2w) = \frac{2}{\sqrt{1 - 4w^2}}.$$

11. $v_1 = \text{Arccos } v_2^2$.
$$\frac{dv_1}{dv_2} = -\frac{1}{\sqrt{1 - (v_2^2)^2}} \frac{d}{dv_2}(v_2^2) = -\frac{2v_2}{\sqrt{1 - v_2^4}}.$$

13. $y = \text{Arcsin } 2x^2$.
$$y' = \frac{1}{\sqrt{1 - (2x^2)^2}} \frac{d}{dx}(2x^2) = \frac{4x}{\sqrt{1 - 4x^4}}.$$

15. $y = \text{Arctan } 7x$.
$$y' = \frac{1}{1 + (7x)^2} \frac{d}{dx}(7x) = \frac{7}{1 + 49x^2}.$$

17. $y = x\,\text{Arctan } x$. By the product rule,
$$y' = x\frac{d}{dx}\text{Arctan } x + \text{Arctan } x \cdot 1 = \frac{x}{1 + x^2} + \text{Arctan } x.$$

19. $y = x\,\text{Arccos } x^2$. By the product rule,
$$\begin{aligned}
y' &= x\frac{d}{dx}(\text{Arccos } x^2) + (\text{Arccos } x^2)\frac{d}{dx}(x) \\
&= x\left(-\frac{2x}{\sqrt{1 - (x^2)^2}}\right) + \text{Arccos } x^2 \cdot 1 \\
&= -\frac{2x^2}{\sqrt{1 - x^4}} + \text{Arccos } x^2.
\end{aligned}$$

21. $r = \theta\,\text{Arcsin } 3\theta$. By the product rule,
$$\begin{aligned}
\frac{dr}{d\theta} &= \theta\frac{d}{d\theta}\text{Arcsin } 3\theta + \text{Arcsin } 3\theta \cdot 1 \\
&= \frac{\theta}{\sqrt{1 - (3\theta)^2}} \frac{d}{d\theta}(3\theta) + \text{Arcsin } 3\theta \\
&= \frac{3\theta}{\sqrt{1 - 9\theta^2}} + \text{Arcsin } 3\theta.
\end{aligned}$$

23. $R = 2V\,\text{Arctan } 3V$. By the product rule,
$$\begin{aligned}
\frac{dR}{dV} &= 2V\frac{d}{dV}\text{Arctan } 3V + (\text{Arctan } 3V)\frac{d}{dV}(2V) \\
&= 2V\frac{3}{1 + (3V)^2} + 2\text{Arctan } 3V \\
&= \frac{6V}{1 + 9V^2} + 2\text{Arctan } 3V.
\end{aligned}$$

25. $y = \dfrac{\text{Arcsin}\,x}{x}$. By the quotient rule,

$$y' = \frac{x\frac{d}{dx}\text{Arcsin}\,x - \text{Arcsin}\,x \cdot 1}{x^2} = \frac{\dfrac{x}{\sqrt{1-x^2}} - \text{Arcsin}\,x}{x^2}.$$

Now simplify the complex fraction by multiplying numerator and denominator by $\sqrt{1-x^2}$:

$$\frac{\dfrac{x}{\sqrt{1-x^2}} - \text{Arcsin}\,x}{x^2} \cdot \frac{\sqrt{1-x^2}}{\sqrt{1-x^2}} = \frac{x - \sqrt{1-x^2}\,\text{Arcsin}\,x}{x^2\sqrt{1-x^2}}.$$

27. $y = \text{Arccos}\,\sqrt{1-x}$.

$$\begin{aligned}
y' &= -\frac{1}{\sqrt{1-(\sqrt{1-x})^2}}\frac{d}{dx}(1-x)^{1/2} = -\frac{1}{\sqrt{1-(1-x)}} \cdot \frac{1}{2}(1-x)^{-1/2} \cdot (-1) \\
&= \frac{1}{\sqrt{x}} \cdot \frac{1}{2(1-x)^{1/2}} = \frac{1}{2\sqrt{x(1-x)}} = \frac{1}{2\sqrt{x-x^2}}.
\end{aligned}$$

29. $y = x^{1/2}\text{Arcsin}\,x$. By the product rule,

$$\begin{aligned}
y' &= x^{1/2}\frac{d}{dx}\text{Arcsin}\,x + (\text{Arcsin}\,x)\frac{d}{dx}(x^{1/2}) = x^{1/2}\frac{1}{\sqrt{1-x^2}} + (\text{Arcsin}\,x)\left(\frac{1}{2}x^{-1/2}\right) \\
&= \frac{\sqrt{x}}{\sqrt{1-x^2}} + \frac{\text{Arcsin}\,x}{2\sqrt{x}}.
\end{aligned}$$

31. $y = (\text{Arcsin}\,x)^2$. By the power rule,

$$y' = 2(\text{Arcsin}\,x)\frac{1}{\sqrt{1-x^2}}.$$

33. $y = (\text{Arccos}\,x)^{1/2}$. By the power rule,

$$\begin{aligned}
y' &= \frac{1}{2}(\text{Arccos}\,x)^{-1/2}\left(-\frac{1}{\sqrt{1-x^2}}\right) = -\frac{1}{2}(\text{Arccos}\,x)^{-1/2}(1-x^2)^{-1/2} \\
&= -\frac{1}{2}\left[(1-x^2)\text{Arccos}\,x\right]^{-1/2}.
\end{aligned}$$

35. $y = \dfrac{\text{Arctan}\,x}{x}$. By the quotient rule,

$$y' = \frac{x\frac{d}{dx}\text{Arctan}\,x - (\text{Arctan}\,x)\frac{d}{dx}(x)}{x^2} = \frac{x \cdot \frac{1}{1+x^2} - \text{Arctan}\,x}{x^2}.$$

Multiply numerator and denominator by $1 + x^2$:

$$y' = \frac{x - (1+x^2)\text{Arctan}\,x}{x^2(1+x^2)}.$$

37. $y = \dfrac{2x}{\text{Arcsin}\,x^2}$. By the quotient rule,

$$y' = \frac{\text{Arcsin}\,x^2\frac{d}{dx}(2x) - 2x\frac{d}{dx}\text{Arcsin}\,x^2}{(\text{Arcsin}\,x^2)^2} = \frac{\text{Arcsin}\,x^2 \cdot 2 - 2x \cdot \dfrac{2x}{\sqrt{1-x^4}}}{(\text{Arcsin}\,x^2)^2}.$$

Now multiply numerator and denominator by $\sqrt{1-x^4}$ to get:

$$y' = \frac{2\sqrt{1-x^4}\,\text{Arcsin}\,x^2 - 4x^2}{\sqrt{1-x^4}(\text{Arcsin}\,x^2)^2}.$$

39.

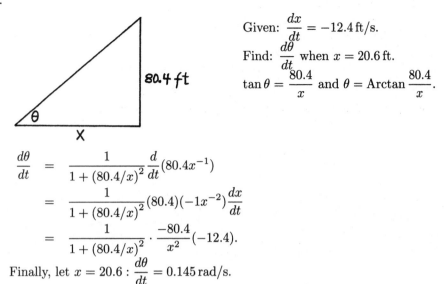

Given: $\dfrac{dx}{dt} = -12.4\,\text{ft/s}$.

Find: $\dfrac{d\theta}{dt}$ when $x = 20.6\,\text{ft}$.

$\tan\theta = \dfrac{80.4}{x}$ and $\theta = \text{Arctan}\,\dfrac{80.4}{x}$.

$$
\begin{aligned}
\frac{d\theta}{dt} &= \frac{1}{1+(80.4/x)^2}\,\frac{d}{dt}(80.4x^{-1}) \\[2mm]
&= \frac{1}{1+(80.4/x)^2}(80.4)(-1x^{-2})\frac{dx}{dt} \\[2mm]
&= \frac{1}{1+(80.4/x)^2}\cdot\frac{-80.4}{x^2}(-12.4).
\end{aligned}
$$

Finally, let $x = 20.6:\ \dfrac{d\theta}{dt} = 0.145\,\text{rad/s}$.

41. Let $y = \text{Arccot}\,u$, so that $u = \cot y$. Then $\dfrac{du}{dx} = -\csc^2 y\,\dfrac{dy}{dx}$.

$$
\frac{dy}{dx} = -\frac{1}{\csc^2 y}\frac{du}{dx} = -\frac{1}{1+\cot^2 y}\frac{du}{dx} = -\frac{1}{1+u^2}\frac{du}{dx}.
$$

6.6   Exponential and Logarithmic Functions

1. $3^3 = 27$; base: 3; exponent: 3. Thus $\log_3 27 = 3$.

5. $(32)^{-1/5} = \dfrac{1}{2}$; base: 32; exponent: $-\dfrac{1}{5}$. Thus $\log_{32}\dfrac{1}{2} = -\dfrac{1}{5}$.

9. $\log_{1/4}\dfrac{1}{16} = 2$; base: $\dfrac{1}{4}$; exponent: 2. Thus $\left(\dfrac{1}{4}\right)^2 = \dfrac{1}{16}$.

13. $\log_{1/3} x = 2$. Changing to exponential form, we get $\left(\dfrac{1}{3}\right)^2 = x$ or $x = \dfrac{1}{9}$.

15. $\log_3 81 = x$; exponential form, $3^x = 81$, so that $x = 4$ by inspection.

17.
$$
\begin{aligned}
\log_x \frac{1}{3} &= -\frac{1}{3} \\[1mm]
x^{-1/3} &= \frac{1}{3} \qquad \text{exponential form} \\[1mm]
\frac{1}{x^{1/3}} &= \frac{1}{3} \\[1mm]
x^{1/3} &= 3 \\[1mm]
\left(x^{1/3}\right)^3 &= 3^3 \\[1mm]
x &= 27.
\end{aligned}
$$

19. $\log_3 x = -2$; in exponential form, $x = 3^{-2} = \dfrac{1}{9}$.

21.

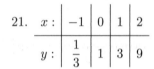

$x:$	-1	0	1	2
$y:$	$\dfrac{1}{3}$	1	3	9

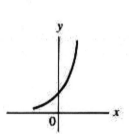

25. $y = \log_3 x$ or $x = 3^y$.

$y:$	-1	0	1	2
$x:$	$\dfrac{1}{3}$	1	3	9

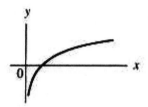

27. $\log_3 4 + \log_3 6 \;=\; \log_3 4 \cdot 6 = \log_3 24.$ $\log_a M + \log_a N = \log_a MN$

29. $\begin{aligned}
5\log_5 2 - 3\log_5 2 \;&=\; \log_5 2^5 - \log_5 2^3 \qquad && k\log_a M = \log_a M^k \\
&=\; \log_5 32 - \log_5 8 \\
&=\; \log_5 \frac{32}{8} && \log_a M - \log_a N = \log_a \frac{M}{N} \\
&=\; \log_5 4.
\end{aligned}$

31. $\begin{aligned}
\frac{1}{2}\log_b 3 - \frac{1}{2}\log_b 9 \;&=\; \log_b 3^{1/2} - \log_b 9^{1/2} \qquad && k\log_a M = \log_a M^k \\
&=\; \log_b \frac{3^{1/2}}{9^{1/2}} = \log_b \frac{\sqrt{3}}{3}. && \log_a M - \log_a N = \log_a \frac{M}{N}
\end{aligned}$

33. $\begin{aligned}
2\log_3 y + \frac{1}{3}\log_3 8 - 2\log_3 5 \;&=\; \log_3 y^2 + \log_3 8^{1/3} - \log_3 5^2 \qquad && k\log_a M = \log_a M^k \\
&=\; \log_3 \frac{y^2 \cdot 8^{1/3}}{5^2} && \text{properties (6.16) and (6.17)} \\
&=\; \log_3 \frac{2y^2}{25}.
\end{aligned}$

35. $\begin{aligned}
\log_3 27 \;&=\; \log_3 3^3 \\
&=\; 3\log_3 3 \qquad && \log_a M^k = k\log_a M \\
&=\; 3 \cdot 1 = 3. && \log_a a = 1
\end{aligned}$

37. $\begin{aligned}
\log_6 \sqrt{6x} \;&=\; \log_6 (6x)^{1/2} = \frac{1}{2}\log_6 6x \qquad && \log_a M^k = k\log_a M \\
&=\; \frac{1}{2}(\log_6 6 + \log_6 x) && \log_a MN = \log_a M + \log_a N \\
&=\; \frac{1}{2}(1 + \log_6 x). && \log_a a = 1
\end{aligned}$

39. $\log_5 \dfrac{1}{25x^2}$ $= \log_5 1 - \log_5 25x^2$ $\log_a \dfrac{M}{N} = \log_a M - \log_a N$

$= 0 - \log_5 (5x)^2$ $\log_a 1 = 0$

$= -2\log_5(5x)$ $\log_a M^k = k\log_a M$

$= -2(\log_5 5 + \log_5 x)$ $\log_a MN = \log_a M + \log_a N$

$= -2(1 + \log_5 x).$ $\log_a a = 1$

41. $\log_3 \dfrac{1}{\sqrt[3]{3x}}$ $= \log_3 \dfrac{1}{(3x)^{1/3}}$

$= \log_3 1 - \log_3 (3x)^{1/3}$ $\log_a \dfrac{M}{N} = \log_a M - \log_a N$

$= 0 - \dfrac{1}{3}\log_3(3x)$ $\log_a 1 = 0,\ \log_a M^k = k\log_a M$

$= -\dfrac{1}{3}(\log_3 3 + \log_3 x)$ $\log_a MN = \log_a M + \log_a N$

$= -\dfrac{1}{3}(1 + \log_3 x).$ $\log_a a = 1$

43. $\log_5 \dfrac{1}{\sqrt{y-2}} = \log_5 (y-2)^{-1/2} = -\dfrac{1}{2}\log_5(y-2).$

45. $\log_{10} \dfrac{x}{\sqrt{x+2}}$ $= \log_{10} \dfrac{x}{(x+2)^{1/2}}$

$= \log_{10} x - \log_{10}(x+2)^{1/2}$ $\log_a \dfrac{M}{N} = \log_a M - \log_a N$

$= \log_{10} x - \dfrac{1}{2}\log_{10}(x+2).$ $\log_a M^k = k\log_a M$

47. $\log_{10} \dfrac{\sqrt{x}}{x+1}$ $= \log_{10} x^{1/2} - \log_{10}(x+1)$ by (6.17)

$= \dfrac{1}{2}\log_{10} x - \log_{10}(x+1).$ by (6.18)

49. $3.62^x = 12.4$

$\log_{10}(3.62)^x = \log_{10} 12.4$

$x\log_{10} 3.62 = \log_{10} 12.4$

$x = \dfrac{\log_{10} 12.4}{\log_{10} 3.62} = 1.96.$

51. $(8.04)^x = 2.85$

$\log_{10}(8.04)^x = \log_{10} 2.85$

$x\log_{10} 8.04 = \log_{10} 2.85$

$x = \dfrac{\log_{10} 2.85}{\log_{10} 8.04} = 0.502.$

53. $(36.4)^x = 0.147$

$\log(36.4)^x = \log 0.147$

$x\log 36.4 = \log 0.147$

$x = \dfrac{\log 0.147}{\log 36.4} = -0.533.$

6.7   Derivative of the Logarithmic Function

1. $y = \ln 2x$.

   $$y' = \frac{1}{2x}\frac{d}{dx}(2x) = \frac{1}{2x}(2) = \frac{1}{x}.$$

3. $y = 4\ln 3x$.

   $$y' = 4\cdot\frac{1}{3x}\frac{d}{dx}(3x) = 4\cdot\frac{1}{3x}(3) = \frac{4}{x}.$$

5. $R = \ln s^2$.

   $$\frac{dR}{ds} = \frac{1}{s^2}\frac{d}{ds}(s^2) = \frac{1}{s^2}(2s) = \frac{2}{s}$$

   or

   $$R = \ln s^2 = 2\ln s; \text{ thus } \frac{dR}{ds} = \frac{2}{s}.$$

7. $y = 2\ln x^3 = 2\cdot 3\ln x$.

   $$y' = \frac{6}{x}.$$

9. $y = \log_{10} x^3 = 3\log_{10} x;\ y' = \frac{3}{x}\log_{10} e.$

11. $R_1 = \ln\sin R_2$.

   $$\frac{dR_1}{dR_2} = \frac{1}{\sin R_2}\frac{d}{dR_2}\sin R_2 = \frac{\cos R_2}{\sin R_2} = \cot R_2.$$

13. $y = x\ln x$. By the product rule,

   $$y' = x\frac{d}{dx}\ln x + \ln x\cdot 1 = x\frac{1}{x} + \ln x = 1 + \ln x.$$

15. $y = \ln\sqrt{x-1} = \ln(x-1)^{1/2} = \frac{1}{2}\ln(x-1)$.

   $$y' = \frac{1}{2(x-1)}.$$

17. $\quad y = \ln\dfrac{1}{\sqrt{x+2}} = \ln(x+2)^{-1/2}$

   $$= -\frac{1}{2}\ln(x+2). \qquad\qquad \log_a M^k = k\log_a M$$

   $$y' = -\frac{1}{2}\frac{1}{x+2} = -\frac{1}{2(x+2)}.$$

19. $z = 3\ln\sqrt[3]{t^2+1} = 3\cdot\frac{1}{3}\ln(t^2+1) = \ln(t^2+1)$.

   $$\frac{dz}{dt} = \frac{1}{t^2+1}\frac{d}{dt}(t^2+1) = \frac{2t}{t^2+1}.$$

21. $\quad y \;=\; \ln\dfrac{x^2}{x+1} = \ln x^2 - \ln(x+1)$ $\qquad\qquad \log_a \dfrac{M}{N} = \log_a M - \log_a N$

$\qquad\;\; =\; 2\ln x - \ln(x+1).$ $\qquad\qquad\qquad\;\; \log_a M^k = k\log_a M$

$\quad y' \;=\; 2\cdot\dfrac{1}{x} - \dfrac{1}{x+1}$

$\qquad\; =\; \dfrac{2}{x}\dfrac{x+1}{x+1} - \dfrac{1}{x+1}\dfrac{x}{x} = \dfrac{2x+2-x}{x(x+1)}$ $\qquad$ adding fractions

$\qquad\; =\; \dfrac{x+2}{x(x+1)}.$

23. $\quad y \;=\; \ln\dfrac{2x}{x^2+1} = \ln 2x - \ln(x^2+1).$ $\qquad \ln\dfrac{M}{N} = \ln M - \ln N$

$\quad y' = \dfrac{1}{x} - \dfrac{2x}{x^2+1} = \dfrac{1}{x}\dfrac{x^2+1}{x^2+1} - \dfrac{2x}{x^2+1}\dfrac{x}{x} = \dfrac{1-x^2}{x(x^2+1)}.$

25. $\quad y \;=\; \ln\dfrac{\sqrt{x}}{2-x^2} = \ln\dfrac{x^{1/2}}{2-x^2}$

$\qquad\;\; =\; \ln x^{1/2} - \ln(2-x^2)$ $\qquad\qquad\qquad\qquad \log_a \dfrac{M}{N} = \log_a M - \log_a N$

$\qquad\;\; =\; \dfrac{1}{2}\ln x - \ln(2-x^2).$ $\qquad\qquad\qquad\quad \log_a M^k = k\log_a M$

$\quad y' \;=\; \dfrac{1}{2}\cdot\dfrac{1}{x} - \dfrac{1}{2-x^2}\dfrac{d}{dx}(2-x^2)$

$\qquad\; =\; \dfrac{1}{2x} - \dfrac{-2x}{2-x^2} = \dfrac{1}{2x}\cdot\dfrac{2-x^2}{2-x^2} + \dfrac{2x}{2-x^2}\cdot\dfrac{2x}{2x}$

$\qquad\; =\; \dfrac{2-x^2+4x^2}{2x(2-x^2)} = \dfrac{2+3x^2}{2x(2-x^2)}.$

27. $\ln\dfrac{\cos x}{\sqrt{x}} = \ln\cos x - \dfrac{1}{2}\ln x.$

$\quad y' = \dfrac{1}{\cos x}(-\sin x) - \dfrac{1}{2x} = -\left(\tan x + \dfrac{1}{2x}\right) = -\dfrac{2x\tan x+1}{2x}.$

29. $\quad y \;=\; \ln\dfrac{\sec^2 x}{\sqrt{x+1}} = \ln\sec^2 x - \ln(x+1)^{1/2}$

$\qquad\;\; =\; 2\ln\sec x - \dfrac{1}{2}\ln(x+1).$

$\quad y' \;=\; 2\dfrac{\sec x\tan x}{\sec x} - \dfrac{1}{2(x+1)}$

$\qquad\; =\; 2\tan x - \dfrac{1}{2(x+1)} = \dfrac{4(x+1)\tan x-1}{2(x+1)}.$

31. $y = \ln^2 x = (\ln x)^2.$ By the power rule,

$\quad y' = 2(\ln x)\dfrac{d}{dx}\ln x = \dfrac{2\ln x}{x}.$

33. $y = (x+1)^{1/2}\ln x.$ By the product rule,

$\quad y' = (x+1)^{1/2}\dfrac{1}{x} + \dfrac{1}{2}(x+1)^{-1/2}\ln x = \dfrac{\sqrt{x+1}}{x} + \dfrac{\ln x}{2\sqrt{x+1}}.$

35. $s = (\sec\theta)\ln\theta.$ By the product rule,

$\quad \dfrac{ds}{d\theta} = \sec\theta\dfrac{d}{d\theta}\ln\theta + \ln\theta\dfrac{d}{d\theta}\sec\theta = \dfrac{\sec\theta}{\theta} + (\sec\theta\tan\theta)\ln\theta.$

37. $y = \ln\left[1 + (x^2 - 1)^{1/2}\right].$

$$y' = \frac{1}{1 + (x^2 - 1)^{1/2}}\frac{d}{dx}\left[1 + (x^2 - 1)^{1/2}\right]$$

$$= \frac{1}{1 + \sqrt{x^2 - 1}}\left[\frac{1}{2}(x^2 - 1)^{-1/2}(2x)\right] = \frac{x}{(1 + \sqrt{x^2 - 1})\sqrt{x^2 - 1}}$$

$$= \frac{x}{\sqrt{x^2 - 1} + (\sqrt{x^2 - 1})^2} = \frac{x}{x^2 - 1 + \sqrt{x^2 - 1}}.$$

39. $y = (\ln x)^2 \ln x^2 = (\ln x)^2 \cdot 2\ln x.$

$$y' = (\ln x)^2 \frac{d}{dx}(2\ln x) + (2\ln x)\frac{d}{dx}(\ln x)^2$$

$$= (\ln x)^2 \frac{2}{x} + (2\ln x)(2\ln x)\frac{1}{x}$$

$$= \frac{2}{x}\ln^2 x + \frac{4}{x}\ln^2 x = \frac{6}{x}\ln^2 x.$$

41. $y = \ln(\ln x).$

$$y' = \frac{1}{\ln x}\frac{d}{dx}(\ln x) = \frac{1}{\ln x}\frac{1}{x} = \frac{1}{x\ln x}.$$

43. $v_1 = \dfrac{\ln v_2}{v_2}.$

$$\frac{dv_1}{dv_2} = \frac{v_2 \cdot \frac{1}{v_2} - \ln v_2}{v_2^2} = \frac{1 - \ln v_2}{v_2^2}.$$

6.8 Derivative of the Exponential Function

1. $y = e^{4x}.$

$$y' = e^{4x}\frac{d}{dx}(4x) = 4e^{4x}.$$

3. $y = e^{x^2}.$

$$y' = e^{x^2}\frac{d}{dx}x^2 = 2xe^{x^2}.$$

5. $y = 2e^{-t^2}.$

$$\frac{dy}{dt} = 2e^{-t^2}\frac{d}{dt}(-t^2) = 2e^{-t^2}(-2t) = -4te^{-t^2}.$$

7. $y = e^{\tan x}.$

$$y' = e^{\tan x}\frac{d}{dx}\tan x = e^{\tan x}\sec^2 x.$$

9. $y = 3^{x^3}.$

$$y' = 3^{x^3}(\ln 3)\frac{d}{dx}(x^3) = 3^{x^3}(\ln 3)(3x^2).$$

11. $y = 4^{x^2}.$ By Formula (6.25),

$$y' = 4^{x^2}(\ln 4)\frac{d}{dx}x^2 = 2x(\ln 4)4^{x^2}.$$

13. $C = 2re^r$. By the product rule,

$$\frac{dC}{dr} = 2r\frac{d}{dr}e^r + e^r\frac{d}{dr}(2r) = 2re^r + 2e^r = 2(r+1)e^r.$$

15. $y = e^{\sin x}$.

$$y' = e^{\sin x}\frac{d}{dx}\sin x = e^{\sin x}\cos x.$$

17. $y = \sin e^x$.

$$y' = (\cos e^x)\frac{d}{dx}e^x = e^x\cos e^x.$$

19. $S = e^{2\omega}\sin\omega$. By the product rule,

$$\frac{dS}{d\omega} = e^{2\omega}\frac{d}{d\omega}\sin\omega + \sin\omega\frac{d}{d\omega}e^{2\omega} = e^{2\omega}\cos\omega + (\sin\omega)e^{2\omega}(2) = e^{2\omega}(\cos\omega + 2\sin\omega).$$

21. $y = \dfrac{\tan x}{e^{x^2}}$. By the quotient rule,

$$\begin{aligned}
y' &= \frac{e^{x^2}\frac{d}{dx}\tan x - \tan x\frac{d}{dx}e^{x^2}}{(e^{x^2})^2} = \frac{e^{x^2}\sec^2 x - (\tan x)e^{x^2}(2x)}{(e^{x^2})^2}\\
&= \frac{e^{x^2}(\sec^2 x - 2x\tan x)}{(e^{x^2})^2} = \frac{\sec^2 x - 2x\tan x}{e^{x^2}}.
\end{aligned}$$

23. $y = (\ln x)e^{\sec x}$. By the product rule,

$$\begin{aligned}
y' &= (\ln x)\frac{d}{dx}e^{\sec x} + e^{\sec x}\frac{d}{dx}\ln x = (\ln x)e^{\sec x}\sec x\tan x + e^{\sec x}\cdot\frac{1}{x}\\
&= e^{\sec x}\left(\sec x\tan x\ln x + \frac{1}{x}\right).
\end{aligned}$$

25. $y = \ln(\sin e^{2x})$.

$$\begin{aligned}
y' &= \frac{1}{\sin e^{2x}}\frac{d}{dx}(\sin e^{2x}) = \frac{1}{\sin e^{2x}}(\cos e^{2x})\frac{d}{dx}e^{2x}\\
&= \frac{1}{\sin e^{2x}}(\cos e^{2x})e^{2x}(2) = 2e^{2x}\frac{\cos e^{2x}}{\sin e^{2x}} = 2e^{2x}\cot e^{2x}.
\end{aligned}$$

27. $y = \dfrac{x}{e^{2x}}$. By the quotient rule,

$$y' = \frac{e^{2x}\cdot 1 - xe^{2x}\cdot 2}{(e^{2x})^2} = \frac{e^{2x}(1 - 2x)}{(e^{2x})^2} = \frac{1 - 2x}{e^{2x}}.$$

29. $y = \dfrac{\sin 2x}{e^x + 1}$. By the quotient rule,

$$\begin{aligned}
y' &= \frac{(e^x + 1)\frac{d}{dx}\sin 2x - \sin 2x\frac{d}{dx}(e^x + 1)}{(e^x + 1)^2} = \frac{(e^x + 1)\cos 2x\cdot 2 - \sin 2x\cdot e^x}{(e^x + 1)^2}\\
&= \frac{2\cos 2x + 2e^x\cos 2x - e^x\sin 2x}{(e^x + 1)^2}.
\end{aligned}$$

31. $\dfrac{d}{dx}\sinh x = \dfrac{d}{dx}\dfrac{1}{2}(e^x - e^{-x}) = \dfrac{1}{2}\left[e^x - e^{-x}(-1)\right] = \dfrac{1}{2}(e^x + e^{-x}) = \cosh x.$

$\dfrac{d}{dx}\cosh x = \dfrac{d}{dx}\dfrac{1}{2}(e^x + e^{-x}) = \dfrac{1}{2}(e^x - e^{-x}) = \sinh x.$

33. $y = \sinh 5x$.

$$y' = (\cosh 5x)\frac{d}{dx}(5x) = 5\cosh 5x.$$

35. $y = \cosh 2x^3$.

$$y' = (\sinh 2x^3)\frac{d}{dx}(2x^3) = 6x^2 \sinh 2x^3.$$

37. $y = x \cosh 3x$. By the product rule,

$$
\begin{aligned}
y' &= \cosh 3x \cdot \frac{d}{dx}(x) + x\frac{d}{dx}\cosh 3x = \cosh 3x \cdot 1 + x \sinh 3x \cdot \frac{d}{dx}(3x)\\
&= \cosh 3x + 3x \sinh 3x.
\end{aligned}
$$

39.
$$
\begin{aligned}
y &= x^{\sin x}\\
\ln y &= \ln x^{\sin x} = \sin x \ln x\\
\frac{1}{y}\frac{dy}{dx} &= \sin x \cdot \frac{1}{x} + \ln x \cos x\\
\frac{dy}{dx} &= y\left(\sin x \cdot \frac{1}{x} + \cos x \ln x\right)\\
&= x^{\sin x}\left(\sin x \cdot \frac{1}{x} + \cos x \ln x\right).
\end{aligned}
$$

Factoring out x^{-1},
$$
\begin{aligned}
\frac{dy}{dx} &= x^{\sin x}x^{-1}(\sin x + x \cos x \ln x)\\
&= x^{(\sin x - 1)}(\sin x + x \cos x \ln x).
\end{aligned}
$$

41.
$$
\begin{aligned}
y &= (\ln x)^x\\
\ln y &= \ln(\ln x)^x\\
&= x\ln(\ln x).\qquad \text{property (6.18)}\\
\frac{1}{y}\frac{dy}{dx} &= x\frac{d}{dx}\ln(\ln x) + \ln(\ln x)\cdot 1 = x\frac{1}{\ln x}\frac{1}{x} + \ln(\ln x)\\
&= \frac{1}{\ln x} + \ln(\ln x).\\
\frac{dy}{dx} &= y\left[\frac{1}{\ln x} + \ln(\ln x)\right],\ \text{where } y = (\ln x)^x.
\end{aligned}
$$

43.
$$
\begin{aligned}
y &= (\tan x)^x\\[4pt]
\ln y &= \ln(\tan x)^x = x \ln \tan x\\[4pt]
\frac{1}{y}\frac{dy}{dx} &= x \cdot \frac{1}{\tan x}\sec^2 x + \ln \tan x\\[4pt]
\frac{dy}{dx} &= y\left(\frac{1}{\tan x}\right)(x \sec^2 x + \tan x \ln \tan x)\\[4pt]
&= (\tan x)^x\left(\frac{1}{\tan x}\right)(x \sec^2 x + \tan x \ln \tan x)\\[4pt]
&= (\tan x)^{x-1}(x \sec^2 x + \tan x \ln \tan x).
\end{aligned}
$$

45. $i = \dfrac{dq}{dt} = \dfrac{d}{dt}\left(2.0\mathrm{e}^{-0.60t}\cos 10\pi t\right)$. By the product rule,

$$
\begin{aligned}
i &= 2.0\left[\mathrm{e}^{-0.60t}\frac{d}{dt}\cos 10\pi t + \cos 10\pi t \frac{d}{dt}\mathrm{e}^{-0.60t}\right]\\[4pt]
&= 2.0\left[\mathrm{e}^{-0.60t}(-\sin 10\pi t)(10\pi) + (\cos 10\pi t)\mathrm{e}^{-0.60t}(-0.60)\right].
\end{aligned}
$$

Substituting $t = 0.20\,\mathrm{s}$, we get
$$i = 2.0\left[0 + 1\cdot \mathrm{e}^{(-0.60)(0.20)}(-0.60)\right] = -1.1\,\mathrm{A}.$$

47. $p(t) = (1 + ae^{-kt})^{-1}$.

$$\frac{dp}{dt} = -1(1 + ae^{-kt})^{-2}[0 + ae^{-kt}(-k)] = \frac{ake^{-kt}}{(1 + ae^{-kt})^2}.$$

6.9 L'Hospital's Rule

1. $\lim\limits_{x \to -2} \dfrac{x^2 - 4}{x + 2} = \lim\limits_{x \to -2} \dfrac{\frac{d}{dx}(x^2 - 4)}{\frac{d}{dx}(x + 2)} = \lim\limits_{x \to -2} \dfrac{2x}{1} = -4.$

3. $\lim\limits_{x \to \infty} \dfrac{3x^2 - 4x}{2x^2 + 1} = \lim\limits_{x \to \infty} \dfrac{\frac{d}{dx}(3x^2 - 4x)}{\frac{d}{dx}(2x^2 + 1)} = \lim\limits_{x \to \infty} \dfrac{6x - 4}{4x} = \lim\limits_{x \to \infty} \dfrac{\frac{d}{dx}(6x - 4)}{\frac{d}{dx}(4x)} = \dfrac{6}{4} = \dfrac{3}{2}.$

5. $\lim\limits_{t \to 3} \dfrac{t^2 + t - 12}{t - 3} = \lim\limits_{t \to 3} \dfrac{\frac{d}{dt}(t^2 + t - 12)}{\frac{d}{dt}(t - 3)} = \lim\limits_{t \to 3} \dfrac{2t + 1}{1} = 7.$

7. $\lim\limits_{m \to 0} \dfrac{\sin 6m}{m} = \lim\limits_{m \to 0} \dfrac{\frac{d}{dm}\sin 6m}{\frac{d}{dm}(m)} = \lim\limits_{m \to 0} \dfrac{6 \cos 6m}{1} = 6.$

9. $\lim\limits_{x \to 0} \dfrac{\tan 3x}{1 - \cos x} = \lim\limits_{x \to 0} \dfrac{3 \sec^2 3x}{\sin x} = \infty$ (limit does not exist).

11. $\lim\limits_{x \to 0} \dfrac{e^x - e^{-x}}{\sin x} = \lim\limits_{x \to 0} \dfrac{e^x - e^{-x}(-1)}{\cos x} = \dfrac{1 + 1}{1} = 2.$

13. $\lim\limits_{x \to 0} \dfrac{1 - e^x}{2x} = \lim\limits_{x \to 0} \dfrac{-e^x}{2} = -\dfrac{1}{2}.$

15. $\lim\limits_{x \to \infty} \dfrac{x + \ln x}{x \ln x} = \lim\limits_{x \to \infty} \dfrac{1 + 1/x}{x \cdot \frac{1}{x} + \ln x \cdot 1} = \lim\limits_{x \to \infty} \dfrac{1 + 1/x}{1 + \ln x} = 0.$

17. $\lim\limits_{x \to 0} \dfrac{x + \sin 2x}{x - \sin 2x} = \lim\limits_{x \to 0} \dfrac{1 + 2\cos 2x}{1 - 2\cos 2x} = \dfrac{1 + 2}{1 - 2} = -3.$

19. $\lim\limits_{x \to \pi/2} \dfrac{\cos x}{\pi - 2x} = \lim\limits_{x \to \pi/2} \dfrac{-\sin x}{-2} = \dfrac{1}{2}.$

21. $\begin{aligned}\lim\limits_{x \to 0+} (\sin x) \ln \sin x &= \lim\limits_{x \to 0+} \dfrac{\ln \sin x}{(\sin x)^{-1}} = \lim\limits_{x \to 0+} \dfrac{\frac{1}{\sin x} \cos x}{-1(\sin x)^{-2} \cos x} \\ &= \lim\limits_{x \to 0+} \left(-\dfrac{\frac{1}{\sin x}}{(\sin x)^{-2}}\right) = \lim\limits_{x \to 0+} (-\sin x) = 0.\end{aligned}$

23. $\lim\limits_{x \to 0+} x \ln x = \lim\limits_{x \to 0+} \dfrac{\ln x}{x^{-1}} = \lim\limits_{x \to 0+} \dfrac{1/x}{-1x^{-2}} = \lim\limits_{x \to 0+} (-x) = 0.$

25. Let $y = \lim\limits_{x \to \infty} \left(1 + \dfrac{1}{x}\right)^{x}$. Then

$$\begin{aligned}
\ln y &= \lim_{x \to \infty} \ln \left(1 + \frac{1}{x}\right)^{x} \\
&= \lim_{x \to \infty} x \ln \left(1 + \frac{1}{x}\right) \qquad \log_a M^k = k \log_a M \\
&= \lim_{x \to \infty} \frac{\ln \left(1 + \dfrac{1}{x}\right)}{x^{-1}}
\end{aligned}$$

which tends to the indeterminate form $\dfrac{0}{0}$. So by L'Hospital's rule,

$$\ln y = \lim_{x \to \infty} \frac{\dfrac{1}{1 + 1/x} \dfrac{d}{dx}\left(1 + \dfrac{1}{x}\right)}{-1x^{-2}} = \lim_{x \to \infty} \frac{\dfrac{1}{1 + 1/x}\left(-\dfrac{1}{x^2}\right)}{-\dfrac{1}{x^2}} = 1.$$

Since $\ln_e y = 1$, $e^1 = y$, and $y = e$.

6.10   Applications

1. $\begin{aligned}
f(x) &= xe^{-x} \\
f'(x) &= xe^{-x}(-1) + e^{-x} = e^{-x} - xe^{-x} \\
f''(x) &= -e^{-x} - xe^{-x}(-1) - e^{-x} = xe^{-x} - 2e^{-x}
\end{aligned}$

Step 1. Critical points:

$f'(x) = e^{-x} - xe^{-x} = e^{-x}(1 - x) = 0$;

thus $x = 1$. From $y = xe^{-x}$, the point $(1, 1/e)$ is the critical point.

Step 2. Test of critical point:

$$f''(1) = xe^{-x} - 2e^{-x}\bigg|_{x=1} = e^{-1} - 2e^{-1} = -\frac{1}{e} < 0.$$

Thus $(1, 1/e)$ is a maximum.

Step 3. Concavity and points of inflection:

$f''(x) = xe^{-x} - 2e^{-x} = 0$.

$e^{-x}(x - 2) = 0$; thus $x = 2$. From $y = xe^{-x}$, the point is $(2, 2/e^2)$.

If $x < 2$, $y'' < 0$ (concave down).

If $x > 2$, $y'' > 0$ (concave up).

Step 4. Other:

$\lim\limits_{x \to +\infty} xe^{-x} = \lim\limits_{x \to +\infty} \dfrac{x}{e^x} = \lim\limits_{x \to +\infty} \dfrac{1}{e^x} = 0$ by L'Hospital's rule. So the positive x-axis is an asymptote.

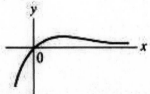

3. $f(x) = x \ln x, \ x > 0$

$$f'(x) = x \cdot \frac{1}{x} + \ln x \cdot 1 = 1 + \ln x$$

$$f''(x) = \frac{1}{x}$$

Step 1. Critical points:

$$
\begin{aligned}
f'(x) = 1 + \ln x &= 0 \\
\ln x &= -1 \\
\log_e x &= -1 \qquad\qquad \ln = \log_e \\
x &= e^{-1} = \frac{1}{e}. \qquad \text{definition of log}
\end{aligned}
$$

From $y = x \ln x$, we have

$$y = \frac{1}{e} \ln \frac{1}{e} = \frac{1}{e}(\ln 1 - \ln e) = \frac{1}{e}(0 - 1) = -\frac{1}{e}.$$

Critical point: $\left(\dfrac{1}{e}, -\dfrac{1}{e}\right)$.

Step 2. Test of critical point:

$$f''(e^{-1}) = \frac{1}{e^{-1}} = e > 0 \ \text{(minimum)}.$$

Step 3. Since $x > 0$, $1/x > 0$. The graph is concave up everywhere.

Other: By L'Hospital's rule, $\displaystyle\lim_{x \to 0+} x \ln x = 0$ (see Exercise 23, Section 6.9, and graph in answer section of book).

5.

$$
\begin{aligned}
f(x) &= \cosh x = \frac{1}{2}(e^x + e^{-x}) \\
f'(x) &= \sinh x = \frac{1}{2}(e^x - e^{-x}) \\
f''(x) &= \cosh x = \frac{1}{2}(e^x + e^{-x})
\end{aligned}
$$

Step 1. Critical points:

$$f'(x) = \frac{1}{2}(e^x - e^{-x}) = 0 \ \text{when} \ x = 0.$$

Step 2. Test of critical point:

$$f''(0) = \frac{1}{2}(1 + 1) = 1 > 0. \ \text{The point } (0, 1) \text{ is a minimum.}$$

Step 3. Concavity:

$$f''(x) = \frac{1}{2}(e^x + e^{-x}) > 0. \ \text{The graph is concave up. (There are no inflection points.)}$$

7. $i = \cos t + \sin t.$

$$
\begin{aligned}
\frac{di}{dt} = -\sin t + \cos t &= 0 \\
-\frac{\sin t}{\cos t} + 1 &= 0 \\
\tan t &= 1 \\
t &= \operatorname{Arctan} 1 = \frac{\pi}{4}.
\end{aligned}
$$

$$i = \cos \frac{\pi}{4} + \sin \frac{\pi}{4} = \frac{1}{\sqrt{2}} + \frac{1}{\sqrt{2}} = \frac{2}{\sqrt{2}} = \sqrt{2} \, \text{A}.$$

Since $\dfrac{d^2 i}{dt^2} = -\cos t - \sin t < 0$ for $t = \dfrac{\pi}{4}$, the critical value corresponds to a maximum.

9. Recall that $v = L\left(\dfrac{di}{dt}\right)$. Thus

$$
\begin{aligned}
v &= 0.0030\frac{d}{dt}[3.0\sin 200t + 2.0\cos 200t] \\
&= (0.0030)[(3.0)(200)\cos 200t - (2.0)(200)\sin 200t]\Big|_{t=0} \\
&= (0.0030)[(3.0)(200)(1) - 0] = 1.8\,\text{V}.
\end{aligned}
$$

11. $N = 12e^{-2.0t}$.

$$
\begin{aligned}
\frac{dN}{dt} &= 12e^{-2.0t}(-2.0) \\
&= -2.0[12e^{-2.0t}] \\
\frac{dN}{dt} &= -2.0N.
\end{aligned}
$$

13.

$$
\begin{aligned}
N &= 6.0e^{-0.25t} \\
\frac{dN}{dt} &= 6.0e^{-0.25t}(-0.25)\Big|_{t=8.5} \\
&= 6.0e^{(-0.25)(8.5)}(-0.25) = -0.18\,\text{g/min}.
\end{aligned}
$$

15. $N = 200e^{(1/2)t}$.

$$
\frac{dN}{dt} = 200\left(\frac{1}{2}\right)e^{(1/2)t}\Big|_{t=3} = 448 \text{ bacteria per hour}.
$$

17. $S = kR^2\ln\dfrac{1}{R} = kR^2(\ln 1 - \ln R) = -kR^2\ln R$, since $\ln 1 = 0$. By the product rule,

$$
\begin{aligned}
\frac{dS}{dR} = -kR^2\frac{1}{R} - 2kR\ln R &= -kR - 2kR\ln R \\
= -kR(1 + 2\ln R) &= 0,\ R \neq 0 \\
2\ln R &= -1 \\
\ln R &= -\frac{1}{2}.
\end{aligned}
$$

By the definition of logarithm, $e^{-1/2} = R$ or $R = \dfrac{1}{\sqrt{e}}$.

19. $x = 5\cos 2t$.

$$
v = \frac{dx}{dt} = -10\sin 2t.
$$

Critical value:

$$
\begin{aligned}
v' = -20\cos 2t &= 0 \\
2t &= \frac{\pi}{2} \\
t &= \frac{\pi}{4}.
\end{aligned}
$$

When $t = \pi/4$, then $x = 5\cos 2\,(\pi/4) = 0$, showing that the particle is at the origin. (At $t = \pi/4$, v is actually a minimum, but its magnitude is a maximum.)

21. In a problem on related rates, we label all quantities that vary by letters, as shown in the figure.

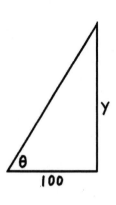

Given $\dfrac{dy}{dt} = 6.0$, find $\dfrac{d\theta}{dt}$ when $y = 150$. From the figure:

$$\tan\theta = \frac{y}{100}$$

$$\theta = \operatorname{Arctan}\frac{y}{100}$$

$$\frac{d\theta}{dt} = \frac{1}{1+(y/100)^2}\left(\frac{1}{100}\right)\frac{dy}{dt}$$

$$= \frac{1}{1+(y/100)^2}\left(\frac{1}{100}\right)(6.0)\Big|_{y=150}$$

$$= 0.018\,\text{rad/s}.$$

23.

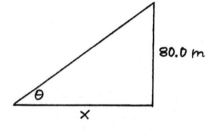

Given: $\dfrac{dx}{dt} = -1.5\,\text{m/s}.$

Find: $\dfrac{d\theta}{dt}$ when $x = 40.0\,\text{m}.$

$$\tan\theta = \frac{80.0}{x}$$

$$\theta = \operatorname{Arctan}\frac{80.0}{x}.$$

$$\frac{d\theta}{dt} = \frac{1}{1+(80.0/x)^2}\frac{d}{dt}(80.0x^{-1})$$

$$= \frac{1}{1+(80.0/x)^2}(80.0)(-1x^{-2})\frac{dx}{dt}$$

$$= \frac{1}{1+(80.0/x)^2}(80.0)\left(-\frac{1}{x^2}\right)(-1.5).$$

When $x = 40.0\,\text{m}$, $\dfrac{d\theta}{dt} = 0.015\,\text{rad/s}.$

25.

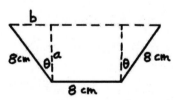

To find an expression for the area, we need to determine a and b in the figure. Observe that
$\cos\theta = \dfrac{a}{8}$ and $\sin\theta = \dfrac{b}{8}$.
Hence $a = 8\cos\theta$ and $b = 8\sin\theta$.

Area of each triangle: $\dfrac{1}{2}ab = \dfrac{1}{2}(64)\sin\theta\cos\theta.$
Area of rectangle: $8a = 64\cos\theta.$
Total area: $A = 64\cos\theta + 64\sin\theta\cos\theta = 64(\cos\theta + \sin\theta\cos\theta).$

$$\frac{dA}{d\theta} = 64[-\sin\theta + \sin\theta(-\sin\theta) + \cos\theta(\cos\theta)]$$

$$= 64(-\sin\theta - \sin^2\theta + \cos^2\theta)$$

$$= 64(-\sin\theta - \sin^2\theta + 1 - \sin^2\theta) \qquad \cos^2\theta = 1 - \sin^2\theta$$

$$= -64(2\sin^2\theta + \sin\theta - 1)$$

$$= -64(2\sin\theta - 1)(\sin\theta + 1) = 0. \qquad \text{factoring}$$

$$2\sin\theta = 1$$

$$\sin\theta = \frac{1}{2} \text{ and } \theta = \frac{\pi}{6}.$$

27.

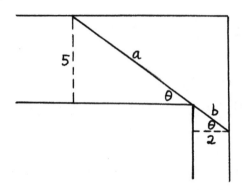

Let $x = a + b$, the length of the pole. Observe that

$$\csc \theta = \frac{a}{5} \qquad \sec \theta = \frac{b}{2}$$
$$a = 5 \csc \theta \qquad b = 2 \sec \theta.$$

So

$$x = 5 \csc \theta + 2 \sec \theta$$
$$\frac{dx}{d\theta} = -5 \csc \theta \cot \theta + 2 \sec \theta \tan \theta$$
$$= -5 \frac{1}{\sin \theta} \frac{\cos \theta}{\sin \theta} + 2 \frac{1}{\cos \theta} \frac{\sin \theta}{\cos \theta} = 0.$$

Multiply by $\sin^2 \theta \cos^2 \theta$:

$$-5 \cos^3 \theta + 2 \sin^3 \theta = 0$$
$$\frac{\sin^3 \theta}{\cos^3 \theta} = \frac{5}{2}$$
$$\tan^3 \theta = \frac{5}{2}$$
$$\tan \theta = \sqrt[3]{\frac{5}{2}}$$
$$\theta = \text{Arctan} \sqrt[3]{\frac{5}{2}} \approx 53.62°.$$

So the length of the ladder is

$$x = 5 \csc 53.62° + 2 \sec 53.62° = 9.58 \, \text{m}.$$

29. $R = k \cos \theta \sin(\theta - \alpha)$, where $k = \dfrac{2 v_0^2}{g \cos^2 \alpha}$. By the product rule,

$\dfrac{dR}{d\theta} = k[\cos \theta \cos(\theta - \alpha) - \sin \theta \sin(\theta - \alpha)]$. From the identity $\cos(A + B) = \cos A \cos B - \sin A \sin B$,

$\dfrac{dR}{d\theta} = k \cos[\theta + (\theta - \alpha)] = k \cos(2\theta - \alpha) = 0$.

Thus, $2\theta - \alpha = \dfrac{\pi}{2}$ and $\theta = \dfrac{\pi}{4} + \dfrac{\alpha}{2}$.

31. $P = Axe^{-x/n}$.

$$\frac{dP}{dx} = Ae^{-x/n} + Axe^{-x/n}\left(-\frac{1}{n}\right) = 0$$
$$Ae^{-x/n}\left(1 - \frac{x}{n}\right) = 0$$
$$x = n.$$

If $x < n$, then $dP/dx > 0$ and if $x > n$, then $dP/dx < 0$. So the critical value corresponds to a maximum by the first-derivative test.

6.11   Newton's Method

1. $f(x) = \cos x + x$

$f'(x) = -\sin x + 1$

By Newton's method, $x_1 = x_0 - \dfrac{\cos x_0 + x_0}{-\sin x_0 + 1} = x_0 + \dfrac{\cos x_0 + x_0}{\sin x_0 - 1}$.

Since the graph of $y = -x$ intersects the graph of $y = \cos x$ between $-\dfrac{\pi}{2}$ and 0, we choose $x_0 = -1$. Thus

$x_1 = -1 + \dfrac{\cos(-1) + (-1)}{\sin(-1) - 1}$.

Setting the calculator in the radian mode, we get $x_1 = -0.75$. It is best to store this value in the memory when computing the next approximation x_2:

$x_2 = -0.75 + \dfrac{\cos(-0.75) + (-0.75)}{\sin(-0.75) - 1} = -0.739$.

A possible sequence is

0.75 $\boxed{+/-}$ $\boxed{\text{STO}}$ $\boxed{\text{COS}}$ $\boxed{+}$ $\boxed{\text{RCL}}$ $\boxed{=}$ $\boxed{\div}$ $\boxed{[}$ $\boxed{\text{RCL}}$ $\boxed{\text{SIN}}$ $\boxed{-}$ 1 $\boxed{]}$ $\boxed{+}$ $\boxed{\text{RCL}}$ $\boxed{=}$ $\boxed{\text{STO}}$

$\rightarrow$ -0.7391111

Having stored the last value, we proceed as before:

$x_3 = -0.739 + \dfrac{\cos(-0.739) + (-0.739)}{\sin(-0.739) - 1} = -0.7390851$.

From this point on, these digits no longer change, so that $x = -0.7390851$ is the root of the equation to 7 decimal places.

3. $f(x) = 4\sin x - x$; $f'(x) = 4\cos x - 1$. The graphs of $y = 4\sin x$ and $y = x$ intersect at the origin and (on the positive side) near $x = 3$. By Newton's method,

$$x_1 = x_0 - \frac{4\sin x_0 - x_0}{4\cos x_0 - 1}$$

$$x_1 = 3 - \frac{4\sin 3 - 3}{4\cos 3 - 1} = 2.509$$

$$x_2 = 2.509 - \frac{4\sin 2.509 - 2.509}{4\cos 2.509 - 1} = 2.4749$$

$$x_3 = 2.4749 - \frac{4\sin 2.4749 - 2.4749}{4\cos 2.4749 - 1} = 2.474577.$$

From this point on these digits no longer change, so that $x = 2.474577$ is the root to 6 decimal places.

5. $f(x) = \sin x - x^2$

$f'(x) = \cos x - 2x$

By Newton's method $x_1 = x_0 - \dfrac{\sin x_0 - x_0^2}{\cos x_0 - 2x_0} = x_0 + \dfrac{\sin x_0 - x_0^2}{2x_0 - \cos x_0}$.

The parabola $y = x^2$ intersects the curve $y = \sin x$ between 0 and $\dfrac{\pi}{2}$. So we choose $x_0 = 1$:

$x_1 = 1 + \dfrac{\sin 1 - 1}{2 - \cos 1} = 0.89$.

Store 0.89 in the memory and evaluate

$x_2 = 0.89 + \dfrac{\sin(0.89) - (0.89)^2}{2(0.89) - \cos(0.89)} = 0.8769$.

Store 0.8769 in the memory and evaluate

$x_3 = 0.8769 + \dfrac{\sin(0.8769) - (0.8769)^2}{2(0.8769) - \cos(0.8769)} = 0.8767262$.

From this point on, these digits no longer change, so that $x = 0.8767262$ is the root of the equation to 7 decimal places.

7. $f(x) = e^x - \dfrac{5}{x}$; $f'(x) = e^x + \dfrac{5}{x^2}$. The graphs of $y = e^x$ and $y = \dfrac{5}{x}$ intersect near $x = 1$. By Newton's method,

$$x_1 = x_0 - \frac{e^{x_0} - 5/x_0}{e^{x_0} + 5/x_0^2}$$

$$x_1 = 1 - \frac{e - 5/1}{e + 5/1^2} = 1.2956$$

$$x_2 = 1.2956 - \frac{e^{1.2956} - 5/1.2956}{e^{1.2956} + 5/(1.2956)^2} = 1.32666$$

$$x_3 = 1.326725.$$

9. $\tan x + x = 2$; since the line $y = 2 - x$ intersects $y = \tan x$ between 0 and $\dfrac{\pi}{2}$, we choose $x_0 = 1$, a convenient value.

Let $f(x) = \tan x + x - 2$; then $f'(x) = \sec^2 x + 1$.

$$x_1 = x_0 - \frac{\tan x_0 + x_0 - 2}{\sec^2 x_0 + 1} = 1 - \frac{\tan 1 + 1 - 2}{\sec^2 1 + 1} = 0.874.$$

$$x_2 = 0.874 - \frac{\tan(0.874) + 0.874 - 2}{\sec^2(0.874) + 1} = 0.85387.$$

$$x_3 = 0.85387 - \frac{\tan(0.85387) + 0.85387 - 2}{\sec^2(0.85387) + 1} = 0.853530.$$

11. $f(x) = x^2 - 2x - 12$; $f'(x) = 2x - 2$. For the negative root use $x_0 = -3$:

$$x_1 = x_0 - \frac{x_0^2 - 2x_0 - 12}{2x_0 - 2}$$

$$x_1 = -3 - \frac{(-3)^2 - 2(-3) - 12}{2(-3) - 2} = -2.625$$

$$x_2 = -2.6056$$

$$x_3 = -2.605551.$$

13. A graphing calculator will quickly locate the roots near -1, between 0 and 1, and near 6. Suppose we find the middle root by letting $x_0 = 1$:

$$f(x) = x^3 - 6x^2 - 4x + 6$$

$$f'(x) = 3x^2 - 12x - 4$$

$$x_1 = x_0 - \frac{x_0^3 - 6x_0^2 - 4x_0 + 6}{3x_0^2 - 12x_0 - 4} = 1 - \frac{1^3 - 6 \cdot 1^2 - 4 \cdot 1 + 6}{3 \cdot 1^2 - 12 \cdot 1 - 4} = 0.7692.$$

$$x_2 = 0.7692 - \frac{(0.7692)^3 - 6(0.7692)^2 - 4(0.7692) + 6}{3(0.7692)^2 - 12(0.7692) - 4} = 0.754211$$

$$x_3 = 0.754211 - \frac{(0.754211)^3 - 6(0.754211)^2 - 4(0.754211) + 6}{3(0.754211)^2 - 12(0.754211) - 4} = 0.754138.$$

15. $f(x) = 2x^3 - 4x^2 - x + 4$; $f'(x) = 6x^2 - 8x - 1$.

$$x_1 = x_0 - \frac{2x_0^3 - 4x_0^2 - x_0 + 4}{6x_0^2 - 8x_0 - 1}$$

Start: $x_0 = -1$

$$x_1 = -1 - \frac{2(-1)^3 - 4(-1)^2 - (-1) + 4}{6(-1)^2 - 8(-1) - 1} = -0.923.$$

$$x_2 = -0.923 - \frac{2(-0.923)^3 - 4(-0.923)^2 - (-0.923) + 4}{6(-0.923)^2 - 8(-0.923) - 1} = -0.9180$$

$$x_3 = -0.917988.$$

Chapter 6 Review

1. In the interval $\left[-\dfrac{\pi}{2}, \dfrac{\pi}{2}\right]$, $\sin\theta = -1$ only for $\theta = -\dfrac{\pi}{2}$.

3. On the interval $[0, \pi]$ the only permissible value is $x = 120° = 2\pi/3$.

5. Let $\theta = \text{Arctan} \dfrac{3}{1}$. Then the length of the hypotenuse is $\sqrt{10}$. Thus

$$\begin{aligned}
\cos(\text{Arctan } 3) &= \cos\theta \\
&= \frac{1}{\sqrt{10}} \\
&= \frac{\sqrt{10}}{10}.
\end{aligned}$$

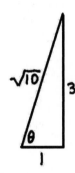

7.

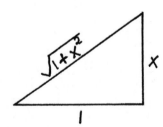

Let $\theta = \text{Arctan} \dfrac{x}{1}$;

then $\sin\theta = \dfrac{x}{\sqrt{1+x^2}}$.

9. $y = x^2 \tan 3x$. By the product rule,

$$\begin{aligned}
y' &= x^2 \frac{d}{dx}\tan 3x + \tan 3x \frac{d}{dx}(x^2) \\
&= x^2 \sec^2 3x \cdot 3 + \tan 3x (2x) \\
&= 3x^2 \sec^2 3x + 2x \tan 3x.
\end{aligned}$$

11. $v = \dfrac{e^{2t}}{t}$. By the quotient rule,

$$\frac{dv}{dt} = \frac{t\frac{d}{dt}e^{2t} - e^{2t}\frac{d}{dt}(t)}{t^2} = \frac{te^{2t}(2) - e^{2t}(1)}{t^2} = \frac{e^{2t}(2t-1)}{t^2}.$$

13. $y = \ln \dfrac{1}{\sqrt{x+3}} = \ln(x+3)^{-1/2}$

$\quad = -\dfrac{1}{2}\ln(x+3)$ $\qquad\qquad\qquad \log_a M^k = k\log_a M$

$\quad y' = -\dfrac{1}{2}\dfrac{1}{x+3} = -\dfrac{1}{2(x+3)}.$

15. $y = \ln \dfrac{\sqrt{2x^2+1}}{x} = \dfrac{1}{2}\ln(2x^2+1) - \ln x.$

$\quad y' = \dfrac{1}{2}\dfrac{1}{2x^2+1}(4x) - \dfrac{1}{x} = \dfrac{2x}{2x^2+1}\cdot\dfrac{x}{x} - \dfrac{1}{x}\dfrac{2x^2+1}{2x^2+1}$

$\qquad = \dfrac{2x^2 - 2x^2 - 1}{x(2x^2+1)} = -\dfrac{1}{x(2x^2+1)}.$

17. $y = \sqrt{\ln 2x} = (\ln 2x)^{1/2}.$ By the power rule,

$\quad y' = \dfrac{1}{2}(\ln 2x)^{-1/2}\dfrac{d}{dx}\ln 2x = \dfrac{1}{2}(\ln 2x)^{-1/2}\dfrac{1}{2x}(2) = \dfrac{1}{2x\sqrt{\ln 2x}}.$

19. $y = \cos(\ln x).$

$\quad y' = -\sin(\ln x)\dfrac{d}{dx}\ln x = -\dfrac{\sin(\ln x)}{x}.$

21. $y = e^{2x}\cot x.$ By the product rule,

$\quad y' = e^{2x}(-\csc^2 x) + (\cot x)e^{2x}(2) = e^{2x}(2\cot x - \csc^2 x).$

23. $y = e^{\text{Arccos}\,3x}.$

$\quad y' = e^{\text{Arccos}\,3x}\dfrac{d}{dx}\text{Arccos}\,3x = e^{\text{Arccos}\,3x}\left(-\dfrac{1}{\sqrt{1-(3x)^2}}\right)\dfrac{d}{dx}(3x) = -\dfrac{3e^{\text{Arccos}\,3x}}{\sqrt{1-9x^2}}.$

25. $y = (\cot x)^x.$

$\quad \ln y = \ln(\cot x)^x$

$\qquad\quad = x\ln\cot x$ $\qquad\qquad\qquad\qquad \log_a M^k = k\log_a M$

$\quad \dfrac{1}{y}\dfrac{dy}{dx} = x\dfrac{d}{dx}\ln\cot x + \ln\cot x\cdot 1$

$\qquad\quad = x\dfrac{1}{\cot x}(-\csc^2 x) + \ln\cot x.$

Note that $\dfrac{1}{\cot x}\csc^2 x = \dfrac{\sin x}{\cos x}\dfrac{1}{\sin^2 x} = \dfrac{1}{\cos x\sin x} = \sec x\csc x.$

Thus

$\quad \dfrac{1}{y}\dfrac{dy}{dx} = -x\sec x\csc x + \ln\cot x$

$\quad \dfrac{dy}{dx} = y(\ln\cot x - x\csc x\sec x)$

$\qquad\quad = (\cot x)^x(\ln\cot x - x\csc x\sec x).$

27.
$$e^{\sin y} + \csc x = 1$$

$$e^{\sin y}\frac{d}{dx}\sin y - \csc x \cot x = 0$$

$$e^{\sin y}\cos y\frac{dy}{dx} = \csc x \cot x$$

$$\frac{dy}{dx} = \frac{\csc x \cot x}{e^{\sin y}\cos y}$$

$$= e^{-\sin y}\sec y \csc x \cot x.$$

29. $\displaystyle\lim_{x\to 0}\frac{\sin 4x}{x} = \lim_{x\to 0}\frac{\frac{d}{dx}\sin 4x}{\frac{d}{dx}(x)}$ by L'Hospital's rule

$$= \lim_{x\to 0}\frac{4\cos 4x}{1} = 4.$$

31. $\displaystyle\lim_{x\to\pi/4}(1-\tan x)\sec 2x = \lim_{x\to\pi/4}\frac{1-\tan x}{\cos 2x}$

$$= \lim_{x\to\pi/4}\frac{-\sec^2 x}{-2\sin 2x} = \frac{-(\sqrt{2})^2}{-2\sin\frac{\pi}{2}} = \frac{2}{2\cdot 1} = 1.$$

33. $\displaystyle\lim_{x\to 0+}\frac{\sin x - x}{x\sin x} = \lim_{x\to 0+}\frac{\cos x - 1}{x\cos x + \sin x}$ L'Hospital's rule

which tends to the indeterminate form $\dfrac{0}{0}$. Applying the rule again, we get

$$\lim_{x\to 0+}\frac{-\sin x}{-x\sin x + \cos x + \cos x} = \frac{0}{0+1+1} = 0.$$

35.

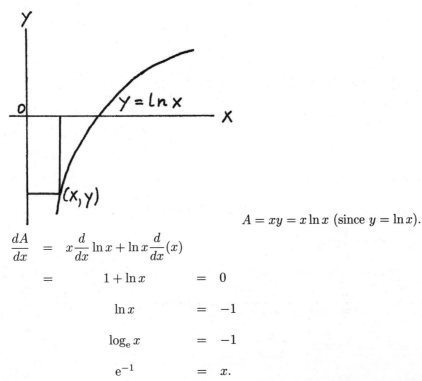

$A = xy = x\ln x$ (since $y = \ln x$).

$$\frac{dA}{dx} = x\frac{d}{dx}\ln x + \ln x\frac{d}{dx}(x)$$

$$= 1 + \ln x = 0$$

$$\ln x = -1$$

$$\log_e x = -1$$

$$e^{-1} = x.$$

So $x = \dfrac{1}{e}$ and $y = \ln e^{-1} = -1\ln e = -1$. The dimensions are $\dfrac{1}{e} \times 1$.

37. (a) From $P = Se^{rt}$, we get $P = 5000e^{(0.10)(15)} = \$22,408.45$.

(b) $\dfrac{dP}{dt} = Se^{rt}\dfrac{d}{dt}(rt) = Se^{rt}r = Pr = rP$. ($P$ grows at a rate that is proportional to the amount present.)

39. Given: $\dfrac{dT}{dt} = 2.00\,\text{K/min}$.

Find: $\dfrac{dp}{dt}$ when $T = 300.0\,\text{K}$.

$$\log_{10} p = -\frac{1706.4}{T} - 7.7760$$

$$\frac{1}{p}(\log_{10} \text{e})\frac{dp}{dt} = \frac{1706.4}{T^2}\frac{dT}{dt}$$

$$= \frac{1706.4}{T^2}(2.00)$$

$$\frac{dp}{dt} = \frac{p}{\log_{10}\text{e}}\frac{1706.4}{T^2}(2.00).$$

When $T = 300.0$, $\log_{10} p = -\dfrac{1706.4}{300} - 7.7760 = -13.464$ and $p = 10^{-13.464}$.

Substituting these values, we get

$$\frac{dp}{dt} = \frac{10^{-13.464}}{\log_{10}\text{e}}\frac{1706.4}{(300.0)^2}(2.00) = 3.00 \times 10^{-15} \text{ mm of mercury per min.}$$

41.

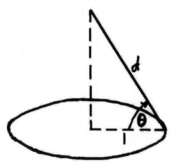

Since $d = \sec\theta$ and $I = k\dfrac{\sin\theta}{d^2}$, we get

$$I = \frac{k\sin\theta}{\sec^2\theta} = k\sin\theta\cos^2\theta.$$

Then

$$\frac{dI}{d\theta} = k\left[\sin\theta(2\cos\theta)(-\sin\theta) + \cos^2\theta\cos\theta\right]$$

$$= k\cos\theta(-2\sin^2\theta + \cos^2\theta)$$

$$= k\cos\theta(-2\sin^2\theta + 1 - \sin^2\theta) \qquad\qquad \cos^2\theta = 1 - \sin^2\theta$$

$$= k\cos\theta(1 - 3\sin^2\theta) = 0.$$

$$\sin^2\theta = \frac{1}{3}$$

$$\sin\theta = \frac{1}{\sqrt{3}}.$$

Hence $\theta = \text{Arcsin}\,\dfrac{1}{\sqrt{3}} \approx 35.3°$.

43. $f(x) = \cos x - x - \dfrac{1}{2}$; $f'(x) = -\sin x - 1$.

$$x_1 = x_0 - \frac{\cos x_0 - x_0 - 1/2}{-\sin x_0 - 1} = x_0 + \frac{\cos x_0 - x_0 - 0.5}{\sin x_0 + 1}.$$

Let $x_0 = 0.5$:

$$x_1 = 0.5 + \frac{\cos 0.5 - 0.5 - 0.5}{\sin 0.5 + 1} = 0.41725$$

$$x_2 = 0.415084$$

$$x_3 = 0.415083.$$

Chapter 7

Integration Techniques

7.1   The Power Formula Again

1. $\displaystyle\int x\sqrt{x^2+1}\,dx = \int (x^2+1)^{1/2}x\,dx \qquad\qquad u = x^2+1;\ du = 2x\,dx.$

   $\displaystyle\frac{1}{2}\int (x^2+1)^{1/2}(2x)\,dx = \frac{1}{2}\int u^{1/2}\,du + C = \frac{1}{2}\cdot\frac{2}{3}u^{3/2} + C = \frac{1}{3}(x^2+1)^{3/2} + C.$

3. $\displaystyle\int \frac{dx}{\sqrt{1-x}} = \int (1-x)^{-1/2}\,dx \qquad\qquad u = 1-x;\ du = -dx.$

   $\displaystyle\qquad\quad = -\int (1-x)^{-1/2}(-dx) = -\int u^{-1/2}\,du$

   $\displaystyle\qquad\quad = -\frac{u^{1/2}}{1/2} + C = -2\sqrt{1-x} + C.$

5. $\displaystyle\int \sin^2 x\cos x\,dx \qquad\qquad u = \sin x;\ du = \cos x\,dx.$

   $\displaystyle\int u^2\,du = \frac{1}{3}u^3 + C = \frac{1}{3}\sin^3 x + C.$

7. $\displaystyle\int \tan^2 2x\sec^2 2x\,dx = \frac{1}{2}\int \tan^2 2x(2\sec^2 2x\,dx) \qquad u = \tan 2x;\ du = 2\sec^2 2x\,dx.$

   $\displaystyle\qquad\qquad = \frac{1}{2}\int u^2\,du = \frac{1}{6}u^3 + C = \frac{1}{6}\tan^3 2x + C.$

9. $\displaystyle\int (1+\tan 3t)^3 \sec^2 3t\,dt \qquad\qquad u = 1+\tan 3t;$

   $\displaystyle\qquad\qquad\qquad\qquad\qquad\qquad\qquad\quad du = 3\sec^2 3t\,dt.$

   $\displaystyle = \frac{1}{3}\int (1+\tan 3t)^3(3\sec^2 3t\,dt) = \frac{1}{3}\int u^3\,du = \frac{1}{3}\frac{u^4}{4} + C = \frac{1}{12}(1+\tan 3t)^4 + C.$

11. $\displaystyle\int (1+4e^r)^3 e^{4r}\,dr = \frac{1}{4}\int (1+e^{4r})^3(4e^{4r}\,dr) \qquad u = 1+e^{4r};\ du = 4e^{4r}\,dr.$

    $\displaystyle\qquad\qquad = \frac{1}{4}\int u^3\,du = \frac{1}{16}u^4 + C = \frac{1}{16}(1+e^{4r})^4 + C.$

13. $\displaystyle\int (1-\cos 5x)^3 \sin 5x\,dx \qquad\qquad u = 1-\cos 5x;\ du = 5\sin 5x\,dx.$

    $\displaystyle\frac{1}{5}\int (1-\cos 5x)^3\cdot 5\sin 5x\,dx = \frac{1}{5}\int u^3\,du = \frac{1}{5}\frac{u^4}{4} + C = \frac{1}{20}(1-\cos 5x)^4 + C.$

15. $\displaystyle\int \frac{\sec^2 3x}{(\tan 3x + 1)^3}\, dx \;=\; \int (\tan 3x + 1)^{-3}\sec^2 3x\, dx \qquad u = \tan 3x + 1;\; du = 3\sec^2 3x\, dx.$

$\displaystyle\qquad\qquad\qquad\qquad =\; \frac{1}{3}\int (\tan 3x + 1)^{-3}(3\sec^2 3x\, dx)$

$\displaystyle\qquad\qquad\qquad\qquad =\; \frac{1}{3}\int u^{-3}\, du = \frac{1}{3}\frac{u^{-2}}{-2} + C = -\frac{1}{6(\tan 3x + 1)^2} + C.$

17. $\displaystyle\int \frac{\ln x\, dx}{x} = \int \ln x\left(\frac{1}{x}\, dx\right). \qquad\qquad\qquad u = \ln x;\; du = \frac{1}{x}\, dx.$

$\displaystyle\int u\, du = \frac{1}{2}u^2 + C = \frac{1}{2}\ln^2 x + C.$

19. $\displaystyle\int \frac{dx}{x\sqrt{\ln x}} \;=\; \int (\ln x)^{-1/2}\frac{1}{x}\, dx \qquad\qquad u = \ln x;\; du = \frac{1}{x}\, dx.$

$\displaystyle\qquad\qquad =\; \int u^{-1/2}\, du = \frac{u^{1/2}}{1/2} + C = 2\sqrt{\ln x} + C.$

21. $\displaystyle\int \frac{\operatorname{Arccos} x\, dx}{\sqrt{1 - x^2}} \qquad\qquad\qquad\qquad u = \operatorname{Arccos} x;\; du = -\frac{dx}{\sqrt{1 - x^2}}.$

$\displaystyle -\int \operatorname{Arccos} x\left(-\frac{dx}{\sqrt{1 - x^2}}\right) = -\int u\, du = -\frac{1}{2}u^2 + C = -\frac{1}{2}(\operatorname{Arccos} x)^2 + C.$

23. $\displaystyle\int_{\pi/6}^{\pi/2} \cos^2 x \sin x\, dx = -\int_{\pi/6}^{\pi/2} (\cos x)^2(-\sin x\, dx).$

$u = \cos x;\; du = -\sin x\, dx$

Lower limit: if $x = \dfrac{\pi}{6}$, then $u = \cos\dfrac{\pi}{6} = \dfrac{\sqrt{3}}{2}$.

Upper limit: if $x = \dfrac{\pi}{2}$, then $u = \cos\dfrac{\pi}{2} = 0$.

$\displaystyle -\int_{\sqrt{3}/2}^{0} u^2\, du = -\frac{1}{3}u^3\Big|_{\sqrt{3}/2}^{0} = 0 + \frac{1}{3}\left(\frac{\sqrt{3}}{2}\right)^3 = \frac{1}{3}\frac{(\sqrt{3})^2\sqrt{3}}{8} = \frac{\sqrt{3}}{8}.$

25. $\displaystyle\int_{1}^{e} \frac{\sqrt{\ln x}}{x}\, dx = \int (\ln x)^{1/2}\left(\frac{1}{x}\, dx\right). \qquad\qquad u = \ln x;\; du = \frac{1}{x}\, dx.$

$\displaystyle\int_{1}^{e} (\ln x)^{1/2}\frac{1}{x}\, dx = \frac{2}{3}(\ln x)^{3/2}\Big|_{1}^{e} = \frac{2}{3}\left[(\ln e)^{3/2} - (\ln 1)^{3/2}\right] = \frac{2}{3}(1 - 0) = \frac{2}{3}$

(since $\ln e = 1$ and $\ln 1 = 0$).

27. $\displaystyle\int \frac{\cot 2x\, dx}{\sin^2 2x} \;=\; \int \cot 2x \csc^2 2x\, dx \qquad\qquad u = \cot 2x;\; du = -2\csc^2 2x\, dx.$

$\displaystyle\qquad\qquad =\; -\frac{1}{2}\int \cot 2x(-2\csc^2 2x\, dx)$

$\displaystyle\qquad\qquad =\; -\frac{1}{2}\int u\, du = -\frac{1}{2}\frac{u^2}{2} + C = -\frac{1}{4}\cot^2 2x + C.$

29. $\displaystyle\int \sqrt{\tan x}\,\sec^2 dx \qquad\qquad\qquad\qquad u = \tan x;\; du = \sec^2 x\, dx.$

$\displaystyle\int u^{1/2}\, du = \frac{2}{3}u^{3/2} + C = \frac{2}{3}(\tan x)^{3/2} + C.$

31. $\displaystyle\int \operatorname{Arctan} 4R\left(\frac{dR}{1 + 16R^2}\right) \;=\; \frac{1}{4}\int \operatorname{Arctan} 4R\frac{4dR}{1 + 16R^2}$

$\displaystyle\qquad\qquad\qquad\qquad u = \operatorname{Arctan} 4R;\; du = \frac{4\, dR}{1 + 16R^2}.$

$\displaystyle\qquad\qquad =\; \frac{1}{4}\int u\, du = \frac{1}{8}u^2 + C = \frac{1}{8}(\operatorname{Arctan} 4R)^2 + C.$

33. $\displaystyle\int \frac{(1+\ln x)^2}{x}\,dx = \int (1+\ln x)^2\left(\frac{1}{x}\,dx\right).$ Let $u = 1 + \ln x$; then $du = \dfrac{1}{x}\,dx.$

$\displaystyle\int u^2\,du = \frac{1}{3}u^3 + C = \frac{1}{3}(1+\ln x)^3 + C.$

7.2 The Logarithmic and Exponential Forms

1. $\displaystyle\int \frac{dx}{x-1}$ $u = x - 1; \ du = dx.$

$\displaystyle\int \frac{dx}{x-1} = \int \frac{du}{u} = \ln|u| + C = \ln|x-1| + C.$

3. $\displaystyle\int \frac{dx}{2+3x} = \frac{1}{3}\int \frac{3\,dx}{2+3x}$ $u = 2 + 3x; \ du = 3\,dx.$

$\displaystyle = \frac{1}{3}\int \frac{du}{u} = \frac{1}{3}\ln|2+3x| + C.$

5. $\displaystyle\int \frac{ds}{1-3s}$ $u = 1 - 3s; \ du = -3\,ds.$

$\displaystyle -\frac{1}{3}\int \frac{-3\,ds}{1-3s} = -\frac{1}{3}\int \frac{du}{u} = -\frac{1}{3}\ln|u| + C = -\frac{1}{3}\ln|1-3s| + C.$

7. $\displaystyle\int_0^1 \frac{x\,dx}{x^2+1} = \frac{1}{2}\int_0^1 \frac{2x\,dx}{x^2+1}$ $u = x^2 + 1; \ du = 2x\,dx.$

$\displaystyle = \frac{1}{2}\ln(x^2+1)\Big|_0^1 = \frac{1}{2}\ln 2 - \frac{1}{2}\ln 1 = \frac{1}{2}\ln 2 - 0$

$\displaystyle = \ln 2^{1/2} = \ln\sqrt{2}.$

9. $\displaystyle\int e^{-x}\,dx$ $u = -x; \ du = -dx.$

$\displaystyle -\int e^{-x}(-dx) = -\int e^u\,du = -e^u + C = -e^{-x} + C.$

11. $\displaystyle\int_0^2 2e^{3x}\,dx = 2\int_0^2 e^{3x}\,dx = \frac{2}{3}\int_0^2 e^{3x}(3\,dx).$ $u = 3x; \ du = 3\,dx.$
Lower limit: if $x = 0$, then $u = 0.$

Upper limit: if $x = 2$, then $u = 6.$

$\displaystyle\frac{2}{3}\int_0^6 e^u\,du = \frac{2}{3}e^u\Big|_0^6 = \frac{2}{3}(e^6 - e^0) = \frac{2}{3}(e^6 - 1).$

13. $\displaystyle\int e^{4x}\,dx$ $u = 4x; \ du = 4\,dx.$

$\displaystyle\frac{1}{4}\int e^{4x}(4\,dx) = \frac{1}{4}\int e^u\,du = \frac{1}{4}e^u + C = \frac{1}{4}e^{4x} + C.$

15. $\displaystyle\int te^{t^2}\,dt = \int e^{t^2}t\,dt$ $u = t^2; \ du = 2t\,dt.$

$\displaystyle = \frac{1}{2}\int e^{t^2}(2t\,dt) = \frac{1}{2}\int e^u\,du = \frac{1}{2}e^u + C = \frac{1}{2}e^{t^2} + C.$

17. $\displaystyle\int e^{\sin R}\cos R\,dR$ $u = \sin R; \ du = \cos R\,dR.$

$\displaystyle\int e^u\,du = e^u + C = e^{\sin R} + C.$

19. $\int \dfrac{\sec^2 2x}{1+\tan 2x}\,dx$ $\qquad\qquad\qquad\qquad u = 1+\tan 2x;\ \ du = 2\sec^2 2x\,dx.$

$\qquad = \dfrac{1}{2}\int \dfrac{2\sec^2 2x\,dx}{1+\tan 2x} = \dfrac{1}{2}\int \dfrac{du}{u} = \dfrac{1}{2}\ln|u| + C = \dfrac{1}{2}\ln|1+\tan 2x| + C.$

21. $\int \dfrac{e^{\text{Arctan}\,x}}{1+x^2}\,dx = \int e^{\text{Arctan}\,x}\left(\dfrac{1}{1+x^2}\,dx\right).$ $\qquad u = \text{Arctan}\,x;\ \ du = \dfrac{1}{1+x^2}\,dx.$

$\quad \int e^u\,du = e^u + C = e^{\text{Arctan}\,x} + C.$

23. $\int \dfrac{dx}{x\ln x} \;=\; \int \dfrac{1}{\ln x}\dfrac{1}{x}\,dx \qquad u = \ln x;\ \ du = \dfrac{1}{x}\,dx.$

$\qquad\qquad\quad = \int \dfrac{du}{u} = \ln|u| + C = \ln|\ln x| + C.$

25. $\int \dfrac{e^x + 1}{e^x}\,dx \;=\; \int \left(\dfrac{e^x}{e^x} + \dfrac{1}{e^x}\right)dx$

$\qquad\qquad\quad = \int (1 + e^{-x})\,dx$

$\qquad\qquad\quad = \int 1\,dx + \int e^{-x}\,dx$

$\qquad\qquad\quad = x + \int e^{-x}\,dx$

$\qquad\qquad\quad = x - \int e^{-x}(-dx)$ $\qquad\qquad u = -x;\ \ du = -dx.$

$\qquad\qquad\quad = x - e^{-x} + C.$

27. $\int \dfrac{e^{-3W}}{1-e^{-3W}}\,dW \;=\; \dfrac{1}{3}\int \dfrac{3e^{-3W}\,dW}{1-e^{-3W}} \qquad u = 1-e^{-3W};\ \ du = 3e^{-3W}\,dW.$

$\qquad\qquad\qquad\quad = \dfrac{1}{3}\int \dfrac{du}{u} = \dfrac{1}{3}\ln|u| + C = \dfrac{1}{3}\ln|1-e^{-3W}| + C.$

29. $\int \dfrac{2x}{(1+x^2)^2}\,dx = \int (1+x^2)^{-2}(2x)\,dx.$ $\qquad u = 1+x^2;\ \ du = 2x\,dx.$

$\quad \int u^{-2}\,du = \dfrac{u^{-1}}{-1} + C = -\dfrac{1}{u} + C = -\dfrac{1}{1+x^2} + C.$

31. $\int (1+e^x)^2\,dx \;=\; \int (1 + 2e^x + e^{2x})\,dx$

$\qquad\qquad\qquad = \int dx + 2\int e^x\,dx + \dfrac{1}{2}\int e^{2x}(2\,dx) \qquad u = 2x;\ \ du = 2\,dx.$

$\qquad\qquad\qquad = x + 2e^x + \dfrac{1}{2}e^{2x} + C.$

33. $\int \dfrac{x+1}{x+2}\,dx = \int \dfrac{(x+2)-1}{x+2}\,dx = \int \left(1 - \dfrac{1}{x+2}\right)dx$

which is equivalent to long division.

$u = x+2;\ \ du = dx.$

$\quad \int 1\,dx - \int \dfrac{dx}{x+2} = \int dx - \int \dfrac{du}{u} = x - \ln|u| + C = x - \ln|x+2| + C.$

35. $\displaystyle\int \frac{\cos x}{(\sin x)^2}\,dx \;=\; \int (\sin x)^{-2}\cos x\,dx$ $u = \sin x;\;\; du = \cos x\,dx.$

$\displaystyle\qquad\qquad = \int u^{-2}\,du = \frac{u^{-1}}{-1} + C = -\frac{1}{u} + C$

$\displaystyle\qquad\qquad = -\frac{1}{\sin x} + C = -\csc x + C.$

37. $\displaystyle\int e^{\sin^2 x}\sin 2x\,dx = \int e^{\sin^2 x}(2\sin x\cos x)\,dx$ by the double-angle formula.

$u = \sin^2 x;\;\; du = 2(\sin x)(\cos x\,dx).$

$\displaystyle\int e^u\,du = e^u + C = e^{\sin^2 x} + C.$

39. $\displaystyle\int_0^{\pi/3} \frac{\sin x}{\cos x}\,dx = -\int_0^{\pi/3} \frac{\sin x}{\cos x}(-dx).$ $u = \cos x;\;\; du = -\sin x\,dx.$

Lower limit: if $x = 0$, then $u = \cos 0 = 1$.

Upper limit: if $x = \dfrac{\pi}{3}$, then $u = \cos\dfrac{\pi}{3} = \dfrac{1}{2}$.

$\displaystyle -\int_1^{1/2} \frac{du}{u} = -\ln|u|\Big|_1^{1/2} = -\ln\left(\frac{1}{2}\right) + \ln 1 = -(\ln 1 - \ln 2) + 0 = \ln 2.$

41. $\displaystyle\int \frac{\cos 2x}{1 + \sin 2x}\,dx \;=\; \frac{1}{2}\int \frac{2\cos 2x\,dx}{1 + \sin 2x}$ $u = 1 + \sin 2x;\;\; du = 2\cos 2x\,dx.$

$\displaystyle\qquad\qquad = \frac{1}{2}\int \frac{du}{u} = \frac{1}{2}\ln|u| + C$

$\displaystyle\qquad\qquad = \frac{1}{2}\ln|1 + \sin 2x| + C$

$\displaystyle\qquad\qquad = \frac{1}{2}\ln(1 + \sin 2x) + C,$

$\qquad\qquad$ since $1 + \sin 2x \geq 0.$

43. $\displaystyle\int \frac{\cos 2x}{(1 + \sin 2x)^2}\,dx$ $u = 1 + \sin 2x;\;\; du = 2\cos 2x\,dx.$

$\displaystyle\frac{1}{2}\int (1 + \sin 2x)^{-2}(2\cos 2x\,dx) \;=\; \frac{1}{2}\int u^{-2}\,du = \frac{1}{2}\frac{u^{-1}}{-1} + C$

$\displaystyle\qquad\qquad = -\frac{1}{2u} + C = -\frac{1}{2(1 + \sin 2x)} + C.$

45. $\displaystyle\int 5^{3x}\,dx$ $u = 3x;\;\; du = 3\,dx.$

$\displaystyle\frac{1}{3}\int 5^{3x}(3\,dx) = \frac{1}{3}\int 5^u\,du = \frac{1}{3}(5^u)\log_5 e = \frac{1}{3}(5^{3x})\log_5 e + C.$

47. $\displaystyle\int 2^{\cot x}\csc^2 x\,dx$ $u = \cot x;\;\; du = -\csc^2 x\,dx.$

$\displaystyle -\int 2^{\cot x}(-\csc^2 x\,dx) \;=\; -\int 2^u\,du = -2^u\log_2 e$

$\displaystyle\qquad\qquad = -2^{\cot x}\log_2 e + C \ \text{ or } \ -\frac{2^{\cot x}}{\ln 2} + C.$

49.

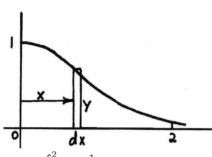

$$I_y = \rho \int_0^2 x^2 \frac{1}{1+x^3}\,dx.$$

Moment of inertia (with respect to y-axis) of typical element:

$$x^2 \cdot \rho y\,dy = x^2 \cdot \rho \cdot \frac{1}{1+x^3}\,dx.$$

Let $u = 1 + x^3$; then $du = 3x^2\,dx$.

For the indefinite integral $\int \dfrac{x^2}{1+x^3}\,dx$ we have

$$\frac{1}{3}\int \frac{3x^2\,dx}{1+x^3} = \frac{1}{3}\int \frac{du}{u} = \frac{1}{3}\ln|u| + C. \text{ Thus}$$

$$
\begin{aligned}
I_y &= \frac{\rho}{3}\int_0^2 \frac{3x^2\,dx}{1+x^3} = \frac{\rho}{3}\ln|1+x^3|\Big|_0^2 = \frac{\rho}{3}[\ln 9 - \ln 1]\\
&= \frac{\rho}{3}[\ln 3^2 - 0] = \frac{\rho}{3}(2\ln 3) = \frac{2}{3}\rho\ln 3.
\end{aligned}
$$

51. $$
\begin{aligned}
i_{av} &= \frac{1}{2-0}\int_0^2 e^{(-1/3)t}\,dt \qquad u = -\frac{1}{3}t;\ \ du = -\frac{1}{3}\,dt.\\
&= \frac{1}{2}(-3)\int_0^2 e^{(-1/3)t}\left(-\frac{1}{3}\,dt\right)\\
&= -\frac{3}{2}e^{(-1/3)t}\Big|_0^2 = -\frac{3}{2}\left(e^{-2/3} - e^0\right)\\
&= -\frac{3}{2}\left(e^{-2/3} - 1\right) = \frac{3}{2}\left(1 - e^{-2/3}\right)\ \text{A.}
\end{aligned}
$$

53.

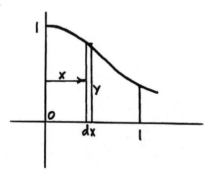

$$V = 2\pi \int_0^1 xe^{-x^2}\,dx.$$

Volume of shell:

$$2\pi \cdot (\text{radius}) \cdot (\text{height}) \cdot (\text{thickness})$$
$$= 2\pi xy\,dx.$$

Let $u = -x^2$; then $du = -2x\,dx$.

$$V = \frac{2\pi}{-2}\int_0^1 e^{-x^2}(-2x)\,dx = -\pi e^{-x^2}\Big|_0^1 = -\pi e^{-1} + \pi \cdot 1 = \pi(-e^{-1} + 1) = \pi\left(1 - \frac{1}{e}\right).$$

55. Volume of shell: $2\pi(\text{radius})(\text{height})(\text{thickness}) = 2\pi x\left(\dfrac{4}{x^2}\right)dx.$

$$
\begin{aligned}
V &= 2\pi \int_1^2 x\left(\frac{4}{x^2}\right)dx = 8\pi \int_1^2 \frac{1}{x}\,dx\\
&= 8\pi \ln|x|\Big|_1^2 = 8\pi(\ln 2 - \ln 1) = 8\pi \ln 2.
\end{aligned}
$$

57. Let $u = 10 - kv$; then $du = -k\,dv$. Multiplying both sides by $-k$, we get

$$\frac{-k\,dv}{10 - kv} = -k\,dt$$

$$\ln|10 - kv| = -kt + C. \qquad \text{integrating}$$

Since $10 - kv > 0$, $\ln|10 - kv| = \ln(10 - kv)$. Thus $\ln(10 - kv) = -kt + C$. If $t = 0$, $v = 0$.

Hence, $\ln 10 = 0 + C$ and

$$\ln(10 - kv) = -kt + \ln 10$$

$$\ln(10 - kv) - \ln 10 = -kt$$

$$\ln\frac{10 - kv}{10} = -kt. \qquad\qquad \log_a M - \log_a N = \log_a \frac{M}{N}$$

By definition of logarithm,

$$e^{-kt} = \frac{10 - kv}{10}$$

$$10e^{-kt} = 10 - kv$$

$$kv = 10(1 - e^{-kt})$$

$$v = \frac{10}{k}(1 - e^{-kt})\,\text{m/s}.$$

59. $\text{mass} = \displaystyle\int (3.00 + 0.0300e^{0.0100x})\,dx = 3.00x + 0.0300\left(\frac{1}{0.0100}\right)e^{0.0100x} + C$

$\qquad\quad = 3.00x + 3e^{0.0100x} + C.$

When $x = 0$, $\text{mass} = 0$; so

$0 = 0 + 3 + C$ and $C = -3$.

$\text{mass} = 3.00x + 3e^{0.0100x} - 3\Big|_{x=6.00} = 18.2\,\text{kg}.$

7.3 Trigonometric Forms

1. $\displaystyle\int \sec^2 2x\,dx \qquad\qquad\qquad\qquad u = 2x;\ du = 2\,dx.$

$\dfrac{1}{2}\displaystyle\int \sec^2 2x(2\,dx) = \dfrac{1}{2}\int \sec^2 u\,du = \dfrac{1}{2}\tan u + C = \dfrac{1}{2}\tan 2x + C.$

3. $\displaystyle\int \sec 3x\tan 3x\,dx \qquad\qquad\qquad u = 3x;\ du = 3\,dx.$

$\dfrac{1}{3}\displaystyle\int \sec 3x\tan 3x(3\,dx) = \dfrac{1}{3}\int \sec u\tan u\,du = \dfrac{1}{3}\sec u + C = \dfrac{1}{3}\sec 3x + C.$

5. $\displaystyle\int \csc 4x\cot 4x\,dx \qquad\qquad\qquad u = 4x;\ du = 4\,dx.$

$\dfrac{1}{4}\displaystyle\int \csc 4x\cot 4x(4\,dx) = \dfrac{1}{4}\int \csc u\cot u\,du = \dfrac{1}{4}(-\csc u) + C = -\dfrac{1}{4}\csc 4x + C.$

7. $\displaystyle\int \tan\frac{1}{2}x\,dx \qquad\qquad\qquad\qquad u = \frac{1}{2}x;\ du = \frac{1}{2}\,dx.$

$2\displaystyle\int \tan\frac{1}{2}x\left(\frac{1}{2}\,dx\right) = 2\int \tan u\,du = 2\ln|\sec u| + C = 2\ln\left|\sec\frac{1}{2}x\right| + C.$

9. $\displaystyle\int \cos 2t\,dt \qquad\qquad\qquad\qquad u = 2t;\ du = 2\,dt.$

$\dfrac{1}{2}\displaystyle\int \cos 2t(2\,dt) = \dfrac{1}{2}\int \cos u\,du = \dfrac{1}{2}\sin u + C = \dfrac{1}{2}\sin 2t + C.$

11. $\int y \csc y^2 \, dy$ $\qquad\qquad u = y^2; \ du = 2y \, dy.$

$\frac{1}{2} \int \csc y^2 (2y \, dy) = \frac{1}{2} \int \csc u \, du = \frac{1}{2} \ln|\csc u - \cot u| + C = \frac{1}{2} \ln|\csc y^2 - \cot y^2| + C.$

13. $\int \frac{\csc^2 y}{1 + \cot y} \, dy$ $\qquad\qquad u = 1 + \cot y; \ du = -\csc^2 y \, dy.$

$-\int \frac{-\csc^2 y \, dy}{1 + \cot y} = -\int \frac{du}{u} = -\ln|u| + C = -\ln|1 + \cot y| + C.$

15. $\int \frac{\sec^2 5t}{4 - \tan 5t} \, dt$ $\qquad\qquad u = 4 - \tan 5t; \ du = -5\sec^2 5t \, dt.$

$-\frac{1}{5} \int \frac{-5\sec^2 5t \, dt}{4 - \tan 5t} = -\frac{1}{5} \int \frac{du}{u} = -\frac{1}{5} \ln|u| + C = -\frac{1}{5} \ln|4 - \tan 5t| + C.$

17. $\int x \sin x^2 \, dx$ $\qquad\qquad u = x^2; \ du = 2x \, dx.$

$\frac{1}{2} \int \sin x^2 (2x) \, dx = \frac{1}{2} \int \sin u \, du = -\frac{1}{2} \cos u + C = -\frac{1}{2} \cos x^2 + C.$

19. $\int T \cot \frac{1}{2}T^2 \, dT$ $\qquad\qquad u = \frac{1}{2}T^2; \ du = T \, dT.$

$\int \cot \frac{1}{2}T^2 (T \, dT) = \int \cot u \, du = \ln|\sin u| + C = \ln\left|\sin \frac{1}{2}T^2\right| + C.$

21. $\int \frac{\cos \sqrt{x}}{\sqrt{x}} \, dx = \int \frac{\cos x^{1/2}}{x^{1/2}} \, dx.$ $\qquad u = x^{1/2}; \ du = \frac{1}{2}x^{-1/2} \, dx = \frac{dx}{2x^{1/2}}.$

$2 \int \cos x^{1/2} \frac{dx}{2x^{1/2}} = 2 \int \cos u \, du = 2 \sin u + C = 2 \sin \sqrt{x} + C.$

23. $\int \tan^3 4x \sec^2 4x \, dx$ $\qquad\qquad u = \tan 4x; \ du = 4\sec^2 4x \, dx.$

$\frac{1}{4} \int \tan^3 4x (4\sec^2 4x \, dx) = \frac{1}{4} \int u^3 \, du = \frac{1}{4}\frac{u^4}{4} + C = \frac{1}{16} \tan^4 4x + C.$

25. $\int \csc e^{3x} (e^{3x}) \, dx$ $\qquad\qquad u = e^{3x}; \ du = 3e^{3x} \, dx.$

$\frac{1}{3} \int \csc e^{3x} (3e^{3x}) \, dx = \frac{1}{3} \int \csc u \, du = \frac{1}{3} \ln|\csc u - \cot u| + C = \frac{1}{3} \ln|\csc e^{3x} - \cot e^{3x}| + C.$

27. $\int \frac{\csc^2 \ln x}{x} \, dx = \int \csc^2 \ln x \left(\frac{dx}{x}\right)$ $\qquad u = \ln x; \ du = \frac{dx}{x}.$

$\qquad\qquad = \int \csc^2 u \, du = -\cot u + C = -\cot(\ln x) + C.$

29. $\int \frac{\sec^2 x}{\tan^2 x} \, dx = \int \sec^2 x \cot^2 x \, dx = \int \frac{1}{\cos^2 x} \frac{\cos^2 x}{\sin^2 x} \, dx$

$\qquad\qquad = \int \frac{dx}{\sin^2 x} = \int \csc^2 x \, dx = -\cot x + C.$

31. $\int \frac{\sin x \, dx}{1 + 2\cos x}$ $\qquad\qquad u = 1 + 2\cos x; \ du = -2\sin x \, dx.$

$-\frac{1}{2} \int \frac{-2\sin x \, dx}{1 + 2\cos x} = -\frac{1}{2} \int \frac{du}{u} = -\frac{1}{2} \ln|u| + C = -\frac{1}{2} \ln|1 + 2\cos x| + C.$

33. $\displaystyle\int_0^{\sqrt{\pi/2}} \omega \cos \omega^2 \, d\omega$ Let $u = \omega^2$; then $du = 2\omega \, d\omega$.

Lower limit: if $\omega = 0$, $u = 0$.

Upper limit: if $\omega = \sqrt{\dfrac{\pi}{2}}$, $u = \omega^2 = \dfrac{\pi}{2}$.

$$\int_0^{\sqrt{\pi/2}} \omega \cos \omega^2 \, d\omega = \frac{1}{2} \int_0^{\sqrt{\pi/2}} (\cos \omega^2)(2\omega \, d\omega) = \frac{1}{2} \int_0^{\pi/2} \cos u \, du$$

$$= \frac{1}{2} \sin u \Big|_0^{\pi/2} = \frac{1}{2} \sin \frac{\pi}{2} - \frac{1}{2} \sin 0 = \frac{1}{2} \cdot 1 - \frac{1}{2} \cdot 0 = \frac{1}{2}.$$

35. $\displaystyle\int_0^{\pi/6} \frac{\cos x}{1 - \sin x} \, dx$ $u = 1 - \sin x$; $du = -\cos x \, dx$.

$$-\int_0^{\pi/6} \frac{-\cos x \, dx}{1 - \sin x} = -\ln|1 - \sin x| \Big|_0^{\pi/6}$$

$$= -\ln\left(1 - \sin \frac{\pi}{6}\right) + \ln(1 - \sin 0)$$

$$= -\ln\left(1 - \frac{1}{2}\right) + \ln 1 = -\ln \frac{1}{2} + 0$$

$$= -(\ln 1 - \ln 2) = \ln 2.$$

37. Recall that $v = \dfrac{q}{C} = \dfrac{1}{C} \displaystyle\int i \, dt$. Thus $v = \dfrac{1}{100 \times 10^{-6}} \displaystyle\int 2.00 \cos 100t \, dt$. Let $u = 100t$; then $du = 100 \, dt$.

$$v = \frac{2.00}{10^{-4}} \int \cos 100t \, dt = \frac{2.00}{10^{-4}} \frac{1}{100} \int \cos 100t (100) \, dt$$

$$= \frac{2.00}{10^{-2}} \int \cos u \, du = 200 \sin u + C = 200 \sin 100t + C.$$

If $t = 0$, $v = 0$. Thus $C = 0$.

$$v = 200 \sin 100t \Big|_{t=0.200} = 183 \, \text{V}, \text{ (calculator set in radian mode)}.$$

39. Volume of shell: $2\pi(\text{radius})(\text{height})(\text{thickness}) = 2\pi x \cos x^2 \, dx$.

$$V = 2\pi \int_0^{\sqrt{\pi/2}} x \cos x^2 \, dx \qquad u = x^2; \ du = 2x \, dx.$$

$$= \pi \int_0^{\sqrt{\pi/2}} \cos x^2 (2x \, dx)$$

$$= \pi \sin x^2 \Big|_0^{\sqrt{\pi/2}} = \pi \sin \frac{\pi}{2} = \pi \cdot 1 = \pi.$$

41. Moment (with respect to y-axis) of typical element: $xy \, dx$.

$$M_y = \int_0^{\sqrt{\pi/2}} x \tan x^2 \, dx. \text{ Let } u = x^2; \text{ then } du = 2x \, dx.$$

$$M_y = \frac{1}{2} \int_0^{\sqrt{\pi/2}} \tan x^2 (2x) \, dx = \frac{1}{2} \ln|\sec x^2| \Big|_0^{\sqrt{\pi/2}} = \frac{1}{2}\left(\ln \sec \frac{\pi}{4} - \ln \sec 0\right)$$

$$= \frac{1}{2}\left(\ln \sqrt{2} - \ln 1\right) = \frac{1}{2}\left(\ln 2^{1/2} - 0\right) = \frac{1}{2} \cdot \frac{1}{2} \ln 2 = \frac{1}{4} \ln 2.$$

43. Period $= \dfrac{2\pi}{\omega}$.

$$
\begin{aligned}
v_{\text{av}} &= \frac{1}{\pi/\omega - 0} \int_0^{\pi/\omega} E \sin \omega t \, dt \qquad\qquad u = \omega t; \ du = \omega \, dt. \\
&= \frac{1}{\pi/\omega} E \left(\frac{1}{\omega}\right) (-\cos \omega t)\Big|_0^{\pi/\omega} \\
&= \frac{E}{\pi}(-\cos \pi + \cos 0) = \frac{E}{\pi}(1+1) = \frac{2E}{\pi} \text{ V.}
\end{aligned}
$$

45. $\displaystyle\int_1^2 \frac{\sin x}{x} \, dx$; $\ h = \dfrac{2-1}{8} = \dfrac{1}{8} = 0.125$.

$x_0 = 1, \ x_1 = 1.125, \ x_2 = 1.125 + 0.125 = 1.25, \ x_3 = 1.25 + 0.125 = 1.375, \ \ldots, \ x_8 = 2$.

The function values are listed next:

$$
\begin{aligned}
f(1) &= \frac{\sin 1}{1} = 0.8415. \\[4pt]
f(1.125) &= \frac{\sin 1.125}{1.125} = 0.8020. \\[4pt]
f(1.25) &= \frac{\sin 1.25}{1.25} = 0.7592. \\[4pt]
f(1.375) &= 0.7134. \\[4pt]
f(1.5) &= 0.6650. \\[4pt]
f(1.625) &= 0.6145. \\[4pt]
f(1.75) &= 0.5623. \\[4pt]
f(1.875) &= 0.5088. \\[4pt]
f(2) &= 0.4546.
\end{aligned}
$$

$$
\int_1^2 \frac{\sin x}{x} \approx 0.125 \left[\frac{1}{2}(0.8415) + 0.8020 + 0.7592 + \cdots + \frac{1}{2}(0.4546) \right] = 0.66.
$$

7.4   Further Trigonometric Forms

1. $\displaystyle\int \sin^2 2x \cos^3 2x \, dx$

TYPE 1, m odd, identity (7.14). $\qquad\qquad\qquad\qquad u = \sin 2x; \ du = 2 \cos 2x \, dx.$

$$
\begin{aligned}
\int \sin^2 2x \cos^2 2x (\cos 2x \, dx) &= \int \sin^2 2x (1 - \sin^2 2x)(\cos 2x \, dx) \\
&= \frac{1}{2} \int \sin^2 2x (1 - \sin^2 2x)(2 \cos 2x \, dx) \\
&= \frac{1}{2} \int u^2 (1 - u^2) \, du = \frac{1}{2} \int (u^2 - u^4) \, du \\
&= \frac{1}{2}\left(\frac{1}{3}u^3 - \frac{1}{5}u^5 \right) + C \\
&= \frac{1}{6} \sin^3 2x - \frac{1}{10} \sin^5 2x + C.
\end{aligned}
$$

3. $\displaystyle\int \sin^3 x \, dx$

TYPE 1, n odd, identity (7.14). $\qquad\qquad\qquad\qquad u = \cos x; \ du = -\sin x \, dx.$

$$
\begin{aligned}
\int \sin^2 x \sin x \, dx &= \int (1 - \cos^2 x) \sin x \, dx \\
&= -\int (1 - \cos^2 x)(-\sin x \, dx) = -\int (1 - u^2) \, du \\
&= -u + \frac{1}{3}u^3 + C = \frac{1}{3} \cos^3 x - \cos x + C.
\end{aligned}
$$

5. $\int \sin^3 x \cos^4 x \, dx$

TYPE 1, n odd, identity (7.14). $\qquad u = \cos x; \;\; du = -\sin x \, dx.$

$$\int \sin^2 x \cos^4 x (\sin x \, dx) \;=\; \int (1 - \cos^2 x) \cos^4 x (\sin x \, dx)$$

$$= \; -\int (1 - \cos^2 x) \cos^4 x (-\sin x \, dx) = -\int (1 - u^2) u^4 \, du$$

$$= \; -\int (u^4 - u^6) \, du = -\frac{1}{5} u^5 + \frac{1}{7} u^7 + C$$

$$= \; \frac{1}{7} \cos^7 x - \frac{1}{5} \cos^5 x + C.$$

7. $\int \sin^3 x \cos^3 x \, dx$

TYPE 1, m odd; so we save $\cos x \, dx$ for du. (Observe that n is also odd, so that $\sin x \, dx$ could be saved for du.)

$$\int \sin^3 x \cos^2 x \cos x \, dx \;=\; \int \sin^3 x (1 - \sin^2 x) \cos x \, dx \qquad u = \sin x; \;\; du = \cos x \, dx.$$

$$= \; \int u^3 (1 - u^2) \, du = \int (u^3 - u^5) \, du$$

$$= \; \frac{1}{4} u^4 - \frac{1}{6} u^6 + C = \frac{1}{4} \sin^4 x - \frac{1}{6} \sin^6 x + C.$$

9. $\int \cos^2 4x \, dx$

TYPE 1, even powers, identity (7.17).

$$\int \cos^2 4x \, dx \;=\; \frac{1}{2} \int (1 + \cos 8x) \, dx$$

$$= \; \frac{1}{2} \int 1 \, dx + \frac{1}{2} \int \cos 8x \, dx \qquad u = 8x; \;\; du = 8 \, dx.$$

$$= \; \frac{1}{2} \int dx + \frac{1}{2} \cdot \frac{1}{8} \int \cos 8x (8 \, dx)$$

$$= \; \frac{1}{2} \int dx + \frac{1}{16} \int \cos u \, du$$

$$= \; \frac{1}{2} x + \frac{1}{16} \sin 8x + C.$$

11. $\int \sin^2 x \cos^2 x \, dx$

TYPE 1, even powers, identities (7.17) and (7.18).

$$\int \frac{1}{2}(1 - \cos 2x) \cdot \frac{1}{2}(1 + \cos 2x) \, dx \;=\; \frac{1}{4} \int (1 - \cos^2 2x) \, dx$$

$$= \; \frac{1}{4} \int dx - \frac{1}{4} \int \cos^2 2x \, dx$$

$$= \; \frac{1}{4} x - \frac{1}{4} \int \frac{1}{2}(1 + \cos 4x) \, dx$$

$$= \; \frac{1}{4} x - \frac{1}{8} \left(x + \frac{1}{4} \sin 4x \right) \qquad u = 4x; \;\; du = 4 \, dx.$$

$$= \; \frac{1}{4} x - \frac{1}{8} x - \frac{1}{32} \sin 4x + C$$

$$= \; \frac{1}{8} x - \frac{1}{32} \sin 4x + C.$$

13. $\displaystyle\int \sin^3 2t \cos^2 2t\, dt$

TYPE 1, n odd. $\qquad\qquad\qquad\qquad\qquad u = \cos 2t;\ \ du = -2\sin 2t\, dt.$

$$\begin{aligned}
\int \sin^2 2t \cos^2 2t(\sin 2t\, dt) &= \int (1 - \cos^2 2t)\cos^2 2t(\sin 2t\, dt) \\
&= -\frac{1}{2}\int (1 - \cos^2 2t)\cos^2 2t(-2\sin 2t\, dt) \\
&= -\frac{1}{2}\int (1 - u^2)u^2\, du = -\frac{1}{2}\int (u^2 - u^4)\, du \\
&= -\frac{1}{2}\left(\frac{u^3}{3} - \frac{u^5}{5}\right) + C = \frac{1}{10}u^5 - \frac{1}{6}u^3 + C. \\
&= \frac{1}{10}\cos^5 2t - \frac{1}{6}\cos^3 2t + C.
\end{aligned}$$

15. $\displaystyle\int \sin^4 4x \cos^3 4x\, dx$

TYPE 1, m odd, identity (7.14). $\qquad\qquad\qquad\qquad u = \sin 4x;\ \ du = 4\cos 4x\, dx.$

$$\begin{aligned}
\int \sin^4 4x \cos^2 4x \cos 4x\, dx &= \int \sin^4 4x(1 - \sin^2 4x)\cos 4x\, dx \\
&= \frac{1}{4}\int \sin^4 4x(1 - \sin^2 4x)(4\cos 4x\, dx) \\
&= \frac{1}{4}\int u^4(1 - u^2)\, du = \frac{1}{4}\left(\frac{1}{5}u^5 - \frac{1}{7}u^7\right) + C \\
&= \frac{1}{20}\sin^5 4x - \frac{1}{28}\sin^7 4x + C.
\end{aligned}$$

17. $\displaystyle\int \tan^3 x\, dx$

TYPE 2, $m = 0$, identity (7.15).

$$\begin{aligned}
\int \tan^3 x\, dx &= \int \tan x \tan^2 x\, dx = \int \tan x(\sec^2 x - 1)\, dx \\
&= \int \tan x \sec^2 x\, dx - \int \tan x\, dx.
\end{aligned}$$

Let $u = \tan x$; then $du = \sec^2 x\, dx$. Using formula (7.10):
$$\int u\, du - \int \tan x\, dx = \frac{1}{2}u^2 - (-\ln|\cos x|) + C = \frac{1}{2}\tan^2 x + \ln|\cos x| + C.$$

19. $\displaystyle\int \tan^2 x \sec^4\, dx$

TYPE 2, m even, identity (7.15). $\qquad\qquad\qquad\qquad u = \tan x;\ \ du = \sec^2 x\, dx.$

$$\begin{aligned}
\int \tan^2 x \sec^2 x \sec^2 x\, dx &= \int \tan^2 x(1 + \tan^2 x)\sec^2 x\, dx \\
&= \int u^2(1 + u^2)\, du = \frac{1}{3}u^3 + \frac{1}{5}u^5 + C \\
&= \frac{1}{3}\tan^3 x + \frac{1}{5}\tan^5 x + C.
\end{aligned}$$

21. $\displaystyle\int \tan y \sec^3 y\, dy$

TYPE 2, n odd.
$$\int \tan y \sec^3 y\, dy = \int \sec^2 y(\sec y \tan y)\, dy.$$
Let $u = \sec y$; then $du = \sec y \tan y\, dy$.
$$\int u^2\, du = \frac{1}{3}u^3 + C = \frac{1}{3}\sec^3 y + C.$$

23. $\int \tan^3 x \sec^3 x \, dx$

TYPE 2, n odd, identity (7.15). $u = \sec x;\ du = \sec x \tan x \, dx.$

$$\int \tan^2 x \sec^2 x (\sec x \tan x \, dx) = \int (\sec^2 x - 1) \sec^2 x (\sec x \tan x \, dx)$$

$$= \int (u^2 - 1) u^2 \, du = \int (u^4 - u^2) \, du$$

$$= \frac{1}{5} u^5 - \frac{1}{3} u^3 + C = \frac{1}{5} \sec^5 x - \frac{1}{3} \sec^3 x + C.$$

25. $\int \cot^6 2x \csc^4 2x \, dx$

TYPE 3, m even, identity (7.16). The technique is the same as for TYPE 2 with m even: set aside $\csc^2 2x$ for du, and change the remaining cosecants to cotangents.

$$\int \cot^6 2x \csc^4 2x \, dx = \int \cot^6 2x \csc^2 2x (\csc^2 2x) \, dx = \int \cot^6 2x (1 + \cot^2 2x)(\csc^2 2x \, dx).$$

Let $u = \cot 2x$; then $du = -2 \csc^2 2x \, dx.$

$$-\frac{1}{2} \int \cot^6 2x (1 + \cot^2 2x)(-2 \csc^2 2x \, dx) = -\frac{1}{2} \int u^6 (1 + u^2) \, du$$

$$= -\frac{1}{2} \int (u^6 + u^8) \, du = -\frac{1}{2} \left(\frac{1}{7} u^7 + \frac{1}{9} u^9 \right) + C$$

$$= -\frac{1}{14} \cot^7 2x - \frac{1}{18} \cot^9 2x + C.$$

27. $\int \csc^6 x \, dx$

TYPE 3, m even, identity (7.16). $u = \cot x;\ du = -\csc^2 x \, dx.$

$$\int \csc^4 x \csc^2 x \, dx = \int (1 + \cot^2 x)^2 \csc^2 x \, dx$$

$$= -\int (1 + \cot^2 x)^2 (-\csc^2 x \, dx)$$

$$= -\int (1 + u^2)^2 \, du = -\int (1 + 2u^2 + u^4) \, du$$

$$= -u - \frac{2}{3} u^3 - \frac{1}{5} u^5 + C$$

$$= -\cot x - \frac{2}{3} \cot^3 x - \frac{1}{5} \cot^5 x + C.$$

29. $\int_0^{\pi/4} (\tan x)^{1/2} \sec^4 \, dx$

TYPE 2, m even, identity (7.15).

$$\int (\tan x)^{1/2} \sec^2 x (\sec^2 x \, dx) = \int (\tan x)^{1/2} (1 + \tan^2 x)(\sec^2 x \, dx).$$

Let $u = \tan x$; $du = \sec^2 x \, dx.$

$$\int u^{1/2} (1 + u^2) \, du = \int (u^{1/2} + u^{5/2}) \, du = \frac{2}{3} u^{3/2} + \frac{2}{7} u^{7/2} + C.$$

$$\int_0^{\pi/4} (\tan x)^{1/2} \sec^4 x \, dx = \frac{2}{3} (\tan x)^{3/2} + \frac{2}{7} (\tan x)^{7/2} \Big|_0^{\pi/4}$$

$$= \frac{2}{3} \left(\tan \frac{\pi}{4} \right)^{3/2} + \frac{2}{7} \left(\tan \frac{\pi}{4} \right)^{7/2} - 0$$

$$= \frac{2}{3} + \frac{2}{7} = \frac{20}{21}.$$

31. $\displaystyle\int_0^\pi (1+\sin x)^2\,dx = \int_0^\pi (1+2\sin x + \sin^2 x)\,dx$

$$= \int_0^\pi dx + 2\int_0^\pi \sin x\,dx + \frac{1}{2}\int_0^\pi (1-\cos 2x)\,dx$$

$$= x\Big|_0^\pi - 2\cos x\Big|_0^\pi + \frac{1}{2}\left(x - \frac{1}{2}\sin 2x\right)\Big|_0^\pi$$

$$= \pi - 2(\cos\pi - \cos 0) + \frac{1}{2}(\pi - 0)$$

$$= \pi - 2(-1-1) + \frac{1}{2}\pi - 0 = 4 + \frac{3}{2}\pi.$$

33. $\displaystyle\int_0^{\pi/4} \frac{\sec^2 x}{1+\tan x}\,dx$

Let $u = 1 + \tan x$; then $du = \sec^2 x\,dx$.

Lower limit: if $x = 0$, $u = 1 + \tan 0 = 1$.

Upper limit: if $x = \dfrac{\pi}{4}$, $u = 1 + \tan\dfrac{\pi}{4} = 1 + 1 = 2$.

$$\int_0^{\pi/4} \frac{\sec^2 x}{1+\tan x}\,dx = \int_1^2 \frac{du}{u} = \ln|u|\Big|_1^2 = \ln 2 - \ln 1 = \ln 2 - 0 = \ln 2.$$

35. $\displaystyle\int \tan^5 4x \sec^4 4x\,dx$

TYPE 2, m even. (This integral can also be treated as TYPE 2 with n odd.)

$u = \tan 4x$; $du = 4\sec^2 4x\,dx$.

$$\int \tan^5 4x \sec^2 4x \sec^2 4x\,dx = \frac{1}{4}\int \tan^5 4x(1+\tan^2 4x)(4\sec^2 4x\,dx)$$

$$= \frac{1}{4}\int u^5(1+u^2)\,du = \frac{1}{4}\left(\frac{1}{6}u^6 + \frac{1}{8}u^8\right) + C$$

$$= \frac{1}{24}\tan^6 4x + \frac{1}{32}\tan^8 4x + C.$$

37. Volume of disk: $\pi(\text{radius})^2 \cdot \text{thickness} = \pi y^2\,dx$.

$V = \pi\displaystyle\int_0^\pi \sin^2 x\,dx$

TYPE 1, even powers, identity (7.18).

$$V = \frac{\pi}{2}\int_0^\pi (1-\cos 2x)\,dx$$

$$= \frac{\pi}{2}\int_0^\pi dx - \frac{\pi}{2}\int_0^\pi \cos 2x\,dx \qquad u = 2x;\ du = 2\,dx.$$

$$= \frac{\pi}{2}\int_0^\pi dx - \frac{\pi}{2}\cdot\frac{1}{2}\int_0^\pi \cos 2x(2\,dx)$$

$$= \frac{\pi}{2}x\Big|_0^\pi - \frac{\pi}{4}\sin 2x\Big|_0^\pi$$

$$= \frac{\pi^2}{2} - 0 = \frac{\pi^2}{2}.$$

39. $i = 20 \cos 100\pi t$. Period: $\dfrac{2\pi}{100\pi} = \dfrac{1}{50}$ second.

$$
\begin{aligned}
i_{\text{rms}}^2 &= \frac{1}{1/50 - 0} \int_0^{1/50} (20)^2 \cos^2 100\pi t \, dt \\
&= 50(400) \int_0^{1/50} \frac{1}{2}(1 + \cos 200\pi t) \, dt \\
&= 10000 \left(t + \frac{1}{200\pi} \sin 200\pi t \right) \Big|_0^{1/50} \\
&= 10000 \left(\frac{1}{50} \right) = 200.
\end{aligned}
$$

$i_{\text{rms}} = \sqrt{200} = \sqrt{100 \cdot 2} = 10\sqrt{2} \, \text{A}.$

41. Since $\sin\left(2t - \dfrac{\pi}{3}\right) = \sin 2t \cos \dfrac{\pi}{3} - \cos 2t \sin \dfrac{\pi}{3} = \dfrac{1}{2} \sin 2t - \dfrac{\sqrt{3}}{2} \cos 2t$, we get, by Exercise 40 with $T = \pi$:

$$
\begin{aligned}
P &= \frac{1}{\pi} \int_0^{\pi} (3 \sin 2t)(5) \left(\frac{1}{2} \sin 2t - \frac{\sqrt{3}}{2} \cos 2t \right) dt \\
&= \frac{15}{2\pi} \int_0^{\pi} \sin^2 2t \, dt - \frac{15\sqrt{3}}{2\pi} \cdot \frac{1}{2} \int_0^{\pi} \sin 2t \cos 2t (2 \, dt).
\end{aligned}
$$

$$
P = \frac{15}{2\pi} \cdot \frac{1}{2} \int_0^{\pi} (1 - \cos 4t) \, dt - \frac{15\sqrt{3}}{2\pi} \cdot \frac{1}{2} \int_0^{\pi} \sin 2t \cos 2t (2 \, dt).
$$

$\qquad u = 4t; \ du = 4 \, dt.$ $\qquad\qquad u = \sin 2t; \ du = 2 \cos 2t \, dt.$

$$
\begin{aligned}
P &= \frac{15}{4\pi} \left(t - \frac{1}{4} \sin 4t \right) \Big|_0^{\pi} - \frac{15\sqrt{3}}{4\pi} \left(\frac{1}{2} \sin^2 2t \right) \Big|_0^{\pi} \\
&= \frac{15}{4\pi} (\pi - 0) - 0 = \frac{15}{4} \, \text{W}.
\end{aligned}
$$

43. $\displaystyle\int_0^{\pi/2} \frac{dx}{\sqrt{1 + \cos x}} \quad (n = 8)$

$h = \dfrac{\pi/2 - 0}{8} = \dfrac{\pi}{16}$

$x_0 = 0, \ x_1 = \dfrac{\pi}{16}, \ x_2 = \dfrac{\pi}{8}, \ x_3 = \dfrac{3\pi}{16}, \ x_4 = \dfrac{\pi}{4},$

$x_5 = \dfrac{5\pi}{16}, \ x_6 = \dfrac{3\pi}{8}, \ x_7 = \dfrac{7\pi}{16}, \ x_8 = \dfrac{\pi}{2}.$

Some of the function values are

$$
\begin{aligned}
f(x_0) &= \frac{1}{\sqrt{1 + \cos 0}} = 0.7071 \\
f(x_1) &= \frac{1}{\sqrt{1 + \cos(\pi/16)}} = 0.7105 \\
f(x_2) &= \frac{1}{\sqrt{1 + \cos(\pi/8)}} = 0.7210 \\
f(x_3) &= \frac{1}{\sqrt{1 + \cos(3\pi/16)}} = 0.7389
\end{aligned}
$$

$\qquad\qquad \cdots$

$$
f(x_8) = \frac{1}{\sqrt{1 + \cos(\pi/2)}} = 1.
$$

$\displaystyle\int_0^{\pi/2} \frac{dx}{\sqrt{1 + \cos x}} \approx \frac{\pi/16}{3} [0.7071 + 4(0.7105) + 2(0.7210) + 4(0.7389) + \cdots + 1] = 1.25.$

45. Since the frequency is 60 cycles per second, the period is $(1/60)\ s$; or:

$$\text{Period} = \frac{120\pi}{2\pi} = \frac{1}{60}\text{ s.}$$

Then

$$\begin{aligned} i_{\text{rms}}^2 &= \frac{1}{1/60 - 0}\int_0^{1/60} 155^2 \sin^2 120\pi t\, dt \\ &= 60 \cdot 155^2 \cdot \frac{1}{2}\int_0^{1/60} (1 - \cos 240\pi t)\, dt \\ &= 60 \cdot 155^2 \cdot \frac{1}{2}\left(t - \frac{1}{240\pi}\sin 240\pi t\right)\Big|_0^{1/60} \qquad u = 240\pi t;\ du = 240\pi\, dt. \\ &= 60 \cdot 155^2 \cdot \frac{1}{2}\left(\frac{1}{60} - 0\right) = 155^2 \cdot \frac{1}{2}. \\ i_{\text{rms}} &= \sqrt{155^2 \cdot \frac{1}{2}} = \frac{155}{\sqrt{2}} \approx 110\,\text{V}. \end{aligned}$$

7.5 Inverse Trigonometric Forms

1. $\displaystyle\int \frac{dx}{\sqrt{1-x^2}} = \text{Arcsin}\, x + C$ by (7.19) with $a = 1$.

3. $\displaystyle\int \frac{dx}{9+4x^2} = \int \frac{dx}{9+(2x)^2}.$ $\qquad u = 2x;\ du = 2\,dx.$

$$\begin{aligned} \frac{1}{2}\int \frac{2\,dx}{9+(2x)^2} &= \frac{1}{2}\int \frac{du}{9+u^2} \\ &= \frac{1}{2}\cdot\frac{1}{3}\text{Arctan}\,\frac{u}{3} + C \qquad \text{by (7.20) with } a = 3 \\ &= \frac{1}{6}\text{Arctan}\,\frac{2}{3}x + C. \end{aligned}$$

5. $\displaystyle\int \frac{x\,dx}{\sqrt{1-x^2}} = \int (1-x^2)^{-1/2} x\,dx.$ $\qquad u = 1 - x^2;\ du = -2x\,dx.$

$$-\frac{1}{2}\int (1-x^2)^{-1/2}(-2x\,dx) = -\frac{1}{2}\int u^{-1/2}\,du = -\frac{1}{2}\frac{u^{1/2}}{1/2} + C = -\sqrt{u} + C = -\sqrt{1-x^2} + C.$$

7. $\displaystyle\int \frac{x\,dx}{16+9x^2}$ $\qquad u = 16 + 9x^2;\ du = 18x\,dx.$

$$\frac{1}{18}\int \frac{18x\,dx}{16+9x^2} = \frac{1}{18}\ln|u| + C = \frac{1}{18}\ln|16+9x^2| + C = \frac{1}{18}\ln(16+9x^2) + C.$$

9. $\displaystyle\int \frac{t\,dt}{4+t^4} = \int \frac{t\,dt}{4+(t^2)^2}.$ $\qquad u = t^2;\ du = 2t\,dt.$

$$\frac{1}{2}\int \frac{2t\,dt}{4+(t^2)^2} = \frac{1}{2}\int \frac{du}{4+u^2} = \frac{1}{2}\cdot\frac{1}{2}\text{Arctan}\,\frac{u}{2} + C \text{ by (7.20) with } a = 2.$$

We now get: $\displaystyle\frac{1}{4}\text{Arctan}\,\frac{1}{2}t^2 + C.$

11. $\displaystyle\int \frac{\csc^2 x\,dx}{\sqrt{4-\cot^2 x}}$ $\qquad u = \cot x;\ du = -\csc^2 x\,dx.$

$$-\int \frac{-\csc^2 x\,dx}{\sqrt{4-\cot^2 x}} = -\int \frac{du}{\sqrt{4-u^2}} = -\text{Arcsin}\,\frac{u}{2} + C = -\text{Arcsin}\left(\frac{1}{2}\cot x\right) + C.$$

13. $\displaystyle\int \frac{dx}{\sqrt{5-3x^2}} = \int \frac{dx}{\sqrt{5-\left(\sqrt{3}x\right)^2}}.$ $\qquad u = \sqrt{3}x; \ du = \sqrt{3}\,dx.$

$\displaystyle\frac{1}{\sqrt{3}}\int \frac{\sqrt{3}\,dx}{\sqrt{5-\left(\sqrt{3}x\right)^2}} = \frac{1}{\sqrt{3}}\int \frac{du}{\sqrt{5-u^2}} = \frac{1}{\sqrt{3}}\text{Arcsin}\,\frac{u}{\sqrt{5}} + C$ by (7.19) with $a = \sqrt{5}$.

We get: $\displaystyle\frac{1}{\sqrt{3}}\text{Arcsin}\,\frac{\sqrt{3}x}{\sqrt{5}} + C = \frac{\sqrt{3}}{3}\text{Arcsin}\,\frac{\sqrt{15}x}{5} + C.$

15. $\displaystyle\int \frac{\cos y \, dy}{2+\sin y} \qquad u = 2 + \sin y; \ du = \cos y \, dy.$

$\displaystyle\int \frac{du}{u} = \ln|u| + C = \ln|2+\sin y| + C = \ln(2+\sin y) + C.$

17. $\displaystyle\int_3^{3\sqrt{3}} \frac{3\,dx}{9+x^2} = 3 \cdot \frac{1}{3}\text{Arctan}\,\frac{x}{3}\Big|_3^{3\sqrt{3}} = \text{Arctan}\,\frac{3\sqrt{3}}{3} - \text{Arctan}\,\frac{3}{3}$

$\displaystyle\qquad\qquad = \text{Arctan}\,\sqrt{3} - \text{Arctan}\,1 = \frac{\pi}{3} - \frac{\pi}{4} = \frac{\pi}{12}.$

19. $\displaystyle\int \frac{dx}{\sqrt{4-3x^2}} = \int \frac{dx}{\sqrt{4-(\sqrt{3}x)^2}}.$ $\qquad u = \sqrt{3}x; \ du = \sqrt{3}\,dx.$

$\displaystyle\frac{1}{\sqrt{3}}\int \frac{\sqrt{3}\,dx}{\sqrt{4-(\sqrt{3}x)^2}} = \frac{1}{\sqrt{3}}\int \frac{du}{\sqrt{4-u^2}} = \frac{1}{\sqrt{3}}\text{Arcsin}\,\frac{u}{2} + C = \frac{1}{\sqrt{3}}\text{Arcsin}\,\frac{\sqrt{3}}{2}x + C.$

21. $\displaystyle\int \frac{x}{\sqrt{4-3x^2}}\,dx = \int (4-3x^2)^{-1/2}x\,dx.$ $\qquad u = 4-3x^2; \ du = -6x\,dx.$

$\displaystyle-\frac{1}{6}\int (4-3x^2)^{-1/2}(-6x\,dx) = -\frac{1}{6}\int u^{-1/2}\,du = -\frac{1}{6}\cdot 2u^{1/2} + C = -\frac{1}{3}\sqrt{4-3x^2} + C.$

23. $\displaystyle\int \frac{dx}{x^2-6x+9} = \int \frac{dx}{(x-3)^2}.$ $\qquad u = x-3; \ du = dx.$

$\displaystyle\int (x-3)^{-2}\,dx = \int u^{-2}\,du = \frac{u^{-1}}{-1} + C = -\frac{1}{x-3} + C.$

25. Completing the square, we have

$\quad x^2 - 6x + 10 = x^2 - 6x + \underline{9} - \underline{9} + 10 = (x^2 - 6x + 9) + 1 = (x-3)^2 + 1.$

$\displaystyle\int \frac{dx}{x^2-6x+10} = \int \frac{dx}{(x-3)^2+1}.$ $\qquad u = x-3; \ du = dx.$

$\displaystyle\int \frac{du}{u^2+1} = \text{Arctan}\,u + C = \text{Arctan}\,(x-3) + C$ by (7.20) with $a = 1$.

27. $\quad 1 - x^2 - 4x = -(x^2 + 4x - 1) = -(x^2 + 4x + 4 - 4 - 1)$

$\displaystyle\qquad\qquad = -(x^2 + 4x + 4) + 5 = 5 - (x+2)^2.$

$\displaystyle\int \frac{dx}{\sqrt{1-x^2-4x}} = \int \frac{dx}{\sqrt{5-(x+2)^2}}.$ $\qquad u = x+2; \ du = dx.$

$\displaystyle\int \frac{du}{\sqrt{5-u^2}} = \text{Arcsin}\,\frac{u}{\sqrt{5}} + C = \text{Arcsin}\,\frac{x+2}{\sqrt{5}} + C.$

29. Completing the square, we have

$$x^2 + 3x + 3 = x^2 + 3x + \frac{9}{4} - \frac{9}{4} + 3 = \left(x^2 + 3x + \frac{9}{4}\right) + \frac{3}{4} = \left(x + \frac{3}{2}\right)^2 + \frac{3}{4}.$$

$$\int \frac{dx}{x^2 + 3x + 3} = \int \frac{dx}{\left(x + \frac{3}{2}\right)^2 + \frac{3}{4}}. \qquad u = x + \frac{3}{2}; \ du = dx.$$

$$\int \frac{du}{u^2 + \frac{3}{4}} = \int \frac{du}{\left(\frac{\sqrt{3}}{2}\right)^2 + u^2} = \frac{2}{\sqrt{3}}\text{Arctan}\,\frac{2}{\sqrt{3}}u + C \text{ by (7.20) with } a = \frac{\sqrt{3}}{2}.$$

We obtain $\dfrac{2}{\sqrt{3}}\text{Arctan}\,\dfrac{2}{\sqrt{3}}\left(x + \dfrac{3}{2}\right) + C = \dfrac{2}{\sqrt{3}}\text{Arctan}\,\dfrac{2x + 3}{\sqrt{3}} + C.$

31. $\displaystyle\int \frac{x + 4}{x^2 + 16}\,dx = \int \frac{x\,dx}{x^2 + 16} + \int \frac{4\,dx}{x^2 + 16}.$

For the first integral, let $u = x^2 + 16$:

$$\frac{1}{2}\int \frac{2x\,dx}{x^2 + 16} + 4\int \frac{dx}{x^2 + 16} \quad = \quad \frac{1}{2}\ln(x^2 + 16) + 4 \cdot \frac{1}{4}\text{Arctan}\,\frac{x}{4} + C$$

$$= \quad \frac{1}{2}\ln(x^2 + 16) + \text{Arctan}\,\frac{1}{4}x + C.$$

33. $\displaystyle\int_0^{\pi/2} \frac{\cos x}{1 + \sin^2 x}\,dx \qquad\qquad u = \sin x; \ du = \cos x\,dx.$

$$\int_0^{\pi/2} \frac{\cos x}{1 + \sin^2 x}\,dx = \text{Arctan}\,(\sin x)\Big|_0^{\pi/2} = \text{Arctan}\,1 - \text{Arctan}\,0 = \frac{\pi}{4}.$$

35. $\displaystyle\int \frac{e^{2x}}{1 + e^{2x}}\,dx \qquad u = 1 + e^{2x}; \ du = 2e^{2x}\,dx.$

$$\frac{1}{2}\int \frac{2e^{2x}\,dx}{1 + e^{2x}} = \frac{1}{2}\int \frac{du}{u} = \frac{1}{2}\ln|u| + C = \frac{1}{2}\ln(1 + e^{2x}) + C.$$

37. $\displaystyle\int \frac{e^\theta\,d\theta}{\sqrt{1 - e^{2\theta}}} = \int \frac{e^\theta\,d\theta}{\sqrt{1 - (e^\theta)^2}}. \qquad u = e^\theta; \ du = e^\theta\,d\theta.$

$$\int \frac{du}{\sqrt{1 - u^2}} = \text{Arcsin}\,u + C = \text{Arcsin}\,e^\theta + C \text{ by (7.19) with } a = 1.$$

39. $\displaystyle A = \int_0^\infty \frac{1}{x^2 + 1}\,dx = \lim_{b \to \infty}\int_0^b \frac{1}{x^2 + 1}\,dx = \lim_{b \to \infty}\text{Arctan}\,x\Big|_0^b = \frac{\pi}{2} - 0 = \frac{\pi}{2}.$

41. Volume of shell: $2\pi \cdot (\text{radius}) \cdot (\text{height}) \cdot (\text{thickness}) = 2\pi xy\,dx.$

$$V \ = \ 2\pi\int_0^\infty x\frac{1}{4 + x^4}\,dx = 2\pi\lim_{b \to \infty}\int_0^b \frac{x\,dx}{4 + (x^2)^2} = 2\pi\lim_{b \to \infty}\frac{1}{2}\int_0^b \frac{2x\,dx}{4 + (x^2)^2}$$

$$= \ \pi\lim_{b \to \infty}\frac{1}{2}\text{Arctan}\,\frac{x^2}{2}\Big|_0^b = \frac{\pi}{2}\lim_{b \to \infty}\left(\text{Arctan}\,\frac{b^2}{2} - 0\right).$$

Since $\tan\theta$ is undefined for $\theta = \dfrac{\pi}{2}$, we obtain:

$$V = \frac{\pi}{2}\left(\frac{\pi}{2}\right) = \frac{\pi^2}{4}.$$

7.6 Integration by Trigonometric Substitution

1. $\int \dfrac{\sqrt{4-x^2}}{x^2}\, dx$. Let $x = 2\sin\theta$, $dx = 2\cos\theta\, d\theta$. Note that

$$\sqrt{4-x^2} = \sqrt{4 - 4\sin^2\theta} = \sqrt{4(1-\sin^2\theta)} = 2\sqrt{\cos^2\theta} = 2\cos\theta.$$

$$\int \frac{\sqrt{4-x^2}\, dx}{x^2} = \int \frac{(2\cos\theta)(2\cos\theta\, d\theta)}{4\sin^2\theta} = \int \frac{\cos^2\theta}{\sin^2\theta}\, d\theta$$

$$= \int \cot^2\theta\, d\theta = \int (\csc^2\theta - 1)\, d\theta = -\cot\theta - \theta + C.$$

From the substituted expression $x = 2\sin\theta$, we get $\sin\theta = \dfrac{x}{2}$ or $\theta = \text{Arcsin}\,\dfrac{x}{2}$. To change $\cot\theta$, we use a diagram. Note that the opposite side has length x and the hypotenuse length 2. The remaining side is of length $\sqrt{4-x^2}$ by the Pythagorean theorem.

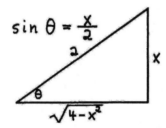

Since $\cot\theta = \dfrac{\sqrt{4-x^2}}{x}$, we get

$$\int \frac{\sqrt{4-x^2}}{x^2}\, dx = -\cot\theta - \theta + C$$

$$= -\frac{\sqrt{4-x^2}}{x} - \text{Arcsin}\,\frac{x}{2} + C.$$

3. $\int x\sqrt{x^2+9}\, dx$. Let $u = x^2 + 9$; then $du = 2x\, dx$ and we simply have

$$\frac{1}{2}\int (x^2+9)^{1/2}(2x\, dx) = \frac{1}{2}\int u^{1/2}\, du = \frac{1}{2}\cdot\frac{2}{3}u^{3/2} + C = \frac{1}{3}(x^2+9)^{3/2} + C.$$

5. $\int \dfrac{dx}{(x^2+25)^{3/2}}$. Let $x = 5\tan\theta$; $dx = 5\sec^2\theta\, d\theta$. Then

$$(x^2+25)^{3/2} = (25\tan^2\theta + 25)^{3/2} = (25)^{3/2}(\tan^2\theta + 1)^{3/2} = 125(\sec^2\theta)^{3/2} = 125\sec^3\theta$$

and we get:

$$\int \frac{5\sec^2\theta\, d\theta}{125\sec^3\theta} = \frac{1}{25}\int \frac{d\theta}{\sec\theta} = \frac{1}{25}\int \cos\theta\, d\theta = \frac{1}{25}\sin\theta + C.$$

The diagram is constructed from the substituted expression $\tan\theta = \dfrac{x}{5}$.

Thus $\dfrac{1}{25}\sin\theta + C$

$$= \frac{1}{25}\frac{x}{\sqrt{x^2+25}} + C.$$

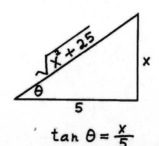

7. $\int \dfrac{dx}{x\sqrt{x^2-4}}$. Let $x = 2\sec\theta$; $dx = 2\sec\theta\tan\theta\,d\theta$. Then

$$\sqrt{x^2-4} = \sqrt{4\sec^2\theta-4} = \sqrt{4(\sec^2\theta-1)} = 2\sqrt{\sec^2\theta-1} = 2\sqrt{\tan^2\theta} = 2\tan\theta \text{ and}$$

$$\int \frac{dx}{x\sqrt{x^2-4}} = \int \frac{2\sec\theta\tan\theta\,d\theta}{2\sec\theta\cdot 2\tan\theta} = \int \frac{1}{2}\,d\theta = \frac{1}{2}\theta + C = \frac{1}{2}\operatorname{Arcsec}\frac{x}{2} + C$$

since $\sec\theta = \dfrac{x}{2}$. This relationship also yields the following diagram:

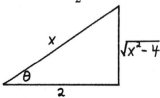

So the answer can also be written: $\dfrac{1}{2}\operatorname{Arctan}\left(\dfrac{1}{2}\sqrt{x^2-4}\right) + C$.

9. $\int \dfrac{dx}{x^2\sqrt{x^2+16}}$. Let $x = 4\tan\theta$; $dx = 4\sec^2\theta\,d\theta$. Then

$$\sqrt{x^2+16} = \sqrt{16\tan^2\theta+16} = 4\sqrt{\tan^2\theta+1} = 4\sqrt{\sec^2\theta} = 4\sec\theta.$$

$$\begin{aligned}
\int \frac{dx}{x^2\sqrt{x^2+16}} &= \int \frac{4\sec^2\theta\,d\theta}{(16\tan^2\theta)(4\sec\theta)} = \frac{1}{16}\int \frac{\sec\theta\,d\theta}{\tan^2\theta} = \frac{1}{16}\int \frac{d\theta}{\cos\theta\tan^2\theta} \\
&= \frac{1}{16}\int \frac{\cos^2\theta\,d\theta}{\cos\theta\sin^2\theta} = \frac{1}{16}\int (\sin\theta)^{-2}\cos\theta\,d\theta
\end{aligned}$$

$$[u = \sin\theta;\ du = \cos\theta\,d\theta.]$$

$$\begin{aligned}
&= \frac{1}{16}\int u^{-2}\,du + C = \frac{1}{16}\frac{u^{-1}}{-1} + C \\
&= -\frac{1}{16}\frac{1}{\sin\theta} + C = -\frac{1}{16}\csc\theta + C.
\end{aligned}$$

From $\tan\theta = \dfrac{x}{4}$, we construct the diagram shown at the right.

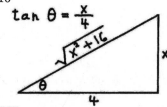

We now have $-\dfrac{1}{16}\csc\theta + C = -\dfrac{1}{16}\dfrac{\sqrt{x^2+16}}{x} + C$.

11. $\int \dfrac{x\,dx}{(x^2-2)^{3/2}}$. Let $u = x^2 - 2$; $du = 2x\,dx$. Then

$$\begin{aligned}
\int (x^2-2)^{-3/2}x\,dx &= \frac{1}{2}\int (x^2-2)^{-3/2}(2x\,dx) \\
&= \frac{1}{2}\int u^{-3/2}\,du = \frac{1}{2}\frac{u^{-1/2}}{-1/2} + C = -\frac{1}{\sqrt{x^2-2}} + C.
\end{aligned}$$

13. $\int \dfrac{x^3\,dx}{\sqrt{x^2-3}}$. Let $x=\sqrt{3}\sec\theta$; $dx=\sqrt{3}\sec\theta\tan\theta\,d\theta$. Then

$$\sqrt{x^2-3}=\sqrt{3\sec^2\theta-3}=\sqrt{3}\sqrt{\sec^2\theta-1}=\sqrt{3}\sqrt{\tan^2\theta}=\sqrt{3}\tan\theta.$$

$$\begin{aligned}
\int \frac{x^3\,dx}{\sqrt{x^2-3}} &= \int \frac{(\sqrt{3})^3\sec^3\theta(\sqrt{3}\sec\theta\tan\theta\,d\theta)}{\sqrt{3}\tan\theta}\\[2mm]
&= 3\sqrt{3}\int \sec^4\theta\,d\theta = 3\sqrt{3}\int \sec^2\theta\sec^2\theta\,d\theta\\[2mm]
&= 3\sqrt{3}\int (1+\tan^2\theta)\sec^2\theta\,d\theta
\end{aligned}$$

$$[u=\tan\theta;\ du=\sec^2\theta\,d\theta.]$$

$$\begin{aligned}
&= 3\sqrt{3}\int (1+u^2)\,du = 3\sqrt{3}\left(u+\frac{1}{3}u^3\right)+C\\[2mm]
&= 3\sqrt{3}\left(\tan\theta+\frac{1}{3}\tan^3\theta\right)+C.
\end{aligned}$$

To construct the diagram, we use
$\sec\theta=\dfrac{x}{\sqrt{3}}.$

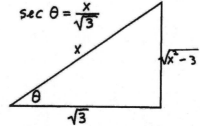

$$\begin{aligned}
3\sqrt{3}\left(\tan\theta+\frac{1}{3}\tan^3\theta\right) &= 3\sqrt{3}\left[\frac{\sqrt{x^2-3}}{\sqrt{3}}+\frac{1}{3}\left(\frac{\sqrt{x^2-3}}{\sqrt{3}}\right)^3\right]+C\\[2mm]
&= 3\sqrt{x^2-3}+\sqrt{3}\frac{(x^2-3)^{3/2}}{3\sqrt{3}}+C\\[2mm]
&= 3\sqrt{x^2-3}+\frac{1}{3}(x^2-3)^{3/2}+C\\[2mm]
&= \sqrt{x^2-3}\left[3+\frac{1}{3}(x^2-3)\right]+C\\[2mm]
&= \frac{1}{3}(x^2+6)\sqrt{x^2-3}+C.
\end{aligned}$$

15. $\int \dfrac{\sqrt{x^2+1}}{x^2}\,dx$. Let $x=\tan\theta$; $dx=\sec^2\theta\,d\theta$. Then,

$$\sqrt{x^2+1}=\sqrt{\tan^2\theta+1}=\sqrt{\sec^2\theta}=\sec\theta.$$

$$\begin{aligned}
\int \frac{\sqrt{x^2+1}\,dx}{x^2} &= \int \frac{\sec\theta\sec^2\theta\,d\theta}{\tan^2\theta} = \int \frac{\sec\theta(1+\tan^2\theta)\,d\theta}{\tan^2\theta}\\[2mm]
&= \int \left(\frac{\sec\theta}{\tan^2\theta}+\sec\theta\right)d\theta = \int \frac{\cos^2\theta}{\sin^2\theta}\frac{1}{\cos\theta}\,d\theta + \int \sec\theta\,d\theta\\[2mm]
&= \int (\sin\theta)^{-2}\cos\theta\,d\theta + \ln|\sec\theta+\tan\theta|+C\\[2mm]
&= -(\sin\theta)^{-1}+\ln|\sec\theta+\tan\theta|+C.
\end{aligned}$$

From $\tan\theta=\dfrac{x}{1}$, we get the following diagram:

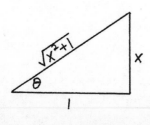

The result is: $-\dfrac{\sqrt{x^2+1}}{x}+\ln|x+\sqrt{x^2+1}|+C.$

17. $\displaystyle\int_0^4 \frac{dx}{(16+x^2)^{3/2}}$. Let $x = 4\tan\theta$; $dx = 4\sec^2\theta\, d\theta$.

Lower limit: if $x = 0$, then $\theta = 0$.

Upper limit: if $x = 4$, then $\tan\theta = 1$ and $\theta = \dfrac{\pi}{4}$.

$$\int_0^{\pi/4} \frac{4\sec^2\theta\, d\theta}{(16+16\tan^2\theta)^{3/2}} = \int_0^{\pi/4} \frac{4\sec^2\theta\, d\theta}{16^{3/2}(\sec^2\theta)^{3/2}} = \frac{4}{16^{3/2}}\int_0^{\pi/4} \frac{d\theta}{\sec\theta}$$

$$= \frac{1}{16}\int_0^{\pi/4} \cos\theta\, d\theta = \frac{1}{16}\sin\theta\Big|_0^{\pi/4} = \frac{1}{16}\cdot\frac{\sqrt{2}}{2} = \frac{\sqrt{2}}{32}.$$

19. $\displaystyle\int \sqrt{9-x^2}\, dx$. Let $x = 3\sin\theta$; $dx = 3\cos\theta\, d\theta$. Then

$$\sqrt{9-x^2} = \sqrt{9 - 9\sin^2\theta} = 3\sqrt{1-\sin^2\theta} = 3\sqrt{\cos^2\theta} = 3\cos\theta.$$

$$\int \sqrt{9-x^2}\, dx = \int (3\cos\theta)(3\cos\theta\, d\theta) = 9\int \cos^2\theta\, d\theta$$

$$= \frac{9}{2}\int (1+\cos 2\theta)\, d\theta = \frac{9}{2}\left(\theta + \frac{1}{2}\sin 2\theta\right) + C.$$

Since $\sin 2\theta = 2\sin\theta\cos\theta$, we get:

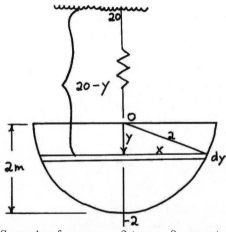

$$\frac{9}{2}(\theta + \sin\theta\cos\theta) + C$$

$$= \frac{9}{2}\left(\text{Arcsin}\,\frac{x}{3} + \frac{x}{3}\frac{\sqrt{9-x^2}}{3}\right) + C$$

$$= \frac{9}{2}\text{Arcsin}\,\frac{x}{3} + \frac{1}{2}x\sqrt{9-x^2} + C.$$

For the definite integral, we now get:

$$\int_0^3 \sqrt{9-x^2}\, dx = \frac{9}{2}\text{Arcsin}\,\frac{x}{3} + \frac{1}{2}x\sqrt{9-x^2}\,\Big|_0^3 = \frac{9}{2}\text{Arcsin}\,1 = \frac{9}{2}\frac{\pi}{2} = \frac{9\pi}{4}.$$

21.

Length of strip: $2x = 2\sqrt{4-y^2}$.

Area of strip: $2\sqrt{4-y^2}\, dy$.

Pressure on strip: $(20-y)w$.

Force against strip:

$$(20-y)w(2\sqrt{4-y^2}\, dy).$$

Summing from $y = -2$ to $y = 0$, we get:

$$F = \int_{-2}^0 2w(20-y)\sqrt{4-y^2}\, dy = 2w\int_{-2}^0 20\sqrt{4-y^2}\, dy - 2w\int_{-2}^0 y\sqrt{4-y^2}\, dy.$$

To evaluate $\displaystyle\int \sqrt{4-y^2}\, dy$, we let $y = 2\sin\theta$, so that $dy = 2\cos\theta\, d\theta$. Thus

$$\sqrt{4-y^2} = \sqrt{4-4\sin^2\theta} = 2\sqrt{1-\sin^2\theta} = 2\cos\theta.$$

$$\int \sqrt{4-y^2}\, dy = \int (2\cos\theta)(2\cos\theta\, d\theta) = 4\int \cos^2\theta\, d\theta$$

$$= 4\int \frac{1}{2}(1+\cos 2\theta)\, d\theta = 2\left(\theta + \frac{1}{2}\sin 2\theta\right)$$

$$= 2(\theta + \sin\theta\cos\theta) = 2\mathrm{Arcsin}\frac{y}{2} + 2\left(\frac{y}{2}\right)\frac{\sqrt{4-y^2}}{2}$$

$$= 2\mathrm{Arcsin}\frac{y}{2} + \frac{1}{2}y\sqrt{4-y^2}.$$

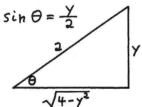

To evaluate $\int y\sqrt{4-y^2}\, dy$, let $u = 4-y^2$; $du = -2y\, dy$. Then

$$-\frac{1}{2}\int (4-y^2)^{1/2}(-2y)\, dy = -\frac{1}{2}\frac{2}{3}(4-y^2)^{3/2} = -\frac{1}{3}(4-y^2)^{3/2}.$$

We now have

$$F = 40w\left[2\mathrm{Arcsin}\frac{y}{2} + \frac{1}{2}y\sqrt{4-y^2}\right] - 2w\left[-\frac{1}{3}(4-y^2)^{3/2}\right]\Big|_{-2}^{0}$$

$$= -40w\left[2\cdot\left(-\frac{\pi}{2}\right)\right] - 2w\left(-\frac{1}{3}\cdot 4^{3/2}\right) \qquad \left(\text{since Arcsin}\,(-1) = -\frac{\pi}{2}\right)$$

$$= 40w\pi + \left(\frac{16}{3}\right)w = \left(40\pi + \frac{16}{3}\right)w\,\mathrm{N}.$$

23.

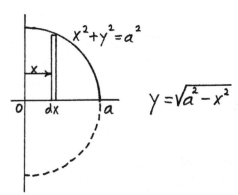

Mass of shell: $\rho\cdot 2\pi(\text{radius})\cdot(\text{height})\cdot(\text{thickness}) = \rho\cdot 2\pi x\sqrt{a^2-x^2}\, dx$.

Moment of inertia of typical shell: $x^2\cdot\rho\cdot 2\pi x\sqrt{a^2-x^2}\, dx$.

$$I_y = 2\int_0^a x^2\cdot\rho\cdot 2\pi x\sqrt{a^2-x^2}\, dx = 4\pi\rho\int_0^a x^3\sqrt{a^2-x^2}\, dx.$$

Let $x = a\sin\theta$; then $dx = a\cos\theta\, d\theta$.

Lower limit: if $x = 0$, then $\theta = 0$.

Upper limit: if $x = a$, $\sin\theta = 1$ and $\theta = \dfrac{\pi}{2}$.

$$4\pi\rho\int_0^{\pi/2} a^3\sin^3\theta\sqrt{a^2-a^2\sin^2\theta}(a\cos\theta\, d\theta) = 4\pi\rho a^5\int_0^{\pi/2}\sin^3\theta\cos^2\theta\, d\theta.$$

The definite integral

$$\int \sin^3\theta\cos^2\theta\, d\theta$$

is an odd power of sine. Thus

$$\int \sin^2\theta \cos^2\theta(\sin\theta\,d\theta) = -\int (1-\cos^2\theta)\cos^2\theta(-\sin^2\theta\,d\theta) \qquad u=\cos\theta;\ \ du=-\sin\theta\,d\theta.$$

$$= -\int (1-u^2)u^2\,du = -\frac{1}{3}u^3 + \frac{1}{5}u^5 = -\frac{1}{3}\cos^3\theta + \frac{1}{5}\cos^5\theta.$$

The result is: $I_y = 4\pi\rho a^5\left(-\dfrac{1}{3}\cos^3\theta + \dfrac{1}{5}\cos^5\theta\right)\Big|_0^{\pi/2} = 0 + 4\pi\rho a^5\left(\dfrac{1}{3}-\dfrac{1}{5}\right) = \dfrac{8a^5\pi\rho}{15}.$

I_y can also be written as follows: since the mass of a sphere is $\rho\cdot\text{volume} = \rho\cdot\dfrac{4}{3}\pi a^3$, we get

$$I_y = \frac{8a^5\pi\rho}{15} = \frac{2}{5}\left(\frac{4}{3}\pi a^3\rho\right)a^2 = \frac{2}{5}ma^2.$$

25.

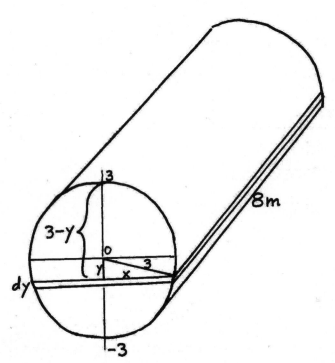

Volume of element: $2x(8\,dy) = 2\sqrt{9-y^2}(8\,dy) = 16\sqrt{9-y^2}\,dy.$

Weight of element: $16w\sqrt{9-y^2}\,dy.$

Work done in moving a typical element to the top of the tank: $(3-y)(16w\sqrt{9-y^2}\,dy).$

Summing from $y=-3$ to $y=0$, we get

$$W = 16w\int_{-3}^{0}(3-y)\sqrt{9-y^2}\,dy = (16w)3\int_{-3}^{0}\sqrt{9-y^2}\,dy - 16w\int_{-3}^{0}y\sqrt{9-y^2}\,dy.$$

The first integral has already been evaluated in Exercise 19. In the second integral, we let $u = 9-y^2;\ \ du = -2y\,dy.$ Thus

$$W = 48w\left(\frac{9}{2}\text{Arcsin}\,\frac{y}{3} + \frac{1}{2}y\sqrt{9-y^2}\right)\Big|_{-3}^{0} - \frac{16w}{-2}\int_{-3}^{0}(9-y^2)^{1/2}(-2y)\,dy$$

$$= 48w\left[0 - \frac{9}{2}\text{Arcsin}\,(-1)\right] + \left[8w\left(\frac{2}{3}\right)(9-y^2)^{3/2}\right]\Big|_{-3}^{0}$$

$$= 48w\left[-\frac{9}{2}\left(-\frac{\pi}{2}\right)\right] + 8w\left(\frac{2}{3}\right)(9)^{3/2}$$

$$= 108\pi w + 144w = (108\pi + 144)w\ \text{J}.$$

7.7 Integration by Parts

1. $\displaystyle\int x e^x \, dx$

$$\begin{aligned} u &= x & dv &= e^x \, dx \\ du &= dx & v &= e^x \end{aligned}$$

$$\int x e^x \, dx = uv - \int v \, du = x e^x - \int e^x \, dx = x e^x - e^x + C.$$

3. $\displaystyle\int x \sin 2x \, dx$

$$\begin{aligned} u &= x & dv &= \sin 2x \, dx \\ du &= dx & v &= -\frac{1}{2} \cos 2x \end{aligned}$$

$$\begin{aligned} \int x \sin 2x \, dx &= uv - \int v \, du = x \left(-\frac{1}{2} \cos 2x \right) - \int \left(-\frac{1}{2} \cos 2x \right) dx \\ &= -\frac{1}{2} x \cos 2x + \frac{1}{2} \int \cos 2x \, dx = -\frac{1}{2} x \cos 2x + \frac{1}{2} \cdot \frac{1}{2} \int \cos 2x (2 \, dx) \\ &= \frac{1}{4} \sin 2x - \frac{1}{2} x \cos 2x + C. \end{aligned}$$

5. $\displaystyle\int x \sec^2 x \, dx$

$$\begin{aligned} u &= x & dv &= \sec^2 x \, dx \\ du &= dx & v &= \tan x \end{aligned}$$

$$\int x \sec^2 x \, dx = x \tan x - \int \tan x \, dx = x \tan x + \ln |\cos x| + C.$$

7. $\displaystyle\int x \ln x \, dx$

$$\begin{aligned} u &= \ln x & dv &= x \, dx \\ du &= \frac{1}{x} \, dx & v &= \frac{1}{2} x^2 \end{aligned}$$

$$\begin{aligned} uv - \int v \, du &= \ln x \left(\frac{1}{2} x^2 \right) - \int \frac{1}{2} x^2 \cdot \frac{1}{x} \, dx = \frac{1}{2} x^2 \ln x - \frac{1}{2} \int x \, dx \\ &= \frac{1}{2} x^2 \ln x - \frac{1}{4} x^2 + C = \frac{1}{4} x^2 (2 \ln x - 1) + C. \end{aligned}$$

9. $\displaystyle\int \mathrm{Arcsin}\, x \, dx$

$$\begin{aligned} u &= \mathrm{Arcsin}\, x & dv &= dx \\ du &= \frac{dx}{\sqrt{1 - x^2}} & v &= x \end{aligned}$$

$$\begin{aligned} \int \mathrm{Arcsin}\, x \, dx &= x \mathrm{Arcsin}\, x - \int \frac{x}{\sqrt{1 - x^2}} \, dx \\ &= x \mathrm{Arcsin}\, x - \int (1 - x^2)^{-1/2} x \, dx \\ &= x \mathrm{Arcsin}\, x + \frac{1}{2} \int (1 - x^2)^{-1/2} (-2x) \, dx \quad (u = 1 - x^2; \ du = -2x \, dx.) \\ &= x \mathrm{Arcsin}\, x + \frac{1}{2} \frac{(1 - x^2)^{1/2}}{1/2} + C \\ &= x \mathrm{Arcsin}\, x + \sqrt{1 - x^2} + C. \end{aligned}$$

11. $\displaystyle\int x \cos 3x \, dx$

$$\begin{aligned} u &= x & dv &= \cos 3x \, dx \\ du &= dx & v &= \frac{1}{3} \sin 3x \end{aligned}$$

$$\begin{aligned} uv - \int v \, du &= x \left(\frac{1}{3} \sin 3x \right) - \int \frac{1}{3} \sin 3x \, dx \\ &= \frac{1}{3} x \sin 3x - \frac{1}{3} \cdot \frac{1}{3} \int \sin 3x (3 \, dx) = \frac{1}{3} x \sin 3x + \frac{1}{9} \cos 3x + C. \end{aligned}$$

13. $\displaystyle\int x^2 e^{-x}\,dx$
$$\begin{aligned} u &= x^2 & dv &= e^{-x}\,dx \\ du &= 2x\,dx & v &= -e^{-x} \end{aligned}$$

$\displaystyle\int x^2 e^{-x}\,dx = -x^2 e^{-x} + 2\int xe^{-x}\,dx.$
$$\begin{aligned} u &= x & dv &= e^{-x}\,dx \\ du &= dx & v &= -e^{-x} \end{aligned}$$

$$\begin{aligned} \int x^2 e^{-x}\,dx &= -x^2 e^{-x} + 2\left[-xe^{-x} + \int e^{-x}\,dx\right] \\ &= -x^2 e^{-x} - 2xe^{-x} - 2e^{-x} + C \\ &= -e^{-x}(x^2 + 2x + 2) + C. \end{aligned}$$

15. $\displaystyle\int \operatorname{Arccot} x\,dx$
$$\begin{aligned} u &= \operatorname{Arccot} x & du &= dx \\ du &= -\frac{1}{1+x^2}\,dx & u &= x \end{aligned}$$

$$\begin{aligned} \int \operatorname{Arccot} x\,dx &= x\operatorname{Arccot} x + \int \frac{x\,dx}{1+x^2} \\ &= x\operatorname{Arccot} x + \frac{1}{2}\int \frac{2x\,dx}{1+x^2} = x\operatorname{Arccot} x + \frac{1}{2}\ln(1+x^2) + C. \end{aligned}$$

17. $\displaystyle\int x\sin x^2\,dx.$ Here integration by parts is not needed: let $u = x^2$; $du = 2x\,dx$. Then
$$\begin{aligned} \int x\sin x^2\,dx &= \frac{1}{2}\int (\sin x^2)(2x)\,dx = \frac{1}{2}\int \sin u\,du \\ &= -\frac{1}{2}\cos u + C = -\frac{1}{2}\cos x^2 + C. \end{aligned}$$

19. $\displaystyle\int e^x \sin x\,dx$
$$\begin{aligned} u &= e^x & dv &= \sin x\,dx \\ du &= e^x\,dx & v &= -\cos x \end{aligned}$$

$\displaystyle\int e^x \sin x\,dx = -e^x \cos x + \int e^x \cos x\,dx.$
$$\begin{aligned} u &= e^x & dv &= \cos x\,dx \\ du &= e^x\,dx & v &= \sin x \end{aligned}$$

$$\begin{aligned} \int e^x \sin x\,dx &= -e^x \cos x + e^x \sin x - \int e^x \sin x\,dx \\ 2\int e^x \sin x\,dx &= -e^x \cos x + e^x \sin x \\ \int e^x \sin x\,dx &= \frac{1}{2}e^x(\sin x - \cos x) + C. \end{aligned}$$

21. $\displaystyle\int e^{-x}\cos \pi x\,dx$
(See also Example 5.)
$$\begin{aligned} u &= e^{-x} & dv &= \cos \pi x\,dx \\ du &= -e^{-x}\,dx & v &= \frac{1}{\pi}\sin \pi x \end{aligned}$$

$\displaystyle\int e^{-x}\cos \pi x\,dx = \frac{1}{\pi}e^{-x}\sin \pi x + \frac{1}{\pi}\left[\int e^{-x}\sin \pi x\,dx\right].$
$$\begin{aligned} u &= e^{-x} & dv &= \sin \pi x\,dx \\ du &= -e^{-x}\,dx & v &= -\frac{1}{\pi}\cos \pi x \end{aligned}$$

$$\begin{aligned} \int e^{-x}\cos \pi x\,dx &= \frac{1}{\pi}e^{-x}\sin \pi x + \frac{1}{\pi}\left[-\frac{1}{\pi}e^{-x}\cos \pi x - \frac{1}{\pi}\int e^{-x}\cos \pi x\,dx\right] \\ &= \frac{1}{\pi}e^{-x}\sin \pi x - \frac{1}{\pi^2}e^{-x}\cos \pi x - \frac{1}{\pi^2}\int e^{-x}\cos \pi x\,dx. \end{aligned}$$

Now solve for the integral:

$$\int e^{-x} \cos \pi x \, dx + \frac{1}{\pi^2} \int e^{-x} \cos \pi x \, dx = \frac{1}{\pi} e^{-x} \sin \pi x - \frac{1}{\pi^2} e^{-x} \cos \pi x$$

$$\left(1 + \frac{1}{\pi^2}\right) \int e^{-x} \cos \pi x \, dx = \frac{1}{\pi^2} e^{-x} (\pi \sin \pi x - \cos \pi x)$$

$$\left(\frac{\pi^2 + 1}{\pi^2}\right) \int e^{-x} \cos \pi x \, dx = \frac{1}{\pi^2} e^{-x} (\pi \sin \pi x - \cos \pi x).$$

We conclude that

$$\int e^{-x} \cos \pi x \, dx = \frac{1}{\pi^2 + 1} \left[e^{-x} (\pi \sin \pi x - \cos \pi x) \right] + C.$$

23. $\displaystyle\int_0^1 x 2^x \, dx$ $\qquad\qquad\qquad\qquad$ $u = x \qquad dv = 2^x \, dx$
$\qquad\qquad\qquad\qquad\qquad\qquad\qquad\qquad\qquad\qquad du = dx \qquad v = \dfrac{2^x}{\ln 2}$

$$\int_0^1 x 2^x \, dx = \frac{x 2^x}{\ln 2}\Big|_0^1 - \int_0^1 \frac{2^x \, dx}{\ln 2} = \frac{x 2^x}{\ln 2}\Big|_0^1 - \frac{2^x}{(\ln 2)^2}\Big|_0^1$$

$$= \frac{2}{\ln 2} - \frac{2}{(\ln 2)^2} + \frac{1}{(\ln 2)^2} = \frac{2\ln 2 - 1}{(\ln 2)^2}.$$

25. $A = \displaystyle\int_0^\pi x \sin x \, dx$ $\qquad\qquad\qquad\qquad$ $u = x \qquad dv = \sin x \, dx$
$\qquad\qquad\qquad\qquad\qquad\qquad\qquad\qquad\qquad\qquad du = dx \qquad v = -\cos x$

$$A = -x \cos x\Big|_0^\pi + \int_0^\pi \cos x \, dx = -x \cos x\Big|_0^\pi + \sin x\Big|_0^\pi$$

$$= -\pi \cos \pi + 0 = -\pi(-1) = \pi.$$

27. $I_y = \displaystyle\int_0^{\pi/2} x^2 \cdot \rho \cos x \, dx.$

For the indefinite integral $\displaystyle\int x^2 \cos x \, dx$:

$$u = x^2 \qquad dv = \cos x \, dx$$
$$du = 2x \, dx \qquad v = \sin x$$

$$\int x^2 \cos x \, dx = x^2 \sin x - 2 \int x \sin x \, dx. \qquad u = x \qquad dv = \sin x \, dx$$
$$du = dx \qquad v = -\cos x$$

$$\int x^2 \cos x \, dx = x^2 \sin x - 2\left[-x \cos x + \int \cos x \, dx\right] = x^2 \sin x + 2x \cos x - 2 \sin x + C.$$

$$I_y = \rho(x^2 \sin x + 2x \cos x - 2 \sin x)\Big|_0^{\pi/2} = \rho\left(\frac{\pi^2}{4} \cdot 1 + 0 - 2 \cdot 1\right) = \left(\frac{\pi^2}{4} - 2\right)\rho.$$

29. Recall that $v = \dfrac{q}{C} = \dfrac{1}{C}\displaystyle\int i \, dt = \dfrac{1}{10 \times 10^{-6}}\int e^{-t} \sin 4t \, dt = 10^5 \int e^{-t} \sin 4t \, dt.$ $\quad$ (1)

$\displaystyle\int e^{-t} \sin 4t \, dt$ $\qquad\qquad\qquad\qquad$ $u = e^{-t} \qquad dv = \sin 4t \, dt$
$\qquad\qquad\qquad\qquad\qquad\qquad\qquad\qquad\qquad\qquad du = -e^{-t} \, dt \qquad v = -\dfrac{1}{4} \cos 4t$

$$\int e^{-t} \sin 4t \, dt = -\frac{1}{4} e^{-t} \cos 4t - \frac{1}{4}\left[\int e^{-t} \cos 4t \, dt\right].$$

$$u = e^{-t} \qquad dv = \cos 4t \, dt$$
$$du = -e^{-t} \, dt \qquad v = \frac{1}{4} \sin 4t$$

$$\int e^{-t} \sin 4t \, dt = -\frac{1}{4} e^{-t} \cos 4t - \frac{1}{4}\left[\frac{1}{4} e^{-t} \sin 4t + \frac{1}{4} \int e^{-t} \sin 4t \, dt\right]$$

$$= -\frac{1}{4} e^{-t} \cos 4t - \frac{1}{16} e^{-t} \sin 4t - \frac{1}{16} \int e^{-t} \sin 4t \, dt.$$

Solving for the integral, we get:

$$\frac{17}{16} \int e^{-t} \sin 4t \, dt \;=\; -\frac{1}{16} e^{-t}(4\cos 4t + \sin 4t)$$

$$\int e^{-t} \sin 4t \, dt \;=\; -\frac{1}{17} e^{-t}(4\cos 4t + \sin 4t)$$

By (1), $v = 10^5 \left(-\dfrac{1}{17} e^{-t}\right)(4\cos 4t + \sin 4t) + C.$

If $t = 0$, $v = 0$. Thus

$$0 = 10^5 \left(-\frac{1}{17}\right)(4) + C \text{ or } C = \frac{4(10^5)}{17},$$

and

$$v = 10^5 \left(-\frac{1}{17} e^{-t}\right)(4\cos 4t + \sin 4t) + \frac{4(10)^5}{17}\bigg|_{t=0.05} = 482 \,\text{V}.$$

7.8   Integration of Rational Functions

1. $\displaystyle \int \frac{dx}{x^2 - 4}$

$$\frac{1}{x^2 - 4} = \frac{1}{(x-2)(x+2)} = \frac{A}{x-2} + \frac{B}{x+2} = \frac{A(x+2) + B(x-2)}{(x-2)(x+2)}$$

$A(x+2) + B(x-2) = 1$

$\underline{x = -2}:\ \ 0 + B(-4) = 1 \text{ or } B = -\dfrac{1}{4}.$

$\underline{x = 2}:\ \ A(4) + 0 = 1 \text{ or } A = \dfrac{1}{4}.$

$$\int \frac{dx}{x^2 - 4} \;=\; \int \left(\frac{1}{4}\frac{1}{x-2} - \frac{1}{4}\frac{1}{x+2}\right) dx$$

$$= \;\frac{1}{4}\ln|x-2| - \frac{1}{4}\ln|x+2| + C = \frac{1}{4}\ln\left|\frac{x-2}{x+2}\right| + C.$$

3. $\displaystyle \int \frac{dx}{x - x^2}$

$$\frac{1}{x - x^2} = \frac{1}{x(1-x)} = \frac{A}{x} + \frac{B}{1-x} = \frac{A(1-x) + Bx}{x(1-x)}$$

$A(1-x) + Bx = 1$

$\underline{x = 1}:\ \ 0 + B(1) = 1 \text{ or } B = 1.$

$\underline{x = 0}:\ \ A(1) + 0 = 1 \text{ or } A = 1.$

$$\int \frac{dx}{x - x^2} \;=\; \int \left(\frac{1}{x} + \frac{1}{1-x}\right) dx \qquad u = 1 - x;\ du = -dx.$$

$$= \;\ln|x| - \ln|1-x| + C = \ln\left|\frac{x}{1-x}\right| + C.$$

5. $\displaystyle \int \frac{5x - 4}{(x-2)(x+1)}\, dx$

$$\frac{5x - 4}{(x-2)(x+1)} = \frac{A}{x-2} + \frac{B}{x+1} = \frac{A(x+1) + B(x-2)}{(x-2)(x+1)}$$

$A(x+1) + B(x-2) = 5x - 4$

$\underline{x = -1}:\ \ 0 + B(-3) = -9 \text{ or } B = 3.$

$\underline{x = 2}:\ \ A(3) + 0 = 6 \text{ or } A = 2.$

$$\int \frac{5x - 4}{(x-2)(x+1)}\, dx \;=\; \int \left(\frac{2}{x-2} + \frac{3}{x+1}\right) dx$$

$$= \;2\ln|x-2| + 3\ln|x+1| + C$$

$$= \;\ln|x-2|^2 + \ln|x+1|^3 + C$$

$$= \;\ln\left|(x-2)^2(x+1)^3\right| + C.$$

7. $\displaystyle\int \frac{x^3\,dx}{x^2-2x-3}$

Since the integrand is not a proper fraction, we must first perform the long division:

$$
\begin{array}{r}
x\ +2\ +\dfrac{7x+6}{x^2-2x-3} \\[2mm]
x^2-2x-3\overline{\smash{\big)}\,x^3} \\
\underline{x^3-2x^2-3x} \\
2x^2\ +3x \\
\underline{2x^2\ -4x-6} \\
7x+6
\end{array}
$$

So the integrand becomes: $\dfrac{x^3}{x^2-2x-3}=x+2+\dfrac{7x+6}{x^2-2x-3}$ and we proceed to split up the fraction:

$$\frac{7x+6}{x^2-2x-3}=\frac{7x+6}{(x-3)(x+1)}=\frac{A}{x-3}+\frac{B}{x+1}=\frac{A(x+1)+B(x-3)}{(x-3)(x+1)}.$$

$A(x+1)+B(x-3)=7x+6$

$\underline{x=-1}:\ 0+B(-4)=-1\ \text{or}\ B=\dfrac{1}{4}.$

$\underline{x=3}:\ A(4)+0=27\ \text{or}\ A=\dfrac{27}{4}.$

$$\int\left(x+2+\frac{27}{4}\frac{1}{x-3}+\frac{1}{4}\frac{1}{x+1}\right)dx=\frac{1}{2}x^2+2x+\frac{27}{4}\ln|x-3|+\frac{1}{4}\ln|x+1|+C.$$

9. $\displaystyle\int \frac{x^2+10x-20}{(x-4)(x-1)(x+2)}\,dx$

$$\frac{x^2+10x-20}{(x-4)(x-1)(x+2)}=\frac{A}{x-4}+\frac{B}{x-1}+\frac{C}{x+2}$$
$$=\frac{A(x-1)(x+2)+B(x-4)(x+2)+C(x-4)(x-1)}{(x-4)(x-1)(x+2)}$$

$A(x-1)(x+2)+B(x-4)(x+2)+C(x-4)(x-1)=x^2+10x-20$

$\underline{x=1}:\ 0+B(-3)(3)+0=-9\ \text{or}\ B=1.$

$\underline{x=-2}:\ 0+0+C(-6)(-3)=-36\ \text{or}\ C=-2.$

$\underline{x=4}:\ A(3)(6)+0+0=36\ \text{or}\ A=2.$

$$\int\frac{x^2+10x-20}{(x-4)(x-1)(x+2)}\,dx=\int\left(\frac{2}{x-4}+\frac{1}{x-1}+\frac{-2}{x+2}\right)dx$$
$$=2\ln|x-4|+\ln|x-1|-2\ln|x+2|+C$$
$$=\ln|x-4|^2+\ln|x-1|-\ln|x+2|^2+C$$
$$=\ln\left|\frac{(x-4)^2(x-1)}{(x+2)^2}\right|+C.$$

11. $\displaystyle\int \frac{3x+5}{(x+1)^2}\,dx$

$$\frac{3x+5}{(x+1)^2}=\frac{A}{x+1}+\frac{B}{(x+1)^2}=\frac{A}{x+1}\frac{x+1}{x+1}+\frac{B}{(x+1)^2}=\frac{A(x+1)+B}{(x+1)^2}$$

$A(x+1)+B=3x+5$

$\underline{x=-1}:\ 0+B=2\ \text{or}\ B=2.$

Thus $A(x+1)+2=3x+5$. Now let $x=$ any value (such as $x=0$):

$\underline{x=0}:\ A(1)+2=5\ \text{or}\ A=3.$

$$\int\frac{3x+5}{(x+1)^2}\,dx=\int\left(\frac{3}{x+1}+\frac{2}{(x+1)^2}\right)dx=\int\left(\frac{3}{x+1}+2(x+1)^{-2}\right)dx$$
$$=3\ln|x+1|+2\frac{(x+1)^{-1}}{-1}+C=3\ln|x+1|-\frac{2}{x+1}+C.$$

13. $\displaystyle\int \frac{-2x^2 + 9x - 7}{(x-2)^2(x+1)}\, dx$

$$\frac{-2x^2 + 9x - 7}{(x-2)^2(x+1)} = \frac{A}{x-2} + \frac{B}{(x-2)^2} + \frac{C}{x+1}$$

$$= \frac{A(x-2)(x+1) + B(x+1) + C(x-2)^2}{(x-2)^2(x+1)}$$

$A(x-2)(x+1) + B(x+1) + C(x-2)^2 = -2x^2 + 9x - 7$

$\underline{x = 2}$: $0 + B(3) + 0 = 3$ or $B = 1$.

$\underline{x = -1}$: $0 + 0 + C(9) = -18$ or $C = -2$.

Using the values obtained for B and C, we get

$A(x-2)(x+1) + 1(x+1) - 2(x-2)^2 = -2x^2 + 9x - 7$.

Now let $x =$ any value (such as $x = 0$):

$\underline{x = 0}$: $A(-2)(1) + (1) - 2(4) = -7$ or $A = 0$.

$\displaystyle\int \left(\frac{1}{(x-2)^2} + \frac{-2}{x+1} \right) dx = \int \left[(x-2)^{-2} - \frac{2}{x+1} \right] dx$ $u = x - 2;\ du = dx.$

$$= \frac{(x-2)^{-1}}{-1} - 2\ln|x+1| + C$$

$$= -\frac{1}{x-2} - 2\ln|x+1| + C.$$

15. $\displaystyle \frac{2x^2 + 1}{(x-2)^3} = \frac{A}{x-2} + \frac{B}{(x-2)^2} + \frac{C}{(x-2)^3}$

$$= \frac{A}{x-2}\frac{(x-2)^2}{(x-2)^2} + \frac{B}{(x-2)^2}\frac{x-2}{x-2} + \frac{C}{(x-2)^3} = \frac{A(x-2)^2 + B(x-2) + C}{(x-2)^3}.$$

$A(x-2)^2 + B(x-2) + C = 2x^2 + 1$

$\underline{x = 2}$: $C = 9$.

Thus

$A(x-2)^2 + B(x-2) + 9 = 2x^2 + 1$

$A(x-2)^2 + B(x-2) = 2x^2 - 8$.

Now let $x =$ any two values (such as $x = 0$ and $x = 1$):

$\underline{x = 0}$: $A(4) + B(-2) = -8$.

$\underline{x = 1}$: $A(1) + B(-1) = -6$.

$2A - B = -4$ (dividing by 2)

$\underline{A - B = -6}$

$A\quad\ \ = 2$ (subtracting)

$\quad B = A + 6 = 8.$

$\displaystyle\int \frac{2x^2 + 1}{(x-2)^3}\, dx = \int \left(\frac{2}{x-2} + \frac{8}{(x-2)^2} + \frac{9}{(x-2)^3} \right) dx$

$$= \int \left[\frac{2}{x-2} + 8(x-2)^{-2} + 9(x-2)^{-3} \right] dx$$

$$= 2\ln|x-2| + 8\frac{(x-2)^{-1}}{-1} + 9\frac{(x-2)^{-2}}{-2} + C$$

$$= 2\ln|x-2| - \frac{8}{x-2} - \frac{9}{2(x-2)^2} + C.$$

17. $\displaystyle\int \frac{x^2 - 3x - 2}{(x+2)(x^2+4)}\,dx$

$$\frac{x^2 - 3x - 2}{(x+2)(x^2+4)} = \frac{A}{x+2} + \frac{Bx+C}{x^2+4} = \frac{A(x^2+4) + (Bx+C)(x+2)}{(x+2)(x^2+4)}$$

$A(x^2+4) + (Bx+C)(x+2) = x^2 - 3x - 2$

$\underline{x = -2}:\ A(8) + 0 = 8 \text{ or } A = 1.$

$1(x^2+4) + (Bx+C)(x+2) = x^2 - 3x - 2$ $\qquad\qquad\qquad A = 1$

Now let $x = 0$ and solve for C:

$(4) + C(2) = -2 \text{ or } C = -3.$ $\qquad\qquad\qquad x = 0$

Since $A = 1$ and $C = -3$, we get

$1(x^2+4) + (Bx - 3)(x+2) = x^2 - 3x - 2.$

At this point we may let x be equal to any value (such as $x = 1$):

$5 + (B - 3)(3) = -4 \text{ or } B = 0.$

So the integral becomes

$$\int\left(\frac{1}{x+2} + \frac{-3}{x^2+4}\right)dx = \ln|x+2| - 3\cdot\frac{1}{2}\text{Arctan}\frac{x}{2} + C = \ln|x+2| - \frac{3}{2}\text{Arctan}\frac{x}{2} + C.$$

19. $\displaystyle\frac{3x^2 + 4x + 3}{(x+1)(x^2+1)} = \frac{A}{x+1} + \frac{Bx+C}{x^2+1} = \frac{A(x^2+1) + (Bx+C)(x+1)}{(x+1)(x^2+1)}$

$A(x^2+1) + (Bx+C)(x+1) = 3x^2 + 4x + 3$

$\underline{x = -1}:\ A(2) = 2 \text{ or } A = 1.$

Thus $1(x^2+1) + (Bx+C)(x+1) = 3x^2 + 4x + 3.$

Let $x = 0$ and solve for C:

$1 + C(1) = 3 \text{ or } C = 2.$

$1(x^2+1) + (Bx+2)(x+1) = 3x^2 + 4x + 3.$

Now let $x = $ any value (such as $x = 1$):

$\underline{x = 1}:\ 2 + (B+2)(2) = 10 \text{ or } B = 2.$

$$\int\frac{3x^2 + 4x + 3}{(x+1)(x^2+1)}\,dx = \int\left(\frac{1}{x+1} + \frac{2x+2}{x^2+1}\right)dx$$

$$= \int\left(\frac{1}{x+1} + \frac{2x}{x^2+1} + 2\cdot\frac{1}{x^2+1}\right)dx$$

$$= \ln|x+1| + \ln|x^2+1| + 2\text{Arctan}\,x + C$$

$$= \ln|(x+1)(x^2+1)| + 2\text{Arctan}\,x + C.$$

21. $\displaystyle\int \frac{x^5\,dx}{(x^2+4)^2} = \int \frac{x^5\,dx}{x^4+8x^2+16}.$

Since the fraction is not a proper fraction, we must first perform the long division:

$$
\begin{array}{r}
x \\
x^4+8x^2+16\,\overline{\smash{\big)}\,x^5 } \\
\underline{x^5+8x^3+16x} \\
-8x^3-16x
\end{array}
$$

The integrand can therefore be written as: $x + \dfrac{-8x^3-16x}{(x^2+4)^2}.$

The fractional part has the form
$$\frac{-8x^3-16x}{(x^2+4)^2} = \frac{Ax+B}{x^2+4} + \frac{Cx+D}{(x^2+4)^2} = \frac{(Ax+B)(x^2+4)+(Cx+D)}{(x^2+4)^2}.$$
Equating numerators:
$$(Ax+B)(x^2+4)+(Cx+D) = -8x^3-16x$$
$$Ax^3+Bx^2+4Ax+4B+Cx+D = -8x^3-16x$$
$$Ax^3+Bx^2+(4A+C)x+(4B+D) = -8x^3-16x.$$
Comparing coefficients:

$A=-8,\; B=0,\; 4A+C=-16,\; 4B+D=0,$ whence $C=16$ and $D=0.$ Thus
$$\int \frac{x^5\,dx}{(x^2+4)^2} = \int\left(x+\frac{-8x}{x^2+4}+\frac{16x}{(x^2+4)^2}\right)dx$$
$$= \int x\,dx - 8\int \frac{x\,dx}{x^2+4} + 16\int (x^2+4)^{-2}x\,dx$$
$$= \int x\,dx - \frac{8}{2}\int \frac{2x\,dx}{x^2+4} + \frac{16}{2}\int (x^2+4)^{-2}(2x\,dx)$$
$$u = x^2+4;\; du = 2x\,dx$$
$$= \frac{1}{2}x^2 - 4\ln(x^2+4) + 8\frac{(x^2+4)^{-1}}{-1} + C$$
$$= \frac{1}{2}x^2 - 4\ln(x^2+4) - \frac{8}{x^2+4} + C.$$

23. $\displaystyle\frac{x^2-3x+5}{x(x^2-2x+5)} = \frac{A}{x} + \frac{Bx+C}{x^2-2x+5} = \frac{A(x^2-2x+5)+(Bx+C)x}{x(x^2-2x+5)}$

$A(x^2-2x+5)+(Bx+C)x = x^2-3x+5$

$\underline{x=0}:\; A(5)=5$ or $A=1.$

$1(x^2-2x+5)+(Bx+C)x = x^2-3x+5.$

Now let $x =$ any two values (such as $x=1$ and $x=-1$):

$\underline{x=1}:\; 4+(B+C)(1)=3$

$\underline{x=-1}:\; 8+(-B+C)(-1)=9.$

$B+C=-1$

$\underline{B-C=1}$

$2B=0,\; B=0$

$C=-1.$

$$\int\left(\frac{1}{x}+\frac{-1}{x^2-2x+5}\right)dx = \int\left(\frac{1}{x}+\frac{-1}{(x-1)^2+4}\right)dx \qquad u=x-1;\; du=dx.$$
$$= \ln|x| - \frac{1}{2}\text{Arctan}\frac{1}{2}(x-1) + C.$$

25. $\displaystyle\int \frac{dx}{x(x^2+2x+2)}$

$$\frac{1}{x(x^2+2x+2)} = \frac{A}{x} + \frac{Bx+C}{x^2+2x+2} = \frac{A(x^2+2x+2)+(Bx+C)x}{x(x^2+2x+2)}$$

$$A(x^2+2x+2) + (Bx+C)x = 1$$

$$Ax^2 + 2Ax + 2A + Bx^2 + Cx = 1$$

$$(A+B)x^2 + (2A+C)x + 2A = 1$$

Comparing coefficients:

$$A + B = 0 \quad \text{coefficients of } x^2$$

$$2A + C = 0 \quad \text{coefficients of } x$$

$$2A = 1 \quad \text{constants}$$

Solution set: $A = \dfrac{1}{2}$, $B = -\dfrac{1}{2}$, $C = -1$.

$$\int \frac{dx}{x(x^2+2x+2)} = \int \left(\frac{1}{2}\frac{1}{x} + \frac{-\frac{1}{2}x-1}{x^2+2x+2} \right)dx = \frac{1}{2}\int \frac{1}{x}\,dx - \frac{1}{2}\int \frac{x+2}{x^2+2x+2}\,dx.$$

For the second integral, if $u = x^2+2x+2$, then $du = (2x+2)\,dx$. So the integral has to be split as follows:

$$\frac{1}{2}\int \frac{1}{x}\,dx - \frac{1}{2}\cdot\frac{1}{2}\int \frac{2(x+2)}{x^2+2x+2}\,dx = \frac{1}{2}\int \frac{1}{x}\,dx - \frac{1}{4}\int \frac{2x+4}{x^2+2x+2}\,dx$$

$$= \frac{1}{2}\int \frac{1}{x}\,dx - \frac{1}{4}\int \frac{(2x+2)+2}{x^2+2x+2}\,dx$$

$$= \frac{1}{2}\int \frac{1}{x}\,dx - \frac{1}{4}\int \frac{2x+2}{x^2+2x+2}\,dx - \frac{1}{4}\int \frac{2\,dx}{x^2+2x+2}$$

$$= \frac{1}{2}\int \frac{1}{x}\,dx - \frac{1}{4}\int \frac{2x+2}{x^2+2x+2}\,dx - \frac{1}{2}\int \frac{dx}{(x+1)^2+1}$$

$$= \frac{1}{2}\ln|x| - \frac{1}{4}\ln|x^2+2x+2| - \frac{1}{2}\text{Arctan}\,(x+1) + C.$$

7.9 Integration by Use of Tables

1. Formula 5; $a = 2$, $b = 1$.
$$\int \frac{dx}{x(2+x)} = \frac{1}{2}\ln \left| \frac{x}{2+x} \right| + C.$$

3. Formula 27; $a = \sqrt{7}$, $a^2 = 7$.
$$\int \sqrt{x^2-7}\,dx = \frac{1}{2}\left(x\sqrt{x^2-7} - 7\ln|x+\sqrt{x^2-7}| \right) + C.$$

5. Formula 16; $a = \sqrt{5}$.
$$\int \frac{dx}{5-x^2} = \frac{1}{2\sqrt{5}}\ln \left| \frac{\sqrt{5}+x}{\sqrt{5}-x} \right| + C.$$

7. $\displaystyle\int \frac{dx}{x^2\sqrt{5x^2+4}} = \int \frac{dx}{x\sqrt{(\sqrt{5}x)^2+4}}$ $\qquad u = \sqrt{5}x; \; du = \sqrt{5}\,dx.$

$$= \sqrt{5}\int \frac{\sqrt{5}\,dx}{(\sqrt{5}x)(\sqrt{5}x)\sqrt{(\sqrt{5}x)^2+4}} = \sqrt{5}\int \frac{du}{u^2\sqrt{u^2+4}}.$$

By Formula 35 with $a = 2$, we get
$$\sqrt{5}\left(-\frac{\sqrt{u^2+4}}{4u} \right) + C = \sqrt{5}\left(-\frac{\sqrt{5x^2+4}}{4(\sqrt{5}x)} \right) + C = -\frac{\sqrt{5x^2+4}}{4x} + C.$$

9. Formula 61; $m = 2$, $n = 1$.

$$\int \sin 2x \sin x \, dx = -\frac{\sin(2+1)x}{2(2+1)} + \frac{\sin(2-1)x}{2(2-1)} + C$$

$$= -\frac{1}{6} \sin 3x + \frac{1}{2} \sin x + C.$$

11. $$\int \frac{dx}{\sqrt{3x^2 + 5}} = \int \frac{dx}{\sqrt{(\sqrt{3}x)^2 + 5}} \qquad u = \sqrt{3}x; \ du = \sqrt{3} \, dx.$$

$$= \frac{1}{\sqrt{3}} \int \frac{\sqrt{3} \, dx}{\sqrt{(\sqrt{3}x)^2 + 5}} = \frac{1}{\sqrt{3}} \int \frac{du}{\sqrt{u^2 + 5}}.$$

By Formula 32 with $a = \sqrt{5}$, we get

$$\frac{1}{\sqrt{3}} \ln \left| u + \sqrt{u^2 + 5} \right| + C = \frac{1}{\sqrt{3}} \ln \left| \sqrt{3}x + \sqrt{3x^2 + 5} \right| + C.$$

13. Formula 42; $n = 2$, $a = 2$.

$$\int x^2 e^{2x} \, dx = \frac{x^2 e^{2x}}{2} - \frac{2}{2} \int x e^{2x} \, dx = \frac{1}{2} x^2 e^{2x} - \int x e^{2x} \, dx.$$

Now by formula 41 with $a = 2$, we get:

$$\frac{1}{2} x^2 e^{2x} - \left[\frac{e^{2x}}{4} (2x - 1) \right] + C = \frac{1}{2} x^2 e^{2x} - \frac{1}{2} x e^{2x} + \frac{1}{4} e^{2x} + C.$$

15. Formula 69

$$\int \tan^6 x \, dx = \frac{\tan^5 x}{5} - \int \tan^4 x \, dx$$

$$= \frac{1}{5} \tan^5 x - \left[\frac{\tan^3 x}{3} - \int \tan^2 x \, dx \right]$$

$$= \frac{1}{5} \tan^5 x - \frac{1}{3} \tan^3 x + \int (\sec^2 x - 1) \, dx$$

$$= \frac{1}{5} \tan^5 x - \frac{1}{3} \tan^3 x + \tan x - x + C.$$

17. $$\int \frac{dx}{4x^2 - 9} = \int \frac{dx}{(2x)^2 - 9} = \frac{1}{2} \int \frac{du}{u^2 - 9}. \qquad u = 2x; \ du = 2 \, dx.$$

By Formula 17 with $a = 3$, we get:

$$\frac{1}{2} \cdot \frac{1}{6} \ln \left| \frac{u - 3}{u + 3} \right| + C = \frac{1}{12} \ln \left| \frac{2x - 3}{2x + 3} \right| + C.$$

19. Formula 12; $a = 3 > 0$, $b = 1$.

$$\int \frac{dx}{x\sqrt{3 + x}} = \frac{1}{\sqrt{3}} \ln \left| \frac{\sqrt{3 + x} - \sqrt{3}}{\sqrt{3 + x} + \sqrt{3}} \right| + C.$$

21. $$\int \frac{dx}{3x^2 - 5} = \int \frac{dx}{(\sqrt{3}x)^2 - 5} = \frac{1}{\sqrt{3}} \int \frac{\sqrt{3} \, dx}{(\sqrt{3}x)^2 - 5} \qquad u = \sqrt{3}x; \ du = \sqrt{3} \, dx.$$

$$= \frac{1}{\sqrt{3}} \int \frac{du}{u^2 - 5}.$$

By formula 17 with $a = \sqrt{5}$, we get:

$$\frac{1}{\sqrt{3}} \frac{1}{2\sqrt{5}} \ln \left| \frac{u - \sqrt{5}}{u + \sqrt{5}} \right| + C = \frac{1}{2\sqrt{15}} \ln \left| \frac{\sqrt{3}x - \sqrt{5}}{\sqrt{3}x + \sqrt{5}} \right| + C.$$

23. Formula 29; $a = \sqrt{10}$.

$$\int \frac{\sqrt{x^2 - 10}}{x} \, dx = \sqrt{x^2 - 10} - \sqrt{10} \, \text{Arccos} \, \frac{\sqrt{10}}{x} + C.$$

25. Formula 63, $m = 3$, $n = 2$.

$$\int \sin 3x \cos 2x \, dx = -\frac{\cos(3+2)x}{2(3+2)} - \frac{\cos(3-2)x}{2(3-2)} = -\frac{1}{10}\cos 5x - \frac{1}{2}\cos x + C.$$

27. $\displaystyle \int \frac{dx}{(4x^2+5)^{3/2}} \quad = \quad \int \frac{dx}{[(2x)^2+5]^{3/2}} \qquad u = 2x;\ \ du = 2\,dx.$

$$= \quad \frac{1}{2}\int \frac{2\,dx}{[(2x)^2+5]^{3/2}} = \frac{1}{2}\int \frac{du}{(u^2+5)^{3/2}}.$$

By Formula 37 with $a = \sqrt{5}$, we get

$$\frac{1}{2}\frac{u}{5\sqrt{u^2+5}} + C = \frac{1}{2}\frac{2x}{5\sqrt{4x^2+5}} + C = \frac{x}{5\sqrt{4x^2+5}} + C.$$

Chapter 7 Review

1. $\displaystyle \int \frac{x\,dx}{x^2+1} \quad = \quad \frac{1}{2}\int \frac{2x\,dx}{x^2+1} \qquad\qquad u = x^2+1;\ \ du = 2x\,dx.$

$$= \quad \frac{1}{2}\ln|u| + C$$

$$= \quad \frac{1}{2}\ln(x^2+1) + C.$$

3. $\displaystyle \int \frac{2\,dx}{x^2+1} = 2\int \frac{dx}{x^2+1} = 2\,\text{Arctan}\,x + C.$

5. $\displaystyle \int \frac{t\,dt}{\sqrt{9-t^2}} \quad = \quad \int (9-t^2)^{-1/2}t\,dt \qquad\qquad u = 9-t^2;\ \ du = -2t\,dt.$

$$= \quad -\frac{1}{2}\int (9-t^2)^{-1/2}(-2t)\,dt$$

$$= \quad -\frac{1}{2}\frac{(9-t^2)^{1/2}}{1/2} + C$$

$$= \quad -\sqrt{9-t^2} + C.$$

7. $\displaystyle \int x\cos 2x^2\,dx \quad = \quad \int \cos 2x^2 \cdot x\,dx \qquad\qquad u = 2x^2;\ \ du = 4x\,dx.$

$$= \quad \frac{1}{4}\int \cos 2x^2 (4x\,dx) = \frac{1}{4}\int \cos u\,du$$

$$= \quad \frac{1}{4}\sin u + C = \frac{1}{4}\sin 2x^2 + C.$$

9. $\displaystyle \int \frac{e^x\,dx}{4+e^{2x}} = \int \frac{e^x\,dx}{4+(e^x)^2}. \qquad\qquad u = e^x;\ \ du = e^x\,dx.$

$$\int \frac{du}{4+u^2} = \frac{1}{2}\text{Arctan}\,\frac{u}{2} + C = \frac{1}{2}\text{Arctan}\,\frac{1}{2}e^x + C.$$

11. $\displaystyle \int \frac{e^x\,dx}{(4+e^x)^2} \quad = \quad \int (4+e^x)^{-2}e^x\,dx \qquad\qquad u = 4+e^x;\ \ du = e^x\,dx.$

$$= \quad \int u^{-2}\,du = \frac{u^{-1}}{-1} + C = \frac{-1}{u} + C = \frac{-1}{4+e^x} + C.$$

13. $\displaystyle \int \frac{\ln x}{x}\,dx \qquad\qquad\qquad\qquad u = \ln x;\ \ du = \frac{1}{x}\,dx.$

$$\int (\ln x)\frac{1}{x}\,dx = \int u\,du = \frac{1}{2}u^2 + C = \frac{1}{2}\ln^2 x + C.$$

15. $\displaystyle\int \frac{x+2}{x^2+4x+5}\,dx$

Let $u = x^2 + 4x + 5$; then $du = (2x+4)\,dx = 2(x+2)\,dx$.

$\displaystyle\frac{1}{2}\int \frac{2(x+2)\,dx}{x^2+4x+5} = \frac{1}{2}\int \frac{du}{u} = \frac{1}{2}\ln|u| + C = \frac{1}{2}\ln|x^2+4x+5| + C.$

17. $\displaystyle\int \frac{dx}{x^2+4x+4} \;=\; \int \frac{dx}{(x+2)^2}$ $\qquad\qquad\qquad u = x+2;\; du = dx.$

$\displaystyle\phantom{\int \frac{dx}{x^2+4x+4}} = \int (x+2)^{-2}\,dx = \frac{(x+2)^{-1}}{-1} + C$

$\displaystyle\phantom{\int \frac{dx}{x^2+4x+4}} = -\frac{1}{x+2} + C.$

19. $\displaystyle\int \ln^2 x\,dx$
$\qquad\qquad\qquad\qquad\qquad\qquad u \;=\; (\ln x)^2 \qquad dv \;=\; dx$
$\qquad\qquad\qquad\qquad\qquad\qquad du \;=\; 2(\ln x)\dfrac{1}{x}\,dx \qquad v \;=\; x$

$\displaystyle\int (\ln x)^2\,dx = uv - \int v\,du = x(\ln x)^2 - \int x\cdot 2(\ln x)\frac{1}{x}\,dx = x\ln^2 x - 2\int \ln x\,dx.$

$\qquad\qquad\qquad\qquad\qquad\qquad u \;=\; \ln x \qquad dv \;=\; dx$
$\qquad\qquad\qquad\qquad\qquad\qquad du \;=\; \dfrac{1}{x}\,dx \qquad v \;=\; x$

$\displaystyle\int \ln^2 x\,dx \;=\; x\ln^2 x - 2\left(x\ln x - \int x\cdot\frac{1}{x}\,dx\right)$

$\displaystyle \;=\; x\ln^2 x - 2x\ln x + 2\int dx = x\ln^2 x - 2x\ln x + 2x + C.$

21. $\displaystyle\int \frac{dx}{x\sqrt{4-x^2}}$. Let $x = 2\sin\theta$; then $dx = 2\cos\theta\,d\theta$.

$\sqrt{4-x^2} = \sqrt{4 - 4\sin^2\theta} = 2\sqrt{1-\sin^2\theta} = 2\sqrt{\cos^2\theta} = 2\cos\theta.$

$\displaystyle\int \frac{2\cos\theta\,d\theta}{(2\sin\theta)(2\cos\theta)} = \frac{1}{2}\int \csc\theta\,d\theta = \frac{1}{2}\ln|\csc\theta - \cot\theta| + C$

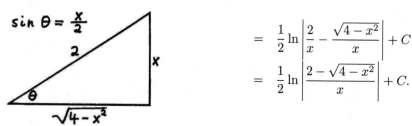

$\displaystyle = \frac{1}{2}\ln\left|\frac{2}{x} - \frac{\sqrt{4-x^2}}{x}\right| + C$

$\displaystyle = \frac{1}{2}\ln\left|\frac{2 - \sqrt{4-x^2}}{x}\right| + C.$

23. $\displaystyle\int \frac{\sin^2 2x\cos 2x}{1 + \sin^3 2x}\,dx$

Let $u = 1 + (\sin 2x)^3$; then $du = 3(\sin 2x)^2(2\cos 2x)\,dx$.

$\displaystyle\frac{1}{6}\int \frac{6\sin^2 2x\cos 2x\,dx}{1 + \sin^3 2x} = \frac{1}{6}\int \frac{du}{u} = \frac{1}{6}\ln|1 + \sin^3 2x| + C.$

25. $\displaystyle\int \sin^2 2x\,dx \;=\; \frac{1}{2}\int (1 - \cos 4x)\,dx = \frac{1}{2}\int dx - \frac{1}{2}\int \cos 4x\,dx$

$\displaystyle \;=\; \frac{1}{2}x - \frac{1}{2}\left(\frac{1}{4}\sin 4x\right) + C \qquad\qquad u = 4x;\; du = 4\,dx$

$\displaystyle \;=\; \frac{1}{2}x - \frac{1}{8}\sin 4x + C.$

27. $\displaystyle\int x\mathrm{e}^{2x}\,dx$

$$u = x \qquad dv = \mathrm{e}^{2x}\,dx$$
$$du = dx \qquad v = \frac{1}{2}\mathrm{e}^{2x}$$

$$
\begin{aligned}
uv - \int v\,du &= x\left(\frac{1}{2}\mathrm{e}^{2x}\right) - \int \frac{1}{2}\mathrm{e}^{2x}\,dx\\
&= \frac{1}{2}x\mathrm{e}^{2x} - \frac{1}{2}\cdot\frac{1}{2}\int \mathrm{e}^{2x}\,(2\,dx)\\
&= \frac{1}{2}x\mathrm{e}^{2x} - \frac{1}{4}\mathrm{e}^{2x} + C = \frac{1}{4}\mathrm{e}^{2x}(2x-1) + C.
\end{aligned}
$$

29. $\displaystyle\int \operatorname{Arctan} 2x\,dx$

$$u = \operatorname{Arctan} 2x \qquad dv = dx$$
$$du = \frac{2\,dx}{1+4x^2} \qquad v = x$$

(Integration by parts.)

$$
\begin{aligned}
x\operatorname{Arctan} 2x - \int \frac{2x\,dx}{1+4x^2} &= x\operatorname{Arctan} 2x - \frac{1}{4}\int \frac{8x\,dx}{1+4x^2} \qquad (u = 1+4x^2;\ du = 8x\,dx.)\\
&= x\operatorname{Arctan} 2x - \frac{1}{4}\ln(1+4x^2) + C.
\end{aligned}
$$

31. $\displaystyle\int \frac{\mathrm{e}^{\tan x}}{\cos^2 x}\,dx = \int \mathrm{e}^{\tan x}\sec^2 x\,dx \qquad\qquad u = \tan x;\ du = \sec^2 x\,dx.$

$$= \int \mathrm{e}^u\,du = \mathrm{e}^u + C = \mathrm{e}^{\tan x} + C.$$

33. $(x^2 - 1) \div (x^2 + 3) = 1 - \dfrac{4}{x^2+3}.$

$$\int \frac{x^2-1}{x^2+3}\,dx = \int \left[1 - \frac{4}{x^2+3}\right]\,dx = x - \frac{4}{\sqrt{3}}\operatorname{Arctan}\frac{x}{\sqrt{3}} + C.$$

35. $\displaystyle\int \frac{1-\cos\omega}{\sin\omega}\,d\omega = \int \left(\frac{1}{\sin\omega} - \frac{\cos\omega}{\sin\omega}\right)\,d\omega$

$$= \int \csc\omega\,d\omega - \int \frac{\cos\omega\,d\omega}{\sin\omega} \qquad\qquad u = \sin\omega;\ du = \cos\omega\,d\omega.$$

$$= \ln|\csc\omega - \cot\omega| - \int \frac{du}{u} = \ln|\csc\omega - \cot\omega| - \ln|\sin\omega| + C.$$

This result can be rewritten as follows:

$$\ln|\csc\omega - \cot\omega| - \ln\frac{1}{|\csc\omega|} + C = \ln|\csc\omega - \cot\omega| - \ln 1 + \ln|\csc\omega| + C$$

$$= \ln|\csc\omega(\csc\omega - \cot\omega)| + C.$$

Another form comes from

$$
\begin{aligned}
\csc^2\omega - \csc\omega\cot\omega &= \frac{1}{\sin^2\omega} - \frac{1}{\sin\omega}\frac{\cos\omega}{\sin\omega} = \frac{1-\cos\omega}{\sin^2\omega} = \frac{1-\cos\omega}{1-\cos^2\omega}\\
&= \frac{1-\cos\omega}{(1-\cos\omega)(1+\cos\omega)} = \frac{1}{1+\cos\omega} = (1+\cos\omega)^{-1}.
\end{aligned}
$$

The result is: $\ln|1+\cos\omega|^{-1} + C = -\ln|1+\cos\omega| + C.$

37.
$$\int \cot^4 x \csc^4 x \, dx = \int \cot^4 x \csc^2 x (\csc^2 x \, dx)$$
$$= \int \cot^4 x (1 + \cot^2 x)(\csc^2 x \, dx)$$
$$= -\int \cot^4 x (1 + \cot^2 x)(-\csc^2 x \, dx) \qquad u = \cot x; \ du = -\csc^2 x \, dx.$$
$$= -\int u^4 (1 + u^2) \, du = -\frac{1}{5} u^5 - \frac{1}{7} u^7 + C$$
$$= -\frac{1}{5} \cot^5 x - \frac{1}{7} \cot^7 x + C.$$

39.
$$\int \sec^3 2x \tan^3 2x \, dx = \int \sec^2 2x \tan^2 2x (\sec 2x \tan 2x \, dx) \qquad \text{(odd power of tangent)}$$
$$= \frac{1}{2} \int \sec^2 2x (\sec^2 2x - 1)(2 \sec 2x \tan 2x \, dx) \quad u = \sec 2x;$$
$$= \frac{1}{2} \int u^2 (u^2 - 1) \, du = \frac{1}{2} \left(\frac{1}{5} u^5 - \frac{1}{3} u^3 \right) + C \quad du = 2 \sec 2x \tan 2x \, dx.$$
$$= \frac{1}{10} \sec^5 2x - \frac{1}{6} \sec^3 2x + C.$$

41.
$$\int \tan^2 x \sec^2 x \, dx \qquad\qquad\qquad\qquad u = \tan x; \ du = \sec^2 x \, dx.$$
$$\int u^2 \, du = \frac{1}{3} u^3 + C = \frac{1}{3} \tan^3 x + C.$$

43.
$$\int \sec^4 3x \, dx = \int \sec^2 3x \sec^2 3x \, dx \qquad\qquad \text{(even power of secant)}$$
$$= \int (1 + \tan^2 3x) \sec^2 3x \, dx \qquad\qquad u = \tan 3x; \ du = 3 \sec^2 3x \, dx.$$
$$= \frac{1}{3} \int (1 + \tan^2 3x)(3 \sec^2 3x \, dx)$$
$$= \frac{1}{3} \int (1 + u^2) \, du = \frac{1}{3} \left(u + \frac{1}{3} u^3 \right) + C$$
$$= \frac{1}{3} \tan 3x + \frac{1}{9} \tan^3 3x + C.$$

45.
$$\int e^x \cos 4x \, dx \qquad\qquad\qquad\qquad\qquad u = e^x \qquad dv = \cos 4x \, dx$$
$$\qquad\qquad\qquad\qquad\qquad\qquad\qquad\qquad du = e^x \, dx \quad v = \frac{1}{4} \sin 4x$$
$$\int e^x \cos 4x \, dx = \frac{1}{4} e^x \sin 4x - \frac{1}{4} \int e^x \sin 4x \, dx.$$
$$\qquad\qquad\qquad\qquad\qquad\qquad\qquad u = e^x \qquad dv = \sin 4x \, dx$$
$$\qquad\qquad\qquad\qquad\qquad\qquad\qquad du = e^x \, dx \quad v = -\frac{1}{4} \cos 4x$$
$$\int e^x \cos 4x \, dx = \frac{1}{4} e^x \sin 4x - \frac{1}{4} \left[-\frac{1}{4} e^x \cos 4x + \frac{1}{4} \int e^x \cos 4x \, dx \right]$$
$$= \frac{1}{4} e^x \sin 4x + \frac{1}{16} e^x \cos 4x - \frac{1}{16} \int e^x \cos 4x \, dx.$$

Solving for the integral:
$$\left(1 + \frac{1}{16} \right) \int e^x \cos 4x \, dx = \frac{1}{4} e^x \sin 4x + \frac{1}{16} e^x \cos 4x$$
$$\frac{17}{16} \int e^x \cos 4x \, dx = \frac{1}{16} e^x (4 \sin 4x + \cos 4x).$$

We conclude that
$$\int e^x \cos 4x \, dx = \frac{1}{17} e^x (4 \sin 4x + \cos 4x) + C.$$

47. $\displaystyle\int \frac{dx}{5x^2+4} = \frac{dx}{(\sqrt{5}x)^2+4}$ $\qquad\qquad u = \sqrt{5}x;\ du = \sqrt{5}\,dx.$

$\displaystyle = \frac{1}{\sqrt{5}}\int \frac{\sqrt{5}\,dx}{(\sqrt{5}x)^2+4} = \frac{1}{\sqrt{5}}\int \frac{du}{u^2+4}$

$\displaystyle = \frac{1}{\sqrt{5}}\cdot\frac{1}{2}\,\text{Arctan}\,\frac{u}{2} + C = \frac{1}{2\sqrt{5}}\,\text{Arctan}\,\frac{\sqrt{5}}{2}x + C.$

49. $\displaystyle\int \frac{x+1}{x^2+2x-8}\,dx = \frac{1}{2}\int \frac{2(x+1)}{x^2+2x-8}\,dx \qquad u = x^2+2x-8;\ du = (2x+2)\,dx.$

$\displaystyle = \frac{1}{2}\ln|x^2+2x-8| + C.$

51. $\displaystyle\frac{3x^2-4x+9}{(x-2)(x^2+9)} = \frac{A}{x-2} + \frac{Bx+C}{x^2+9} = \frac{A(x^2+9)+(Bx+C)(x-2)}{(x-2)(x^2+9)}$

$A(x^2+9)+(Bx+C)(x-2) = 3x^2-4x+9$

$\underline{x=2}:\ A(13) = 13$ or $A = 1.$

$1(x^2+9)+(Bx+C)(x-2) = 3x^2-4x+9.$ Now let $x = 0$ and find C:

$9+C(-2) = 9$ or $C = 0.$

$1(x^2+9)+(Bx+0)(x-2) = 3x^2-4x+9$

Finally, let $x =$ any value (such as $x = 1$):

$\underline{x=1}:\ 10+B(-1) = 8$ or $B = 2.$

$\displaystyle\int \frac{3x^2-4x+9}{(x-2)(x^2+9)}\,dx = \int \left(\frac{1}{x-2} + \frac{2x}{x^2+9}\right)\,dx$

$\displaystyle = \ln|x-2| + \ln|x^2+9| + C = \ln|(x-2)(x^2+9)| + C.$

Chapter 8

Parametric Equations, Vectors, and Polar Coordinates

8.1   Vectors and Parametric Equations

1. $x = 3t$, $y = t + 1$. From the second equation, $t = y - 1$. Substituting in the first equation we get $x = 3(y - 1)$, and $x - 3y + 3 = 0$.

3.
$$
\begin{aligned}
x &= t^2 + 1 \\
\underline{y} &= \underline{t} \\
x &= y^2 + 1 \quad \text{(substitution)} \\
y^2 &= x - 1
\end{aligned}
$$

5.
$$
\begin{aligned}
x &= \cos\theta \\
\underline{y} &= \underline{\sin^2\theta} \\
x^2 &= \cos^2\theta \\
\underline{y} &= \underline{\sin^2\theta} \\
x^2 + \quad y &= \cos^2\theta + \sin^2\theta = 1 \quad \text{(adding)} \\
y &= 1 - x^2
\end{aligned}
$$

7.
$$
\begin{aligned}
y &= \sec^2\theta \\
\underline{x} &= \underline{3\tan\theta} \\
y &= \sec^2\theta \\
\frac{x^2}{9} &= \tan^2\theta \\
y - \frac{x^2}{9} &= \sec^2\theta - \tan^2\theta = 1 \quad \text{(basic identity)} \\
y &= 1 + \frac{1}{9}x^2
\end{aligned}
$$

9. From $x = \ln t$, we get $e^x = t$. Substituting in $y = t + 2$, we have $y = e^x + 2$.

11. $x = t^2$; $v_x = 2t \big|_{t=2} = 4$.
    $y = 2 - t$; $v_y = -1$.
    Thus $\vec{v} = 4\vec{i} - \vec{j}$.

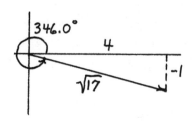

13. $x = t^2 - 2$; $v_x = 2t \big|_{t=-1} = -2$.
    $y = \dfrac{1}{3}t^3 + 2t + 1$; $v_y = t^2 + 2 \big|_{t=-1} = 3$.
    Thus $\vec{v} = -2\vec{i} + 3\vec{j}$.

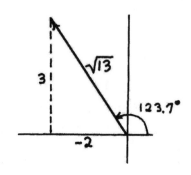

15. $x = (t-3)^2$; $v_x = 2(t-3) \big|_{t=2} = -2$.
    $y = 2t$; $v_y = 2$.
    Thus $\vec{v} = -2\vec{i} + 2\vec{j}$.

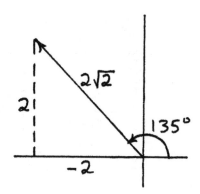

17. $x = e^t$; $v_x = e^t \big|_{t=0} = 1$.
    $y = e^{-t}$; $v_y = -e^{-t} \big|_{t=0} = -1$.
    Thus $\vec{v} = \vec{i} - \vec{j}$.

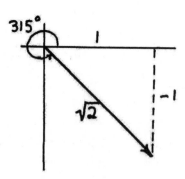

19. $x = \sec t$; $v_x = \sec t \tan t \big|_{t=\pi/6} = \dfrac{2}{\sqrt{3}} \cdot \dfrac{1}{\sqrt{3}} = \dfrac{2}{3}$.

    $y = 2\tan t$; $v_y = 2\sec^2 t \big|_{t=\pi/6} = 2\left(\dfrac{2}{\sqrt{3}}\right)^2 = \dfrac{8}{3}$.

    Thus $\vec{v} = \dfrac{2}{3}\vec{i} + \dfrac{8}{3}\vec{j}$.

    $|\vec{v}| = \sqrt{\left(\dfrac{2}{3}\right)^2 + \left(\dfrac{8}{3}\right)^2} = \sqrt{\dfrac{4 + 64}{9}} = \dfrac{\sqrt{4 \cdot 17}}{3} = \dfrac{2}{3}\sqrt{17}$; $\theta = 76.0°$.

21. $x = 2t^2$; $v_x = 4t$, $a_x = 4$.

$y = \frac{1}{3}t^3$; $v_y = t^2$, $a_y = 2t$.

When $t = -1$, $v_x = -4$, $a_x = 4$ and $v_y = 1$, $a_y = -2$.

Thus $\vec{v} = -4\vec{i} + \vec{j}$ and $\vec{a} = 4\vec{i} - 2\vec{j}$.

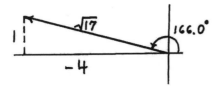

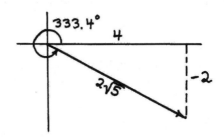

23. $x = \frac{1}{4}t^4 - 2t^2 + t$; $v_x = t^3 - 4t + 1$, $a_x = 3t^2 - 4$.

$y = \frac{1}{3}t^3 - t$; $v_y = t^2 - 1$, $a_y = 2t$.

When $t = 0$, $v_x = 1$, $a_x = -4$ and $v_y = -1$, $a_y = 0$.

Thus $\vec{v} = \vec{i} - \vec{j}$ and $\vec{a} = -4\vec{i}$.

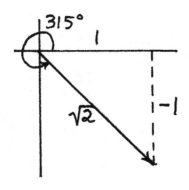

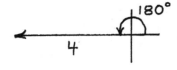

25. $x = 4\cos t$; $v_x = -4\sin t$, $a_x = -4\cos t$.

$y = 4\sin t$; $v_y = 4\cos t$, $a_y = -4\sin t$.

When $t = \frac{3\pi}{4}$, $v_x = -\frac{4}{\sqrt{2}}$, $a_x = \frac{4}{\sqrt{2}}$ and $v_y = -\frac{4}{\sqrt{2}}$, $a_y = -\frac{4}{\sqrt{2}}$.

Thus $\vec{v} = -\frac{4}{\sqrt{2}}\vec{i} - \frac{4}{\sqrt{2}}\vec{j}$ and $\vec{a} = \frac{4}{\sqrt{2}}\vec{i} - \frac{4}{\sqrt{2}}\vec{j}$.

$|\vec{v}| = |\vec{a}| = \sqrt{\frac{16}{2} + \frac{16}{2}} = \sqrt{16} = 4$.

The direction of $\vec{v}$ is 225° and the direction of $\vec{a}$ is 315°. The curve is a circle of radius 4

centered at the origin.

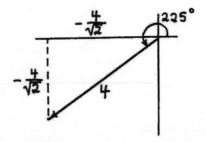

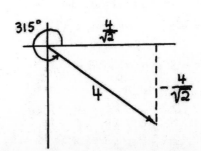

27. $x = (\tan t)^2;\ v_x = 2\tan t\sec^2 t\big|_{t=\pi/3} = 2\sqrt{3}(2^2) = 8\sqrt{3}.$

$y = \csc t;\ v_y = -\csc t\cot t\big|_{t=\pi/3} = -\dfrac{2}{\sqrt{3}}\cdot\dfrac{1}{\sqrt{3}} = -\dfrac{2}{3}.$

Thus $\vec{v} = 8\sqrt{3}\,\vec{i} - \dfrac{2}{3}\vec{j}$ (4th quadrant).

$|\vec{v}| = \sqrt{\left(8\sqrt{3}\right)^2 + \left(-\dfrac{2}{3}\right)^2} \approx 13.87\,\text{m/s};\ \theta = \text{Arctan}\,\dfrac{-2/3}{8\sqrt{3}} = -2.75°.$

29. By Galileo's principle, the horizontal and vertical components of the velocity can be treated separately.

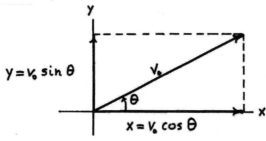

(a) In the horizontal direction, we need only the relationship distance = rate × time to obtain
$x = (v_0\cos\theta)t = v_0 t\cos\theta.$

In the vertical direction, we have $g = -10\,\text{m/s}^2$ by the assumption that the upward direction is positive. So if v denotes the velocity, then $v = -10t + C$. If $t = 0$, $v = v_0\sin\theta$, so that $v_0\sin\theta = 0 + C$. Thus $v = -10t + v_0\sin\theta.$

Integrating, we get $y = -5t^2 + (v_0\sin\theta)t + k.$

If $t = 0$, then $0 = k$, so that $y = v_0 t\sin\theta - 5t^2.$

(b) If $v_0 = 40\,\text{m/s}$ and $\theta = 30°$, then

$x = 40t\cos 30°,\ y = 40t\sin 30° - 5t^2$

$x = 40t\left(\dfrac{\sqrt{3}}{2}\right),\ y = 40t\left(\dfrac{1}{2}\right) - 5t^2$

$x = 20\sqrt{3}t,\ y = 20t - 5t^2$

$v_x = 20\sqrt{3},\ v_y = 20 - 10t\big|_{t=1} = 10.$

Magnitude: $\sqrt{(20\sqrt{3})^2 + (10)^2} = \sqrt{1200 + 100} = 10\sqrt{13}\,\text{m/s}.$

Direction: $\text{Arctan}\left[\dfrac{10}{20\sqrt{3}}\right] = 16.1°.$

Also, $a_x = 0,\ a_y = -10.$

Magnitude: $\sqrt{0 + 100} = 10\,\text{m/s}^2.$ Direction: $-90°.$

(c) The range is the value of x when $y = 0$:

$0 = v_0 t\sin\theta - 5t^2$

$t(v_0\sin\theta - 5t) = 0$

$t = 0,\ t = \dfrac{1}{5}v_0\sin\theta.$

Substituting the nonzero value in the equation $x = v_0 t\cos\theta$, we get:

$R = x = v_0\left(\dfrac{1}{5}v_0\sin\theta\right)\cos\theta$

$R = \dfrac{1}{5}v_0^2\sin\theta\cos\theta.$

8.2   Arc Length

1. $y = \frac{2}{3}x^{3/2}$; $y' = \frac{2}{3} \cdot \frac{3}{2}x^{1/2} = x^{1/2}$; $(y')^2 = x$.

$$s = \int_0^3 \sqrt{1+x}\,dx = \frac{2}{3}(1+x)^{3/2}\bigg|_0^3 = \frac{2}{3}\left[4^{3/2}-1\right] = \frac{14}{3}.$$

3. $y = \frac{1}{2}(e^x + e^{-x})$; $y' = \frac{1}{2}(e^x - e^{-x})$.

$$1 + (y')^2 = 1 + \frac{1}{4}(e^{2x} - 2 + e^{-2x}) = \frac{1}{4}(e^{2x} + 2 + e^{-2x}) = \left[\frac{1}{2}(e^x + e^{-x})\right]^2.$$

$$\begin{aligned}
s &= \int_0^1 \sqrt{\left[\frac{1}{2}(e^x + e^{-x})\right]^2}\,dx = \int_0^1 \frac{1}{2}(e^x + e^{-x})\,dx \\
&= \frac{1}{2}(e^x - e^{-x})\bigg|_0^1 = \frac{1}{2}(e - e^{-1}) \approx 1.18.
\end{aligned}$$

5. $y = \frac{1}{6}x^3 + \frac{1}{2x} = \frac{1}{6}x^3 + \frac{1}{2}x^{-1}$; $y' = \frac{1}{2}x^2 - \frac{1}{2}x^{-2}$; $(y')^2 = \frac{1}{4}x^4 - \frac{1}{2} + \frac{1}{4}x^{-4}$;

$$1 + (y')^2 = 1 + \frac{1}{4}x^4 - \frac{1}{2} + \frac{1}{4}x^{-4} = \frac{1}{4}x^4 + \frac{1}{2} + \frac{1}{4}x^{-4} = \left(\frac{1}{2}x^2 + \frac{1}{2}x^{-2}\right)^2.$$

$$\begin{aligned}
s &= \int_1^3 \sqrt{\left(\frac{1}{2}x^2 + \frac{1}{2}x^{-2}\right)^2}\,dx = \int_1^3 \left(\frac{1}{2}x^2 + \frac{1}{2}x^{-2}\right)dx \\
&= \frac{1}{2}\int_1^3 (x^2 + x^{-2})\,dx = \frac{1}{2}\left(\frac{1}{3}x^3 + \frac{x^{-1}}{-1}\right)\bigg|_1^3 = \frac{1}{2}\left(\frac{1}{3}x^3 - \frac{1}{x}\right)\bigg|_1^3 \\
&= \frac{1}{2}\left[\left(9 - \frac{1}{3}\right) - \left(\frac{1}{3} - 1\right)\right] = \frac{1}{2}\left(10 - \frac{2}{3}\right) = \frac{14}{3}.
\end{aligned}$$

7. $x = 3t^2$, $y = t^3$; $\dfrac{dx}{dt} = 6t$, $\dfrac{dy}{dt} = 3t^2$.

$$\begin{aligned}
s &= \int_0^{\sqrt{5}} \sqrt{(6t)^2 + (3t^2)^2}\,dt = \int_0^{\sqrt{5}} \sqrt{36t^2 + 9t^4}\,dt \\
&= \int_0^{\sqrt{5}} \sqrt{9t^2(4 + t^2)}\,dt = 3\int_0^{\sqrt{5}} \sqrt{4 + t^2}\, t\,dt \\
&= \frac{3}{2}\int_0^{\sqrt{5}} (4 + t^2)^{1/2}(2t\,dt) \qquad u = 4 + t^2;\ du = 2t\,dt. \\
&= \frac{3}{2} \cdot \frac{2}{3}(4 + t^2)^{3/2}\bigg|_0^{\sqrt{5}} = (4 + 5)^{3/2} - 4^{3/2} = 27 - 8 = 19.
\end{aligned}$$

9. $x = \cos^3\theta$, $y = \sin^3\theta$; $\dfrac{dx}{d\theta} = 3\cos^2\theta(-\sin\theta)$, $\dfrac{dy}{d\theta} = 3\sin^2\theta\cos\theta$;

$$\left(\frac{dx}{d\theta}\right)^2 = 9\cos^4\theta\sin^2\theta; \left(\frac{dy}{d\theta}\right)^2 = 9\sin^4\theta\cos^2\theta.$$

$$\begin{aligned}
s &= \int_0^{\pi/2} \sqrt{9\cos^4\theta\sin^2\theta + 9\sin^4\theta\cos^2\theta}\,d\theta \\
&= \int_0^{\pi/2} \sqrt{9\cos^2\theta\sin^2\theta(\cos^2\theta + \sin^2\theta)}\,d\theta \qquad \sin^2\theta + \cos^2\theta = 1 \\
&= \int_0^{\pi/2} 3\cos\theta\sin\theta\,d\theta = 3\int_0^{\pi/2} \sin\theta\cos\theta\,d\theta \qquad u = \sin\theta;\ du = \cos\theta. \\
&= \frac{3}{2}\sin^2\theta\bigg|_0^{\pi/2} = \frac{3}{2}.
\end{aligned}$$

11. $x = e^t \sin t; \quad \dfrac{dx}{dt} = e^t \cos t + e^t \sin t.$

$y = e^t \cos t; \quad \dfrac{dy}{dt} = -e^t \sin t + e^t \cos t.$

$$\left(\dfrac{dx}{dt}\right)^2 = e^{2t} \cos^2 t + 2e^{2t} \sin t \cos t + e^{2t} \sin^2 t$$

$$= e^{2t}(\cos^2 t + \sin^2 t) + 2e^{2t} \sin t \cos t = e^{2t} + 2e^{2t} \sin t \cos t.$$

$$\left(\dfrac{dy}{dt}\right)^2 = e^{2t} \sin^2 t - 2e^{2t} \sin t \cos t + e^{2t} \cos^2 t$$

$$= e^{2t}(\sin^2 t + \cos^2 t) - 2e^{2t} \sin t \cos t = e^{2t} - 2e^{2t} \sin t \cos t.$$

By Formula (8.5),

$$s = \int_0^1 \sqrt{2e^{2t}}\, dt = \sqrt{2} \int_0^1 e^t\, dt = \sqrt{2} e^t \Big|_0^1 = \sqrt{2}(e - 1).$$

13. $y = \left(a^{2/3} - x^{2/3}\right)^{3/2}; \ y' = \dfrac{3}{2}\left(a^{2/3} - x^{2/3}\right)^{1/2}\left(-\dfrac{2}{3}x^{-1/3}\right) = -x^{-1/3}\left(a^{2/3} - x^{2/3}\right)^{1/2};$

$(y')^2 = \left(-x^{-1/3}\right)^2\left[\left(a^{2/3} - x^{2/3}\right)^{1/2}\right]^2 = x^{-2/3}\left(a^{2/3} - x^{2/3}\right) = x^{-2/3}a^{2/3} - 1.$

$1 + (y')^2 = x^{-2/3}a^{2/3}; \ \sqrt{1 + (y')^2} = x^{-1/3}a^{1/3}.$

$$\text{Arc length} = 4\int_0^a x^{-1/3}a^{1/3}\, dx = 4a^{1/3}\int_0^a x^{-1/3}\, dx$$

$$= 4a^{1/3} \cdot \dfrac{3}{2}x^{2/3}\Big|_0^a = 6a^{1/3}a^{2/3} = 6a.$$

15. $y = 2x^2, \ y' = 4x.$

$\text{Arclength} = \displaystyle\int_0^3 \sqrt{1 + (4x)^2}\, dx = 18.4599.$

17. $y = \ln x, \ y' = \dfrac{1}{x}.$

$\text{Arc length} = \displaystyle\int_1^4 \sqrt{1 + (1/x)^2}\, dx = 3.3428.$

19. $y = e^{-x}, \ y' = -e^{-x}.$

$s = \displaystyle\int_1^5 \sqrt{1 + (-e^{-x})^2}\, dx = \int_1^5 \sqrt{1 + e^{-2x}}\, dx = 4.0333.$

8.3 Polar Coordinates

1.

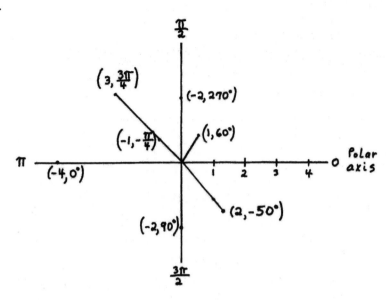

3. $x = r\cos\theta = 2\cos 120° = 2\left(-\dfrac{1}{2}\right) = -1.$

   $y = r\sin\theta = 2\sin 120° = 2\left(\dfrac{\sqrt{3}}{2}\right) = \sqrt{3}.$

5. $x = r\cos\theta = -6\cos\left(\dfrac{\pi}{3}\right) = -6\left(\dfrac{1}{2}\right) = -3.$

   $y = r\sin\theta = -6\sin\left(\dfrac{\pi}{3}\right) = -6\left(\dfrac{\sqrt{3}}{2}\right) = -3\sqrt{3}.$

7. $x = r\cos\theta = -2\cos 170° = 1.97.$

   $y = r\sin\theta = -2\sin 170° = -0.35.$

9. $r^2 = x^2 + y^2 = 1^2 + (-1)^2 = 2,\; r = \sqrt{2},\; \tan\theta = \dfrac{-1}{1} = -1.$

   Since θ is in the fourth quadrant, $\theta = 315° = \dfrac{7\pi}{4}.$

11. $r = \sqrt{3^2 + 4^2} = 5,\; \tan\theta = \dfrac{4}{3}$; since θ is in the first quadrant, $\theta = \text{Arctan}\,\dfrac{4}{3} = 0.93.$

13.
$$
\begin{aligned}
x &= 2 \\
r\cos\theta &= 2 \\
r &= \dfrac{2}{\cos\theta} \\
r &= 2\sec\theta.
\end{aligned}
$$

15. $x^2 + y^2 = r^2 = 2$, or $r = \pm\sqrt{2}$. To generate the circle, it is sufficient to take the positive square root: $r = \sqrt{2}.$

17. $2x - 4y = 5$. Since $x = r\cos\theta$ and $y = r\sin\theta$, we get
$$2r\cos\theta - 4r\sin\theta = 5.$$

19. $x^2 - 2x + y = 2$. Substituting $x = r\cos\theta$ and $y = r\sin\theta$, we get
$$r^2\cos^2\theta - 2r\cos\theta + r\sin\theta = 2.$$

21.
$$
\begin{aligned}
x^2 - y^2 &= 1 \\
(r\cos\theta)^2 - (r\sin\theta)^2 &= 1 \\
r^2\cos^2\theta - r^2\sin^2\theta &= 1 \\
r^2(\cos^2\theta - \sin^2\theta) &= 1 \\
r^2(\cos 2\theta) &= 1 \qquad \text{double-angle formula} \\
r^2 &= \frac{1}{\cos 2\theta} = \sec 2\theta.
\end{aligned}
$$

23. $2x^2 + 4y^2 = 1$; $2(r\cos\theta)^2 + 4(r\sin\theta)^2 = 1$, or $2r^2\cos^2\theta + 4r^2\sin^2\theta = 1$. This equation can be simplified by replacing $\cos^2\theta$ by $1 - \sin^2\theta$ and solving for r^2:
$$
\begin{aligned}
2r^2(1 - \sin^2\theta) + 4r^2\sin^2\theta &= 1 \\
r^2(2 - 2\sin^2\theta + 4\sin^2\theta) &= 1 \\
r^2 &= \frac{1}{2 + 2\sin^2\theta} = \frac{1}{2(1 + \sin^2\theta)}.
\end{aligned}
$$

25. $r\sin\theta = 2$ leads directly to $y = 2$.

27.
$$
\begin{aligned}
\theta &= \frac{\pi}{4} \\
\tan\theta &= \tan\frac{\pi}{4} \\
\frac{y}{x} &= 1 \text{ and } y = x.
\end{aligned}
$$

29.
$$
\begin{aligned}
r &= \cos\theta \\
r^2 &= r\cos\theta \qquad \text{multiplying by } r \\
x^2 + y^2 &= x.
\end{aligned}
$$
Alternatively, $r = \cos\theta = \dfrac{x}{r}$.
From $r = \dfrac{x}{r}$ we have $r^2 = x$ or $x^2 + y^2 = x$.

31.
$$
\begin{aligned}
r &= 1 + \cos\theta \\
r^2 &= r + r\cos\theta \qquad \text{multiplying by } r \\
(1)\quad x^2 + y^2 &= \pm\sqrt{x^2 + y^2} + x \\
x^2 + y^2 - x &= \pm\sqrt{x^2 + y^2} \\
(x^2 + y^2 - x)^2 &= x^2 + y^2. \qquad \text{squaring both sides}
\end{aligned}
$$
Alternatively,
$$
\begin{aligned}
r &= 1 + \cos\theta \\
r &= 1 + \frac{x}{r} \qquad \text{since } \cos\theta = \frac{x}{r} \\
\pm\sqrt{x^2 + y^2} &= 1 + \frac{x}{\pm\sqrt{x^2 + y^2}} \\
x^2 + y^2 &= \pm\sqrt{x^2 + y^2} + x, \text{ which agrees with (1).}
\end{aligned}
$$

33.
$$r = 1 - \cos\theta$$
$$r^2 = r - r\cos\theta \qquad \text{multiplying by } r$$
$$x^2 + y^2 = \pm\sqrt{x^2+y^2} - x$$
$$x^2 + y^2 + x = \pm\sqrt{x^2+y^2}$$
$$(x^2+y^2+x)^2 = x^2+y^2. \qquad \text{squaring both sides}$$

Alternatively,
$$r = 1 - \cos\theta$$
$$r = 1 - \frac{x}{r} \qquad \cos\theta = \frac{x}{r}$$
$$\pm\sqrt{x^2+y^2} = 1 - \frac{x}{\pm\sqrt{x^2+y^2}}$$
$$x^2+y^2 = \pm\sqrt{x^2+y^2} - x \qquad \text{multiplying by } \pm\sqrt{x^2+y^2}$$
$$x^2+y^2+x = \pm\sqrt{x^2+y^2}$$
$$(x^2+y^2+x)^2 = x^2+y^2.$$

35.
$$r = 1 - 2\sin\theta$$
$$r^2 = r - 2r\sin\theta \qquad \text{multiplying by } r$$
$$x^2+y^2 = \pm\sqrt{x^2+y^2} - 2y$$
$$x^2+y^2+2y = \pm\sqrt{x^2+y^2}$$
$$(x^2+y^2+2y)^2 = x^2+y^2. \qquad \text{squaring both sides}$$

37.
$$r = -2 + 4\sin\theta$$
$$r^2 = -2r + 4r\sin\theta \qquad \text{multiplying by } r$$
$$x^2+y^2 = -2\left(\pm\sqrt{x^2+y^2}\right) + 4y$$
$$(x^2+y^2-4y)^2 = 4(x^2+y^2).$$

39. By the double-angle formula,
$$r^2 = \sin 2\theta = 2\sin\theta\cos\theta$$
$$r^4 = 2(r\sin\theta)(r\cos\theta) \qquad \text{multiplying by } r^2$$
$$(x^2+y^2)^2 = 2xy.$$

41. $r = \dfrac{2}{1-\cos\theta}$. Dividing both sides by r, we get
$$1 = \frac{2}{r(1-\cos\theta)} = \frac{2}{r - r\cos\theta}$$
$$1 = \frac{2}{\pm\sqrt{x^2+y^2} - x}$$
$$\pm\sqrt{x^2+y^2} - x = 2$$
$$\pm\sqrt{x^2+y^2} = x + 2$$
$$x^2+y^2 = x^2 + 4x + 4 \qquad \text{squaring both sides}$$
$$y^2 = 4(x+1).$$

43.
$$r = \frac{4}{1-\sin\theta}$$
$$r - r\sin\theta = 4 \qquad \text{multipling by } (1-\sin\theta)$$
$$\pm\sqrt{x^2+y^2} - y = 4$$
$$\pm\sqrt{x^2+y^2} = y + 4$$
$$x^2+y^2 = y^2 + 8y + 16$$
$$x^2 = 8(y+2).$$

45.
$$
\begin{aligned}
r^2 &= 3\cos 2\theta = 3(\cos^2\theta - \sin^2\theta) \qquad \text{double-angle formula}\\
r^4 &= 3(r^2\cos^2\theta - r^2\sin^2\theta) \qquad \text{multiplying by } r^2\\
(x^2+y^2)^2 &= 3(x^2-y^2).
\end{aligned}
$$

47. $r = a\sin 3\theta = a\sin(2\theta + \theta) = a(\sin 2\theta\cos\theta + \cos 2\theta\sin\theta)$.
$$
\begin{aligned}
r &= a\left[(2\sin\theta\cos\theta)\cos\theta + (\cos^2\theta - \sin^2\theta)\sin\theta\right]\\
&= a(2\sin\theta\cos^2\theta + \cos^2\theta\sin\theta - \sin^3\theta)\\
&= a(3\sin\theta\cos^2\theta - \sin^3\theta)\\
r^4 &= a\left[3(r\sin\theta)(r^2\cos^2\theta) - r^3\sin^3\theta\right] \qquad \text{multiplying by } r^3\\
(x^2+y^2)^2 &= a(3x^2y - y^3).
\end{aligned}
$$

8.4   Curves in Polar Coordinates

1. $r = 2$ is a circle of radius 2.

3. $r = 2\cos\theta$ is a circle but may be viewed as a special case of a rose with one leaf.

5. $r\cos\theta = 4$ is the line $x = 4$.

7. $r = 2(1 + \cos\theta)$.
   Cardioid.
   If $\theta = 0°$, $r = 4$.
   If $\theta = 90°$ or $270°$, $r = 2$.
   If $\theta = 180°$, $r = 0$.

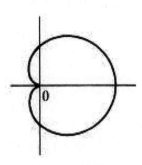

9. Limaçon.
   If $\theta = 90°$ or $270°$, $r = 9$.
   If $\theta = 0°$, $r = 14$.
   If $\theta = 180°$, $r = 4$.

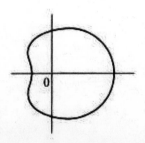

11. $r = 3 - 2\cos\theta$.

    Limaçon.

    If $\theta = 0°$, $r = 1$.

    If $\theta = 90°$ or $270°$, $r = 3$.

    If $\theta = 180°$, $r = 5$.

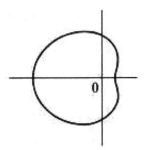

13. Limaçon with a loop.

    If $\theta = 0°$ or $180°$, $r = 1$.

    If $\theta = 90°$, $r = -1$.

    If $\theta = 270°$, $r = 3$.

    Also, $r = 0$ when

    $$1 - 2\sin\theta = 0$$
    $$\sin\theta = \frac{1}{2}$$
    $$\theta = 30°, \ 150°.$$

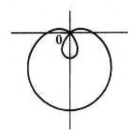

15. $r = -1 - 2\sin\theta$.

    Limaçon with a loop.

    If $\theta = 0°$ or $180°$, $r = -1$.

    If $\theta = 90°$, $r = -3$.

    If $\theta = 270°$, $r = 1$.

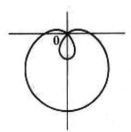

17. Four-leaf rose.

    $r = 0$ when $\theta = 0°$, $90°$, $180°$, $270°$.

    $|r| = 4$ when $\theta = 45°$, $135°$, $225°$, $315°$.

19. $r = 3\cos 3\theta$, three-leaf rose. When $\theta = 0°$, $r = 1$; when $\theta = \pm 30°$, $r = \cos(\pm 90°) = 0$. So one leaf is traced out in the interval $\theta = -30°$ to $\theta = +30°$. The location of the other two leaves follows from the symmetry.

21. $r^2 = 4 \sin 2\theta$, lemniscate. When $\theta = 0°$ or $90°$, we get $r = 0$. When $\theta = 45°$, $r^2 = 4 \sin 90° = 4$ and $r = \pm 2$. So the upper portion of the figure is traced out in the interval $\theta = 0°$ to $\theta = 90°$ with $r = 2$ at $\theta = 45°$.

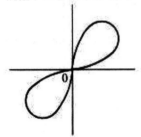

23. $r = 3\sqrt{\sin 2\theta}$, $0 \le \theta \le 90°$; $r = 0$ when $\theta = 0°$ or $90°$; r attains its largest value when $\theta = 45°$: $r = 3$.

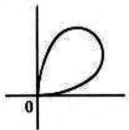

25. $r = \cos 3\left(\theta - \dfrac{\pi}{6}\right)$ is the three-leaf rose $r = \cos 3\theta$ rotated $\dfrac{\pi}{6}$ in the counterclockwise direction.

27. $r = \sin 3\theta$, three-leaf rose. When $\theta = 0°$ or $60°$, we get $r = 0$. When $\theta = 30°$, $r = \sin 3 \cdot 30° = \sin 90° = 1$. So one leaf is traced out in the interval $\theta = 0°$ to $\theta = 60°$, attaining a maximum value of $r = 1$ at $\theta = 30°$. The location of the other two leaves follows from the symmetry.

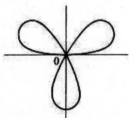

29. $r = \dfrac{2}{\theta}$; as $\theta \to 0$, $r \to \infty$; as $\theta \to \infty$, $r \to 0$.

(In other words, as the curve circles the pole, r gets closer to 0; see curve in answer section.)

31. Multiply both sides of the equation by $2 - \cos\theta$:

$$
\begin{aligned}
2r - r\cos\theta &= 3 \\
\pm 2\sqrt{x^2 + y^2} - x &= 3 \\
\pm 2\sqrt{x^2 + y^2} &= x + 3 \\
4(x^2 + y^2) &= (x+3)^2 = x^2 + 6x + 9 \\
3x^2 + 4y^2 - 6x - 9 &= 0
\end{aligned}
$$

which is the equation of an ellipse. To plot the curve, it is sufficient to get the four "intercepts":

θ:	0°	90°	180°	270°
r:	3	$\dfrac{3}{2}$	1	$\dfrac{3}{2}$

33. If $\theta = 0°$ or $180°$, $r = 3$.

If $\theta = 90°$, $r = \dfrac{6}{5}$.

If $\theta = 270°$, $r = -6$.

Some other points are given in the following table:

θ:	−10°	190°	250°	260°	280°	290°
r:	4.1	4.1	−7.3	−6.3	−6.3	−7.3

The resulting curve is a hyperbola. (See the sketch in the answer section.)

8.5 Areas in Polar Coordinates

1. $A = \dfrac{1}{2}\displaystyle\int_0^{\pi/6} 2^2\, d\theta = 2\theta \Big|_0^{\pi/6} = \dfrac{\pi}{3}.$

3. $A = \dfrac{1}{2}\displaystyle\int_0^{\pi/4} \sec^2\theta\, d\theta = \dfrac{1}{2}\tan\theta \Big|_0^{\pi/4} = \dfrac{1}{2}.$

5. $A = \dfrac{1}{2}\displaystyle\int_0^{2} \theta^2\, d\theta = \dfrac{1}{2}\dfrac{1}{3}\theta^3 \Big|_0^{2} = \dfrac{4}{3}.$

7. 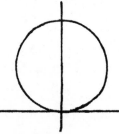 $r = 2\sin\theta$ is a circle (one-leaf rose), similar to Figure 8.19. The curve is traced out in the interval $\theta = 0$ to $\theta = \pi$:

$A = \dfrac{1}{2}\displaystyle\int_0^{\pi} (2\sin\theta)^2\, d\theta = 2\int_0^{\pi} \dfrac{1}{2}(1 - \cos 2\theta)\, d\theta = \int_0^{\pi} 1\, d\theta - \int_0^{\pi} \cos 2\theta\, d\theta.$

For the second integral, let $u = 2\theta$; $du = 2\, d\theta$. So we get

$\displaystyle\int_0^{\pi} d\theta - \dfrac{1}{2}\int_0^{\pi} \cos 2\theta (2\, d\theta) = \theta - \dfrac{1}{2}\sin 2\theta \Big|_0^{\pi} = \pi - 0 = \pi.$

9.

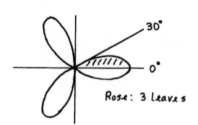

Rose: 3 Leaves

Since $r = 4\cos 3\theta = 0$ when $\theta = 30°$, it follows that half a leaf is traced out in the interval $\theta = 0°$ to $\theta = 30°$.

So integrating from $\theta = 0$ to $\theta = \dfrac{\pi}{6}$ yields one-sixth of the area:

$$\frac{1}{6}A = \frac{1}{2}\int_0^{\pi/6} (4\cos 3\theta)^2 \, d\theta = 8\int_0^{\pi/6} \cos^2 3\theta \, d\theta$$

$$= 8\int_0^{\pi/6} \frac{1}{2}(1 + \cos 6\theta)\, d\theta = 4\int_0^{\pi/6}(1 + \cos 6\theta)\, d\theta.$$

To integrate $\displaystyle\int \cos 6\theta \, d\theta$, let $u = 6\theta$; $du = 6\, d\theta$.

$$\frac{1}{6}\int \cos 6\theta (6\, d\theta) = \frac{1}{6}\sin 6\theta + C.$$

It follows that $\dfrac{1}{6}A = 4\left(\theta + \dfrac{1}{6}\sin 6\theta\right)\Big|_0^{\pi/6} = 4\left(\dfrac{\pi}{6}\right) = \dfrac{2\pi}{3}.$

So $A = 6\left(\dfrac{2\pi}{3}\right) = 4\pi.$

11. $r = 3\sin 2\theta$, four-leaf rose.

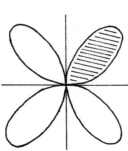

When $\theta = 0°$, $r = 0$ and when $\theta = 90°$,
$r = 3\sin 2(90°) = 3\sin 180° = 0$.
So one leaf is traced out in the interval $\theta = 0$ to $\theta = \dfrac{\pi}{2}$. The entire area is equal to four times the area of one leaf.

$$A = 4 \cdot \frac{1}{2}\int_0^{\pi/2}(3\sin 2\theta)^2 \, d\theta = 18\int_0^{\pi/2}\sin^2 2\theta \, d\theta$$

$$= 18\int_0^{\pi/2}\frac{1}{2}(1 - \cos 4\theta)\, d\theta$$

$$= 9\int_0^{\pi/2} 1\, d\theta - 9\int_0^{\pi/2}\cos 4\theta \, d\theta \qquad u = 4\theta;\ du = 4\, d\theta.$$

$$= 9\theta\Big|_0^{\pi/2} - \frac{9}{4}\sin 4\theta\Big|_0^{\pi/2} = \frac{9\pi}{2}.$$

13.

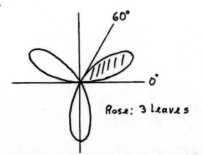

Rose: 3 Leaves

Since $r = 6\sin 3\theta = 0$ when $\theta = 0°$ and $\theta = 60°$, it follows that one leaf is traced out in the interval from $\theta = 0°$ to $\theta = 60°$.

So integrating from $\theta = 0$ to $\theta = \dfrac{\pi}{3}$ yields one-third of the area:

$$\frac{1}{3}A = \frac{1}{2}\int_0^{\pi/3} (6\sin 3\theta)^2 \, d\theta = \frac{1}{2}\int_0^{\pi/3} 36\sin^2 3\theta \, d\theta$$

$$= 18\int_0^{\pi/3} \frac{1}{2}(1 - \cos 6\theta) \, d\theta = 9\int_0^{\pi/3}(1 - \cos 6\theta) \, d\theta$$

$$= 9\left(\theta - \frac{1}{6}\sin 6\theta\right)\Big|_0^{\pi/3} = 9\left(\frac{\pi}{3}\right) = 3\pi. \qquad u = 6\theta; \; du = 6\, d\theta.$$

So $A = 9\pi$.

15. $r = \sqrt{\sin\theta}$.

This curve is traced out in the interval $\theta = 0$ to $\theta = \pi$.

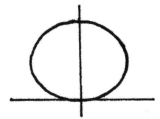

$$A = \frac{1}{2}\int_0^{\pi}\left(\sqrt{\sin\theta}\right)^2 \, d\theta = \frac{1}{2}\int_0^{\pi}\sin\theta \, d\theta$$

$$= -\frac{1}{2}\cos\theta\Big|_0^{\pi} = -\frac{1}{2}(\cos\pi - \cos 0) = -\frac{1}{2}(-1 - 1) = 1.$$

17. For a limaçon without a loop, the limits of integration are 0 and 2π:

$$A = \frac{1}{2}\int_0^{2\pi}(2 + \cos\theta)^2 \, d\theta = \frac{1}{2}\int_0^{2\pi}(4 + 4\cos\theta + \cos^2\theta) \, d\theta.$$

Since $\cos^2\theta = \frac{1}{2}(1 + \cos 2\theta)$, we get

$$A = \frac{1}{2}\int_0^{2\pi}\left[4 + 4\cos\theta + \frac{1}{2}(1 + \cos 2\theta)\right] d\theta$$

$$= \frac{1}{2}\left[4\theta + 4\sin\theta + \frac{1}{2}\left(\theta + \frac{1}{2}\sin 2\theta\right)\right]\Big|_0^{2\pi} \qquad u = 2\theta; \; du = 2\, d\theta.$$

$$= \frac{1}{2}\left[4(2\pi) + \frac{1}{2}(2\pi)\right] = \frac{9\pi}{2}.$$

19. For any cardioid, the limits of integration are 0 and 2π:

$$A = \frac{1}{2}\int_0^{2\pi}[2(1 - \cos\theta)]^2 \, d\theta = 2\int_0^{2\pi}(1 - 2\cos\theta + \cos^2\theta) \, d\theta$$

$$= 2\int_0^{2\pi}\left[1 - 2\cos\theta + \frac{1}{2}(1 + \cos 2\theta)\right] d\theta = 2\left[\theta - 2\sin\theta + \frac{1}{2}\left(\theta + \frac{1}{2}\sin 2\theta\right)\right]\Big|_0^{2\pi}$$

$$= 2\left[2\pi + \frac{1}{2}(2\pi)\right] = 6\pi.$$

21. For a limaçon without a loop, the limits of integration are 0 and 2π:

$$A = \frac{1}{2}\int_0^{2\pi}(2 - \sin\theta)^2 \, d\theta = \frac{1}{2}\int_0^{2\pi}(4 - 4\sin\theta + \sin^2\theta) \, d\theta$$

$$= \frac{1}{2}\int_0^{2\pi}\left[4 - 4\sin\theta + \frac{1}{2}(1 - \cos 2\theta)\right] d\theta$$

$$= \frac{1}{2}\left[4\theta + 4\cos\theta + \frac{1}{2}\left(\theta - \frac{1}{2}\sin 2\theta\right)\right]\Big|_0^{2\pi} \qquad u = 2\theta; \; du = 2\, d\theta.$$

$$= \frac{1}{2}\left[8\pi + 4 + \frac{1}{2}(2\pi)\right] - \frac{1}{2}[4] = \frac{1}{2}(9\pi + 4) - 2 = \frac{9\pi}{2}.$$

23. For any cardioid, the limits of integration are 0 and 2π:

$$
\begin{aligned}
A &= \frac{1}{2}\int_0^{2\pi}(3+3\cos\theta)^2\,d\theta = \frac{9}{2}\int_0^{2\pi}(1+2\cos\theta+\cos^2\theta)\,d\theta \\
&= \frac{9}{2}\int_0^{2\pi}\left[1+2\cos\theta+\frac{1}{2}(1+\cos 2\theta)\right]d\theta \\
&= \frac{9}{2}\left[\theta+2\sin\theta+\frac{1}{2}\left(\theta+\frac{1}{2}\sin 2\theta\right)\right]\Bigg|_0^{2\pi} = \frac{9}{2}\left[2\pi+\frac{1}{2}(2\pi)\right] = \frac{27\pi}{2}.
\end{aligned}
$$

25.

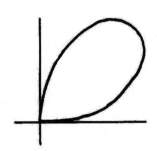

If $r^2 = 4\sin 2\theta$, then $r = 2\sqrt{\sin 2\theta}$, which is only the upper half of the lemniscate. Note that $r = 0$ when $\theta = 0°$ and $\theta = 90°$.

The total area is therefore given by:

$$
\begin{aligned}
A &= 2\cdot\frac{1}{2}\int_0^{\pi/2}\left(2\sqrt{\sin 2\theta}\right)^2 d\theta = \int_0^{\pi/2}4\sin 2\theta\,d\theta \qquad u = 2\theta;\ du = 2\,d\theta. \\
&= 4\left(-\frac{1}{2}\cos 2\theta\right)\Bigg|_0^{\pi/2} = -2\cos 2\theta\Bigg|_0^{\pi/2} = -2(-1-1) = 4.
\end{aligned}
$$

27. $r^2 = 2\cos 2\theta$, lemniscate.

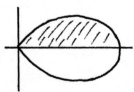

The function $r = \sqrt{2\cos 2\theta}$ is one-half of the lemniscate, shown in the figure. When $\theta = 0$, $r = \sqrt{2}$, the largest possible value. When $\theta = 45°$,

$$r = \sqrt{2\cos 90°} = 0.$$

So integrating from $\theta = 0$ to $\theta = \dfrac{\pi}{4}$ yields the area of the shaded region. By symmetry, the area enclosed by the entire lemniscate is four times this value.

$$
\begin{aligned}
A &= 4\cdot\frac{1}{2}\int_0^{\pi/4}\left(\sqrt{2\cos 2\theta}\right)^2 d\theta = 2\int_0^{\pi/4}2\cos 2\theta\,d\theta \\
&= 2\int_0^{\pi/4}\cos 2\theta(2\,d\theta) = 2\sin 2\theta\Bigg|_0^{\pi/4} = 2\sin 2\left(\frac{\pi}{4}\right) = 2.
\end{aligned}
$$

29. Similar to Exercise 25. Since $r = \sqrt{6}\sqrt{\sin 2\theta}$,

$$
\begin{aligned}
A &= 2\cdot\frac{1}{2}\int_0^{\pi/2}\left(\sqrt{6}\sqrt{\sin 2\theta}\right)^2 d\theta = \int_0^{\pi/2}6\sin 2\theta\,d\theta \qquad u = 2\theta;\ du = 2\,d\theta. \\
&= -3\cos 2\theta\Bigg|_0^{\pi/2} = -3(-1-1) = 6.
\end{aligned}
$$

31. $r = 1 - 2\sin\theta$.

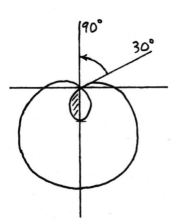

$$r = 0 \text{ when } \quad 1 - 2\sin\theta \;=\; 0$$
$$\sin\theta \;=\; \frac{1}{2}$$
$$\theta \;=\; 30°, \; 150°.$$

Half the loop is traced out in the interval $\dfrac{\pi}{6}$ to $\dfrac{\pi}{2}$. So

$$
\begin{aligned}
A &= 2 \cdot \frac{1}{2} \int_{\pi/6}^{\pi/2} (1 - 2\sin\theta)^2 \, d\theta = \int_{\pi/6}^{\pi/2} (1 - 4\sin\theta + 4\sin^2\theta) \, d\theta \\
&= \int_{\pi/6}^{\pi/2} [1 - 4\sin\theta + 2(1 - \cos 2\theta)] \, d\theta = \theta + 4\cos\theta + 2\left(\theta - \frac{1}{2}\sin 2\theta\right)\Big|_{\pi/6}^{\pi/2} \\
&= \left(\frac{\pi}{2} + \pi\right) - \left(\frac{\pi}{6} + 4\cos\frac{\pi}{6} + \frac{\pi}{3} - \sin\frac{\pi}{3}\right) \\
&= \frac{3\pi}{2} - \frac{\pi}{6} - 4\left(\frac{\sqrt{3}}{2}\right) - \frac{\pi}{3} + \frac{\sqrt{3}}{2} = \pi - 3\left(\frac{\sqrt{3}}{2}\right) = \frac{1}{2}(2\pi - 3\sqrt{3}).
\end{aligned}
$$

33.

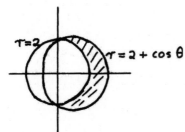

To determine where the curves intersect, we substitute $r = 2$ in $r = 2 + \cos\theta$ and solve for θ:

$$2 \;=\; 2 + \cos\theta$$
$$\cos\theta \;=\; 0.$$

So $\theta = -90°,\ 90°$. We can therefore obtain the upper half of the shaded region by integrating from $\theta = 0$ to $\theta = \dfrac{\pi}{2}$:

$$
\begin{aligned}
\frac{1}{2}A &= \frac{1}{2}\int_0^{\pi/2} (2 + \cos\theta)^2 \, d\theta - \frac{1}{2}\int_0^{\pi/2} 2^2 \, d\theta \\
A &= \int_0^{\pi/2} (2 + \cos\theta)^2 \, d\theta - \int_0^{\pi/2} 4 \, d\theta \\
&= \int_0^{\pi/2} (4 + 4\cos\theta + \cos^2\theta) \, d\theta - 4\theta\Big|_0^{\pi/2} \\
&= \int_0^{\pi/2} \left[4 + 4\cos\theta + \frac{1}{2}(1 + \cos 2\theta)\right] d\theta - 4\left(\frac{\pi}{2}\right) \\
&= 4\theta + 4\sin\theta + \frac{1}{2}\left(\theta + \frac{1}{2}\sin 2\theta\right)\Big|_0^{\pi/2} - 2\pi \\
&= \left[2\pi + 4 + \frac{1}{2}\left(\frac{\pi}{2}\right)\right] - 0 - 2\pi = 4 + \frac{\pi}{4} = \frac{16 + \pi}{4}.
\end{aligned}
$$

35.

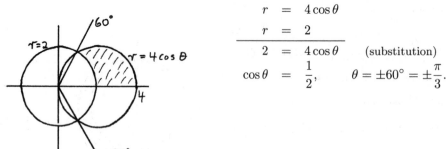

$$r = 4\cos\theta$$
$$r = 2$$
$$\overline{\quad\quad}$$
$$2 = 4\cos\theta \quad \text{(substitution)}$$
$$\cos\theta = \frac{1}{2}, \qquad \theta = \pm 60° = \pm\frac{\pi}{3}.$$

$$
\begin{aligned}
A &= 2\cdot\frac{1}{2}\int_0^{\pi/3}\left[(4\cos\theta)^2 - 2^2\right]d\theta = \int_0^{\pi/3}(16\cos^2\theta - 4)\,d\theta \\
&= \int_0^{\pi/3}\left[8(1+\cos 2\theta) - 4\right]d\theta = \int_0^{\pi/3}(4 + 8\cos 2\theta)\,d\theta \qquad u = 2\theta;\ du = 2\,d\theta. \\
&= 4\theta + 4\sin 2\theta\Big|_0^{\pi/3} = \frac{4\pi}{3} + 4\sin\frac{2\pi}{3} \\
&= \frac{4\pi}{3} + 4\left(\frac{\sqrt{3}}{2}\right) = 4\left(\frac{\pi}{3} + \frac{\sqrt{3}}{2}\right).
\end{aligned}
$$

37.

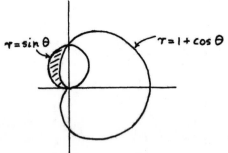

By inspection, we see that the curves intersect at $\theta = 90°$ and $\theta = 180°$.

$$
\begin{aligned}
A &= \frac{1}{2}\int_{\pi/2}^{\pi}\sin^2\theta\,d\theta - \frac{1}{2}\int_{\pi/2}^{\pi}(1+\cos\theta)^2\,d\theta \\
&= \frac{1}{2}\int_{\pi/2}^{\pi}\sin^2\theta\,d\theta - \frac{1}{2}\int_{\pi/2}^{\pi}(1 + 2\cos\theta + \cos^2\theta)\,d\theta \\
&= \frac{1}{4}\int_{\pi/2}^{\pi}(1 - \cos 2\theta)\,d\theta - \frac{1}{2}\int_{\pi/2}^{\pi}\left[1 + 2\cos\theta + \frac{1}{2}(1+\cos 2\theta)\right]d\theta \\
&= \frac{1}{4}\left(\theta - \frac{1}{2}\sin 2\theta\right)\Big|_{\pi/2}^{\pi} - \frac{1}{2}\left[\theta + 2\sin\theta + \frac{1}{2}\left(\theta + \frac{1}{2}\sin 2\theta\right)\right]\Big|_{\pi/2}^{\pi} \\
&= \frac{1}{4}\left(\pi - \frac{\pi}{2}\right) - \frac{1}{2}\left[\left(\pi + \frac{\pi}{2}\right) - \left(\frac{\pi}{2} + 2 + \frac{\pi}{4}\right)\right] \\
&= \frac{1}{4}\left(\frac{\pi}{2}\right) - \frac{1}{2}\left(\frac{3\pi}{4} - 2\right) = 1 - \frac{\pi}{4}.
\end{aligned}
$$

39. (Partial solution)

Shaded area:
$$\frac{1}{2}\int_0^{\pi/2} 16(1+\sin\theta)^2\,d\theta - \frac{1}{2}\int_0^{\pi/2} 64\sin^2\theta\,d\theta.$$
Area below polar axis (right side):
$$\frac{1}{2}\int_{-\pi/2}^0 16(1+\sin\theta)^2\,d\theta.$$
Total area $= (32-4\pi)+(12\pi-32)=8\pi.$

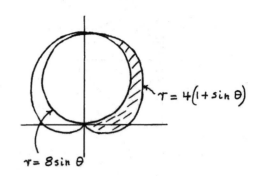

41.

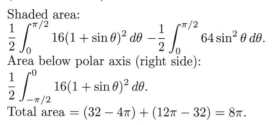

Substituting $r=3\cos\theta$ in $r=1+\cos\theta$, we get:

$$3\cos\theta = 1+\cos\theta$$

$$\cos\theta = \frac{1}{2}$$

$$\theta = \frac{\pi}{3},\ -\frac{\pi}{3}.$$

Area 1:
$$\frac{1}{2}\int_{\pi/3}^{\pi/2}(3\cos\theta)^2\,d\theta$$
$$=\frac{9}{2}\int_{\pi/3}^{\pi/2}\cos^2\theta\,d\theta$$
$$=\frac{9}{4}\int_{\pi/3}^{\pi/2}(1+\cos 2\theta)\,d\theta$$
$$=\frac{9}{4}\left(\theta+\frac{1}{2}\sin 2\theta\right)\Big|_{\pi/3}^{\pi/2}$$

$$=\frac{9}{4}\left[\left(\frac{\pi}{2}\right)-\left(\frac{\pi}{3}+\frac{1}{2}\frac{\sqrt{3}}{2}\right)\right]=\frac{9}{4}\left(\frac{\pi}{2}-\frac{\pi}{3}-\frac{\sqrt{3}}{4}\right)=\frac{9}{4}\left(\frac{\pi}{6}-\frac{\sqrt{3}}{4}\right).$$

Area 2:
$$\frac{1}{2}\int_0^{\pi/3}(1+\cos\theta)^2\,d\theta = \frac{1}{2}\int_0^{\pi/3}(1+2\cos\theta+\cos^2\theta)\,d\theta$$
$$=\frac{1}{2}\int_0^{\pi/3}\left[1+2\cos\theta+\frac{1}{2}(1+\cos 2\theta)\right]d\theta$$
$$=\frac{1}{2}\left[\theta+2\sin\theta+\frac{1}{2}\left(\theta+\frac{1}{2}\sin 2\theta\right)\right]\Big|_0^{\pi/3}$$
$$=\frac{1}{2}\left(\frac{\pi}{3}+\frac{2\sqrt{3}}{2}+\frac{1}{2}\frac{\pi}{3}+\frac{1}{4}\frac{\sqrt{3}}{2}\right)$$
$$=\frac{1}{2}\left(\frac{\pi}{2}+\frac{9\sqrt{3}}{8}\right).$$

$$
\begin{aligned}
\text{Total area} &= 2(\text{Area 1}+\text{Area 2})\\
&=\frac{9}{2}\left(\frac{\pi}{6}-\frac{\sqrt{3}}{4}\right)+\left(\frac{\pi}{2}+\frac{9\sqrt{3}}{8}\right)\\
&=\frac{9\pi}{12}-\frac{9\sqrt{3}}{8}+\frac{\pi}{2}+\frac{9\sqrt{3}}{8}=\frac{3\pi}{4}+\frac{2\pi}{4}=\frac{5\pi}{4}.
\end{aligned}
$$

Chapter 8 Review

1.
$$x = 4\cos^2\theta$$

$$\underline{y = 4\sin\theta}$$

$$\frac{x}{4} = \cos^2\theta$$

$$\frac{y^2}{16} = \sin^2\theta$$

$$\frac{x}{4} + \frac{y^2}{16} = \cos^2\theta + \sin^2\theta = 1$$

$$4x + y^2 = 16.$$

3. (a) $x = 5\cos t$; $v_x = -5\sin t$, $a_x = -5\cos t$.

   $y = 5\sin t$; $v_y = 5\cos t$, $a_y = -5\sin t$.

   When $t = \dfrac{\pi}{4}$, we have

   $$\vec{v} = -\frac{5}{\sqrt{2}}\vec{i} + \frac{5}{\sqrt{2}}\vec{j},\ |\vec{v}| = \sqrt{\left(-\frac{5}{\sqrt{2}}\right)^2 + \left(\frac{5}{\sqrt{2}}\right)^2} = \sqrt{25} = 5\,\text{m/s};\ \theta = 135°;$$

   $$\vec{a} = -\frac{5}{\sqrt{2}}\vec{i} - \frac{5}{\sqrt{2}}\vec{j},\ |\vec{a}| = 5\,\text{m/s}^2;\ \theta = 225°.$$

   (b) $x = 5\sin t$; $v_x = 5\cos t$, $a_x = -5\sin t$.

   $y = 5\cos t$; $v_y = -5\sin t$, $a_y = -5\cos t$.

   When $t = \dfrac{\pi}{4}$, we get

   $$\vec{v} = \frac{5}{\sqrt{2}}\vec{i} - \frac{5}{\sqrt{2}}\vec{j},\ |\vec{v}| = 5\,\text{m/s};\ \theta = 315°;$$

   $$\vec{a} = -\frac{5}{\sqrt{2}}\vec{i} - \frac{5}{\sqrt{2}}\vec{j}\ \text{(same as in part (a))}.$$

5. $y = \ln\sec x$; $y' = \dfrac{\sec x\tan x}{\sec x} = \tan x$; $(y')^2 = \tan^2 x$.

   $1 + (y')^2 = 1 + \tan^2 x = \sec^2 x$.

   $$s = \int_0^{\pi/4}\sqrt{\sec^2 x}\,dx = \int_0^{\pi/4}\sec x\,dx = \ln|\sec x + \tan x|\Big|_0^{\pi/4}$$

   $$= \ln(\sqrt{2} + 1) - \ln 1 = \ln(1 + \sqrt{2}).\qquad\text{(since }\ln 1 = 0\text{)}$$

7. Since $y = r\sin\theta$ and $x = r\cos\theta$, $y = 3x$ becomes

   $$(r\sin\theta)^2 = 3(r\cos\theta)$$

   $$r^2\sin^2\theta = 3r\cos\theta,$$

   which can be rewritten as follows:

   $$r = \frac{3\cos\theta}{\sin^2\theta} = 3\frac{\cos\theta}{\sin\theta}\frac{1}{\sin\theta} = 3\cot\theta\csc\theta.$$

9.
$$r^2 = 4\tan\theta = 4\frac{\sin\theta}{\cos\theta} = 4\frac{r\sin\theta}{r\cos\theta}$$

$$x^2 + y^2 = 4\frac{y}{x}$$

$$x(x^2 + y^2) = 4y.$$

11. $r = \sin\theta$ is a circle (one-leaf rose).

13. Four-leaf rose.

   $r = \pm 2$ when $\theta = 0°,\ 90°,\ 180°,\ 270°$.

   $r = 0$ when $\theta = 45°,\ 135°,\ 225°,\ 315°$.

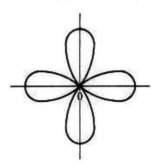

15. $r^2 = 9 \sin 2\theta$ is a lemniscate: $r = 0$ when $\theta = 0°$ and $\theta = 90°$, $r = \pm\sqrt{9} = \pm 3$ when $\theta = 45°$.

17. $\begin{aligned} A &= \frac{1}{2}\int_0^{\pi/8} \tan^2 2\theta\, d\theta = \frac{1}{2}\int_0^{\pi/8} (\sec^2 2\theta - 1)\, d\theta \\ &= \frac{1}{2}\left(\frac{1}{2}\tan 2\theta - \theta\right)\Big|_0^{\pi/8} = \frac{1}{2}\left(\frac{1}{2} - \frac{\pi}{8}\right) = \frac{1}{4} - \frac{\pi}{16}. \end{aligned}$

19.

The function $r = \sqrt{a^2 \cos 2\theta}$ is the right half of the lemniscate, shown in the figure. When $\theta = 0$, $r = a$ and when $\theta = 45°$, $r = 0$. So the shaded region is traced out in the interval $\theta = 0$ to $\theta = \dfrac{\pi}{4}$. The area enclosed by the entire lemniscate is four times the shaded area. Thus

$\begin{aligned} A &= 4 \cdot \frac{1}{2}\int_0^{\pi/4}\left(\sqrt{a^2\cos 2\theta}\right)^2 d\theta = 2\int_0^{\pi/4} a^2\cos 2\theta\, d\theta \\ &= a^2 \int_0^{\pi/4} \cos 2\theta\,(2\, d\theta) = a^2\sin 2\theta\Big|_0^{\pi/4} = a^2. \end{aligned}$

21.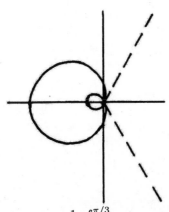

$\begin{aligned} r = 0 \text{ when } 1 - 2\cos\theta &= 0 \\ \cos\theta &= \frac{1}{2} \\ \theta &= -\frac{\pi}{3},\ \frac{\pi}{3}. \end{aligned}$

$\begin{aligned} A &= 2 \cdot \frac{1}{2}\int_0^{\pi/3}(1 - 2\cos\theta)^2\, d\theta = \int_0^{\pi/3}(1 - 4\cos\theta + 4\cos^2\theta)\, d\theta \\ &= \int_0^{\pi/3}[1 - 4\cos\theta + 2(1 + \cos 2\theta)]\, d\theta = \theta - 4\sin\theta + 2\theta + \sin 2\theta\Big|_0^{\pi/3} \\ &= \frac{\pi}{3} - 4\frac{\sqrt{3}}{2} + \frac{2\pi}{3} + \frac{\sqrt{3}}{2} = \pi - \frac{3}{2}\sqrt{3} = \frac{1}{2}\left(2\pi - 3\sqrt{3}\right). \end{aligned}$

23.

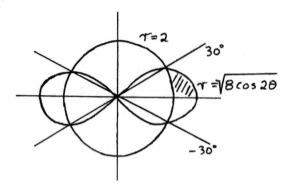

Substituting $r = 2$ in $r^2 = 8 \cos 2\theta$, we get

$$
\begin{aligned}
4 &= 8 \cos 2\theta \\
\cos 2\theta &= \frac{1}{2} \\
2\theta &= \pm 60° \\
\theta &= \pm 30° = \pm \frac{\pi}{6}.
\end{aligned}
$$

The desired area is equal to four times the shaded area:

$$
\begin{aligned}
A &= 4 \cdot \frac{1}{2} \int_0^{\pi/6} \left[\left(\sqrt{8 \cos 2\theta} \right)^2 - 2^2 \right] d\theta = 2 \int_0^{\pi/6} (8 \cos 2\theta - 4) \, d\theta \\
&= 8 \sin 2\theta - 8\theta \Big|_0^{\pi/6} = 8 \sin \frac{\pi}{3} - 8 \left(\frac{\pi}{6} \right) \\
&= 8 \left(\frac{\sqrt{3}}{2} \right) - \frac{4\pi}{3} = 4 \left(\sqrt{3} - \frac{\pi}{3} \right).
\end{aligned}
$$

25. If $x = F(\theta)$ and $y = G(\theta)$ are a set of parametric equations, then $\dfrac{dy}{dx} = \dfrac{G'(\theta)}{F'(\theta)}$ by (8.2). Now use the product rule:

$$
\begin{array}{ll}
x = f(\theta) \cos \theta & \qquad y = f(\theta) \sin \theta \\[2mm]
\dfrac{dx}{d\theta} = f'(\theta) \cos \theta - f(\theta) \sin \theta & \qquad \dfrac{dy}{d\theta} = f'(\theta) \sin \theta + f(\theta) \cos \theta
\end{array}
$$

$$
\frac{dy}{dx} = \frac{f'(\theta) \sin \theta + f(\theta) \cos \theta}{f'(\theta) \cos \theta - f(\theta) \sin \theta}.
$$

27. $f(\theta) = 2\theta$, $f'(\theta) = 2$, $\dfrac{dy}{dx} = \dfrac{2 \sin \theta + 2\theta \cos \theta}{2 \cos \theta - 2\theta \sin \theta}$.

At $\theta = \dfrac{\pi}{2}$,

$$
\frac{dy}{dx} = \frac{2 \sin \frac{\pi}{2} + 2 \left(\frac{\pi}{2} \right) \cos \frac{\pi}{2}}{2 \cos \frac{\pi}{2} - 2 \left(\frac{\pi}{2} \right) \sin \frac{\pi}{2}} = \frac{2 + 0}{0 - \pi} = -\frac{2}{\pi}.
$$

Chapter 9

Three Dimensional Space; Partial Derivatives; Multiple Integrals

9.1 Surfaces in Three Dimensions

1. The trace in the xy-plane is the circle $x^2 + y^2 = 4$.

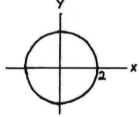

In three-space, $x^2 + y^2 = 4$ represents a cylinder (note the missing z-variable) whose axis is the z-axis.

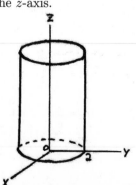

3. The trace in the xy-plane is the parabola $y^2 = 9x$.

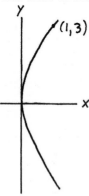

In three-space, $y^2 = 9x$ represents a cylinder extending along the z-axis: every cross-section parallel to the xy-plane is the parabola $y^2 = 9x$.

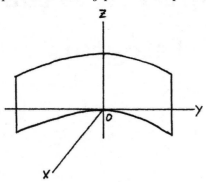

5. The trace in the yz-plane is the ellipse $y^2 + 4z^2 = 4$.

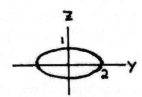

In three-space, $y^2 + 4z^2 = 4$ represents a cylinder (note the missing x-variable). The cylinder extends along the x-axis: every cross-section parallel to the yz-plane is the ellipse $y^2 + 4z^2 = 4$.

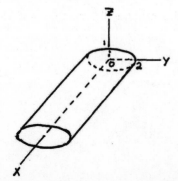

7. The trace in the xz-plane is the parabola $z = 3x - 3x^2$.

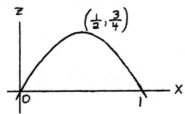

In three-space, $z = 3x - 3x^2$ represents a cylinder extending along the y-axis: every cross-section parallel to the xz-plane is the parabola $z = 3x - 3x^2$.

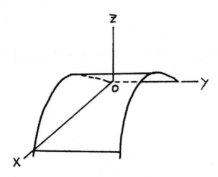

9. Trace in the xy-plane: $y = e^x$.

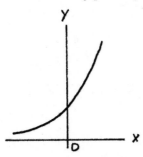

In three-space, $y = e^x$ represents a cylinder extending along the z-axis. (See graph in Answer Section.)

11. The plane $2x - 3y + z = 6$ can be sketched from the intercepts, since 3 noncollinear points determine a plane. (See graph in Answer Section.)

13. $z = x$ is a line in the xz-plane. Because of the missing y-variable the plane $z = x$ extends along the y-axis.

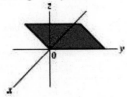

15. $x^2 + y^2 + z^2 = 9$ is a sphere of radius 3.

17. $9x^2 + 4y^2 + z^2 = 36$ (ellipsoid)

  1. Trace in xy-plane $(z = 0)$: $9x^2 + 4y^2 = 36$.

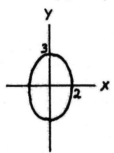

  2. Trace in xz-plane $(y = 0)$: $9x^2 + z^2 = 36$.

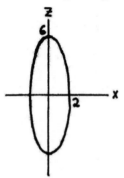

  3. Trace in yz-plane $(x = 0)$: $4y^2 + z^2 = 36$.

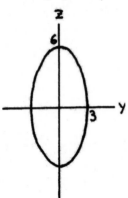 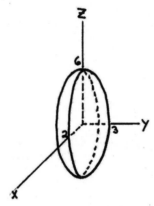

19. $z = x^2 + y^2$ is a paraboloid similar to Example 1.

21. $z = 1 + 2x^2 + 4y^2$ (paraboloid)

  1. Trace in xy-plane $(z = 0)$: $0 = 1 + 2x^2 + 4y^2$ or $2x^2 + 4y^2 = -1$ (imaginary locus).

  2. Trace in xz-plane $(y = 0)$: $z = 1 + 2x^2$.

3. Trace in yz-plane $(x = 0)$: $z = 1 + 4y^2$.

Cross-section: let $z = 2$ in the given equation. The resulting ellipse, $2x^2 + 4y^2 = 1$, is parallel to the xy-plane and two units above.

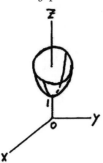

23. $4x^2 + y^2 - z^2 = 4$ (hyperboloid of one sheet)

1. Trace in xy-plane $(z = 0)$: $4x^2 + y^2 = 4$.

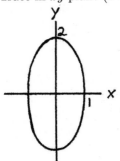

2. Trace in yz-plane $(x = 0)$: $y^2 - z^2 = 4$.

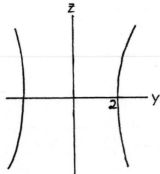

3. Trace in xz-plane $(y = 0)$: $4x^2 - z^2 = 4$, another hyperbola.

Cross-section: let $z = \pm 2 : 4x^2 + y^2 = 8$ (ellipse).

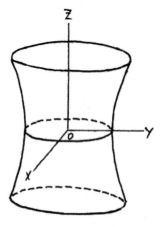

25. $-2x^2 - y^2 + z^2 = 6$ (hyperboloid of two sheets)

    1. Trace in xy-plane $(z = 0)$: $-2x^2 - y^2 = 6$ (imaginary locus).

    2. Trace in xz-plane $(y = 0)$: $-2x^2 + z^2 = 6$.

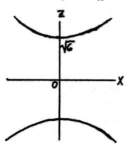

    3. Trace in yz-plane $(x = 0)$: $-y^2 + z^2 = 6$.

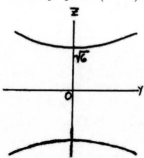

Cross-section: let $z = 3$ and $z = -3$ in the given equation. In each case, the result is an ellipse:

$$-2x^2 - y^2 + 9 = 6 \text{ or } 2x^2 + y^2 = 3.$$

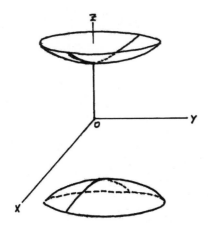

27. $x^2 - 4y^2 + z^2 = 4$ (hyperboloid of one sheet)

    1. Trace in xz-plane $(y = 0)$: $x^2 + z^2 = 4$ (circle).

    2. Trace in yz-plane $(x = 0)$: $-4y^2 + z^2 = 4$.

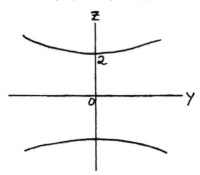

    3. Trace in xy-plane $(z = 0)$: $x^2 - 4y^2 = 4$, another hyperbola.

Cross-section: $y = \pm 1 : x^2 + z^2 = 8$ (circle). (See graph in Answer Section.)

29. $9y^2 - 36x^2 - 16z^2 = 144$ (hyperboloid of two sheets)

    1. Trace in xy-plane $(z = 0)$: $9y^2 - 36x^2 = 144$ or $y^2 - 4x^2 = 16$.

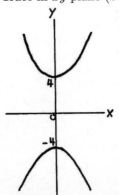

2. Trace in xz-plane $(y = 0)$: $-36x^2 - 16z^2 = 144$ (imaginary locus).

3. Trace in yz-plane $(x = 0)$: $9y^2 - 16z^2 = 144$.

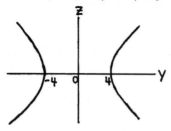

Cross-section: let $y = 8$ and $y = -8$ in the given equation. In each case, the resulting ellipse,

$$9(8)^2 - 36x^2 - 16z^2 = 144 \text{ or } 9x^2 + 4z^2 = 108$$

is parallel to the xz-plane and 8 units to the right (respectively left).

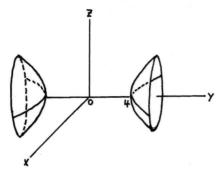

31. Saddle shape. See graph in Answer Section.

9.2 Partial Derivatives

1. $f(x, y) = 2x^2 + 5y^2 + 1$

   (a) $\dfrac{\partial f}{\partial x} = 4x + 0 = 4x$, since y is a constant.

   (b) $\dfrac{\partial f}{\partial y} = 0 + 10y = 10y$, since x is a constant.

3. $f(x, y) = x + \sin 2y$

   (a) $\dfrac{\partial f}{\partial x} = 1 + 0 = 1$, since $\sin 2y$ is a constant.

   (b) $\dfrac{\partial f}{\partial y} = 0 + \cos 2y \cdot 2 = 2 \cos 2y$.

5. $f(x, y) = \dfrac{1}{3} x^3 - 3 \cos y$

   (a) Since y is a constant, $\dfrac{\partial}{\partial x}(-3 \cos y) = 0$; so $\dfrac{\partial f}{\partial x} = \dfrac{1}{3} \cdot 3x^2 - 0 = x^2$.

   (b) Since x is now treated as a constant, $\dfrac{\partial}{\partial y}\left(\dfrac{1}{3} x^3\right) = 0$; so $\dfrac{\partial f}{\partial y} = 0 - 3(-\sin y) = 3 \sin y$.

7. $f(x,y) = e^{2x} + \ln y$

   (a) Since y is a constant, $\dfrac{\partial}{\partial x} \ln y = 0$; so $\dfrac{\partial f}{\partial x} = e^{2x} \cdot 2 + 0 = 2e^{2x}$.

   (b) Since x is a constant, $\dfrac{\partial}{\partial y} e^{2x} = 0$; so $\dfrac{\partial f}{\partial y} = 0 + \dfrac{1}{y} = \dfrac{1}{y}$.

9. $f(x,y) = x^2 \ln y$

   (a) Since $\ln y$ is a constant coefficient, $\dfrac{\partial}{\partial x}(x^2 \ln y) = \ln y \dfrac{\partial}{\partial x}(x^2) = \ln y \cdot 2x = 2x \ln y$.

   (b) Since x^2 is now treated as a constant coefficient, $\dfrac{\partial}{\partial y}(x^2 \ln y) = x^2 \dfrac{\partial}{\partial y} \ln y = x^2 \cdot \dfrac{1}{y} = \dfrac{x^2}{y}$.

11. $f(x,y) = 3x^2 \tan 2y$

   (a) $f(x,y) = \tan 2y \cdot 3x^2$. Since $\tan 2y$ is a constant coefficient,
   $\dfrac{\partial f}{\partial x} = \tan 2y \cdot \dfrac{\partial}{\partial x}(3x^2) = 6x \tan 2y$.

   (b) Now $3x^2$ is a constant coefficient: $\dfrac{\partial f}{\partial y} = 3x^2 \cdot \sec^2 2y \cdot 2 = 6x^2 \sec^2 2y$.

13. $f(x,y) = x\sqrt{x+y} = x(x+y)^{1/2}$

   (a) Since x and $(x+y)^{1/2}$ are both functions of x, we need the product rule:
   $$
   \begin{aligned}
   \frac{\partial f}{\partial x} &= x\frac{\partial}{\partial x}(x+y)^{1/2} + (x+y)^{1/2}\frac{\partial}{\partial x}(x) \\
   &= x \cdot \frac{1}{2}(x+y)^{-1/2} + (x+y)^{1/2} \cdot 1 \\
   &= \frac{x}{2(x+y)^{1/2}} + \frac{(x+y)^{1/2}}{1} \cdot \frac{2(x+y)^{1/2}}{2(x+y)^{1/2}} \\
   &= \frac{x + 2(x+y)^{1/2}(x+y)^{1/2}}{2(x+y)^{1/2}} \\
   &= \frac{x + 2(x+y)}{2(x+y)^{1/2}} = \frac{3x + 2y}{2\sqrt{x+y}}.
   \end{aligned}
   $$

   (b) This time x is a constant coefficient:
   $$
   \frac{\partial f}{\partial y} = x\frac{\partial}{\partial y}(x+y)^{1/2} = x \cdot \frac{1}{2}(x+y)^{-1/2} = \frac{x}{2\sqrt{x+y}}.
   $$

15. $f(x,y) = y^2 \tan xy$

   (a) Since y^2 is a constant coefficient,
   $$
   \frac{\partial f}{\partial x} = y^2 \frac{\partial}{\partial x} \tan xy = y^2 \sec^2 xy \cdot y = y^3 \sec^2 xy.
   $$

   (b) For this case, we need the product rule:
   $$
   \begin{aligned}
   \frac{\partial f}{\partial y} &= y^2 \frac{\partial}{\partial y} \tan xy + \tan xy \frac{\partial}{\partial y} y^2 \\
   &= y^2 \sec^2 xy \cdot x + \tan xy \cdot 2y \\
   &= xy^2 \sec^2 xy + 2y \tan xy.
   \end{aligned}
   $$

17. $f(x,y) = \dfrac{\cos x^2 y}{y}$

   (a) $f(x,y) = \dfrac{1}{y} \cos x^2 y$. Since y is a constant coefficient, we get
   $$
   \begin{aligned}
   \frac{\partial f}{\partial x} &= \frac{1}{y}\frac{\partial}{\partial x} \cos x^2 y = \frac{1}{y}(-\sin x^2 y)\frac{\partial}{\partial x}(x^2 y) \\
   &= \frac{1}{y}(-\sin x^2 y)(2xy) = -2x \sin x^2 y.
   \end{aligned}
   $$

(b) Since y and $\cos x^2 y$ are both function of y, we need the quotient rule:

$$\frac{\partial f}{\partial y} = \frac{y\frac{\partial}{\partial y}\cos x^2 y - \cos x^2 y \frac{\partial}{\partial y}(y)}{y^2}$$

$$= \frac{y(-\sin x^2 y)\frac{\partial}{\partial y}(x^2 y) - \cos x^2 y \cdot 1}{y^2}$$

$$= \frac{y(-\sin x^2 y)(x^2) - \cos x^2 y}{y^2}$$

$$= \frac{-x^2 y \sin x^2 y - \cos x^2 y}{y^2}.$$

19. $f(x, y) = \dfrac{(x+y^2)^{1/2}}{x}$

(a) By the quotient rule,

$$\frac{\partial f}{\partial x} = \frac{x\frac{\partial}{\partial x}(x+y^2)^{1/2} - (x+y^2)^{1/2}\frac{\partial}{\partial x}(x)}{x^2} = \frac{x \cdot \frac{1}{2}(x+y^2)^{-1/2}(1) - (x+y^2)^{1/2} \cdot 1}{x^2}.$$

To clear fractions, multiply numerator and denominator by $2(x+y^2)^{1/2}$:

$$\frac{\partial f}{\partial x} = \frac{x - 2(x+y^2)}{2x^2(x+y^2)^{1/2}} = \frac{-x - 2y^2}{2x^2\sqrt{x+y^2}}.$$

(b) Writing $f(x) = \dfrac{1}{x}(x+y^2)^{1/2}$, $\dfrac{1}{x}$ is now a constant coefficient.

$$\frac{\partial f}{\partial y} = \frac{1}{x} \cdot \frac{1}{2}(x+y^2)^{-1/2}\frac{\partial}{\partial y}(x+y^2) = \frac{1}{2x(x+y^2)^{1/2}}(0 + 2y) = \frac{y}{x\sqrt{x+y^2}}.$$

21. $f(x, y) = 5x - 2y - 3$

(a) $\dfrac{\partial f}{\partial x} = 5$ (b)$\dfrac{\partial f}{\partial y} = -2$

(c) $\dfrac{\partial^2 f}{\partial x^2} = 0$ (d)$\dfrac{\partial^2 f}{\partial y^2} = 0$

(e) $\dfrac{\partial^2 f}{\partial x \partial y} = 0$

23. $f(x, y) = 2x^2 + 3y^2 + 3x^2 y^2 + 5xy$

(a) $\dfrac{\partial f}{\partial x} = 4x + 6xy^2 + 5y$ (b)$\dfrac{\partial f}{\partial y} = 6y + 6x^2 y + 5x$

(c) $\dfrac{\partial^2 f}{\partial x^2} = 4 + 6y^2$ (d)$\dfrac{\partial^2 f}{\partial y^2} = 6 + 6x^2$

(e) $\dfrac{\partial^2 f}{\partial x \partial y} = \dfrac{\partial}{\partial x}\left(\dfrac{\partial f}{\partial y}\right) = \dfrac{\partial}{\partial x}(6y + 6x^2 y + 5x) = 12xy + 5$

or

$$\frac{\partial^2 f}{\partial y \partial x} = \frac{\partial}{\partial y}\left(\frac{\partial f}{\partial x}\right) = \frac{\partial}{\partial y}(4x + 6xy^2 + 5y) = 12xy + 5$$

25. $f(x, y) = \sin(2x + y)$

(a) $\dfrac{\partial f}{\partial x} = 2\cos(2x + y)$ (b)$\dfrac{\partial f}{\partial y} = \cos(2x + y)$

(c) $\dfrac{\partial^2 f}{\partial x^2} = \dfrac{\partial}{\partial x}[2\cos(2x + y)] = -4\sin(2x + y)$

(d) $\dfrac{\partial^2 f}{\partial y^2} = \dfrac{\partial}{\partial y}[\cos(2x + y)] = -\sin(2x + y)$

(e) $\dfrac{\partial^2 f}{\partial x \partial y} = \dfrac{\partial}{\partial x}\left(\dfrac{\partial f}{\partial y}\right) = \dfrac{\partial}{\partial x}[\cos(2x + y)] = -2\sin(2x + y)$

or

$$\frac{\partial^2 f}{\partial x \partial y} = \frac{\partial^2 f}{\partial y \partial x} = \frac{\partial}{\partial y}\left(\frac{\partial f}{\partial x}\right) = \frac{\partial}{\partial y}[2\cos(2x + y)] = -2\sin(2x + y)$$

27. $f(x,y) = \ln(x^2 y) + \tan y$

(a) $\dfrac{\partial f}{\partial x} = \dfrac{1}{x^2 y} \dfrac{\partial}{\partial x}(x^2 y) + 0 = \dfrac{2xy}{x^2 y} = \dfrac{2}{x}$

(b) $\dfrac{\partial f}{\partial y} = \dfrac{1}{x^2 y} \dfrac{\partial}{\partial y}(x^2 y) + \sec^2 y = \dfrac{1}{y} + \sec^2 y$

(c) $\dfrac{\partial^2 f}{\partial x^2} = -\dfrac{2}{x^2}$

(d) $\dfrac{\partial^2 f}{\partial y^2} = -\dfrac{1}{y^2} + 2(\sec y)\dfrac{\partial}{\partial y}\sec y = -\dfrac{1}{y^2} + 2\sec^2 y \tan y$

(e) $\dfrac{\partial^2 f}{\partial x \partial y} = \dfrac{\partial}{\partial x}\left(\dfrac{\partial f}{\partial y}\right) = \dfrac{\partial}{\partial x}\left(\dfrac{1}{y} + \sec^2 y\right) = 0$

or

$\dfrac{\partial^2 f}{\partial y \partial x} = \dfrac{\partial}{\partial y}\left(\dfrac{\partial f}{\partial x}\right) = \dfrac{\partial}{\partial y}\left(\dfrac{2}{x}\right) = 0$

29. $f(x,y) = \operatorname{Arctan} \dfrac{y}{x}$

(a) $\dfrac{\partial f}{\partial x} = \dfrac{1}{1 + \left(\dfrac{y}{x}\right)^2}\left(-\dfrac{y}{x^2}\right) = -\dfrac{y}{x^2 + y^2}$

(b) $\dfrac{\partial f}{\partial y} = \dfrac{1}{1 + \left(\dfrac{y}{x}\right)^2}\left(\dfrac{1}{x}\right) = \dfrac{1}{1 + \dfrac{y^2}{x^2}} \cdot \dfrac{1}{x} \cdot \dfrac{x}{x} = \dfrac{x}{x^2 + y^2}$

(c) $\dfrac{\partial^2 f}{\partial x^2} = \dfrac{\partial}{\partial x}\left[-y(x^2 + y^2)^{-1}\right] = y(x^2 + y^2)^{-2}(2x) = \dfrac{2xy}{(x^2 + y^2)^2}$

(d) $\dfrac{\partial^2 f}{\partial y^2} = \dfrac{\partial}{\partial y}\left[x(x^2 + y^2)^{-1}\right] = -x(x^2 + y^2)^{-2}(2y) = -\dfrac{2xy}{(x^2 + y^2)^2}$

(e)
$$\begin{aligned}
\dfrac{\partial^2 f}{\partial x \partial y} &= \dfrac{\partial}{\partial x}\left(\dfrac{\partial f}{\partial y}\right) = \dfrac{\partial}{\partial x}\dfrac{x}{x^2 + y^2} \\
&= \dfrac{(x^2 + y^2)\cdot 1 - x \cdot 2x}{(x^2 + y^2)^2} \qquad \text{quotient rule} \\
&= \dfrac{-x^2 + y^2}{(x^2 + y^2)^2} \quad \text{or} \\
\dfrac{\partial^2 f}{\partial x \partial y} &= \dfrac{\partial^2 f}{\partial y \partial x} = \dfrac{\partial}{\partial y}\left(\dfrac{\partial f}{\partial x}\right) = \dfrac{\partial}{\partial y}\left(-\dfrac{y}{x^2 + y^2}\right) \\
&= -\dfrac{(x^2 + y^2)\cdot 1 - y \cdot 2y}{(x^2 + y^2)^2} \qquad \text{quotient rule} \\
&= -\dfrac{x^2 - y^2}{(x^2 + y^2)^2} = \dfrac{-x^2 + y^2}{(x^2 + y^2)^2}
\end{aligned}$$

31. $f(x,y,z) = \sqrt{xyz} = x^{1/2}y^{1/2}z^{1/2}$

(a) $\dfrac{\partial f}{\partial x} = \dfrac{1}{2}x^{-1/2}y^{1/2}z^{1/2} = \dfrac{\sqrt{yz}}{2\sqrt{x}} \cdot \dfrac{\sqrt{x}}{\sqrt{x}} = \dfrac{\sqrt{xyz}}{2x}$

(b) $\dfrac{\partial f}{\partial y} = x^{1/2} \cdot \dfrac{1}{2}y^{-1/2}z^{1/2} = \dfrac{\sqrt{xz}}{2\sqrt{y}} \cdot \dfrac{\sqrt{y}}{\sqrt{y}} = \dfrac{\sqrt{xyz}}{2y}$

(c) $\dfrac{\partial^2 f}{\partial x^2} = \dfrac{\partial}{\partial x}\left(\dfrac{1}{2}x^{-1/2}y^{1/2}z^{1/2}\right) = \dfrac{1}{2}\left(-\dfrac{1}{2}\right)x^{-3/2}y^{1/2}z^{1/2} = -\dfrac{\sqrt{yz}}{4x^{3/2}} \cdot \dfrac{\sqrt{x}}{\sqrt{x}} = -\dfrac{\sqrt{xyz}}{4x^2}$

(d) $\dfrac{\partial^2 f}{\partial y^2} = \dfrac{\partial}{\partial y}\left(x^{1/2} \cdot \dfrac{1}{2}y^{-1/2}z^{1/2}\right) = -\dfrac{1}{4}y^{-3/2}x^{1/2}z^{1/2} = -\dfrac{\sqrt{xz}}{4y^{3/2}} \cdot \dfrac{\sqrt{y}}{\sqrt{y}} = -\dfrac{\sqrt{xyz}}{4y^2}$

(e) $\dfrac{\partial^2 f}{\partial x \partial y} = \dfrac{\partial}{\partial x}\left(x^{1/2} \cdot \dfrac{1}{2}y^{-1/2}z^{1/2}\right) = \dfrac{1}{2}x^{-1/2} \cdot \dfrac{1}{2}y^{-1/2}z^{1/2} = \dfrac{\sqrt{z}}{4\sqrt{xy}} \cdot \dfrac{\sqrt{xy}}{\sqrt{xy}} = \dfrac{\sqrt{xyz}}{4xy}$

33. (a) $\dfrac{\partial}{\partial x}(ye^x) = ye^x\Big|_{(1,-1,-e)} = (-1)e = -e$

 (b) $\dfrac{\partial}{\partial y}(ye^x) = e^x\Big|_{(1,-1,-e)} = e$

35. $z = \ln(x^2 + y)$ at $\left(\dfrac{1}{\sqrt{2}}, \dfrac{1}{2}, 0\right)$

 (a) $\dfrac{\partial}{\partial x}\ln(x^2 + y) = \dfrac{1}{x^2 + y}(2x) = \dfrac{2\left(1/\sqrt{2}\right)}{\left(1/\sqrt{2}\right)^2 + 1/2} = \dfrac{2}{\sqrt{2}} \cdot \dfrac{\sqrt{2}}{\sqrt{2}} = \sqrt{2}$

 (b) $\dfrac{\partial}{\partial y}\ln(x^2 + y) = \dfrac{1}{x^2 + y} = \dfrac{1}{\left(1/\sqrt{2}\right)^2 + 1/2} = 1$

37. $Z = \dfrac{RX}{R + X}$. By the quotient rule,

 $\dfrac{\partial Z}{\partial R} = \dfrac{(R + X)(X) - RX(1)}{(R + X)^2} = \dfrac{RX + X^2 - RX}{(R + X)^2} = \dfrac{X^2}{(R + X)^2}.$

 If $R = 8.0$ and $X = 5.0$, $\dfrac{\partial Z}{\partial R} = \dfrac{(5.0)^2}{(8.0 + 5.0)^2} = 0.15.$

39. $\dfrac{1}{R} = \dfrac{R_1 + R_2}{R_1 R_2}$ and $R = \dfrac{R_1 R_2}{R_1 + R_2}$. By the quotient rule,

 $\dfrac{\partial R}{\partial R_1} = \dfrac{(R_1 + R_2)(R_2) - R_1 R_2(1)}{(R_1 + R_2)^2} = \dfrac{R_2^2}{(R_1 + R_2)^2}.$

 When $R_1 = 10.0\,\Omega$ and $R_2 = 20.0\,\Omega$,

 $\dfrac{\partial R}{\partial R_1} = \dfrac{(20.0)^2}{(10.0 + 20.0)^2} = 0.444.$

41. $g_m = \dfrac{\partial}{\partial v_g}\left[0.50(v_g + 0.1v_p)^{4/3}\right] = 0.50\left(\dfrac{4}{3}\right)(v_g + 0.1v_p)^{1/3}(1) = 0.67(v_g + 0.1v_p)^{1/3}\Omega^{-1}.$

43. $\dfrac{\partial T}{\partial x} = \dfrac{\partial}{\partial x}(1 + x^3 y) = 3x^2 y\Big|_{(2,1)} = 3(4)(1) = 12°\,\text{C/m}.$

45. $y = A\cos\omega\left(t - \dfrac{x}{v}\right).$

 $\begin{aligned}
\dfrac{\partial y}{\partial x} &= -A\sin\omega\left(t - \dfrac{x}{v}\right)\dfrac{\partial}{\partial x}\omega\left(t - \dfrac{x}{v}\right) = \left[-A\sin\omega\left(t - \dfrac{x}{v}\right)\right]\left(-\dfrac{\omega}{v}\right) \\
&= \left(\dfrac{\omega}{v}\right)A\sin\omega\left(t - \dfrac{x}{v}\right). \\
\dfrac{\partial^2 y}{\partial x^2} &= \left[\left(\dfrac{\omega}{v}\right)A\cos\omega\left(t - \dfrac{x}{v}\right)\right]\left(-\dfrac{\omega}{v}\right) = \left(-\dfrac{\omega^2}{v^2}\right)A\cos\omega\left(t - \dfrac{x}{v}\right) \\
&= \dfrac{1}{v^2}\dfrac{\partial^2 y}{\partial t^2} \text{ since } \dfrac{\partial^2 y}{\partial t^2} = -\omega^2\left[A\cos\omega\left(t - \dfrac{x}{v}\right)\right].
\end{aligned}$

9.3 Applications of Partial Derivatives

1. $P = \dfrac{2L}{V^2} = 2LV^{-2}$. Pattern: $dP = \dfrac{\partial}{\partial L}(\quad)dL + \dfrac{\partial}{\partial V}(\quad)dV.$

 $dP = \dfrac{\partial}{\partial L}(2LV^{-2})dL + \dfrac{\partial}{\partial V}(2LV^{-2})dV = 2V^{-2}dL + (2L)(-2V^{-3})dV = \dfrac{2}{V^2}dL - \dfrac{4L}{V^3}dV.$

3. $L = \dfrac{1}{\sqrt{X^2 + Y^2}} = (X^2 + Y^2)^{-1/2}$. Pattern: $dL = \dfrac{\partial}{\partial X}(\quad)dX + \dfrac{\partial}{\partial Y}(\quad)dY.$

 $\begin{aligned}
dL &= -\dfrac{1}{2}(X^2 + Y^2)^{-3/2}(2X)\,dX - \dfrac{1}{2}(X^2 + Y^2)^{-3/2}(2Y)\,dY \\
&= -\dfrac{X\,dX}{(X^2 + Y^2)^{3/2}} - \dfrac{Y\,dY}{(X^2 + Y^2)^{3/2}}.
\end{aligned}$

5. $M = \dfrac{\sin\theta_1}{\sin\theta_2} = \sin\theta_1\csc\theta_2$. Pattern: $dM = \dfrac{\partial}{\partial\theta_1}(\quad)d\theta_1 + \dfrac{\partial}{\partial\theta_2}(\quad)d\theta_2$.

$$\begin{aligned}
dM &= \frac{\partial}{\partial\theta_1}(\sin\theta_1\csc\theta_2)d\theta_1 + \frac{\partial}{\partial\theta_2}(\sin\theta_1\csc\theta_2)d\theta_2 \\
&= \cos\theta_1\csc\theta_2\, d\theta_1 + (\sin\theta_1)(-\csc\theta_2\cot\theta_2)d\theta_2 \\
&= \frac{\cos\theta_1}{\sin\theta_2}\,d\theta_1 - \frac{\sin\theta_1}{\sin\theta_2\tan\theta_2}\,d\theta_2.
\end{aligned}$$

7. $f = r + \tan r\omega$

$$df = \frac{\partial}{\partial r}(r + \tan r\omega)\,dr + \frac{\partial}{\partial\omega}(r + \tan r\omega)\,d\omega = (1 + \omega\sec^2 r\omega)\,dr + (r\sec^2 r\omega)\,d\omega.$$

9. $V = \dfrac{L}{P^2} = LP^{-2}$.

Differential:
$$\begin{aligned}
dV &= \frac{\partial}{\partial L}(LP^{-2})dL + \frac{\partial}{\partial P}(LP^{-2})dP \\
dV &= P^{-2}dL + L(-2P^{-3})dP \\
dV &= \frac{1}{P^2}dL - \frac{2L}{P^3}dP.
\end{aligned}$$
$L = 8.0,\ dL = \pm 0.5$
$P = 3.0,\ dP = \pm 0.2$

Approximate maximum error: either
$$dV = \frac{1}{(3.0)^2}(+0.5) - \frac{2(8.0)}{(3.0)^3}(-0.2)$$
or
$$dV = \frac{1}{(3.0)^2}(-0.5) - \frac{2(8.0)}{(3.0)^3}(+0.2).$$
Thus $dV = \pm 0.17$.

11. $V = \pi r^2 h$.

$$dV = \frac{\partial}{\partial r}(\pi r^2 h)\,dr + \frac{\partial}{\partial h}(\pi r^2 h)\,dh = 2\pi r h\,dr + \pi r^2\,dh;\quad h = 10.0\,\text{cm},\ dh = 0.1\,\text{cm}$$
$r = 5.00\,\text{cm},\ dr = 0.08\,\text{cm}$

Approximate maximum error: $\Delta V \approx dV = 2\pi(5.00)(10.0)(0.08) + \pi(5.00)^2(0.1) = 33\,\text{cm}^3$.

13. $i = 1 - e^{-R/L}$.

Differential:
$$\begin{aligned}
di &= \frac{\partial}{\partial L}\left(1 - e^{-R/L}\right)dL + \frac{\partial}{\partial R}\left(1 - e^{-R/L}\right)dR \\
&= -e^{-R/L}\left(\frac{R}{L^2}\right)dL + \left(-e^{-R/L}\right)\left(-\frac{1}{L}\right)dR \\
di &= e^{-R/L}\left[-\frac{R}{L^2}\,dL + \frac{1}{L}\,dR\right].
\end{aligned}$$
$R = 1.2,\ dR = \pm 0.05$

$L = 0.70,\ dL = \pm 0.01$

Approximate maximum error: either
$$di = e^{-1.2/0.70}\left[-\frac{1.2}{(0.70)^2}(-0.01) + \frac{1}{0.70}(0.05)\right]$$
or
$$di = e^{-1.2/0.70}\left[-\frac{1.2}{(0.70)^2}(0.01) + \frac{1}{0.70}(-0.05)\right].$$
Thus $di = \pm 0.017\,\text{A}$.

Approximate maximum percentage error: $\dfrac{di}{i} \times 100 = \dfrac{0.017}{1 - e^{-1.2/0.70}} \times 100 = 2.1\%$.

15. $T = 2\pi\sqrt{L/g} = 2\pi L^{1/2}g^{-1/2}$.

$$
\begin{aligned}
dT &= \frac{\partial}{\partial L}\left(2\pi L^{1/2}g^{-1/2}\right)dL + \frac{\partial}{\partial g}\left(2\pi L^{1/2}g^{-1/2}\right)dg \\
&= \pi L^{-1/2}g^{-1/2}\,dL - \pi L^{1/2}g^{-3/2}\,dg \\
&= \pi\left(\frac{1}{\sqrt{Lg}}\,dL - \frac{\sqrt{L}}{g^{3/2}}\,dg\right).
\end{aligned}
$$

$L = 15.0\,\text{cm}, \ dL = \pm 0.2\,\text{cm}$

$g = 980\,\text{cm/s}^2, \ dg = \pm 6,\text{cm/s}^2$

Approximate maximum error: either

$$
dT = \pi\left[\frac{1}{\sqrt{(15.0)(980)}}(+0.2) - \frac{\sqrt{15.0}}{980^{3/2}}(-6)\right]
$$

or

$$
dT = \pi\left[\frac{1}{\sqrt{(15.0)(980)}}(-0.2) - \frac{\sqrt{15.0}}{980^{3/2}}(+6)\right].
$$

Thus, $dT = \pm 0.0076\,\text{s}$.

Approximate maximum percentage error: $\dfrac{dT}{T} \times 100 = \dfrac{0.0076}{2\pi\sqrt{15.0/980}} \times 100 = 0.98\%$.

17. $V = \dfrac{1}{3}\pi r^2 h$.

Given: $\dfrac{dr}{dt} = 1.0\,\dfrac{\text{cm}}{\text{min}}$, $\dfrac{dh}{dt} = 1.0\,\dfrac{\text{cm}}{\text{min}}$.

Find: $\dfrac{dV}{dt}$ when $r = 10\,\text{cm}$ and $h = 20\,\text{cm}$.

$$
\begin{aligned}
\frac{dV}{dt} &= \frac{\partial V}{\partial r}\frac{dr}{dt} + \frac{\partial V}{\partial h}\frac{dh}{dt} \\
&= \left(\frac{2}{3}\pi rh\right)\frac{dr}{dt} + \left(\frac{1}{3}\pi r^2\right)\frac{dh}{dt} \\
&= \left(\frac{2}{3}\pi rh\right)(1.0) + \left(\frac{1}{3}\pi r^2\right)(1.0)\Big|_{r=10,\,h=20} \\
&= \frac{500\pi}{3} = 520\,\text{cm}^3/\text{min}.
\end{aligned}
$$

19. $P = i^2 R$.

Given: $\dfrac{di}{dt} = 2.0\,\text{A/s}$ and $\dfrac{dR}{dt} = 3.0\,\Omega/\text{s}$.

Find: $\dfrac{dP}{dt}$ when $i = 10\,\text{A}$ and $R = 50\,\Omega$.

$\dfrac{dP}{dt} = \dfrac{\partial}{\partial i}(i^2 R)\dfrac{di}{dt} + \dfrac{\partial}{\partial R}(i^2 R)\dfrac{dR}{dt} = 2iR\dfrac{di}{dt} + i^2\dfrac{dR}{dt} = 2iR(2.0) + i^2(3.0)$.

When $i = 10$ and $R = 50$, we get

$\dfrac{dP}{dt} = 2(10)(50)(2.0) + (10)^2(3.0) = 2300\,\text{W/s}$.

21. $z = 3x^2 + 2y^2 + 4x - 4y - 1$

Critical points:

$\dfrac{\partial z}{\partial x} = 6x + 4 = 0$ or $x = -\dfrac{2}{3}$

$\dfrac{\partial z}{\partial y} = 4y - 4 = 0$ or $y = 1$.

The critical point is at $\left(-\dfrac{2}{3}, 1\right)$.

$\dfrac{\partial^2 z}{\partial x^2} = 6, \ \dfrac{\partial^2 z}{\partial y^2} = 4, \ \dfrac{\partial^2 z}{\partial x \partial y} = 0$.

Thus $A = 6 \cdot 4 - 0^2 = 24 > 0$.

Since $A > 0$ and $\dfrac{\partial^2 z}{\partial x^2} > 0$, $f(x, y)$ has a minimum at $\left(-\dfrac{2}{3}, 1\right)$.

Also, $\quad z \;=\; 3\left(-\dfrac{2}{3}\right)^2 + 2(1)^2 + 4\left(-\dfrac{2}{3}\right) - 4(1) - 1$

$$\qquad\qquad = \dfrac{4}{3} + 2 - \dfrac{8}{3} - 4 - 1 = -3 - \dfrac{4}{3} = -\dfrac{13}{3}.$$

23. $z = 2xy - x^2 - 2y^2 + 3x + 5$

    Critical points:

$$\dfrac{\partial z}{\partial x} \;=\; 2y - 2x + 3 \;=\; 0$$

$$\dfrac{\partial z}{\partial y} \;=\; 2x - 4y \;=\; 0$$

$$\underline{\hspace{6cm}}$$

$$\qquad\quad -2y + 3 \;=\; 0 \qquad\qquad \text{(adding)}$$

$$y \;=\; \dfrac{3}{2}, \; x = 2y = 3.$$

The critical point is at $\left(3, \dfrac{3}{2}\right)$.

$\dfrac{\partial^2 z}{\partial x^2} = -2, \; \dfrac{\partial^2 z}{\partial y^2} = -4, \; \dfrac{\partial^2 z}{\partial x \partial y} = \dfrac{\partial}{\partial x}\left(\dfrac{\partial z}{\partial y}\right) = \dfrac{\partial}{\partial x}(2x - 4y) = 2.$

Thus $A = (-2)(-4) - (2)^2 = 4 > 0.$

Since $A > 0$ and $\dfrac{\partial^2 z}{\partial x^2} < 0$, $f(x, y)$ has a maximum at $\left(3, \dfrac{3}{2}\right)$.

Finally, $z = 2(3)\left(\dfrac{3}{2}\right) - (3)^2 - 2\left(\dfrac{3}{2}\right)^2 + 3(3) + 5 = \dfrac{19}{2}$,

so that the maximum point is $\left(3, \dfrac{3}{2}, \dfrac{19}{2}\right)$.

25. $z = y^2 - x^2 - 2xy - 4y$

    Critical points:

$$\dfrac{\partial z}{\partial x} \;=\; -2x - 2y \;= 0$$

$$\dfrac{\partial z}{\partial y} \;=\; 2y - 2x - 4 \;= 0$$

$$\underline{\hspace{5cm}}$$

$$\qquad\quad -4x - 4 \;= 0 \qquad \text{adding}$$

$$x \;= -1$$

$$y \;= 1.$$

The critical point is at $(-1, 1)$; $z = 1^2 - (-1)^2 - 2(-1)(1) - 4(1) = -2.$

$\dfrac{\partial^2 z}{\partial x^2} = -2, \; \dfrac{\partial^2 z}{\partial y^2} = 2, \; \dfrac{\partial^2 z}{\partial x \partial y} = -2.$

Thus $A = (-2)(2) - (-2)^2 = -8 < 0.$

Since $A < 0$, the point $(-1, 1, -2)$ is a saddle point.

27. $z = x^2 + 2y^3 - x - 12y - 4$

Critical points:

$\dfrac{\partial z}{\partial x} = 2x - 1 = 0$ or $x = \dfrac{1}{2}$.

$\dfrac{\partial z}{\partial y} = 6y^2 - 12 = 0$ or $y = \pm\sqrt{2}$.

Substituting in the given equation, we find that the critical points are $\left(\dfrac{1}{2}, \sqrt{2}, -15.6\right)$ and

$\left(\dfrac{1}{2}, -\sqrt{2}, 7.1\right)$.

$\dfrac{\partial^2 z}{\partial x^2} = 2$, $\dfrac{\partial^2 z}{\partial y^2} = 12y$, $\dfrac{\partial^2 z}{\partial x \partial y} = 0$. $A = 2(12y) - 0^2 = 24y$.

For the first critical point, $A > 0$ and $\dfrac{\partial^2 z}{\partial x^2} > 0$. So $\left(\dfrac{1}{2}, \sqrt{2}, -15.6\right)$ is a minimum.

For the other critical point, $A < 0$. So $\left(\dfrac{1}{2}, -\sqrt{2}, 7.1\right)$ is a saddle point.

29. $z = 3x^3 - xy^2 + x$

Critical points:

$\dfrac{\partial z}{\partial x} = 9x^2 - y^2 + 1 = 0$, $\dfrac{\partial z}{\partial y} = -2xy = 0$, which implies that $x = 0$ or $y = 0$.

Substituting $y = 0$ in the first equation, we get $9x^2 + 1 = 0$ which has no real roots. If we substitute $x = 0$ in the first equation, we get

$-y^2 + 1 = 0$, whence $y = \pm 1$;

so the critical points are at $(0, 1)$ and $(0, -1)$.

$\dfrac{\partial^2 z}{\partial x^2} = 18x$, $\dfrac{\partial^2 z}{\partial y^2} = -2x$, $\dfrac{\partial^2 z}{\partial x \partial y} = -2y$.

So $A = (18x)(-2x) - (-2y)^2$ or $A = -36x^2 - 4y^2$.

For both critical points, $A = -4 < 0$.

We conclude that $f(x, y)$ has saddle points at $(0, 1)$ and $(0, -1)$.

31. $z = 9xy - x^3 - y^3$

$\dfrac{\partial z}{\partial x} = 9y - 3x^2 = 0$ or $3y - x^2 = 0$.

$\dfrac{\partial z}{\partial y} = 9x - 3y^2 = 0$ or $3x - y^2 = 0$.

Substituting $x = \dfrac{1}{3}y^2$ in the first equation, we get

$$3y - \left(\dfrac{1}{3}y^2\right)^2 = 0$$

$$3y - \dfrac{1}{9}y^4 = 0$$

$$27y - y^4 = 0$$

$$y(27 - y^3) = 0$$

$$y = 0, 3.$$

When $y = 0$, $x = 0$ and when $y = 3$, $x = 3$. So the critical points are $(0, 0, 0)$ and $(3, 3, 27)$.

$\dfrac{\partial^2 z}{\partial x^2} = -6x$, $\dfrac{\partial^2 z}{\partial y^2} = -6y$, $\dfrac{\partial^2 z}{\partial x \partial y} = 9$.

$A = (-6x)(-6y) - 9^2 = 36xy - 81$.

We conclude that $(0, 0, 0)$ is a saddle point since $A < 0$. For the other critical point,

$A = 36(3 \cdot 3) - 81 > 0$ and $\dfrac{\partial^2 z}{\partial x^2} < 0$. So $(3, 3, 27)$ is a maximum.

33. Let x and y be the first two numbers; the $60 - x - y$ is the third. The quantity to be maximized is the product P:

$$P \;=\; xy(60 - x - y)$$

$$P \;=\; 60xy - x^2y - xy^2$$

$$\frac{\partial P}{\partial x} \;=\; 60y - 2xy - y^2 \;=\; 0$$

$$\frac{\partial P}{\partial y} \;=\; 60x - x^2 - 2xy \;=\; 0$$

$$-2xy \;=\; x^2 - 60x \qquad \text{second equation}$$

$$y \;=\; -\frac{1}{2}x + 30.$$

Substituting in first equation:

$$60\left(-\frac{1}{2}x + 30\right) - 2x\left(-\frac{1}{2}x + 30\right) - \left(-\frac{1}{2}x + 30\right)^2 \;=\; 0$$

$$-30x + 1800 + x^2 - 60x - \frac{1}{4}x^2 + 30x - 900 \;=\; 0$$

$$\frac{3}{4}x^2 - 60x + 900 \;=\; 0$$

$$x^2 - 80x + 1200 \;=\; 0$$

$$(x - 20)(x - 60) \;=\; 0$$

$$x \;=\; 20$$

$$y \;=\; 20$$

$$\text{third number} \;=\; 60 - 20 - 20 = 20.$$

35. x units of W_1 and y units of W_2 selling at $\$P_1$ and $\$P_2$, respectively, produce a total revenue of $xP_1 + yP_2$ (dollars). So the profit P is given by

$$P(x, y) \;=\; xP_1 + yP_2 - C(x, y)$$

$$\frac{\partial P}{\partial x} \;=\; P_1 - \frac{\partial C(x, y)}{\partial x} = 0$$

$$\frac{\partial P}{\partial y} \;=\; P_2 - \frac{\partial C(x, y)}{\partial y} = 0.$$

So if the profit is indeed a maximum at (a, b), the critical point, then

$$\frac{\partial C(a, b)}{\partial x} = P_1 \text{ and } \frac{\partial C(a, b)}{\partial y} = P_2.$$

37.
$$f(x, y) \;=\; 0$$

$$\frac{\partial f}{\partial x}dx + \frac{\partial f}{\partial y}dy \;=\; 0$$

$$\frac{\partial f}{\partial x} + \frac{\partial f}{\partial y}\frac{dy}{dx} \;=\; 0$$

$$\frac{dy}{dx} \;=\; -\frac{\dfrac{\partial f}{\partial x}}{\dfrac{\partial f}{\partial y}}$$

39. Let $f = x^6 + 2x^5y^2 - 6x^3y^3 - 4y^4 + 10$;
$$\frac{\partial f}{\partial x} = 6x^5 + 10x^4y^2 - 18x^2y^3;$$
$$\frac{\partial f}{\partial y} = 4x^5y - 18x^3y^2 - 16y^3.$$
By Exercise 37,
$$\frac{dy}{dx} = -\frac{\partial f/\partial x}{\partial f/\partial y} = -\frac{6x^5 + 10x^4y^2 - 18x^2y^3}{4x^5y - 18x^3y^2 - 16y^3}.$$

41. Let $f = 2x^5 + 3x^4y - 4x^3y^2 + 7x^2y^2 + 1$;
$$\frac{\partial f}{\partial x} = 10x^4 + 12x^3y - 12x^2y^2 + 14xy^2; \quad \frac{\partial f}{\partial y} = 3x^4 - 8x^3y + 14x^2y.$$
By Exercise 37,
$$\frac{dy}{dx} = -\frac{\partial f/\partial x}{\partial f/\partial y} = -\frac{10x^4 + 12x^3y - 12x^2y^2 + 14xy^2}{3x^4 - 8x^3y + 14x^2y}.$$

43. Let $f = 7x^4y^8 + 16x^3y^5 + 25x^2 - 7y^2 + 2$;
$$\frac{\partial f}{\partial x} = 28x^3y^8 + 48x^2y^5 + 50x;$$
$$\frac{\partial f}{\partial y} = 56x^4y^7 + 80x^3y^4 - 14y.$$
$$\frac{dy}{dx} = -\frac{\partial f/\partial x}{\partial f/\partial y} = -\frac{28x^3y^8 + 48x^2y^5 + 50x}{56x^4y^7 + 80x^3y^4 - 14y}.$$

45. Let $f = y^2 + y\cos x - 2$;
$$\frac{\partial f}{\partial x} = -y\sin x; \quad \frac{\partial f}{\partial y} = 2y + \cos x.$$
By Exercise 37,
$$\frac{dy}{dx} = -\frac{\partial f/\partial x}{\partial f/\partial y} = -\frac{-y\sin x}{2y + \cos x} = \frac{y\sin x}{2y + \cos x}.$$

47. Let $f = \cot(x^2 - y^2) - 3xy + 7$;
$$\frac{\partial f}{\partial x} = -\csc^2(x^2 - y^2)(2x) - 3y;$$
$$\frac{\partial f}{\partial y} = -\csc^2(x^2 - y^2)(-2y) - 3x.$$
$$\frac{dy}{dx} = -\frac{-2x\csc^2(x^2 - y^2) - 3y}{2y\csc^2(x^2 - y^2) - 3x} = \frac{2x\csc^2(x^2 - y^2) + 3y}{2y\csc^2(x^2 - y^2) - 3x}.$$

49. Let $f = \ln y - x\sin y + 3y^2$;
$$\frac{\partial f}{\partial x} = -\sin y; \quad \frac{\partial f}{\partial y} = \frac{1}{y} - x\cos y + 6y.$$
By Exercise 37,
$$\frac{dy}{dx} = -\frac{\partial f/\partial x}{\partial f/\partial y} = -\frac{-\sin y}{1/y - x\cos y + 6y} = \frac{\sin y}{1/y - x\cos y + 6y} = \frac{y\sin y}{1 - xy\cos x + 6y^2}.$$

9.4 Curve Fitting

1.

x_i	y_i	x_i^2	x_iy_i
2	14	4	28
4	36	16	144
6	53	36	318
8	78	64	624
10	92	100	920

Totals 30 273 220 2034 $n = 5$

$$D = \begin{vmatrix} 30 & 220 \\ 5 & 30 \end{vmatrix} = (30)(30) - (5)(220) = -200.$$

$$A = \begin{vmatrix} 30 & 2034 \\ 5 & 273 \end{vmatrix} = (30)(273) - (5)(2034) = -1980.$$

$$B = \begin{vmatrix} 2034 & 220 \\ 273 & 30 \end{vmatrix} = (2034)(30) - (273)(220) = 960.$$

$a = \dfrac{A}{D} = \dfrac{-1980}{-200} = 9.9$, $b = \dfrac{B}{D} = \dfrac{960}{-200} = -4.8.$

Since $y = ax + b$, we get $y = 9.9x - 4.8$.

3.

x_i	y_i	x_i^2	x_iy_i
1	0.70	1	0.70
2	1.65	4	3.3
3	2.00	9	6
4	3.15	16	12.6
5	3.80	25	19
6	4.25	36	25.5

Totals 21 15.55 91 67.1 $n = 6$

$$D = \begin{vmatrix} 21 & 91 \\ 6 & 21 \end{vmatrix} = (21)(21) - (91)(6) = -105.$$

$$A = \begin{vmatrix} 21 & 67.1 \\ 6 & 15.55 \end{vmatrix} = (21)(15.55) - (67.1)(6) = -76.05.$$

$$B = \begin{vmatrix} 67.1 & 91 \\ 15.55 & 21 \end{vmatrix} = (67.1)(21) - (91)(15.55) = -5.95.$$

$a = \dfrac{A}{D} = \dfrac{-76.05}{-105} = 0.724$, $b = \dfrac{B}{D} = \dfrac{-5.95}{-105} = 0.057.$

Equation: $y = 0.724x + 0.057$.

5. We let $T_1' = T_1^2 = (20)^2$, $T_2' = T_2^2 = (30)^2$, and so on:

T_i'	R_i	$(T_i')^2$	$T_i' R_i$
$20^2 = 400$	7	160000	2800
$30^2 = 900$	8	810000	7200
$40^2 = 1600$	11	2560000	17600
$50^2 = 2500$	16	6250000	40000
$60^2 = 3600$	22	12960000	79200
$70^2 = 4900$	27	24010000	132300
Totals	13900	91	46750000	279100	$n = 6$

$$D = \begin{vmatrix} 13900 & 46750000 \\ 6 & 13900 \end{vmatrix} = -87290000.$$

$$A = \begin{vmatrix} 13900 & 279100 \\ 6 & 91 \end{vmatrix} = -409700.$$

$$B = \begin{vmatrix} 279100 & 46750000 \\ 91 & 13900 \end{vmatrix} = -3.7476 \times 10^8.$$

$$a = \frac{A}{D} = \frac{-409700}{-87290000} = 0.0047, \; b = \frac{B}{D} = 4.2933.$$

Thus $R = 0.0047T^2 + 4.2933$.

7. $y = a \ln x + b$

x_i	x_i'	y_i	$(x_i')^2$	$x_i' y_i$
3	1.09861	-0.81	1.20695	-0.88987
4	1.38629	-0.22	1.92181	-0.30498
5	1.60944	0.22	2.59029	0.35408
6	1.79176	0.57	3.21040	1.02130
7	1.94591	0.88	3.78657	1.71240
Totals	7.83201	0.64	12.71602	1.89293	$n = 5$

$$D = \begin{vmatrix} 7.83201 & 12.71602 \\ 5 & 7.83201 \end{vmatrix} = -2.23972.$$

$$A = \begin{vmatrix} 7.83201 & 1.89293 \\ 5 & 0.64 \end{vmatrix} = -4.45216.$$

$$B = \begin{vmatrix} 1.89293 & 12.71602 \\ 0.64 & 7.83201 \end{vmatrix} = 6.68719.$$

$$a = \frac{A}{D} = 1.988, \; b = \frac{B}{D} = -2.986.$$

Equation: $y = 1.988 \ln x - 2.986$.

9.5 Iterated Integrals

1. $\displaystyle\int_0^1 \left(\int_0^x x^2 y^2 \, dy \right) dx \;=\; \int_0^1 x^2 \frac{y^3}{3} \Big|_0^x dx = \int_0^1 x^2 \left(\frac{x^3}{3} - 0 \right) dx$

$\displaystyle\qquad\qquad =\; \int_0^1 \frac{1}{3} x^5 \, dx = \frac{1}{3} \frac{x^6}{6} \Big|_0^1 = \frac{1}{18}.$

3. $\displaystyle\int_0^1 \left(\int_{\sqrt{x}}^1 y \, dy \right) dx \;=\; \int_0^1 \frac{1}{2} y^2 \Big|_{\sqrt{x}}^1 dx = \int_0^1 \frac{1}{2} \left[1^2 - \left(\sqrt{x} \right)^2 \right] dx$

$\displaystyle\qquad\qquad =\; \frac{1}{2} \int_0^1 (1 - x) \, dx = \frac{1}{2} \left(x - \frac{1}{2} x^2 \right) \Big|_0^1 = \frac{1}{2} \left(1 - \frac{1}{2} \right) = \frac{1}{4}.$

5. $\displaystyle\int_0^2 \int_0^{\sqrt{y-1}} xy \, dx \, dy \;=\; \int_0^2 y \cdot \frac{1}{2} x^2 \Big|_0^{\sqrt{y-1}} dy = \int_0^2 y \cdot \frac{1}{2} \left(\sqrt{y-1} \right)^2 dy$

$\displaystyle\qquad\qquad =\; \frac{1}{2} \int_0^2 y(y-1) \, dy = \frac{1}{2} \int_0^2 (y^2 - y) \, dy$

$\displaystyle\qquad\qquad =\; \frac{1}{2} \left(\frac{1}{3} y^3 - \frac{1}{2} y^2 \right) \Big|_0^2 = \frac{1}{2} \left(\frac{8}{3} - 2 \right) = \frac{1}{3}.$

7. $\displaystyle\int_1^3 \left(\int_0^{\sqrt{9-y^2}} y \, dx \right) dy \;=\; \int_1^3 yx \Big|_0^{\sqrt{9-y^2}} dy = \int_1^3 y \left(\sqrt{9 - y^2} - 0 \right) dy$

$\displaystyle\qquad\qquad =\; \int_1^3 (9 - y^2)^{1/2} y \, dy \qquad\qquad u = 9 - y^2; \; du = -2y \, dy.$

$\displaystyle\qquad\qquad =\; -\frac{1}{2} \int_1^3 (9 - y^2)^{1/2}(-2y) \, dy = -\frac{1}{2} \frac{(9 - y^2)^{3/2}}{3/2} \Big|_1^3$

$\displaystyle\qquad\qquad =\; -\frac{1}{3} (9 - y^2)^{3/2} \Big|_1^3 = 0 + \frac{1}{3} (8)^{3/2} = \frac{1}{3} \left(\sqrt{8} \right)^3$

$\displaystyle\qquad\qquad =\; \frac{1}{3} \left(\sqrt{8} \right)^2 \left(\sqrt{8} \right) = \frac{1}{3} (8) \cdot 2\sqrt{2} = \frac{16\sqrt{2}}{3}.$

9. $\displaystyle\int_0^{\sqrt{\pi/6}} \left(\int_0^x \cos x^2 \, dy \right) dx \;=\; \int_0^{\sqrt{\pi/6}} (\cos x^2) y \Big|_0^x dx = \int_0^{\sqrt{\pi/6}} (\cos x^2) x \, dx$

$\displaystyle\qquad\qquad =\; \frac{1}{2} \int_0^{\sqrt{\pi/6}} \cos x^2 (2x) \, dx \qquad\qquad u = x^2; \; du = 2x \, dx.$

$\displaystyle\qquad\qquad =\; \frac{1}{2} \sin x^2 \Big|_0^{\sqrt{\pi/6}} = \frac{1}{2} \sin \frac{\pi}{6} = \frac{1}{4}.$

11. $\displaystyle\int_0^{\pi/4} \left(\int_0^{\sec y} 2x \, dx \right) dy = \int_0^{\pi/4} x^2 \Big|_0^{\sec y} dy = \int_0^{\pi/4} \sec^2 y \, dy = \tan y \Big|_0^{\pi/4} = 1.$

13. $\displaystyle\int_0^3 \left(\int_0^x e^x \, dy \right) dx = \int_0^3 y e^x \Big|_0^x dx = \int_0^3 x e^x \, dx$

$\qquad\qquad\qquad\qquad\qquad u = x \qquad dv = e^x \, dx$

$\qquad\qquad\qquad\qquad\qquad du = dx \qquad v = e^x$

$\displaystyle x e^x \Big|_0^3 - \int_0^3 e^x \, dx = x e^x \Big|_0^3 - e^x \Big|_0^3 = 3e^3 - (e^3 - 1) = 2e^3 + 1.$

15. $\displaystyle\int_2^3 \int_1^x \left(\int_0^{6y} xy \, dz \right) dy \, dx \;=\; \int_2^3 \int_1^x xyz \Big|_0^{6y} dy \, dx$

$\displaystyle\qquad\qquad =\; \int_2^3 \int_1^x xy(6y) \, dy \, dx = \int_2^3 x \cdot 2y^3 \Big|_1^x dx$

$\displaystyle\qquad\qquad =\; \int_2^3 2x(x^3 - 1) \, dx = \int_2^3 (2x^4 - 2x) \, dx = \frac{397}{5}.$

17.

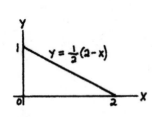

$$A = \int_0^2 \int_0^{(2-x)/2} dy\,dx = \int_0^2 y \Big|_0^{(2-x)/2} dx$$

$$= \int_0^2 \frac{1}{2}(2-x)\,dx = x - \frac{1}{4}x^2 \Big|_0^2 = 1.$$

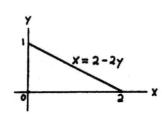

$$A = \int_0^1 \int_0^{2-2y} dx\,dy = \int_0^1 x \Big|_0^{2-2y} dy$$

$$= \int_0^1 (2-2y)\,dy = 2y - y^2 \Big|_0^1 = 2 - 1 = 1.$$

19.

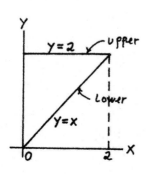

$$A = \int_0^2 \int_x^2 dy\,dx = \int_0^2 y \Big|_x^2 dx$$

$$= \int_0^2 (2-x)\,dx = 2x - \frac{1}{2}x^2 \Big|_0^2 = 4 - 2 = 2.$$

$$A = \int_0^2 \int_0^y dx\,dy = \int_0^2 x \Big|_0^y dy$$

$$= \int_0^2 y\,dy = \frac{1}{2}y^2 \Big|_0^2 = 2.$$

21.

$$A = \int_0^4 \int_0^{2x} dy\,dx = \int_0^4 y \Big|_0^{2x} dx$$

$$= \int_0^4 2x\,dx = x^2 \Big|_0^4 = 16.$$

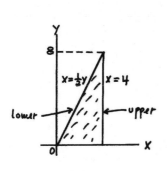

$$A = \int_0^8 \int_{(1/2)y}^4 dx\,dy = \int_0^8 x\Big|_{(1/2)y}^4 dy$$
$$= \int_0^8 \left(4 - \frac{1}{2}y\right) dy = 4y - \frac{1}{4}y^2\Big|_0^8$$
$$= 32 - 16 = 16.$$

23.

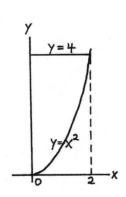

$$A = \int_0^2 \int_{x^2}^4 dy\,dx = \int_0^2 y\Big|_{x^2}^4 dx$$
$$= \int_0^2 (4 - x^2)\,dx = 4x - \frac{1}{3}x^3\Big|_0^2$$
$$= 8 - \frac{8}{3} = \frac{24}{3} - \frac{8}{3} = \frac{16}{3}.$$

$$A = \int_0^4 \int_0^{\sqrt{y}} dx\,dy = \int_0^4 \sqrt{y}\,dy$$
$$= \frac{2}{3}y^{3/2}\Big|_0^4 = \frac{2}{3}(8 - 0) = \frac{16}{3}.$$

25.

$$A = \int_0^1 \int_x^{\sqrt{x}} dy\,dx = \int_0^1 y\Big|_x^{\sqrt{x}} dx$$
$$= \int_0^1 \left(\sqrt{x} - x\right)\,dx = \frac{2}{3}x^{3/2} - \frac{x^2}{2}\Big|_0^1$$
$$= \frac{2}{3} - \frac{1}{2} = \frac{1}{6}.$$

$$A = \int_0^1 \int_{y^2}^y dx\,dy = \int_0^1 x\Big|_{y^2}^y dy$$
$$= \int_0^1 (y - y^2)\,dy = \frac{1}{2}y^2 - \frac{1}{3}y^3\Big|_0^1$$
$$= \frac{1}{2} - \frac{1}{3} = \frac{1}{6}.$$

27.

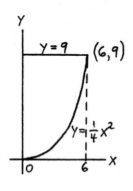

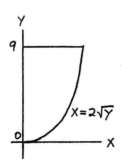

$$A = \int_0^6 \int_{x^2/4}^9 dy\, dx = \int_0^6 y\Big|_{x^2/4}^9 dx$$

$$= \int_0^6 \left(9 - \frac{1}{4}x^2\right) dx = 36.$$

$$A = \int_0^9 \int_0^{2\sqrt{y}} dx\, dy = \int_0^9 x\Big|_0^{2\sqrt{y}} dy$$

$$= \int_0^9 2\sqrt{y}\, dy = 2 \cdot \frac{2}{3} x^{3/2}\Big|_0^9 = 36.$$

29.

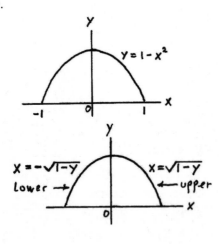

$$A = \int_{-1}^1 \int_0^{1-x^2} dy\, dx = \int_{-1}^1 y\Big|_0^{1-x^2} dx$$

$$= \int_{-1}^1 (1-x^2)\, dx = x - \frac{1}{3}x^3\Big|_{-1}^1$$

$$= \left(1 - \frac{1}{3}\right) - \left(-1 + \frac{1}{3}\right) = \frac{4}{3}.$$

$$A = \int_0^1 \int_{-\sqrt{1-y}}^{\sqrt{1-y}} dx\, dy = \int_0^1 x\Big|_{-\sqrt{1-y}}^{\sqrt{1-y}} dy$$

$$= \int_0^1 \left(\sqrt{1-y} + \sqrt{1-y}\right) dy$$

$$= 2\int_0^1 \sqrt{1-y}\, dy \quad u = 1-y;\ du = -dy.$$

$$= -2\int_0^1 (1-y)^{1/2}(-dy)$$

$$= -2 \cdot \frac{2}{3}(1-y)^{3/2}\Big|_0^1$$

$$= 0 + \frac{4}{3} \cdot 1^{3/2} = \frac{4}{3}.$$

31.

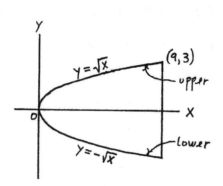

$$A = \int_0^9 \int_{-\sqrt{x}}^{\sqrt{x}} dy\, dx = \int_0^9 y \Big|_{-\sqrt{x}}^{\sqrt{x}} dx$$

$$= \int_0^9 2\sqrt{x}\, dx = 36.$$

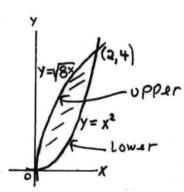

$$A = \int_{-3}^3 \int_{y^2}^9 dx\, dy = \int_{-3}^3 x \Big|_{y^2}^9 dy$$

$$= \int_{-3}^3 (9 - y^2)\, dy = 36.$$

33.

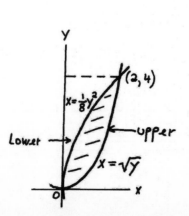

$$
\begin{array}{rcl}
y &=& x^2 \\
y^2 &=& 8x \\
\hline
y^2 &=& x^4 \\
y^2 &=& 8x \\
\hline
0 &=& x^4 - 8x \quad \text{(subtracting)} \\
x(x^3 - 8) &=& 0 \\
x &=& 0,\ 2
\end{array}
$$

$$A = \int_0^2 \int_{x^2}^{\sqrt{8x}} dy\, dx = \int_0^2 y \Big|_{x^2}^{\sqrt{8x}} dx = \int_0^2 \left(\sqrt{8x} - x^2 \right) dx$$

$$= \int_0^2 \left(2\sqrt{2} x^{1/2} - x^2 \right) dx = 2\sqrt{2} \cdot \frac{2}{3} x^{3/2} - \frac{1}{3} x^3 \Big|_0^2 = 2\sqrt{2} \cdot \frac{2}{3} \cdot 2^{3/2} - \frac{1}{3} \cdot 8$$

$$= \frac{4}{3} \cdot 2^{1/2} \cdot 2^{3/2} - \frac{8}{3} = \frac{4}{3} \cdot 2^2 - \frac{8}{3} = \frac{16}{3} - \frac{8}{3} = \frac{8}{3}.$$

$$A = \int_0^4 \int_{y^2/8}^{\sqrt{y}} dx\, dy = \int_0^4 x \Big|_{y^2/8}^{\sqrt{y}} dy$$

$$= \int_0^4 \left(\sqrt{y} - \frac{1}{8} y^2 \right) dy = \left(\frac{2}{3} y^{3/2} - \frac{1}{24} y^3 \right) \Big|_0^4$$

$$= \frac{2}{3} \cdot 4^{3/2} - \frac{1}{24} \cdot 64$$

$$= \frac{2}{3} \cdot 8 - \frac{8}{3} = \frac{16}{3} - \frac{8}{3} = \frac{8}{3}.$$

35.

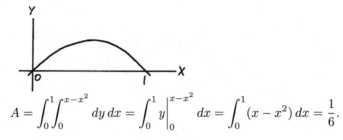

$$A = \int_0^1 \int_0^{x-x^2} dy\,dx = \int_0^1 y\Big|_0^{x-x^2} dx = \int_0^1 (x-x^2)\,dx = \frac{1}{6}.$$

37.

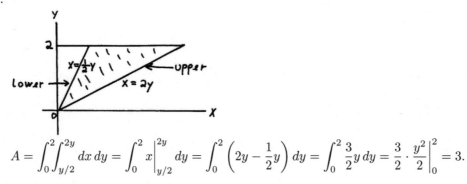

$$A = \int_0^2 \int_{y/2}^{2y} dx\,dy = \int_0^2 x\Big|_{y/2}^{2y} dy = \int_0^2 \left(2y - \frac{1}{2}y\right) dy = \int_0^2 \frac{3}{2}y\,dy = \frac{3}{2}\cdot\frac{y^2}{2}\Big|_0^2 = 3.$$

9.6 Volumes by Double Integration

1.

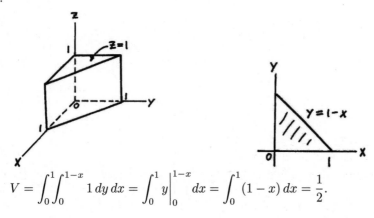

$$V = \int_0^1 \int_0^{1-x} 1\,dy\,dx = \int_0^1 y\Big|_0^{1-x} dx = \int_0^1 (1-x)\,dx = \frac{1}{2}.$$

3.

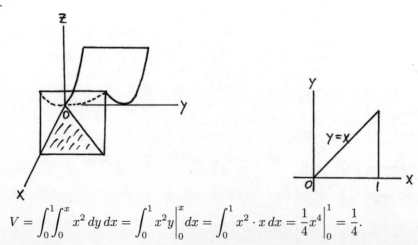

$$V = \int_0^1 \int_0^x x^2\,dy\,dx = \int_0^1 x^2 y\Big|_0^x dx = \int_0^1 x^2 \cdot x\,dx = \frac{1}{4}x^4\Big|_0^1 = \frac{1}{4}.$$

5.

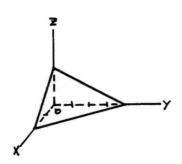 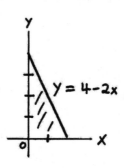

If $z = 0$, we get $y = 4 - 2x$, which is the trace in the xy-plane. Thus $y = 0$ is the lower function and $y = 4 - 2x$ is the upper function. The integrand is $z = \dfrac{1}{2}(4 - 2x - y)$.

$$
\begin{aligned}
V &= \int_0^2 \int_0^{4-2x} \left(\frac{1}{2}\right)(4 - 2x - y)\,dy\,dx = \frac{1}{2}\int_0^2 \left(4y - 2xy - \frac{1}{2}y^2\right)\Bigg|_0^{4-2x} dx \\
&= \frac{1}{2}\int_0^2 \left[4(4 - 2x) - 2x(4 - 2x) - \frac{1}{2}(4 - 2x)^2\right] dx \\
&= \frac{1}{2}\int_0^2 (16 - 8x - 8x + 4x^2 - 8 + 8x - 2x^2)\,dx \\
&= \frac{1}{2}\int_0^2 (2x^2 - 8x + 8)\,dx = \frac{1}{2}\left(\frac{2}{3}x^3 - 4x^2 + 8x\right)\Bigg|_0^2 \\
&= \frac{1}{2}\left(\frac{16}{3} - 16 + 16\right) = \frac{8}{3}.
\end{aligned}
$$

7.

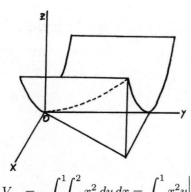 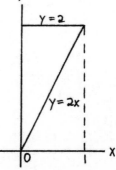

$$
\begin{aligned}
V &= \int_0^1 \int_{2x}^2 x^2\,dy\,dx = \int_0^1 x^2 y\Big|_{2x}^2 dx = \int_0^1 x^2(2 - 2x)\,dx \\
&= \int_0^1 (2x^2 - 2x^3)\,dx = \frac{2}{3}x^3 - \frac{1}{2}x^4\Big|_0^1 = \frac{2}{3} - \frac{1}{2} = \frac{1}{6}.
\end{aligned}
$$

9. From $x^2 + y^2 = 9$, we get $y = \sqrt{9 - x^2}$.

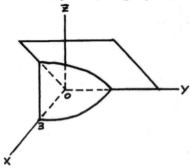
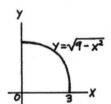

$$V = \int_0^3 \int_0^{\sqrt{9-x^2}} x \, dy \, dx = \int_0^3 xy \Big|_0^{\sqrt{9-x^2}} dx$$

$$= \int_0^3 x\sqrt{9 - x^2} \, dx = -\frac{1}{2} \int_0^3 (9 - x^2)^{1/2}(-2x) \, dx \qquad u = 9 - x^2; \ du = -2x \, dx.$$

$$= -\frac{1}{2} \cdot \frac{2}{3}(9 - x^2)^{3/2} \Big|_0^3 = 0 + \frac{1}{3}(9)^{3/2} = 9.$$

11. Since the surface $z = xy$ lies entirely above the region in the xy-plane, a two-dimensional figure is sufficient for obtaining the limits of integration.

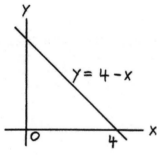

$$V = \int_0^4 \int_0^{4-x} xy \, dy \, dx$$

$$= \int_0^4 x \cdot \frac{1}{2} y^2 \Big|_0^{4-x} dx$$

$$= \int_0^4 \frac{1}{2} x(4 - x)^2 \, dx = \frac{1}{2} \int_0^4 (16x - 8x^2 + x^3) \, dx$$

$$= \frac{1}{2} \left(8x^2 - \frac{8}{3} x^3 + \frac{1}{4} x^4 \right) \Big|_0^4 = \frac{1}{2} \left(8 \cdot 4^2 - \frac{8}{3} \cdot 4^3 + \frac{1}{4} \cdot 4^4 \right)$$

$$= \frac{1}{2} \cdot 4^3 \left(2 - \frac{8}{3} + 1 \right) = 32 \left(\frac{9}{3} - \frac{8}{3} \right) = \frac{32}{3}.$$

13.

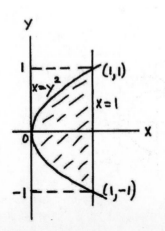

Since the paraboloid lies above the region, a two-dimensional figure is sufficient for obtaining the limits of integration. Since x is given as a function of y, it is convenient to use (9.14) with $h_1(y) = y^2$ and $h_2(y) = 1$.

Thus

$$V = \int_{-1}^{1}\int_{y^2}^{1}(x^2 + 3y^2)\,dx\,dy = \int_{-1}^{1}\left(\frac{1}{3}x^3 + 3xy^2\right)\Big|_{y^2}^{1}\,dy$$

$$= \int_{-1}^{1}\left[\left(\frac{1}{3} + 3y^2\right) - \left(\frac{1}{3}y^6 + 3y^4\right)\right]\,dy$$

$$= \int_{-1}^{1}\left(\frac{1}{3} + 3y^2 - \frac{1}{3}y^6 - 3y^4\right)\,dy$$

$$= \frac{1}{3}y + y^3 - \frac{1}{21}y^7 - \frac{3}{5}y^5\Big|_{-1}^{1} = \left(\frac{1}{3} + 1 - \frac{1}{21} - \frac{3}{5}\right) - \left(-\frac{1}{3} - 1 + \frac{1}{21} + \frac{3}{5}\right)$$

$$= 2\left(\frac{1}{3} + 1 - \frac{1}{21} - \frac{3}{5}\right) = (2)\frac{35 + 105 - 5 - 63}{105}$$

$$= \frac{144}{105} = \frac{48}{35}.$$

15.

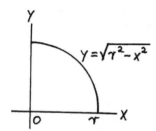

Equation of sphere: $x^2 + y^2 + z^2 = r^2$.
Upper surface: $z = \sqrt{r^2 - x^2 - y^2}$.
Trace in xy-plane: $x^2 + y^2 = r^2$.

The first-quadrant region is shown in the figure. The volume above this region is one-eighth of the total:

$$V = 8\int_{0}^{r}\int_{0}^{\sqrt{r^2-x^2}}\sqrt{r^2 - x^2 - y^2}\,dy\,dx.$$

17.

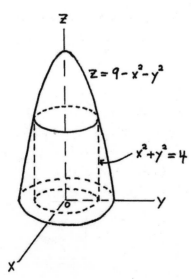

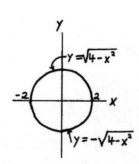

$$V = \int_{-2}^{2}\int_{-\sqrt{4-x^2}}^{\sqrt{4-x^2}}(9 - x^2 - y^2)\,dy\,dx.$$

19.

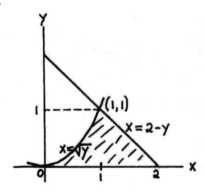

By (9.14) with $h_1(y) = \sqrt{y}$ and $h_2(y) = 2 - y$, we get:

$$V = \int_0^1 \int_{\sqrt{y}}^{2-y} \sqrt{8 - 2x^2 - y^2} \, dx \, dy.$$

21.

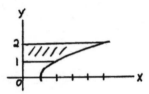

By (9.14) with $h_1(y) = 0$ and $h_2(y) = y^2 + 1$, we get:

$$V = \int_1^2 \int_0^{y^2+1} xy \, dx \, dy.$$

23.

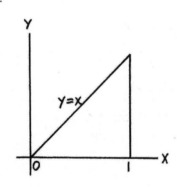 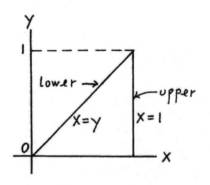

$$\int_0^1 \int_y^1 F(x,y) \, dx \, dy$$

25.

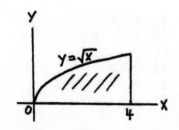 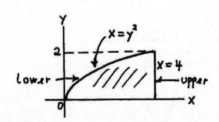

$$\int_0^2 \int_{y^2}^4 F(x,y) \, dx \, dy$$

27.

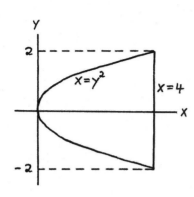

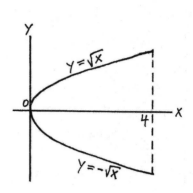

$$\int_0^4 \int_{-\sqrt{x}}^{\sqrt{x}} F(x,y)\, dy\, dx$$

29.

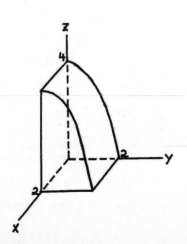

From $x = 4y^2$, we get $y = \dfrac{1}{2}\sqrt{x}$. So by (9.13) with $g_1(x) = 0$ and $g_2(x) = \dfrac{1}{2}\sqrt{x}$, we get:

$$V = \int_0^{16} \int_0^{\sqrt{x}/2} F(x,y)\, dy\, dx.$$

31.

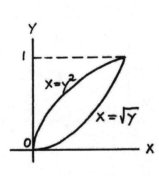

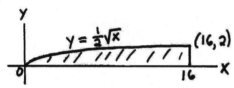

$$\int_0^1 \int_{x^2}^{\sqrt{x}} F(x,y)\, dy\, dx$$

33.

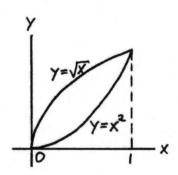

$$
\begin{aligned}
V &= \int_0^2 \int_0^2 (4 - y^2)\, dy\, dx \\
&= \int_0^2 \left(4y - \frac{1}{3}y^3 \right) \Big|_0^2 dx \\
&= \int_0^2 \left(8 - \frac{8}{3} \right) dx = \frac{16}{3} \int_0^2 dx \\
&= \frac{16}{3}(2) = \frac{32}{3}.
\end{aligned}
$$

35. Consider the figure for the first-octant volume:

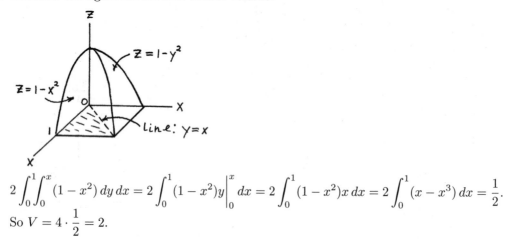

$$2 \int_0^1 \int_0^x (1 - x^2) \, dy \, dx = 2 \int_0^1 (1 - x^2) y \Big|_0^x \, dx = 2 \int_0^1 (1 - x^2) x \, dx = 2 \int_0^1 (x - x^3) \, dx = \frac{1}{2}.$$

So $V = 4 \cdot \dfrac{1}{2} = 2.$

9.7   Mass, Centroids, and Moments of Inertia

1. Area of typical element: $dy \, dx$.

$$A = \int_0^4 \int_0^{x/2} dy \, dx = \int_0^4 y \Big|_0^{x/2} dx = \int_0^4 \frac{1}{2} x \, dx = 4.$$

Mass of typical element: $\dfrac{1}{4} x \, dy \, dx$.

$$
\begin{aligned}
m &= \int_0^4 \int_0^{x/2} \frac{1}{4} x \, dy \, dx \\
&= \int_0^4 \frac{1}{4} xy \Big|_0^{x/2} dx \\
&= \int_0^4 \frac{1}{4} x \left(\frac{1}{2} x \right) dx = \frac{8}{3}.
\end{aligned}
$$

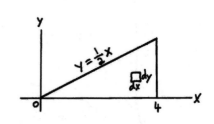

3. Area of typical element: $dy \, dx$.

$$
\begin{aligned}
A &= \int_0^4 \int_{\sqrt{x}}^2 dy \, dx = \int_0^4 y \Big|_{\sqrt{x}}^2 dx = \int_0^4 (2 - \sqrt{x}) \, dx \\
&= 2x - \frac{2}{3} x^{3/2} \Big|_0^4 = 8 - \frac{16}{3} = \frac{24}{3} - \frac{16}{3} = \frac{8}{3}.
\end{aligned}
$$

Mass of typical element: $xy \, dy \, dx$.

$$
\begin{aligned}
m &= \int_0^4 \int_{\sqrt{x}}^2 xy \, dy \, dx = \int_0^4 x \cdot \frac{1}{2} y^2 \Big|_{\sqrt{x}}^2 dx \\
&= \int_0^4 \frac{1}{2} x (4 - x) \, dx = \frac{1}{2} \int_0^4 (4x - x^2) \, dx \\
&= \frac{16}{3}.
\end{aligned}
$$

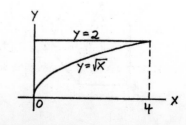

5. Area of typical element: $dy\,dx$.

$$A = \int_{-2}^{2}\int_{0}^{4-x^2} dy\,dx = \int_{-2}^{2}(4-x^2)\,dx = \frac{32}{3}.$$

Mass of typical element: $x^2\,dy\,dx$.

$$\begin{aligned}
m &= \int_{-2}^{2}\int_{0}^{4-x^2} x^2\,dy\,dx = \int_{-2}^{2} x^2 y\Big|_{0}^{4-x^2}\,dx \\
&= \int_{-2}^{2} x^2(4-x^2)\,dx = \frac{4}{3}x^3 - \frac{1}{5}x^5\Big|_{-2}^{2} \\
&= \left(\frac{32}{3} - \frac{32}{5}\right) - \left(-\frac{32}{3} + \frac{32}{5}\right) \\
&= 2\left(\frac{32}{3} - \frac{32}{5}\right) = 64\left(\frac{1}{3} - \frac{1}{5}\right) \\
&= 64\left(\frac{2}{15}\right) = \frac{128}{15}.
\end{aligned}$$

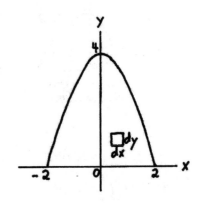

7.

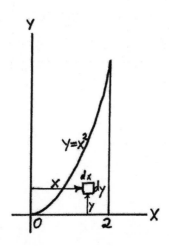

Moment of typical element with respect to the y-axis: $x\,dy\,dx$.

$$\bar{x} = \frac{M_y}{A} = \frac{\int_0^2\int_0^{x^2} x\,dy\,dx}{\int_0^2\int_0^{x^2} dy\,dx} = \frac{\int_0^2 xy\Big|_0^{x^2}\,dx}{\int_0^2 y\Big|_0^{x^2}\,dx} = \frac{\int_0^2 x\cdot x^2\,dx}{\int_0^2 x^2\,dx} = \frac{4}{8/3} = \frac{3}{2}.$$

Moment of typical element with respect to the x-axis: $y\,dy\,dx$.

$$\bar{y} = \frac{M_x}{A} = \frac{\int_0^2\int_0^{x^2} y\,dy\,dx}{8/3} = \frac{3}{8}\int_0^2 \frac{1}{2}y^2\Big|_0^{x^2}\,dx = \frac{3}{8}\int_0^2 \frac{1}{2}x^4\,dx = \frac{6}{5}.$$

9.

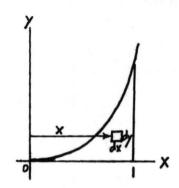

Moment of inertia of typical element: $x^2 \cdot \rho \, dy \, dx = \rho x^2 \, dy \, dx$.

$$I_y = \rho \int_0^1 \int_0^{x^3} x^2 \, dy \, dx = \rho \int_0^1 x^2 y \Big|_0^{x^3} dx = \rho \int_0^1 x^5 \, dx = \frac{\rho}{6}.$$

Mass: $\rho \int_0^1 \int_0^{x^3} dy \, dx = \rho \int_0^1 x^3 \, dx = \frac{\rho}{4}.$ $R_y = \sqrt{\frac{\rho}{6} \frac{4}{\rho}} = \sqrt{\frac{2}{3}} = \frac{\sqrt{6}}{3}.$

11.

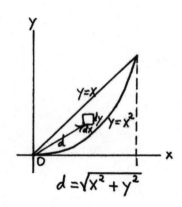

$d = \sqrt{x^2 + y^2}$

Moment of inertia of typical element: $\rho \, d^2 \, dy \, dx = \rho(x^2 + y^2) \, dy \, dx$. [See Equation (9.24).]

$$
\begin{aligned}
I_o &= \int_0^1 \int_{x^2}^x \rho(x^2 + y^2) \, dy \, dx = \rho \int_0^1 \left(x^2 y + \frac{1}{3} y^3 \right) \Big|_{x^2}^x dx \\
&= \rho \int_0^1 \left[\left(x^3 + \frac{1}{3} x^3 \right) - \left(x^4 + \frac{1}{3} x^6 \right) \right] dx = \rho \int_0^1 \left(\frac{4}{3} x^3 - x^4 - \frac{1}{3} x^6 \right) dx = \frac{3\rho}{35}.
\end{aligned}
$$

13.

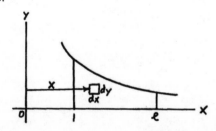

Moment of inertia of typical element: $x^2 \cdot \rho y \, dx = \rho x^2 \, dy \, dx$.

$$I_y = \rho \int_1^e \int_0^{1/x} x^2 \, dy \, dx = \rho \int_1^e x^2 y \Big|_0^{1/x} dx = \rho \int_1^e x \, dx = \frac{\rho}{2} x^2 \Big|_1^e = \frac{\rho}{2}(e^2 - 1).$$

Mass: $\rho \int_1^e \int_0^{1/x} dy \, dx = \rho \int_1^e \frac{1}{x} \, dx = \rho \ln|x| \Big|_1^e = \rho(\ln e - \ln 1) = \rho(1 - 0) = \rho.$

$$R_y = \sqrt{\frac{\rho(e^2 - 1)}{2} \frac{1}{\rho}} = \sqrt{\frac{e^2 - 1}{2}}.$$

15. Moment of typical element with respect to the y-axis: $x \, dy \, dx$.

$$\bar{x} = \frac{M_y}{A} = \frac{\int_1^e \int_0^{1/x} x \, dy \, dx}{\int_1^e \int_0^{1/x} dy \, dx}.$$

By Exercise 13, $A = 1$. So $\bar{x} = \int_1^e xy \Big|_0^{1/x} dx = \int_1^e 1 \, dx = e - 1$.

17.

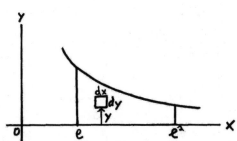

$$y = \frac{1}{\sqrt[3]{x}} = x^{-1/3}$$

Moment of inertia of typical element: $y^2 \cdot \rho \, dy \, dx = \rho y^2 \, dy \, dx$.

$$\begin{aligned}
I_x &= \rho \int_e^{e^2} \int_0^{x^{-1/3}} y^2 \, dy \, dx = \rho \int_e^{e^2} \frac{1}{3} y^3 \Big|_0^{x^{-1/3}} dx \\
&= \rho \int_e^{e^2} \frac{1}{3} (x^{-1/3})^3 \, dx = \frac{\rho}{3} \int_e^{e^2} x^{-1} \, dx = \frac{\rho}{3} \ln |x| \Big|_e^{e^2} \\
&= \frac{\rho}{3}(\ln e^2 - \ln e) = \frac{\rho}{3}(2 \ln e - \ln e) \\
&= \frac{\rho}{3}(2 - 1) = \frac{\rho}{3}. \qquad\qquad \ln e = 1
\end{aligned}$$

19.

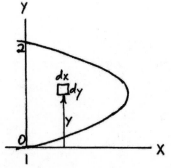

$$\begin{aligned}
x = 4y - 2y^2 &= 0 \\
2y(2 - y) &= 0 \\
y &= 0, \ 2
\end{aligned}$$

Moment of typical element: $y^2 \cdot \rho \, dx \, dy$.

$$I_x = \int_0^2 \int_0^{4y - 2y^2} y^2 \cdot \rho \, dx \, dy = \rho \int_0^2 y^2 x \Big|_0^{4y - 2y^2} dy = \rho \int_0^2 y^2 (4y - 2y^2) \, dy = \frac{16\rho}{5}.$$

21.

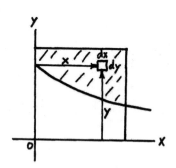

Moment (with respect to y-axis) of typical element: $x\,dy\,dx$.

$$M_y = \int_0^1\!\!\int_{e^{-x}}^1 x\,dy\,dx = \int_0^1 xy\Big|_{e^{-x}}^1 dx = \int_0^1 x(1-e^{-x})\,dx = \int_0^1 x\,dx - \int_0^1 xe^{-x}\,dx.$$

[Integration by parts]
$$u = x \qquad dv = e^{-x}\,dx$$
$$du = dx \qquad v = -e^{-x}$$

$$\frac{1}{2}x^2\Big|_0^1 - \left[-xe^{-x}\Big|_0^1 + \int_0^1 e^{-x}\,dx\right] = \frac{1}{2}x^2\Big|_0^1 - \left[-xe^{-x}\Big|_0^1 - e^{-x}\Big|_0^1\right]$$

$$= \frac{1}{2}x^2\Big|_0^1 + xe^{-x}\Big|_0^1 + e^{-x}\Big|_0^1 = \frac{1}{2} + e^{-1} + e^{-1} - 1 = \frac{2}{e} - \frac{1}{2}.$$

$$A = \int_0^1\!\!\int_{e^{-x}}^1 dy\,dx = \int_0^1 (1-e^{-x})\,dx = x + e^{-x}\Big|_0^1 = 1 + e^{-1} - 1 = \frac{1}{e}.$$

$$\overline{x} = \frac{2/e - 1/2}{1/e} = 2 - \frac{1}{2}e = \frac{4-e}{2}.$$

$$M_x = \int_0^1\!\!\int_{e^{-x}}^1 y\,dy\,dx = \int_0^1 \frac{1}{2}y^2\Big|_{e^{-x}}^1 dx = \frac{1}{2}\int_0^1 (1-e^{-2x})\,dx \qquad u = -2x; \ du = -2\,dx.$$

$$= \frac{1}{2}\left(x + \frac{1}{2}e^{-2x}\right)\Big|_0^1 = \frac{1}{2}\left(1 + \frac{1}{2}e^{-2}\right) - \frac{1}{2}\left(\frac{1}{2}\right)$$

$$= \frac{1}{2} + \frac{1}{4}e^{-2} - \frac{1}{4} = \frac{1}{4e^2} + \frac{1}{4}.$$

$$\overline{y} = \frac{1/(4e^2) + 1/4}{1/e} = \frac{1}{4e} + \frac{e}{4} = \frac{e^2+1}{4e}.$$

23.

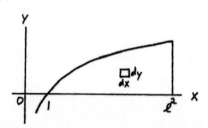

Mass of typical element: $x\,dy\,dx$.

$$m = \int_1^{e^2}\!\!\int_0^{\ln x} x\,dy\,dx = \int_1^{e^2} xy\Big|_0^{\ln x} dx = \int_1^{e^2} x\ln x\,dx.$$

Integration by parts:

$$u = \ln x \qquad dv = x\,dx$$

Recall: $\ln e = 1$; $\ln 1 = 0$

$$du = \frac{1}{x}\,dx \qquad v = \frac{1}{2}x^2$$

$$\int_1^{e^2} x \ln x \, dx = \frac{1}{2} x^2 \ln x \Big|_1^{e^2} - \int_1^{e^2} \frac{1}{2} x^2 \cdot \frac{1}{x} \, dx$$

$$= \frac{1}{2} x^2 \ln x \Big|_1^{e^2} - \frac{1}{4} x^2 \Big|_1^{e^2} = \frac{1}{2} e^4 \ln e^2 - 0 - \frac{1}{4} e^4 + \frac{1}{4}$$

$$= \frac{1}{2} e^4 (2 \ln e) - \frac{1}{4} e^4 + \frac{1}{4} = e^4 - \frac{1}{4} e^4 + \frac{1}{4} = \frac{1}{4} (3e^4 + 1).$$

25.

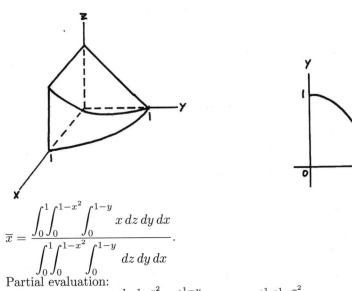

$$\bar{x} = \frac{\displaystyle\int_0^1 \int_0^{1-x^2} \int_0^{1-y} x \, dz \, dy \, dx}{\displaystyle\int_0^1 \int_0^{1-x^2} \int_0^{1-y} dz \, dy \, dx}.$$

Partial evaluation:

$$\text{Numerator} = \int_0^1 \int_0^{1-x^2} xz \Big|_0^{1-y} dy \, dx = \int_0^1 \int_0^{1-x^2} x(1-y) \, dy \, dx$$

$$= \int_0^1 x \left(y - \frac{1}{2} y^2 \right) \Big|_0^{1-x^2} dx = \int_0^1 x \left[(1-x^2) - \frac{1}{2} (1-x^2)^2 \right] dx$$

$$= \frac{1}{2} \int_0^1 (x - x^5) \, dx = \frac{1}{6}.$$

27.

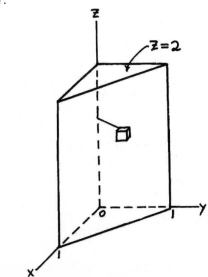

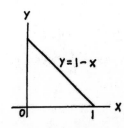

Moment of inertia (with respect to z-axis) of typical element: $\rho(x^2 + y^2) \, dz \, dy \, dx$.

$$
\begin{aligned}
I_z &= \rho \int_0^1 \int_0^{1-x} \int_0^2 (x^2 + y^2)\, dz\, dy\, dx \\
&= \rho \int_0^1 \int_0^{1-x} (x^2 + y^2)z \Big|_0^2 dy\, dx = 2\rho \int_0^1 \int_0^{1-x} (x^2 + y^2)\, dy\, dx \\
&= 2\rho \int_0^1 \left(x^2 y + \frac{1}{3}y^3 \right) \Big|_0^{1-x} dx = 2\rho \int_0^1 \left[x^2(1-x) + \frac{1}{3}(1-x)^3 \right] dx \\
&= 2\rho \int_0^1 x^2(1-x)\, dx + \frac{2\rho}{3} \int_0^1 (1-x)^3\, dx \qquad u = 1 - x; \;\; du = -dx. \\
&= 2\rho \left(\frac{1}{3}x^3 - \frac{1}{4}x^4 \right) \Big|_0^1 - \frac{2\rho}{3} \frac{(1-x)^4}{4} \Big|_0^1 \\
&= 2\rho \left(\frac{1}{3} - \frac{1}{4} \right) + \frac{2\rho}{12} = 2\rho \left(\frac{1}{12} \right) + \frac{2\rho}{12} = \frac{1}{3}\rho.
\end{aligned}
$$

29. $I_y = \rho \displaystyle\int_0^1 \int_{x^2}^x \int_0^{xy} (x^2 + z^2)\, dz\, dy\, dx.$

31. Moment of inertia of typical element with respect to the x-axis: $(y^2 + z^2)\rho\, dz\, dy\, dx.$

$$
\begin{aligned}
I_x &= \int_0^1 \int_0^2 \int_0^{x^2} (y^2 + z^2)\rho\, dz\, dy\, dx = \rho \int_0^1 \int_0^2 \left(y^2 z + \frac{1}{3}z^3 \right) \Big|_0^{x^2} dy\, dx \\
&= \rho \int_0^1 \int_0^2 \left(y^2 x^2 + \frac{1}{3}x^6 \right) dy\, dx = \rho \int_0^1 \left(x^2 \cdot \frac{1}{3}y^3 + \frac{1}{3}x^6 y \right) \Big|_0^2 dx \\
&= \rho \int_0^1 \left(\frac{8}{3}x^2 + \frac{2}{3}x^6 \right) dx = \frac{62\rho}{63}.
\end{aligned}
$$

33.
$$
\begin{aligned}
V &= \int_0^1 \int_0^{1-x} \int_0^2 dz\, dy\, dx = \int_0^1 \int_0^{1-x} z \Big|_0^2 dy\, dx = 2 \int_0^1 \int_0^{1-x} dy\, dx \\
&= 2 \int_0^1 y \Big|_0^{1-x} dx = 2 \int_0^1 (1-x)\, dx = 1.
\end{aligned}
$$

$$
\begin{aligned}
M_{yz} &= \int_0^1 \int_0^{1-x} \int_0^2 x\, dz\, dy\, dx = \int_0^1 \int_0^{1-x} xz \Big|_0^2 dy\, dx \\
&= 2 \int_0^1 \int_0^{1-x} x\, dy\, dx = 2 \int_0^1 xy \Big|_0^{1-x} dx = 2 \int_0^1 x(1-x)\, dx = \frac{1}{3}.
\end{aligned}
$$

$$
\bar{x} = \frac{M_{yz}}{V} = \frac{1}{3}.
$$

$$
\begin{aligned}
M_{xy} &= \int_0^1 \int_0^{1-x} \int_0^2 z\, dz\, dy\, dx = \int_0^1 \int_0^{1-x} \frac{1}{2}z^2 \Big|_0^2 dy\, dx \\
&= 2 \int_0^1 \int_0^{1-x} dy\, dx = 2 \int_0^1 (1-x)\, dx = 1.
\end{aligned}
$$

$$
\bar{z} = \frac{M_{xy}}{V} = 1.
$$

35.

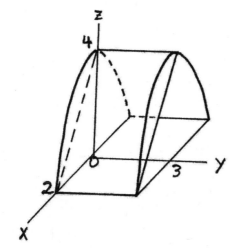

Lower surface: $z = 4 - 2x$.

Upper surface: $z = 4 - x^2$.

$$V = \int_0^2 \int_0^3 \int_{4-2x}^{4-x^2} (1 + x)\, dz\, dy\, dx.$$

9.8   Volumes in Cylindrical Coordinates

1.

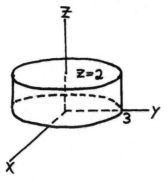 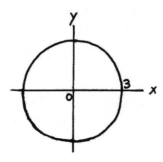

The limits of integration depend on the region in the xy-plane. The circle $x^2 + y^2 = 9$ becomes $r = 3$, while θ ranges from $\theta = 0$ to $\theta = 2\pi$. (See figure on the right.) The equation of the surface is $z = 2$.

$$V = \int_0^{2\pi} \int_0^3 2r\, dr\, d\theta = \int_0^{2\pi} r^2 \Big|_0^3 d\theta = \int_0^{2\pi} 9\, d\theta = 9\theta \Big|_0^{2\pi} = 18\pi.$$

3. The paraboloid $z = 9 - x^2 - y^2$ is shown in Exercise 17, Section 9.6. The region in the xy-plane is the trace $0 = 9 - x^2 - y^2$, which is a circle of radius 3, that is, $r = 3$ in cylindrical coordinates. The equation of the surface becomes $z = 9 - (x^2 + y^2) = 9 - r^2$.

$$\begin{aligned} V &= \int_0^{2\pi} \int_0^3 (9 - r^2)r\, dr\, d\theta = \int_0^{2\pi} \int_0^3 (9r - r^3)\, dr\, d\theta \\ &= \int_0^{2\pi} \left(\frac{9}{2}r^2 - \frac{1}{4}r^4 \right) \Big|_0^3 d\theta = \int_0^{2\pi} \left(\frac{81}{2} - \frac{81}{4} \right) d\theta = \frac{81}{4}\theta \Big|_0^{2\pi} = \frac{81\pi}{2}. \end{aligned}$$

5.

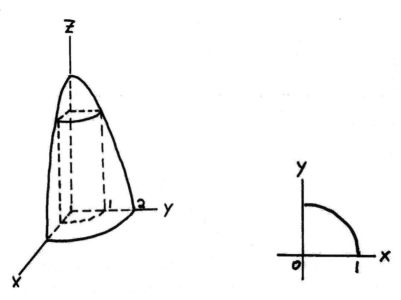

The limits of integration depend on the region in the xy-plane, shown in the figure on the right: the circle is $r = 1$, while θ ranges from $\theta = 0$ to $\theta = \pi/2$. The surface is $z = 4 - x^2 - y^2 = 4 - (x^2 + y^2) = 4 - r^2$. So

$$V = \int_0^{\pi/2}\!\!\int_0^1 (4 - r^2)r\,dr\,d\theta = \int_0^{\pi/2}\!\!\int_0^1 (4r - r^3)\,dr\,d\theta$$

$$= \int_0^{\pi/2} \left(2r - \frac{1}{4}r^4\right)\bigg|_0^1 dr = \int_0^{\pi/2} \frac{7}{4}\,d\theta = \frac{7\pi}{8}.$$

7.

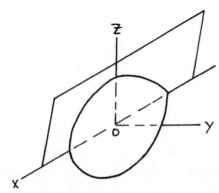

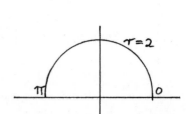

The limits of integration depend on the region in the xy-plane, shown in the figure on the right. The circle is $r = 2$ and θ ranges from 0 to π. The equation of the surface is $z = 3y = 3r\sin\theta$.

$$V = \int_0^\pi\!\!\int_0^2 3r\sin\theta\, r\,dr\,d\theta = 3\int_0^\pi (\sin\theta)\frac{1}{3}r^3\bigg|_0^2 d\theta = 8\int_0^\pi \sin\theta\,d\theta$$

$$= -8\cos\theta\bigg|_0^\pi = -8\cos\pi + 8\cos 0 = 8 + 8 = 16.$$

9.

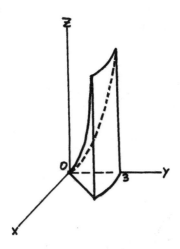

The region bounded by $y = x$ and $x^2 + y^2 = 9$ is shown in the figure on the right: the circle is $r = 3$, while θ ranges from $\theta = \pi/4$ to $\theta = \pi/2$. The surface is $z = x^2 + y^2 = r^2$. So

$$V = \int_{\pi/4}^{\pi/2} \int_0^3 r^2\, r\, dr\, d\theta = \int_{\pi/4}^{\pi/2} \frac{1}{4} r^4 \Big|_0^3 d\theta = \frac{1}{4}(81) \int_{\pi/4}^{\pi/2} d\theta = \frac{1}{4}(81)\theta \Big|_{\pi/4}^{\pi/2}$$

$$= \frac{1}{4}(81)\left(\frac{\pi}{2} - \frac{\pi}{4}\right) = \frac{1}{4}(81)\frac{\pi}{4} = \frac{81\pi}{16}.$$

11. Region in xy-plane: the circle $r = 1$ (first quadrant); $z = xy = (r\cos\theta)(r\sin\theta) = r^2 \cos\theta \sin\theta$.

$$V = \int_0^{\pi/2} \int_0^1 r^2 \sin\theta \cos\theta\, r\, dr\, d\theta = \int_0^{\pi/2} (\sin\theta \cos\theta)\frac{1}{4} r^4 \Big|_0^1 d\theta$$

$$= \frac{1}{4} \int_0^{\pi/2} \sin\theta \cos\theta\, d\theta \qquad\qquad u = \sin\theta;\ du = \cos\theta\, d\theta.$$

$$= \frac{1}{4} \cdot \frac{1}{2}(\sin\theta)^2 \Big|_0^{\pi/2} = \frac{1}{8}.$$

13.

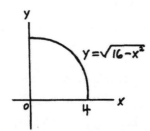

$\int_0^4 \int_0^{\sqrt{16-x^2}} x\, dy\, dx$. The first-quadrant region, shown in the figure, is bounded by a circle of radius 4 ($r = 4$); θ ranges from $\theta = 0$ to $\theta = \pi/2$. The integrand x is replaced by $r\cos\theta$.

$$\int_0^{\pi/2} \int_0^4 (r\cos\theta) r\, dr\, d\theta = \int_0^{\pi/2} \int_0^4 (\cos\theta) r^2\, dr\, d\theta$$

$$= \int_0^{\pi/2} (\cos\theta)\frac{1}{3} r^3 \Big|_0^4 d\theta = \frac{4^3}{3} \int_0^{\pi/2} \cos\theta\, d\theta$$

$$= \frac{64}{3} \sin\theta \Big|_0^{\pi/2} = \frac{64}{3}.$$

15.

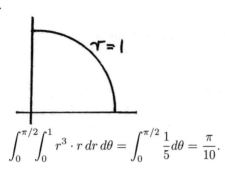

The first-quadrant region is bounded by the circle $r = 1$. The integrand is $(x^2 + y^2)^{3/2} = (r^2)^{3/2} = r^3$.

$$\int_0^{\pi/2} \int_0^1 r^3 \cdot r \, dr \, d\theta = \int_0^{\pi/2} \frac{1}{5} d\theta = \frac{\pi}{10}.$$

17.

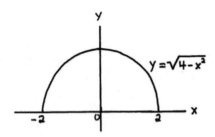

The region, which is bounded by a semicircle of radius 2, is shown in the figure: $r = 2$ and $\theta = 0$ to $\theta = \pi$. Since $x = r \cos \theta$, we have

$$\int_0^{\pi} \int_0^2 (r \cos \theta) r \, dr \, d\theta = \int_0^{\pi} (\cos \theta) \frac{1}{3} r^3 \Big|_0^2 \, d\theta = \frac{8}{3} \sin \theta \Big|_0^{\pi} = 0.$$

Chapter 9 Review

1. Plane; the intercepts are $x = 2$, $y = 2$, and $z = 1$. (See graph in Answer Section.)

3. Cylinder; $2y^2 + z^2 = 4$ is an ellipse in the yz-plane. (See graph in Answer Section.)

5. Trace in xy-plane: $y = 3x^2$.

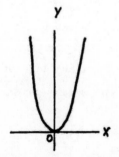

The cylinder $y = 3x^2$ extends along the z-axis (note the missing z-variable).

7. $z^2 - 4x^2 - y^2 = 4$ (hyperboloid of two sheets)

    1. Trace in xy-plane ($z = 0$): $-4x^2 - y^2 = 4$ (no trace; imaginary locus).

    2. Trace in yz-plane ($x = 0$): $z^2 - y^2 = 4$.

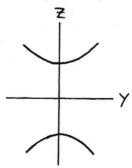

    3. Trace in xz-plane ($y = 0$): $z^2 - 4x^2 = 4$ (another hyperbola).

Cross-section: let $z = 3$ in the original equation:

$$9 - 4x^2 - y^2 = 4 \text{ or } 4x^2 + y^2 = 5.$$

The resulting ellipse is in a plane parallel to the xy-plane and 3 units above it. (See graph in Answer Section.)

9. $4x^2 + 2y^2 - z^2 = 9$ (hyperboloid of one sheet)

    1. Trace in xy-plane ($z = 0$): $4x^2 + 2y^2 = 9$.

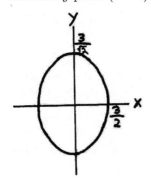

    2. Trace in xz-plane ($y = 0$): $4x^2 - z^2 = 9$.

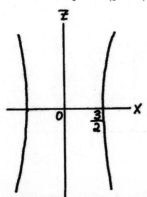

3. Trace in yz-plane ($x = 0$): $2y^2 - z^2 = 9$.

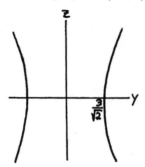

Cross-section: let $z = 4$ in the original equation:

$$4x^2 + 2y^2 - 4^2 = 9 \text{ or } 4x^2 + 2y^2 = 25.$$

The resulting ellipse is parallel to the xy-plane and 4 units above.

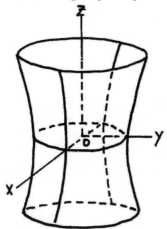

11. $z = \ln(x^2 + y^2)^{1/2} + \sin xy = \dfrac{1}{2}\ln(x^2 + y^2) + \sin xy$

(a) $\dfrac{\partial z}{\partial x} = \dfrac{1}{2}\dfrac{1}{x^2 + y^2}(2x) + (\cos xy)y = \dfrac{x}{x^2 + y^2} + y\cos xy$

(b) $\dfrac{\partial z}{\partial y} = \dfrac{1}{2}\dfrac{1}{x^2 + y^2}(2y) + (\cos xy)x = \dfrac{y}{x^2 + y^2} + x\cos xy$

(c) $\dfrac{\partial^2 z}{\partial x^2} = \dfrac{\partial}{\partial x}\left(\dfrac{\partial z}{\partial x}\right) = \dfrac{(x^2 + y^2)(1) - x(2x)}{(x^2 + y^2)^2} + y(-\sin xy)y = \dfrac{y^2 - x^2}{(x^2 + y^2)^2} - y^2\sin xy$

(d) $\dfrac{\partial^2 z}{\partial y^2} = \dfrac{\partial}{\partial y}\left(\dfrac{\partial z}{\partial y}\right) = \dfrac{(x^2 + y^2)(1) - y(2y)}{(x^2 + y^2)^2} + x(-\sin xy)x = \dfrac{x^2 - y^2}{(x^2 + y^2)^2} - x^2\sin xy$

(e) $\dfrac{\partial^2 z}{\partial x\partial y} = \dfrac{\partial}{\partial x}\left(\dfrac{\partial z}{\partial y}\right) = \dfrac{\partial}{\partial x}\left[y(x^2 + y^2)^{-1} + x\cos xy\right]$

$\qquad = -y(x^2 + y^2)^{-2}(2x) + x(-\sin xy)y + \cos xy$

$\qquad = -\dfrac{2xy}{(x^2 + y^2)^2} - xy\sin xy + \cos xy$

13. $T = 2xy - x^2 - 2y^2 + 3x + 5$

Critical point:

$$\frac{\partial T}{\partial x} = 2y - 2x + 3 = 0$$

$$\frac{\partial T}{\partial y} = 2x - 4y = 0$$

$$\overline{\qquad\qquad\qquad\qquad}$$

$$-2y + 3 = 0 \qquad\qquad \text{adding}$$

$$y = \frac{3}{2}, \quad x = 3.$$

$$\frac{\partial^2 T}{\partial x^2} = -2, \ \frac{\partial^2 T}{\partial y^2} = -4, \ \frac{\partial^2 T}{\partial x \partial y} = 2$$

$$A = (-2)(-4) - (2)^2 = 4 > 0$$

Since $A > 0$ and $\dfrac{\partial^2 T}{\partial x^2} < 0$, T has a maximum at $\left(3, \dfrac{3}{2}\right)$. Finally, substituting in T, we get

$$T\Big|_{(3,3/2)} = \frac{19}{2} = 9.5° \, \text{C (warmest point)}.$$

15. $T = 2xy - x^2 - 2y^2 + 3x + 5$; $\dfrac{\partial T}{\partial x} = 2y - 2x + 3\Big|_{(-3,1)} = 11° \, \text{C/cm}.$

17.

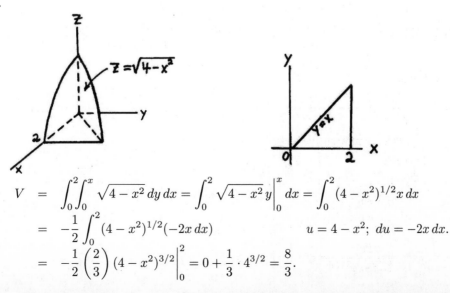

$$V = \int_0^2 \int_0^x \sqrt{4 - x^2} \, dy \, dx = \int_0^2 \sqrt{4 - x^2} \, y \Big|_0^x \, dx = \int_0^2 (4 - x^2)^{1/2} x \, dx$$

$$= -\frac{1}{2} \int_0^2 (4 - x^2)^{1/2}(-2x \, dx) \qquad\qquad u = 4 - x^2; \ du = -2x \, dx.$$

$$= -\frac{1}{2}\left(\frac{2}{3}\right)(4 - x^2)^{3/2}\Big|_0^2 = 0 + \frac{1}{3} \cdot 4^{3/2} = \frac{8}{3}.$$

19.

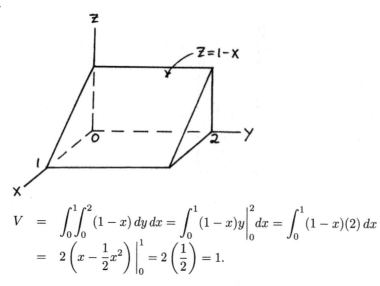

$$V = \int_0^1 \int_0^2 (1-x)\,dy\,dx = \int_0^1 (1-x)y\Big|_0^2\,dx = \int_0^1 (1-x)(2)\,dx$$

$$= 2\left(x - \frac{1}{2}x^2\right)\Big|_0^1 = 2\left(\frac{1}{2}\right) = 1.$$

21.

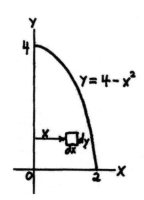

Moment of inertia of typical element: $x^2 \cdot \rho\,dy\,dx$.

$$I_y = \rho \int_0^2 \int_0^{4-x^2} x^2\,dy\,dx = \frac{64\rho}{15}.$$

23.

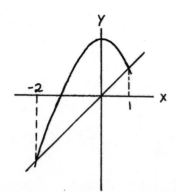

$$y = x$$
$$\underline{y = 2 - x^2}$$
$$0 = x - 2 + x^2 \quad \text{(subtracting)}$$
$$x^2 + x - 2 = 0$$
$$(x+2)(x-1) = 0$$
$$x = -2,\ 1$$

Moment of typical element with respect to the y-axis: $x\,dy\,dx$.

$$\bar{x} = \frac{M_y}{A} = \frac{\int_{-2}^1 \int_x^{2-x^2} x\,dy\,dx}{\int_{-2}^1 \int_x^{2-x^2} dy\,dx} = \frac{\int_{-2}^1 xy\Big|_x^{2-x^2}\,dx}{\int_{-2}^1 y\Big|_x^{2-x^2}\,dx} = \frac{\int_{-2}^1 x(2 - x^2 - x)\,dx}{\int_{-2}^1 (2 - x^2 - x)\,dx} = \frac{-9/4}{9/2} = -\frac{1}{2}.$$

Moment of typical element with respect to the x-axis: $y\,dy\,dx$.

$$\bar{y} = \frac{M_x}{A} = \frac{\int_{-2}^1 \int_x^{2-x^2} y\,dy\,dx}{\int_{-2}^1 \int_x^{2-x^2} dy\,dx} = \frac{\int_{-2}^1 \frac{1}{2}y^2\Big|_x^{2-x^2}\,dx}{9/2} = \frac{2}{9} \cdot \frac{1}{2} \int_{-2}^1 \left[(2 - x^2)^2 - x^2\right]\,dx = \frac{2}{5}.$$

25.

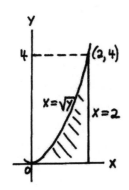 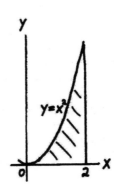

$$\int_0^2 \int_0^{x^2} F(x,y)\, dy\, dx$$

27. $V = \pi r^2 h$

$r = 10.00\,\text{cm}, \ dr = \pm 0.1\,\text{mm} = \pm 0.01\,\text{cm}$

$h = 15.000\,\text{cm}, \ dh = \pm 0.05\,\text{mm} = \pm 0.005\,\text{cm}.$

$$
\begin{aligned}
dV &= \frac{\partial}{\partial r}(\pi r^2 h)\, dr + \frac{\partial}{\partial h}(\pi r^2 h)\, dh \\
&= 2\pi r h\, dr + \pi r^2\, dh \\
&= 2\pi(10.00)(15.000)(\pm 0.01) + \pi(10.00)^2(\pm 0.005) \\
&= \pm 11.0\,\text{cm}^3.
\end{aligned}
$$

$$\frac{dV}{V} \times 100 = \frac{11.0}{\pi(10.00)^2(15.000)} \times 100 = 0.23\%.$$

29.

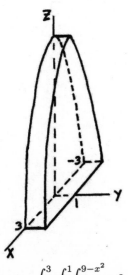 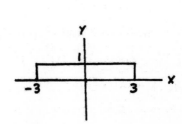

$$I_z = \rho \int_{-3}^3 \int_0^1 \int_0^{9-x^2} (x^2 + y^2)\, dz\, dy\, dx = \frac{384\rho}{5}.$$

31.

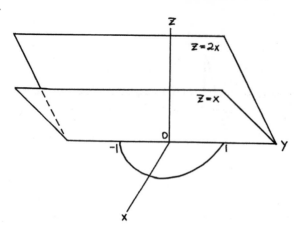

Lower surface: $z = x$.

Upper surface: $z = 2x$.

$$I_x = \int_{-1}^{1} \int_{0}^{\sqrt{1-y^2}} \int_{x}^{2x} \rho(y^2 + z^2)\, dz\, dx\, dy.$$

33.

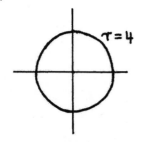

The trace in the xy-plane ($z = 0$) is the circle $x^2 + y^2 = 16$. So the region is bounded by $r = 4$, while θ ranges from 0 to 2π. The integrand is $z = 16 - (x^2 + y^2) = 16 - r^2$. So

$$V = \int_{0}^{2\pi} \int_{0}^{4} (16 - r^2) r\, dr\, d\theta = \int_{0}^{2\pi} \left(8r^2 - \frac{1}{4} r^4 \right) \Big|_{0}^{4} d\theta = \int_{0}^{2\pi} 64\, d\theta = 128\pi.$$

Chapter 10

Infinite Series

10.1 Introduction to Infinite Series

1. $a = 1$, $r = \dfrac{1}{3}$; $S = \dfrac{a}{1-r} = \dfrac{1}{1-(1/3)} = \dfrac{3}{2}$.

3. $a = \dfrac{2}{3}$, $ar = \dfrac{2}{3^2}$, $\dfrac{2}{3}r = \dfrac{2}{3^2}$ and $r = \dfrac{1}{3}$;
   $$S = \dfrac{a}{1-r} = \dfrac{2/3}{1-1/3} = \dfrac{2/3}{2/3} = 1.$$

5. $a = 1$, $r = \dfrac{3}{4}$; $S = \dfrac{1}{1-(3/4)} = \dfrac{1}{(1/4)} = 4$.

7. $a = \dfrac{4}{9}$, $ar = \dfrac{4}{9^2}$, $\dfrac{4}{9}a = \dfrac{4}{9^2}$ and $r = \dfrac{1}{9}$;
   $$S = \dfrac{a}{1-r} = \dfrac{4/9}{1-1/9} = \dfrac{4/9}{8/9} = \dfrac{1}{2}.$$

9. $\dfrac{2}{3} - \dfrac{4}{9} + \dfrac{8}{27} - \ldots$

   $$a = \dfrac{2}{3},\ ar = -\dfrac{4}{9}$$
   $$\dfrac{2}{3}r = -\dfrac{4}{9}$$
   $$r = -\dfrac{2}{3}.$$
   $$S = \dfrac{a}{1-r} = \dfrac{2/3}{1-(-2/3)} = \dfrac{2/3}{1+2/3} = \dfrac{2/3}{5/3} = \dfrac{2}{3} \cdot \dfrac{3}{5} = \dfrac{2}{5}.$$

   Alternatively, the series can be written $S = \dfrac{2}{3} + \dfrac{2}{3}\left(-\dfrac{2}{3}\right) + \dfrac{2}{3}\left(-\dfrac{2}{3}\right)^2 - \cdots$, showing that
   $a = \dfrac{2}{3}$, $r = -\dfrac{2}{3}$.

11. $a = 1$ and $r = -\dfrac{3}{4}$; $S = \dfrac{1}{1-r} = \dfrac{1}{1-(-3/4)} = \dfrac{1}{7/4} = \dfrac{4}{7}$.

13. Since $a = 1$ and $r = -\dfrac{2}{7}$, $S = \dfrac{a}{1-r} = \dfrac{1}{1-(-2/7)} = \dfrac{1}{1+2/7} = \dfrac{7}{9}$.

15. $0.212121\ldots = \dfrac{21}{10^2} + \dfrac{21}{10^4} + \dfrac{21}{10^6} + \cdots$

$a = \dfrac{21}{10^2}, \; ar = \dfrac{21}{10^4},$

$\dfrac{21}{10^2}r = \dfrac{21}{10^4} \text{ and } r = \dfrac{1}{10^2};$

$S = \dfrac{a}{1-r} = \dfrac{21/10^2}{1 - 1/10^2} = \dfrac{21}{10^2 - 1} = \dfrac{21}{99} = \dfrac{7}{33}.$

17. $0.757575\ldots \;=\; 0.75 + 0.0075 + 0.000075 + \cdots$

$= \;\dfrac{75}{100} + \dfrac{75}{10000} + \dfrac{75}{1000000} + \cdots$

$= \;\dfrac{75}{10^2} + \dfrac{75}{10^4} + \dfrac{75}{10^6} + \cdots$

$a = \dfrac{75}{10^2}, \; ar \;=\; \dfrac{75}{10^4}$

$\dfrac{75}{10^2}r \;=\; \dfrac{75}{10^4}$

$r \;=\; \dfrac{1}{10^2}.$

$S = \dfrac{a}{1-r} = \dfrac{75/10^2}{1 - 1/10^2} = \dfrac{75}{10^2 - 1} = \dfrac{75}{99} = \dfrac{25}{33}.$

19. $0.001001001\ldots = 0.001 + 0.000001 + 0.000000001 + \cdots = \dfrac{1}{10^3} + \dfrac{1}{10^6} + \dfrac{1}{10^9} + \cdots$

$a = \dfrac{1}{10^3}, \; ar = \dfrac{1}{10^6},$

$\dfrac{1}{10^3}r = \dfrac{1}{10^6} \text{ and } r = \dfrac{1}{10^3};$

$S = \dfrac{a}{1-r} = \dfrac{1/10^3}{1 - 1/10^3} = \dfrac{1}{10^3 - 1} = \dfrac{1}{999}.$

21. $0.50707\ldots \;=\; 0.5 + 0.007 + 0.00007 + 0.0000007 + \ldots$

$= \;\dfrac{1}{2} + \dfrac{7}{10^3} + \dfrac{7}{10^5} + \dfrac{7}{10^7} + \cdots$

$= \;\dfrac{1}{2} + \left(\dfrac{7}{10^3} + \dfrac{7}{10^3}\dfrac{1}{10^2} + \dfrac{7}{10^3}\dfrac{1}{10^4} + \cdots \right)$

$a = \dfrac{7}{10^3}, \; r = \dfrac{1}{10^2}.$

$S = \dfrac{1}{2} + \dfrac{7/10^3}{1 - 1/10^2} = \dfrac{1}{2} + \dfrac{7}{10^3 - 10} = \dfrac{1}{2} + \dfrac{7}{990} = \dfrac{495 + 7}{990} = \dfrac{502}{990} = \dfrac{251}{495}.$

10.2   Tests for Convergence

1. $\lim\limits_{n\to\infty} \dfrac{n}{2n+2} = \lim\limits_{n\to\infty} \dfrac{1}{2 + 2/n} = \dfrac{1}{2}$

(n^{th} term does not go to 0).

3. Dividing numerator and denominator by n^2, we get for the limit

$\lim\limits_{n\to\infty} \dfrac{5n^2}{2n^2 - 2} = \lim\limits_{n\to\infty} \dfrac{5}{2 - 2/n^2} = \dfrac{5}{2}$

(n^{th} term does not go to 0).

5. $\displaystyle\int_1^\infty \frac{dx}{(x+1)^2}\, dx = \lim_{b\to\infty}\int_1^b (x+1)^{-2}\, dx \qquad u = x+1;\ du = dx.$

$\displaystyle \qquad = \lim_{b\to\infty} \frac{(x+1)^{-1}}{-1}\Big|_1^b = \lim_{b\to\infty}\left(-\frac{1}{x+1}\right)\Big|_1^b$

$\displaystyle \qquad = \lim_{b\to\infty}\left[-\frac{1}{b+1}+\frac{1}{2}\right] = \frac{1}{2}. \qquad \text{(convergent)}$

7. $\displaystyle\int_1^\infty \frac{x}{x^2+1}\, dx = \lim_{b\to\infty}\frac{1}{2}\int_1^b \frac{2x\, dx}{x^2+1} \qquad u = x^2+1;\ du = 2x\, dx.$

$\displaystyle \qquad = \lim_{b\to\infty}\frac{1}{2}\ln(x^2+1)\Big|_1^b = \lim_{b\to\infty}\frac{1}{2}\left[\ln(b^2+1)-\ln 2\right].$

This limit does not exist; so the series diverges.

9. $\displaystyle\int_0^\infty \frac{dx}{(2x+2)^2} = \lim_{b\to\infty}\int_0^b (2x+2)^{-2}\, dx \qquad u = 2x+2;\ du = 2\, dx.$

$\displaystyle \qquad = \lim_{b\to\infty}\frac{1}{2}\int_0^b (2x+2)^{-2}(2\, dx) = \lim_{b\to\infty}\frac{1}{2}\frac{(2x+2)^{-1}}{-1}\Big|_0^b$

$\displaystyle \qquad = \lim_{b\to\infty}\left(-\frac{1}{2(2x+2)}\right)\Big|_0^b$

$\displaystyle \qquad = \lim_{b\to\infty}\left(-\frac{1}{2(2b+2)}+\frac{1}{4}\right) = \frac{1}{4}. \qquad \text{(convergent)}$

11. $\displaystyle\int_2^\infty \frac{x^2}{x^3-2}\, dx = \lim_{b\to\infty}\frac{1}{3}\int_2^b \frac{3x^2\, dx}{x^3-2} \qquad u = x^3-2;\ du = 3x^2\, dx.$

$\displaystyle \qquad = \lim_{b\to\infty}\frac{1}{3}\ln(x^3-2)\Big|_2^b = \lim_{b\to\infty}\frac{1}{3}\left[\ln(b^3-2)-\ln 6\right].$

Since this limit does not exist, the series diverges.

13. $\displaystyle\int_1^\infty \frac{x}{e^x}\, dx = \int_1^\infty xe^{-x}dx \qquad\qquad \text{Integration by parts:}$

$$\begin{aligned} u &= x & dv &= e^{-x}dx \\ du &= dx & v &= -e^{-x} \end{aligned}$$

$\displaystyle \lim_{b\to\infty}\left[-xe^{-x}\Big|_1^b + \int_1^b e^{-x}dx\right] = \lim_{b\to\infty}\left[-xe^{-x}\Big|_1^b - e^{-x}\Big|_1^b\right]$

$\displaystyle \qquad = \lim_{b\to\infty}\left[\left(-be^{-b}+e^{-1}\right)-\left(e^{-b}-e^{-1}\right)\right]$

$\displaystyle \qquad = \lim_{b\to\infty}\left[\left(-\frac{b}{e^b}+\frac{1}{e}\right)-\left(\frac{1}{e^b}-\frac{1}{e}\right)\right]$

$\displaystyle \qquad = \frac{2}{e}. \qquad \text{(convergent)}$

15. $\displaystyle\int_1^\infty \frac{x\, dx}{(x^2+1)^{3/2}} = \lim_{b\to\infty}\frac{1}{2}\int_1^b \frac{2x\, dx}{(x^2+1)^{3/2}} \qquad u = x^2+1;\ du = 2x\, dx.$

$\displaystyle \qquad = \lim_{b\to\infty}\frac{1}{2}\int_1^b (x^2+1)^{-3/2}(2x\, dx) = \lim_{b\to\infty}\frac{1}{2}\frac{(x^2+1)^{-1/2}}{-1/2}\Big|_1^b$

$\displaystyle \qquad = \lim_{b\to\infty}\frac{1}{2}\left(\frac{-2}{\sqrt{x^2+1}}\right)\Big|_1^b = \lim_{b\to\infty}\left(-\frac{1}{\sqrt{b^2+1}}+\frac{1}{\sqrt{2}}\right)$

$\displaystyle \qquad = \frac{1}{\sqrt{2}}. \qquad \text{(convergent)}$

17. $\displaystyle\int_0^\infty \frac{x\, dx}{(x^2+2)^2} = \frac{1}{4}. \qquad \text{(convergent)}$

19. $\displaystyle\int_2^\infty \frac{x}{\sqrt{x^2+2}}\, dx = \lim_{b\to\infty} \int_2^b (x^2+2)^{-1/2}\, dx$

$\displaystyle\qquad\qquad\qquad = \lim_{b\to\infty} \frac{1}{2} \int_2^b (x^2+2)^{-1/2}(2x)\, dx \qquad\qquad u = x^2+2;\ du = 2x\, dx.$

$\displaystyle\qquad\qquad\qquad = \lim_{b\to\infty} \frac{1}{2} \frac{(x^2+2)^{1/2}}{1/2}\Big|_2^b = \lim_{b\to\infty}\left(\sqrt{b^2+2}-\sqrt{6}\right).$

Since this limit does not exist, the integral diverges.

21. $\displaystyle\frac{1}{n^2+1} < \frac{1}{n^2}$ \qquad\qquad (converges by comparison to p-series with $p = 2$).

23. $\displaystyle\frac{1}{n-5} > \frac{1}{n}$ \qquad\qquad (diverges by comparison to the harmonic series).

25. $\displaystyle\frac{1}{n^3+2} < \frac{1}{n^3}$ \qquad\qquad (converges by comparison to p-series with $p = 3$).

27. $\displaystyle\frac{1}{n^2+n} < \frac{1}{n^2}$ \qquad\qquad (converges by comparison to p-series with $p = 2$).

29. $\displaystyle\frac{1}{3^n+1} < \frac{1}{3^n}$ \qquad\qquad (converges by comparison to geometric series).

31. $\displaystyle\frac{1}{3^n-1} < \frac{1}{2^n}$ for $n \geq 2$ \quad (converges by comparison to the geometric series with $r = \frac{1}{2}$).

33. $\displaystyle\frac{1}{\sqrt{n}-1} = \frac{1}{n^{1/2}-1} > \frac{1}{n^{1/2}}$ \qquad\qquad (diverges by comparison to p-series with $p = 1/2$).

35. $\displaystyle\frac{1}{n^3-1} < \frac{1}{n^2}$ for $n \geq 2$ \qquad\qquad (converges by comparison to p-series with $p = 2$).

37. $\displaystyle\frac{1+\sin n}{n^3} \leq \frac{2}{n^3} < \frac{1}{n^2}$ for $n > 2$ \qquad\qquad (comparison to p-series).

39. $\displaystyle\frac{1}{\ln n} > \frac{1}{n}$ \qquad\qquad (diverges by comparison to the harmonic series).

41. $\displaystyle\lim_{n\to\infty}\frac{a_{n+1}}{a_n} = \lim_{n\to\infty}\frac{1/3^{n+1}}{1/3^n} = \lim_{n\to\infty}\frac{1}{3^{n+1}}\frac{3^n}{1} = \lim_{n\to\infty}\frac{1}{3\cdot 3^n}\frac{3^n}{1} = \lim_{n\to\infty}\frac{1}{3} = \frac{1}{3} < 1.$ (convergent)

43. $\displaystyle\lim_{n\to\infty}\frac{a_{n+1}}{a_n} = \lim_{n\to\infty}\frac{\dfrac{1}{(n+1)!}}{\dfrac{1}{n!}}$

Now observe that $(n+1)! = (n+1)n!$ (for example, $5! = 5\cdot 4!$).

$\displaystyle\lim_{n\to\infty}\frac{1}{(n+1)n!}\frac{n!}{1} = \lim_{n\to\infty}\frac{1}{n+1} = 0 < 1. \qquad$ (convergent)

45. $\lim_{n\to\infty} \dfrac{a_{n+1}}{a_n} = \lim_{n\to\infty} \dfrac{\dfrac{4^{n+1}}{(n+1)!}}{\dfrac{4^n}{n!}} = \lim_{n\to\infty} \dfrac{4^{n+1}}{(n+1)!}\dfrac{n!}{4^n}$

Now observe that $(n+1)! = (n+1)\cdot n!$. (For example $6! = 6\cdot 5!$.) Also, $4^{n+1} = 4^n \cdot 4$.

$\lim_{n\to\infty} \dfrac{4^n \cdot 4}{(n+1)n!}\dfrac{n!}{4^n} = \lim_{n\to\infty} \dfrac{4}{n+1} = 0 < 1.$ (convergent)

47. $\lim_{n\to\infty} \dfrac{a_{n+1}}{a_n} = \lim_{n\to\infty} \dfrac{\dfrac{(n+1)^2}{2^{n+1}}}{\dfrac{n^2}{2^n}} = \lim_{n\to\infty} \dfrac{(n+1)^2}{2^{n+1}}\cdot\dfrac{2^n}{n^2}$

$= \lim_{n\to\infty} \dfrac{(n+1)^2}{2^n\cdot 2}\dfrac{2^n}{n^2} = \lim_{n\to\infty} \dfrac{n^2+2n+1}{2n^2}$

$= \lim_{n\to\infty} \dfrac{1+2/n+1/n^2}{2} = \dfrac{1}{2} < 1.$ (convergent)

49. $\lim_{n\to\infty} \dfrac{a_{n+1}}{a_n} = \lim_{n\to\infty} \dfrac{(n+1)(2/3)^{n+1}}{n(2/3)^n} = \lim_{n\to\infty} \dfrac{(n+1)(2/3)^n\cdot(2/3)}{n(2/3)^n}$

$= \lim_{n\to\infty} \dfrac{2}{3}\cdot\dfrac{n+1}{n} = \lim_{n\to\infty} \dfrac{2}{3}\cdot\dfrac{1+1/n}{1}$ dividing numerator and denominator by n

$= \dfrac{2}{3} < 1.$ (convergent)

51. $\lim_{n\to\infty} \dfrac{a_{n+1}}{a_n} = \lim_{n\to\infty} \dfrac{(n+1)\left(\dfrac{3}{2}\right)^{n+1}}{n\left(\dfrac{3}{2}\right)^n} = \lim_{n\to\infty} \dfrac{(n+1)\left(\dfrac{3}{2}\right)^n\left(\dfrac{3}{2}\right)}{n\left(\dfrac{3}{2}\right)^n}$

$= \lim_{n\to\infty} \dfrac{3(n+1)}{2n} = \lim_{n\to\infty} \dfrac{3+3/n}{2} = \dfrac{3}{2} > 1.$ (divergent)

53. $\lim_{n\to\infty} \dfrac{a_{n+1}}{a_n} = \lim_{n\to\infty} \dfrac{(n+1)!/7^{n+1}}{n!/7^n} = \lim_{n\to\infty} \dfrac{(n+1)!}{7^{n+1}}\dfrac{7^n}{n!}$

$= \lim_{n\to\infty} \dfrac{(n+1)n!}{7\cdot 7^n}\dfrac{7^n}{n!} = \lim_{n\to\infty} \dfrac{n+1}{7} = \infty.$ (divergent)

55. $a_{n+1} = \dfrac{(n+1)!}{1\cdot 3\cdot 5\cdots(2n-1)(2n+1)}$

$\lim_{n\to\infty} \dfrac{a_{n+1}}{a_n} = \lim_{n\to\infty} \dfrac{(n+1)n!}{1\cdot 3\cdot 5\cdots(2n-1)(2n+1)}\cdot\dfrac{1\cdot 3\cdot 5\cdots(2n-1)}{n!}$

$= \lim_{n\to\infty} \dfrac{n+1}{2n+1} = \lim_{n\to\infty} \dfrac{1+1/n}{2+1/n} = \dfrac{1}{2} < 1.$ (convergent)

57. $a_{n+1} = \dfrac{1\cdot 4\cdot 7\cdots(3n-2)(3n+1)}{2\cdot 4\cdot 6\cdots(2n)(2n+2)}$

$\dfrac{a_{n+1}}{a_n} = \dfrac{1\cdot 4\cdot 7\cdots(3n-2)(3n+1)}{2\cdot 4\cdot 6\cdots(2n)(2n+2)}\cdot\dfrac{2\cdot 4\cdot 6\cdots(2n)}{1\cdot 4\cdot 7\cdots(3n-2)}$

$\lim_{n\to\infty} \dfrac{3n+1}{2n+2} = \lim_{n\to\infty} \dfrac{3+1/n}{2+2/n} = \dfrac{3}{2} > 1.$ (divergent)

59. $$\lim_{n\to\infty} \frac{a_{n+1}}{a_n} = \frac{\dfrac{(n+1)^2-1}{(n+1)^3}}{\dfrac{n^2-1}{n^3}}$$

$$= \lim_{n\to\infty} \frac{n^2+2n}{n^3+3n^2+3n+1} \cdot \frac{n^3}{n^2-1}$$

$$= \lim_{n\to\infty} \frac{n^5+2n^4}{n^5+3n^4+2n^3-2n^2-3n-1} = 1. \qquad \text{(test fails)}$$

By the integral test,

$$\int_2^\infty \frac{x^2-1}{x^3}\,dx = \int_2^\infty \left(\frac{1}{x}-\frac{1}{x^3}\right)\,dx = \int_2^\infty \frac{1}{x}\,dx - \int_2^\infty \frac{1}{x^3}\,dx.$$

Consider the first integral:

$$\lim_{b\to\infty} \int_2^b \frac{1}{x}\,dx = \lim_{b\to\infty} \ln x \Big|_2^b = \lim_{b\to\infty} (\ln b - \ln 2).$$

Since this limit does not exist, the series diverges.

10.3   Maclaurin Series

1. $$\begin{aligned} f(x) &= \sin x & f(0) &= 0 \\ f'(x) &= \cos x & f'(0) &= 1 \\ f''(x) &= -\sin x & f''(0) &= 0 \\ f'''(x) &= -\cos x & f'''(0) &= -1 \\ f^{(4)}(x) &= \sin x & f^{(4)}(0) &= 0 \\ f^{(5)}(x) &= \cos x & f^{(5)}(0) &= 1 \end{aligned}$$

$$\begin{aligned} \sin x &= 0 + 1x + \frac{0}{2!}x^2 + \frac{-1}{3!}x^3 + \frac{0}{4!}x^4 + \frac{1}{5!}x^5 + \cdots \\ &= x - \frac{x^3}{3!} + \frac{x^5}{5!} - \cdots \end{aligned}$$

3. $$\begin{aligned} f(x) &= \sin 2x & f(0) &= 0 \\ f'(x) &= 2\cos 2x & f'(0) &= 2 \\ f''(x) &= -2^2 \sin 2x & f''(0) &= 0 \\ f'''(x) &= -2^3 \cos 2x & f'''(0) &= -2^3 \\ f^{(4)}(x) &= 2^4 \sin 2x & f^{(4)}(0) &= 0 \\ f^{(5)}(x) &= 2^5 \cos 2x & f^{(5)}(0) &= 2^5 \end{aligned}$$

$$\begin{aligned} \sin 2x &= 0 + 2x + \frac{0}{2!}x^2 + \frac{-2^3}{3!}x^3 + \frac{0}{4!}x^4 + \frac{2^5}{5!}x^5 + \cdots \\ &= 2x - \frac{2^3}{3!}x^3 + \frac{2^5}{5!}x^5 - \cdots \end{aligned}$$

5. $$\begin{aligned} f(x) &= e^{-x} & f(0) &= 1 \\ f'(x) &= -e^{-x} & f'(0) &= -1 \\ f''(x) &= e^{-x} & f''(0) &= 1 \\ f'''(x) &= -e^{-x} & f'''(0) &= -1 \\ f^{(4)}(x) &= e^{-x} & f^{(4)}(0) &= 1 \end{aligned}$$

$$\begin{aligned} e^{-x} &= 1 - 1x + \frac{1}{2!}x^2 + \frac{-1}{3!}x^3 + \frac{1}{4!}x^4 - \cdots \\ &= 1 - x + \frac{x^2}{2!} - \frac{x^3}{3!} + \frac{x^4}{4!} - \cdots \end{aligned}$$

7.
$$\begin{aligned}
f(x) &= \ln(1+x) & f(0) &= 0 \\
f'(x) &= \frac{1}{x+1} = (x+1)^{-1} & f'(0) &= 1 \\
f''(x) &= -1(x+1)^{-2} & f''(0) &= -1 \\
f'''(x) &= 1 \cdot 2(x+1)^{-3} & f'''(0) &= 2! \\
f^{(4)}(x) &= -1 \cdot 2 \cdot 3(x+1)^{-4} & f^{(4)}(0) &= -3! \\
f^{(5)}(x) &= 1 \cdot 2 \cdot 3 \cdot 4(x+1)^{-5} & f^{(5)}(0) &= 4! \\
f^{(6)}(x) &= -1 \cdot 2 \cdot 3 \cdot 4 \cdot 5(x+1)^{-6} & f^{(6)}(0) &= -5!
\end{aligned}$$

$$\begin{aligned}
\ln(x+1) &= 0 + 1x + \frac{-1}{2!}x^2 + \frac{2!}{3!}x^3 + \frac{-3!}{4!}x^4 + \frac{4!}{5!}x^5 + \cdots \\
&= x - \frac{x^2}{2} + \frac{x^3}{3} - \frac{x^4}{4} + \frac{x^5}{5} - \cdots
\end{aligned}$$

9.
$$\begin{aligned}
f(x) &= \sinh x = \frac{1}{2}(e^x - e^{-x}) & f(0) &= 0 \\
f'(x) &= \frac{1}{2}(e^x + e^{-x}) & f'(0) &= 1 \\
f''(x) &= \frac{1}{2}(e^x - e^{-x}) & f''(0) &= 0 \\
f'''(x) &= \frac{1}{2}(e^x + e^{-x}) & f'''(0) &= 1 \\
f^{(4)}(x) &= \frac{1}{2}(e^x - e^{-x}) & f^{(4)}(0) &= 0 \\
f^{(5)}(x) &= \frac{1}{2}(e^x + e^{-x}) & f^{(5)}(0) &= 1
\end{aligned}$$

$$\begin{aligned}
\sinh x &= 0 + 1x + \frac{0}{2!}x^2 + \frac{1}{3!}x^3 + \frac{0}{4!}x^4 + \frac{1}{5!}x^5 + \cdots \\
&= x + \frac{x^3}{3!} + \frac{x^5}{5!} + \cdots
\end{aligned}$$

11.
$$\begin{aligned}
f(x) &= \operatorname{Arctan} x & f(0) &= 0 \\
f'(x) &= \frac{1}{1+x^2} = (1+x^2)^{-1} & f'(0) &= 1 \\
f''(x) &= -(1+x^2)(2x) = -\frac{2x}{(1+x^2)^2} & f''(0) &= 0 \\
f'''(x) &= \frac{6x^2 - 2}{(1+x^2)^3} & f'''(0) &= -2 \\
f^{(4)}(x) &= \frac{4!(x - x^3)}{(1+x^2)^4} & f^{(4)}(0) &= 0 \\
f^{(5)}(x) &= \frac{4!(5x^4 - 10x^2 + 1)}{(1+x^2)^5} & f^{(5)}(0) &= 4!
\end{aligned}$$

$$\begin{aligned}
\operatorname{Arctan} x &= 0 + 1x + \frac{0}{2!}x^2 + \frac{-2}{3!}x^3 + \frac{0}{4!}x^4 + \frac{4!}{5!}x^5 + \cdots \\
&= x - \frac{x^3}{3} + \frac{x^5}{5} - \cdots
\end{aligned}$$

13. This difficult series is best obtained indirectly by using the binomial theorem to expand the derivative of Arcsin x and integrating the resulting series:

$$
\begin{aligned}
\frac{d}{dx}\text{Arcsin}\,x \;=\;& \frac{1}{\sqrt{1-x^2}} = (1-x^2)^{-1/2} = \left[1+(-x^2)\right]^{-1/2} \\[2mm]
=\;& 1 - \frac{1}{2}(-x^2) + \frac{\left(-\frac{1}{2}\right)\left(-\frac{3}{2}\right)}{2!}(-x^2)^2 + \frac{\left(-\frac{1}{2}\right)\left(-\frac{3}{2}\right)\left(-\frac{5}{2}\right)}{3!}(-x^2)^3 \\[2mm]
& + \frac{\left(-\frac{1}{2}\right)\left(-\frac{3}{2}\right)\left(-\frac{5}{2}\right)\left(-\frac{7}{2}\right)}{4!}(-x^2)^4 \\[2mm]
& + \cdots \\[2mm]
=\;& 1 + \frac{1}{2}x^2 + \frac{\frac{1}{2}\cdot\frac{3}{2}}{2!}x^4 + \frac{\frac{1}{2}\cdot\frac{3}{2}\cdot\frac{5}{2}}{3!}x^6 + \frac{\frac{1}{2}\cdot\frac{3}{2}\cdot\frac{5}{2}\cdot\frac{7}{2}}{4!}x^8 + \cdots
\end{aligned}
$$

Integrating term by term yields the desired series:

$$
\begin{aligned}
\int_0^x \frac{d}{dx}\text{Arcsin}\,x\,dx = \text{Arcsin}\,x& \\[2mm]
= x + \frac{1}{2}\cdot\frac{1}{3}x^3 + \frac{\frac{1}{2}\cdot\frac{3}{2}}{2!}\cdot\frac{1}{5}x^5 &+ \frac{\frac{1}{2}\cdot\frac{3}{2}\cdot\frac{5}{2}}{3!}\cdot\frac{1}{7}x^7 + \frac{\frac{1}{2}\cdot\frac{3}{2}\cdot\frac{5}{2}\cdot\frac{7}{2}}{4!}\cdot\frac{1}{9}x^9 + \cdots \\[2mm]
= x + \frac{1\cdot x^3}{2\cdot 3} + \frac{1\cdot 3\cdot x^5}{2\cdot 4\cdot 5} &+ \frac{1\cdot 3\cdot 5\cdot x^7}{2\cdot 4\cdot 6\cdot 7} + \frac{1\cdot 3\cdot 5\cdot 7\cdot x^9}{2\cdot 4\cdot 6\cdot 8\cdot 9} + \cdots
\end{aligned}
$$

15. In $\sum$-form, $\sin x = \displaystyle\sum_{n=0}^{\infty}\frac{(-1)^n x^{2n+1}}{(2n+1)!}$. By the ratio test,

$$
\lim_{n\to\infty}\left|\frac{a_{n+1}}{a_n}\right| = \lim_{n\to\infty}\left|\frac{\dfrac{x^{2n+3}}{(2n+3)!}}{\dfrac{x^{2n+1}}{(2n+1)!}}\right| = \lim_{n\to\infty}\left|\frac{x^{2n+1}x^2}{(2n+3)(2n+2)(2n+1)!}\cdot\frac{(2n+1)!}{x^{2n+1}}\right|
$$

$$
= \lim_{n\to\infty}\left|\frac{x^2}{(2n+3)(2n+2)}\right| = 0 < 1 \text{ for all } x.
$$

So the series converges for all x. Similarly, $\cos x = \displaystyle\sum_{n=0}^{\infty}\frac{(-1)^n x^{2n}}{(2n)!}$.

$$
\lim_{n\to\infty}\left|\frac{a_{n+1}}{a_n}\right| = \lim_{n\to\infty}\left|\frac{\dfrac{x^{2(n+1)}}{[2(n+1)]!}}{\dfrac{x^{2n}}{(2n)!}}\right| = \lim_{n\to\infty}\left|\frac{x^{2n+2}}{(2n+2)!}\cdot\frac{(2n)!}{x^{2n}}\right|
$$

$$
= \lim_{n\to\infty}\left|\frac{x^{2n}x^2}{(2n+2)(2n+1)(2n)!}\cdot\frac{(2n)!}{x^{2n}}\right|
$$

$$
= \lim_{n\to\infty}\left|\frac{x^2}{(2n+2)(2n+1)}\right| = 0 < 1 \text{ for all } x.
$$

So the series converges for all x.

10.4 Operations with Series

1. $\sin x = x - \dfrac{x^3}{3!} + \dfrac{x^5}{5!} - \cdots$. Replacing x by $3x$,

$$\sin 3x = 3x - \frac{(3x)^3}{3!} + \frac{(3x)^5}{5!} - \cdots = 3x - \frac{3^3 x^3}{3!} + \frac{3^5 x^5}{5!} - \cdots.$$

3. $e^x = 1 + x + \dfrac{x^2}{2!} + \dfrac{x^3}{3!} + \dfrac{x^4}{4!} + \cdots$. Replacing x by $-x$,

$$e^{-x} = 1 - x + \frac{x^2}{2!} - \frac{x^3}{3!} + \frac{x^4}{4!} - \cdots.$$

5. $\cos x = 1 - \dfrac{x^2}{2!} + \dfrac{x^4}{4!} \cdots$. Replacing x by $\sqrt{x}$,

$$\cos \sqrt{x} = 1 - \frac{(\sqrt{x})^2}{2!} + \frac{(\sqrt{x})^4}{4!} - \cdots = 1 - \frac{x}{2!} + \frac{x^2}{4!} - \cdots.$$

7. $\cos x = 1 - \dfrac{x^2}{2!} + \dfrac{x^4}{4!} - \dfrac{x^6}{6!} + \cdots$. Multiplying the series by x, we get

$$x \cos x = x - \frac{x^3}{2!} + \frac{x^5}{4!} - \frac{x^7}{6!} + \cdots.$$

9. $\ln(1 + x) = x - \dfrac{x^2}{2} + \dfrac{x^3}{3} - \dfrac{x^4}{4} + \cdots$. Replacing x by x^2,

$$\begin{aligned}
\ln(1 + x^2) &= x^2 - \frac{(x^2)^2}{2} + \frac{(x^2)^3}{3} - \frac{(x^2)^4}{4} + \cdots \\
&= x^2 - \frac{x^4}{2} + \frac{x^6}{3} - \frac{x^8}{4} + \cdots.
\end{aligned}$$

11. Using the expansion for $\text{Arctan}\, x$ from Exercise 11, Section 10.3, we have

$$\begin{aligned}
\frac{\text{Arctan}\, x}{x} &= \frac{x - \dfrac{x^3}{3} + \dfrac{x^5}{5} - \cdots}{x} \\
&= 1 - \frac{x^2}{3} + \frac{x^4}{5} - \cdots.
\end{aligned}$$

13. $$\begin{aligned}
\frac{\ln(1 + x)}{x} &= \frac{x - \dfrac{x^2}{2} + \dfrac{x^3}{3} - \dfrac{x^4}{4} + \cdots}{x} \\
&= 1 - \frac{x}{2} + \frac{x^2}{3} - \frac{x^3}{4} + \cdots
\end{aligned}$$

15. $$\begin{aligned}
\frac{d}{dx} \sin x &= \frac{d}{dx} \left(x - \frac{x^3}{3!} + \frac{x^5}{5!} - \frac{x^7}{7!} + \cdots \right) \\
&= 1 - \frac{3x^2}{3 \cdot 2!} + \frac{5x^4}{5 \cdot 4!} - \frac{7x^6}{7 \cdot 6!} + \cdots \\
&= 1 - \frac{x^2}{2!} + \frac{x^4}{4!} - \frac{x^6}{6!} + \cdots = \cos x
\end{aligned}$$

17. $$\begin{aligned}
\frac{d}{dx} \ln(1 + x) &= \frac{1}{1 + x} \\
&= 1 + (-x) + (-x)^2 + (-x)^3 + (-x)^4 + \cdots \\
&= 1 - x + x^2 - x^3 + x^4 - \cdots
\end{aligned}$$
$$\text{(geometric series with } r = -x)$$
$$\begin{aligned}
\ln(1 + x) &= \int_0^x \frac{dx}{1 + x} = \int_0^x (1 - x + x^2 - x^3 + x^4 - \cdots)\, dx \\
&= x - \frac{x^2}{2} + \frac{x^3}{3} - \frac{x^4}{4} + \frac{x^5}{5} - \cdots
\end{aligned}$$

19. Carrying only powers up to the fourth power and using the expansion for e^{-x} from Exercise 3, we get

$$
\begin{aligned}
e^{-x} &= 1 - x + \frac{x^2}{2} - \frac{x^3}{6} + \frac{x^4}{24} - \cdots \\
\cos x &= 1 - \frac{x^2}{2} + \frac{x^4}{24} - \cdots
\end{aligned}
$$

$$
\begin{array}{l}
\underline{} \\
1 - x + \dfrac{x^2}{2} - \dfrac{x^3}{6} + \dfrac{x^4}{24} - \cdots \\
\qquad\quad - \dfrac{x^2}{2} + \dfrac{x^3}{2} - \dfrac{x^4}{4} + \cdots \\
\qquad\qquad\qquad\qquad + \dfrac{x^4}{24} - \cdots \\
\underline{} \\
1 - x \qquad\quad + \dfrac{1}{3}x^3 - \dfrac{1}{6}x^4 + \cdots
\end{array}
$$

21. $\ln(1 + x^2)^3 = 3\ln(1 + x^2)$

$$
= 3\left(x^2 - \frac{x^4}{2} + \frac{x^6}{3} - \frac{x^8}{4} + \cdots \right) \qquad \text{by Exercise 9}
$$

23. $\ln(1+x) + \operatorname{Arctan} x = x - \dfrac{x^2}{2} + \dfrac{x^3}{3} - \dfrac{x^4}{4} + \dfrac{x^5}{5} - \dfrac{x^6}{6} + \dfrac{x^7}{7} - \dfrac{x^8}{8} + \dfrac{x^9}{9} - \cdots$

$$
+ x - \frac{x^3}{3} + \frac{x^5}{5} - \frac{x^7}{7} + \frac{x^9}{9} - \cdots
$$

$$
= 2x - \frac{x^2}{2} - \frac{x^4}{4} + \frac{2x^5}{5} - \frac{x^6}{6} - \frac{x^8}{8} + \frac{2x^9}{9} - \cdots
$$

25. $-\sqrt{3} + j : \ r = 2, \ \theta = 150° = \dfrac{5\pi}{6}$. Thus $-\sqrt{3} + j = 2e^{5\pi j/6}$.

27. $3j : \ r = 3, \ \theta = 90° = \dfrac{\pi}{2}$. Thus $3j = 3e^{\pi j/2}$.

29. $-2 + 2j : \ r = \sqrt{8} = 2\sqrt{2}, \ \theta = 135° = \dfrac{3\pi}{4}$. Thus $-2 + 2j = 2\sqrt{2}e^{3\pi j/4}$.

10.5 Computations with Series; Applications

1. $\sin x = x - \dfrac{x^3}{3!} + \dfrac{x^5}{5!} - \dfrac{x^7}{7!} + \cdots$

$\sin(0.7) = 0.7 - \dfrac{(0.7)^3}{3!} + \dfrac{(0.7)^5}{5!} = 0.644234 \qquad \text{(three terms)}$

max. error (fourth term): $-\dfrac{(0.7)^7}{7!} = -0.000016$

(a)   0.644234   sum of first three terms

$\underline{\quad -0.000016 \quad}$ error

(b)   0.644218

The values of (a) and (b) agree to four decimal places: 0.6442.

3. $10° = \dfrac{10°\pi}{180°} = \dfrac{\pi}{18}$

$$\cos x = 1 - \frac{x^2}{2!} + \frac{x^4}{4!} - \frac{x^6}{6!} + \cdots$$

$$\cos \frac{\pi}{18} = 1 - \frac{1}{2!}\left(\frac{\pi}{18}\right)^2 = 0.98477 \qquad \text{(two terms)}$$

max. error (third term): $\dfrac{1}{4!}\left(\dfrac{\pi}{18}\right)^4 = 0.00004$

  (a)   0.98477   sum of first two terms

         0.00004   error

  (b)   0.98481

The values of (a) and (b) agree to four decimal places (after rounding off): 0.9848.

5. $\quad e^x = 1 + x + \dfrac{x^2}{2!} + \dfrac{x^3}{3!} + \dfrac{x^4}{4!} + \cdots$

$$e^{-0.2} = 1 + (-0.2) + \frac{(-0.2)^2}{2!} + \frac{(-0.2)^3}{3!} \qquad \text{(four terms)}$$

$$= 0.818667$$

max. error (fifth term): $\dfrac{(-0.2)^4}{4!} = 0.000067$

  (a)   0.818667   sum of first four terms

         0.000067   error

  (b)   0.818734

The values of (a) and (b) agree to four decimal places (after rounding off): 0.8187.

7. $\quad \cos x = 1 - \dfrac{x^2}{2!} + \dfrac{x^4}{4!} - \dfrac{x^6}{6!} + \dfrac{x^8}{8!} - \cdots$

$$\cos 1.2 = 1 - \frac{(1.2)^2}{2!} + \frac{(1.2)^4}{4!} - \frac{(1.2)^6}{6!} = 0.36225 \quad \text{(four terms)}$$

max. error (fifth term): $\dfrac{(1.2)^8}{8!} = 0.0001$

  (a)   0.36225   sum of first four terms

         0.0001   error

  (b)   0.36235

The values of (a) and (b) agree to three decimal places: 0.362.

9. $\ln(1 + x) = x - \dfrac{x^2}{2} + \dfrac{x^3}{3} - \dfrac{x^4}{4} + \cdots$

$$\ln 1.1 = \ln(1 + 0.1) = 0.1 - \frac{(0.1)^2}{2} + \frac{(0.1)^3}{3} \qquad \text{(three terms)}$$

$$= 0.095333$$

max. error (fourth term): $-\dfrac{(0.1)^4}{4} = -0.000025$

  (a)   0.095333   sum of first three terms

         -0.000025 error

  (b)   0.095308

The values of (a) and (b) agree to four decimal places: 0.0953.

11. $\text{Arcsin } \dfrac{1}{2} = \dfrac{\pi}{6}$; so

$$\frac{\pi}{6} = \frac{1}{2} + \frac{1 \cdot (1/2)^3}{2 \cdot 3} + \frac{1 \cdot 3 \cdot (1/2)^5}{2 \cdot 4 \cdot 5} = 0.5232 \text{ and } \pi \approx 6(0.5232) = 3.14.$$

13.
$$\frac{1-\cos x}{x} = \left[1-\left(1-\frac{x^2}{2!}+\frac{x^4}{4!}-\frac{x^6}{6!}+\frac{x^8}{8!}-\cdots\right)\right]/x$$
$$= \frac{x}{2!}-\frac{x^3}{4!}+\frac{x^5}{6!}-\frac{x^7}{8!}+\cdots$$
$$\int_0^{1/2}\frac{1-\cos x}{x}\,dx = \int_0^{1/2}\left(\frac{x}{2!}-\frac{x^3}{4!}+\frac{x^5}{6!}-\frac{x^7}{8!}+\cdots\right)dx$$
$$= \frac{x^2}{2\cdot2!}-\frac{x^4}{4\cdot4!}+\frac{x^6}{6\cdot6!}-\frac{x^8}{8\cdot8!}+\cdots\Big|_0^{1/2}$$
$$= 0.06185 \quad\text{using three terms}$$

error: $-\dfrac{(0.5)^8}{8\cdot8!}=-1.2\times10^{-8}$ (fourth term)

15.
$$\cos x = 1-\frac{x^2}{2!}+\frac{x^4}{4!}-\frac{x^6}{6!}+\frac{x^8}{8!}-\frac{x^{10}}{10!}+\cdots$$
$$\cos\sqrt{x} = 1-\frac{x}{2!}+\frac{x^2}{4!}-\frac{x^3}{6!}+\frac{x^4}{8!}-\frac{x^5}{10!}+\cdots$$
$$\int_0^1\cos\sqrt{x}\,dx = \int_0^1\left(1-\frac{x}{2!}+\frac{x^2}{4!}-\frac{x^3}{6!}+\frac{x^4}{8!}-\frac{x^5}{10!}+\cdots\right)dx$$
$$= x-\frac{x^2}{2\cdot2!}+\frac{x^3}{3\cdot4!}-\frac{x^4}{4\cdot6!}+\frac{x^5}{5\cdot8!}-\frac{x^6}{6\cdot10!}+\cdots\Big|_0^1$$
$$= 1-\frac{1}{2\cdot2!}+\frac{1}{3\cdot4!}-\frac{1}{4\cdot6!}+\frac{1}{5\cdot8!}$$
$$= 0.76354663 \quad\text{(five terms)}$$

max. error (sixth term): $-\dfrac{1}{6\cdot10!}=-0.00000005$

(a)   0.76354663   sum of first five terms

    −0.00000005   error

(b)   0.76354658

Result: 0.76355 to five decimal places.

17.
$$e^x = 1+x+\frac{x^2}{2!}+\frac{x^3}{3!}+\frac{x^4}{4!}+\cdots$$
$$e^{-x^2} = 1-x^2+\frac{x^4}{2!}-\frac{x^6}{3!}+\frac{x^8}{4!}-\cdots$$
$$\int_0^{0.3}e^{-x^2}\,dx = x-\frac{x^3}{3}+\frac{x^5}{5\cdot2!}-\frac{x^7}{7\cdot3!}\Big|_0^{0.3}=0.29124\quad\text{using four terms}$$

error: $\dfrac{(0.3)^9}{9\cdot4!}=9.1\times10^{-8}$ (fifth term)

19. $f(x)=\sin x \qquad f\left(\frac{\pi}{6}\right)=\frac{1}{2}$

$f'(x)=\cos x \qquad f'\left(\frac{\pi}{6}\right)=\frac{\sqrt3}{2}$

$f''(x)=-\sin x \qquad f''\left(\frac{\pi}{6}\right)=-\frac{1}{2}$

$\sin x=\frac{1}{2}+\frac{\sqrt3}{2}\left(x-\frac{\pi}{6}\right)-\frac{1}{2}\frac{1}{2!}\left(x-\frac{\pi}{6}\right)^2+\cdots$

$29°=30°-1°=\frac{\pi}{6}-\frac{\pi}{180}$

Thus $x-\frac{\pi}{6}=\left(\frac{\pi}{6}-\frac{\pi}{180}\right)-\frac{\pi}{6}=-\frac{\pi}{180}$.

$\sin29°=\frac{1}{2}+\frac{\sqrt3}{2}\left(-\frac{\pi}{180}\right)-\frac{1}{2}\frac{1}{2!}\left(-\frac{\pi}{180}\right)^2=0.4848.$

21. $f(x) = \cos x \qquad f\left(\dfrac{\pi}{6}\right) = \dfrac{\sqrt{3}}{2}$

$f'(x) = -\sin x \qquad f'\left(\dfrac{\pi}{6}\right) = -\dfrac{1}{2}$

$f''(x) = -\cos x \qquad f''\left(\dfrac{\pi}{6}\right) = -\dfrac{\sqrt{3}}{2}$

$f'''(x) = \sin x \qquad f'''\left(\dfrac{\pi}{6}\right) = \dfrac{1}{2}$

$\cos x = \dfrac{\sqrt{3}}{2} + \left(-\dfrac{1}{2}\right)\left(x - \dfrac{\pi}{6}\right) + \dfrac{-\sqrt{3}/2}{2!}\left(x - \dfrac{\pi}{6}\right)^2 + \dfrac{1/2}{3!}\left(x - \dfrac{\pi}{6}\right)^3 + \cdots$

$31° = 30° + 1° = \dfrac{\pi}{6} + \dfrac{\pi}{180} = x$

Thus $x - \dfrac{\pi}{6} = \left(\dfrac{\pi}{6} + \dfrac{\pi}{180}\right) - \dfrac{\pi}{6} = \dfrac{\pi}{180}.$

$\cos 31° = \dfrac{\sqrt{3}}{2} - \dfrac{1}{2}\left(\dfrac{\pi}{180}\right) - \dfrac{\sqrt{3}}{4}\left(\dfrac{\pi}{180}\right)^2 + \dfrac{1}{12}\left(\dfrac{\pi}{180}\right)^3 = 0.85717.$

23. $f(x) = \cos x \qquad f\left(\dfrac{\pi}{3}\right) = \dfrac{1}{2}$

$f'(x) = -\sin x \qquad f'\left(\dfrac{\pi}{3}\right) = -\dfrac{\sqrt{3}}{2}$

$f''(x) = -\cos x \qquad f''\left(\dfrac{\pi}{3}\right) = -\dfrac{1}{2}$

$\cos x = \dfrac{1}{2} - \dfrac{\sqrt{3}}{2}\left(x - \dfrac{\pi}{3}\right) - \dfrac{1}{2}\dfrac{1}{2!}\left(x - \dfrac{\pi}{3}\right)^2 + \cdots$

$58° = 60° - 2° = \dfrac{\pi}{3} - \dfrac{2°\pi}{180°} = \dfrac{\pi}{3} - \dfrac{\pi}{90}$

Thus $x - \dfrac{\pi}{3} = \left(\dfrac{\pi}{3} - \dfrac{\pi}{90}\right) - \dfrac{\pi}{3} = -\dfrac{\pi}{90}.$

$\cos 58° = \dfrac{1}{2} - \dfrac{\sqrt{3}}{2}\left(-\dfrac{\pi}{90}\right) - \dfrac{1}{2}\dfrac{1}{2!}\left(-\dfrac{\pi}{90}\right)^2 = 0.5299.$

25. Expand $\ln x$ about $c = 1$.

$f(x) = \ln x \qquad\qquad f(1) = 0$

$f'(x) = \dfrac{1}{x} = x^{-1} \qquad f'(1) = 1$

$f''(x) = -1x^{-2} \qquad\quad f''(1) = -1$

$f'''(x) = 1 \cdot 2x^{-3} \qquad f'''(1) = 2!$

$f^{(4)}(x) = -1 \cdot 2 \cdot 3x^{-3} \qquad f^{(4)}(1) = -3!$

$\ln x = 0 + 1(x - 1) + \dfrac{-1}{2!}(x - 1)^2 + \dfrac{2!}{3!}(x - 1)^3 + \dfrac{-3!}{4!}(x - 1)^4 + \cdots$

$\qquad = (x - 1) - \dfrac{(x - 1)^2}{2} + \dfrac{(x - 1)^3}{3} - \dfrac{(x - 1)^4}{4} + \cdots$

27. Expand the given function about $x = 1$:

$f(x) = 5x^5 + 10x^4 - 2x^3 + x^2 + 5 \qquad f(1) = 19$

$f'(x) = 25x^4 + 40x^3 - 6x^2 + 2x \qquad f'(1) = 61$

$f''(x) = 100x^3 + 120x^2 - 12x + 2 \qquad f''(1) = 210$

$f'''(x) = 300x^2 + 240x - 12 \qquad f'''(1) = 528$

$f^{(4)}(x) = 600x + 240 \qquad f^{(4)}(1) = 840$

$f^{(5)}(x) = 600 \qquad f^{(5)}(1) = 600$

$f(x) = 19 + 61(x - 1) + \dfrac{210}{2!}(x - 1)^2 + \dfrac{528}{3!}(x - 1)^3 + \dfrac{840}{4!}(x - 1)^4 + \dfrac{600}{5!}(x - 1)^5$

$\qquad = 5(x - 1)^5 + 35(x - 1)^4 + 88(x - 1)^3 + 105(x - 1)^2 + 61(x - 1) + 19$

29. (a) $\sin\theta = \theta - \dfrac{\theta^3}{3!} + \dfrac{\theta^5}{5!} - \dfrac{\theta^7}{7!} + \cdots$

If $\theta \approx 0$, then the higher powers of θ become negligible, so that $\sin\theta \approx \theta$.

(b) If $\sin\theta$ is replaced by θ, then
$$\frac{d^2\theta}{dt^2} = -\frac{g}{L}\theta,$$
that is, the acceleration is directly proportional to the displacement and oppositely directed. By Example 5, Section 6.10, θ has the form
$$\theta = a\cos\sqrt{\frac{g}{L}}\,t.$$

(c) The period of the cosine function is
$$\frac{2\pi}{\sqrt{\dfrac{g}{L}}} = 2\pi\sqrt{\frac{L}{g}}.$$

31. $e^x = 1 + x + \dfrac{x^2}{2!} + \dfrac{x^3}{3!} + \cdots$

$i = 2e^{-0.5t^2} = 2\left[1 - 0.5t^2 + \dfrac{1}{2!}(-0.5t^2)^2 + \dfrac{1}{3!}(-0.5t^2)^3 + \cdots\right]$

Letting $t = 0.2$ and using three terms, we get 1.9604.

max. error (fourth term): -0.000003

1.960400

-0.000003

1.960397

So $i = 1.96040\,\text{A}$ accurate to five decimal places.

33. $\cos u = 1 - \dfrac{u^2}{2!} + \dfrac{u^4}{4!} - \cdots$

$\cos u^2 = 1 - \dfrac{u^4}{2!} + \dfrac{u^8}{4!} - \cdots$

$\displaystyle\int_0^t \cos u^2\,du = \int_0^t \left(1 - \dfrac{u^4}{2!} + \dfrac{u^8}{4!} - \cdots\right)du$

$= u - \dfrac{u^5}{5\cdot 2!} + \dfrac{u^9}{9\cdot 4!} - \cdots \Big|_0^t$

$= t - \dfrac{t^5}{5\cdot 2!} + \dfrac{t^9}{9\cdot 4!} - \cdots$

35. $e^x = 1 + x + \dfrac{x^2}{2!} + \dfrac{x^3}{3!} + \dfrac{x^4}{4!} + \dfrac{x^5}{5!} + \dfrac{x^6}{6!} + \cdots$

$e^{-x^2/2} = 1 - \dfrac{1}{2}x^2 + \dfrac{1}{4}\dfrac{x^4}{2!} - \dfrac{1}{8}\dfrac{x^6}{3!} + \dfrac{1}{16}\dfrac{x^8}{4!} - \dfrac{1}{32}\dfrac{x^{10}}{5!} + \dfrac{1}{64}\dfrac{x^{12}}{6!} - \cdots$

$\dfrac{1}{\sqrt{2\pi}}\displaystyle\int_0^1 e^{-x^2/2}\,dx = \dfrac{1}{\sqrt{2\pi}}\left(x - \dfrac{1}{2}\dfrac{x^3}{3} + \dfrac{1}{4}\dfrac{x^5}{5\cdot 2!} - \dfrac{1}{8}\dfrac{x^7}{7\cdot 3!} + \dfrac{1}{16}\dfrac{x^9}{9\cdot 4!} - \dfrac{1}{32}\dfrac{x^{11}}{11\cdot 5!}\right)\Big|_0^1$

$= \dfrac{1}{\sqrt{2\pi}}\left(1 - \dfrac{1}{2}\dfrac{1}{3} + \dfrac{1}{4}\dfrac{1}{5\cdot 2!} - \dfrac{1}{8}\dfrac{1}{7\cdot 3!} + \dfrac{1}{16}\dfrac{1}{9\cdot 4!} - \dfrac{1}{32}\dfrac{1}{11\cdot 5!}\right)$

$= 0.3413.$

error: $\dfrac{1}{\sqrt{2\pi}}\dfrac{1}{64}\dfrac{1}{13\cdot 6!} = 6.7\times 10^{-7}$

37. $q(t) = \dfrac{\ln(1+t^2)}{t^2} = \dfrac{t^2 - \dfrac{t^4}{2} + \dfrac{t^6}{3} - \dfrac{t^8}{4} + \dfrac{t^{10}}{5} - \dfrac{t^{12}}{6} + \cdots}{t^2}$

$\qquad\qquad = 1 - \dfrac{t^2}{2} + \dfrac{t^4}{3} - \dfrac{t^6}{4} + \dfrac{t^8}{5} - \dfrac{t^{10}}{6} + \cdots$

$\quad i = \dfrac{dq}{dt} = 0 - \dfrac{2t}{2} + \dfrac{4t^3}{3} - \dfrac{6t^5}{4} + \dfrac{8t^7}{5} - \dfrac{10t^9}{6} + \cdots$

$\qquad\qquad = -t + \dfrac{4}{3}t^3 - \dfrac{3}{2}t^5 + \dfrac{8}{5}t^7 - \dfrac{5}{3}t^9 + \cdots$

Let $t = 0.40$:

$\quad i = -0.40 + \dfrac{4}{3}(0.40)^3 - \dfrac{3}{2}(0.40)^5 + \dfrac{8}{5}(0.40)^7$

$\qquad = -0.327405.$ (using 4 terms)

error: $-\dfrac{5}{3}(0.40)^9 = -0.0004$ (fifth term)

(a) -0.327405 sum of first four terms

$\quad\dfrac{-0.0004}{\qquad}$ error

(b) -0.327805

The values of (a) and (b) agree to two decimal places: $-0.33\,\text{A}$.

39. $\dfrac{P_\theta}{P} = 1 + \dfrac{1}{4}\left(\sin\dfrac{\theta}{2}\right)^2 + \dfrac{9}{64}\left(\sin\dfrac{\theta}{2}\right)^4 + \cdots$

When $\theta = 5°$, $\dfrac{P_\theta}{P} = 1.0005$.

10.6   Fourier Series

1.

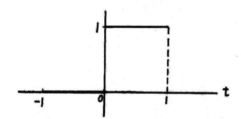

(a) Period: $2p = 2$, so that $p = 1$.

$a_0 = \dfrac{1}{1}\displaystyle\int_{-1}^{1} f(t)\,dt = \int_{-1}^{0} 0\,dt + \int_{0}^{1} 1\,dt = 0 + t\Big|_0^1 = 1$

$\dfrac{a_0}{2} = \dfrac{1}{2}.$

$a_1 = \dfrac{1}{1}\displaystyle\int_{-1}^{1} f(t)\cos\dfrac{1\pi t}{1}\,dt = \int_{-1}^{0} 0\,dt + \int_{0}^{1} 1\cdot\cos\pi t\,dt$

$\quad = \dfrac{1}{\pi}\sin\pi t\Big|_0^1 = 0.$ $u = \pi t;\ du = \pi\,dt.$

$a_2 = \dfrac{1}{1}\displaystyle\int_{-1}^{1} f(t)\cos\dfrac{2\pi t}{1}\,dt = \int_{-1}^{0} 0\,dt + \int_{0}^{1} 1\cdot\cos 2\pi t\,dt$

$\quad = \dfrac{1}{2\pi}\sin 2\pi t\Big|_0^1 = 0.$ $u = 2\pi t;\ du = 2\pi\,dt.$

$a_n = \dfrac{1}{1}\displaystyle\int_{-1}^{1} f(t)\cos\dfrac{n\pi t}{p}\,dt = \int_{-1}^{0} 0\,dt + \int_{0}^{1} 1\cdot\cos\dfrac{n\pi t}{1}\,dt$

$\quad = \dfrac{1}{n\pi}\sin n\pi t\Big|_0^1 = \dfrac{1}{n\pi}\sin n\pi = 0.$ $u = n\pi t;\ du = n\pi\,dt.$

(b) $b_1 = \dfrac{1}{1}\displaystyle\int_{-1}^{1} f(t)\sin\dfrac{1\pi t}{1}\,dt = \int_{-1}^{0} 0\,dt + \int_{0}^{1} 1\cdot\sin\pi t\,dt$

$\qquad = -\dfrac{1}{\pi}\cos\pi t\Big|_{0}^{1} = -\dfrac{1}{\pi}(\cos\pi - \cos 0) \qquad u = \pi t;\ du = \pi\,dt.$

$\qquad = -\dfrac{1}{\pi}(-1-1) = \dfrac{2}{\pi}.$

$b_2 = \dfrac{1}{1}\displaystyle\int_{-1}^{1} f(t)\sin\dfrac{2\pi t}{1}\,dt = \int_{-1}^{0} 0\,dt + \int_{0}^{1} 1\cdot\sin 2\pi t\,dt$

$\qquad = -\dfrac{1}{2\pi}\cos 2\pi t\Big|_{0}^{1} = -\dfrac{1}{2\pi}(\cos 2\pi - \cos 0) \qquad u = 2\pi t;\ du = 2\pi\,dt.$

$\qquad = -\dfrac{1}{2\pi}(1-1) = 0.$

$b_3 = \dfrac{1}{1}\displaystyle\int_{-1}^{1} f(t)\sin\dfrac{3\pi t}{1}\,dt = \int_{-1}^{0} 0\,dt + \int_{0}^{1} 1\cdot\sin 3\pi t\,dt$

$\qquad = -\dfrac{1}{3\pi}\cos 3\pi t\Big|_{0}^{1} \qquad u = 3\pi t;\ du = 3\pi\,dt.$

$\qquad = -\dfrac{1}{3\pi}(\cos 3\pi - \cos 0) = -\dfrac{1}{3\pi}(-1-1) = \dfrac{2}{3\pi}.$

$b_n = \dfrac{1}{1}\displaystyle\int_{-1}^{1} f(t)\sin\dfrac{n\pi t}{p}\,dt = \int_{-1}^{0} 0\,dt + \int_{0}^{1} 1\cdot\sin\dfrac{n\pi t}{1}\,dt \qquad u = n\pi t;\ du = n\pi\,dt.$

$\qquad = -\dfrac{1}{n\pi}\cos n\pi t\Big|_{0}^{1} = -\dfrac{1}{n\pi}(\cos n\pi - 1) = \dfrac{1}{n\pi}(1 - \cos n\pi).$

Now recall that

$\cos 2\pi = \cos 4\pi = \cos 6\pi = \ldots = 1,$

$\cos \pi = \cos 3\pi = \cos 5\pi = \ldots = -1.$

Hence $1 - \cos n\pi = 0$ whenever n is even and $1 - \cos n\pi = 1 - (-1) = 2$ whenever n is odd.

Thus

$$b_n = \begin{cases} 0, & n \text{ even} \\[2mm] \dfrac{2}{n\pi}, & n \text{ odd} \end{cases}$$

It follows that (since $p = 1$)

$f(t) = \dfrac{1}{2} + \dfrac{2}{\pi}\sin\dfrac{1\pi t}{1} + 0 + \dfrac{2}{3\pi}\sin\dfrac{3\pi t}{1} + 0 + \dfrac{2}{5\pi}\sin\dfrac{5\pi t}{1} + \cdots$

$\qquad = \dfrac{1}{2} + \dfrac{2}{\pi}\left(\sin\pi t + \dfrac{1}{3}\sin 3\pi t + \dfrac{1}{5}\sin 5\pi t + \cdots\right).$

3.

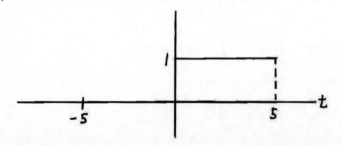

Period: $2p = 10$, so that $p = 5$.

$$a_0 = \frac{1}{5}\int_{-5}^{5} f(t)\,dt = \frac{1}{5}\int_{-5}^{0} 0\,dt + \frac{1}{5}\int_{0}^{5} 1\,dt = \frac{1}{5}t\Big|_{0}^{5} = 1$$

$$\frac{a_0}{2} = \frac{1}{2}.$$

$$a_1 = \frac{1}{5}\int_{-5}^{5} f(t)\cos\frac{1\pi t}{5}\,dt = \frac{1}{5}\int_{-5}^{0} 0\cos\frac{\pi t}{5}\,dt + \frac{1}{5}\int_{0}^{5} 1\cdot\cos\frac{\pi t}{5}\,dt$$

$$= \frac{1}{5}\frac{5}{\pi}\int_{0}^{5}\cos\frac{\pi t}{5}\left(\frac{\pi}{5}\,dt\right) = \frac{1}{\pi}\sin\frac{\pi t}{5}\Big|_{0}^{5} = 0. \qquad u = \frac{\pi t}{5};\ du = \frac{\pi}{5}\,dt.$$

$$a_2 = \int_{-5}^{5} f(t)\cos\frac{2\pi t}{5}\,dt = \frac{1}{5}\int_{-5}^{0} 0\cos\frac{2\pi t}{5}\,dt + \frac{1}{5}\int_{0}^{5} 1\cdot\cos\frac{2\pi t}{5}\,dt$$

$$= \frac{1}{5}\frac{5}{2\pi}\int_{0}^{5}\cos\frac{2\pi}{5}\left(\frac{2\pi}{5}\,dt\right) = \frac{1}{2\pi}\sin\frac{2\pi t}{5}\Big|_{0}^{5} = 0. \qquad u = \frac{2\pi t}{5};\ du = \frac{2\pi}{5}\,dt.$$

$$a_n = \frac{1}{5}\int_{-5}^{5} f(t)\cos\frac{n\pi t}{5}\,dt = \frac{1}{5}\int_{-5}^{0} 0\cos\frac{n\pi t}{5}\,dt + \frac{1}{5}\int_{0}^{5} 1\cdot\cos\frac{n\pi t}{5}\,dt$$

$$= \frac{1}{5}\frac{5}{n\pi}\int_{0}^{5}\cos\frac{n\pi t}{5}\left(\frac{n\pi}{5}\,dt\right) = \frac{1}{n\pi}\sin\frac{n\pi t}{5}\Big|_{0}^{5} = 0. \qquad u = \frac{n\pi t}{5};\ du = \frac{n\pi}{5}\,dt.$$

$$b_1 = \frac{1}{5}\int_{-5}^{5} f(t)\sin\frac{1\pi t}{5}\,dt = \frac{1}{5}\int_{-5}^{0} 0\sin\frac{\pi t}{5}\,dt + \frac{1}{5}\int_{0}^{5} 1\cdot\sin\frac{\pi t}{5}\,dt$$

$$= \frac{1}{5}\frac{5}{\pi}\int_{0}^{5}\sin\frac{\pi t}{5}\left(\frac{\pi}{5}\,dt\right) = -\frac{1}{\pi}\cos\frac{\pi t}{5}\Big|_{0}^{5} \qquad u = \frac{\pi t}{5};\ du = \frac{\pi}{5}\,dt.$$

$$= -\frac{1}{\pi}(\cos\pi - \cos 0) = -\frac{1}{\pi}(-1 - 1) = \frac{2}{\pi}.$$

$$b_2 = \frac{1}{5}\int_{-5}^{5} f(t)\sin\frac{2\pi t}{5}\,dt = \frac{1}{5}\int_{-5}^{0} 0\sin\frac{2\pi t}{5}\,dt + \frac{1}{5}\int_{0}^{5} 1\cdot\sin\frac{2\pi t}{5}\,dt$$

$$= \frac{1}{5}\frac{5}{2\pi}\int_{0}^{5}\sin\frac{2\pi t}{5}\left(\frac{2\pi}{5}\,dt\right) = -\frac{1}{2\pi}\cos\frac{2\pi t}{5}\Big|_{0}^{5} \qquad u = \frac{2\pi t}{5};\ du = \frac{2\pi}{5}\,dt.$$

$$= -\frac{1}{2\pi}(\cos 2\pi - \cos 0) = 0.$$

$$b_3 = \frac{1}{5}\int_{-5}^{5} f(t)\sin\frac{3\pi t}{5}\,dt = \frac{1}{5}\int_{-5}^{0} 0\sin\frac{3\pi t}{5}\,dt + \frac{1}{5}\int_{0}^{5} 1\cdot\sin\frac{3\pi t}{5}\,dt$$

$$= \frac{1}{5}\frac{5}{3\pi}\int_{0}^{5}\sin\frac{3\pi t}{5}\left(\frac{3\pi}{5}\,dt\right) = -\frac{1}{3\pi}\cos\frac{3\pi t}{5}\Big|_{0}^{5} \qquad u = \frac{3\pi t}{5};\ du = \frac{3\pi}{5}\,dt.$$

$$= -\frac{1}{3\pi}(\cos 3\pi - \cos 0) = -\frac{1}{3\pi}(-1 - 1) = \frac{2}{3\pi}.$$

$$b_n = \frac{1}{5}\int_{-5}^{5} f(t)\sin\frac{n\pi t}{5}\,dt = \frac{1}{5}\int_{-5}^{0} 0\sin\frac{n\pi t}{5}\,dt + \frac{1}{5}\int_{0}^{5} 1\cdot\sin\frac{n\pi t}{5}\,dt$$

$$= -\frac{1}{5}\frac{5}{n\pi}\int_{0}^{5}\sin\frac{n\pi t}{5}\left(\frac{n\pi}{5}\,dt\right) = -\frac{1}{n\pi}\left(\cos\frac{n\pi t}{5}\right)\Big|_{0}^{5}$$

$$= -\frac{1}{n\pi}(\cos n\pi - \cos 0) = -\frac{1}{n\pi}(\cos n\pi - 1) = \frac{1}{n\pi}(1 - \cos n\pi).$$

Now recall that

$$\cos 2\pi = \cos 4\pi = \cos 6\pi = \cdots = 1,$$

$$\cos\pi = \cos 3\pi = \cos 5\pi = \cdots = -1.$$

So $1 - \cos n\pi = 0$ whenever n is even and $1 - \cos n\pi = 1 - (-1) = 2$ whenever n is odd. Thus

$$b_n = \begin{cases} 0, & n \text{ even} \\ \dfrac{2}{n\pi}, & n \text{ odd} \end{cases}$$

It now follows that (since $p = 5$)

$$f(t) = \frac{1}{2} + \frac{2}{\pi}\sin\frac{1\pi t}{5} + 0 + \frac{2}{3\pi}\sin\frac{3\pi t}{5} + 0 + \frac{2}{5\pi}\sin\frac{5\pi t}{5} + \cdots$$

$$= \frac{1}{2} + \frac{2}{\pi}\left(\sin\frac{\pi t}{5} + \frac{1}{3}\sin\frac{3\pi t}{5} + \frac{1}{5}\sin\frac{5\pi t}{5} + \cdots\right).$$

5.

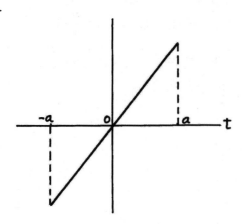

(a) Since the period $2p = 2a,\ p = a$.

$$a_0 = \frac{1}{a}\int_{-a}^{a} t\,dt = \frac{1}{a}\cdot\frac{1}{2}t^2\Big|_{-a}^{a} = 0.$$

$$a_1 = \frac{1}{a}\int_{-a}^{a} t\cos\frac{1\pi t}{a}\,dt = \frac{1}{a}\int_{-a}^{a} t\cos\frac{\pi t}{a}\,dt.$$

$$u = t \qquad dv = \cos\frac{\pi t}{a}\,dt$$
$$du = dt \qquad v = \frac{a}{\pi}\sin\frac{\pi t}{a}$$

$$a_1 = \frac{1}{a}\left[t\left(\frac{a}{\pi}\right)\sin\frac{\pi t}{a}\Big|_{-a}^{a} - \frac{a}{\pi}\int_{-a}^{a}\sin\frac{\pi t}{a}\,dt\right]$$

$$= \frac{1}{a}\left[0 - \frac{a}{\pi}\left(-\frac{a}{\pi}\right)\cos\frac{\pi t}{a}\Big|_{-a}^{a}\right]$$

$$= \frac{a}{\pi^2}(\cos\pi - \cos(-\pi)) = 0. \qquad \cos(-\theta) = \cos\theta$$

$$a_2 = \frac{1}{a}\int_{-a}^{a} t\cos\frac{2\pi t}{a}\,dt$$

$$u = t \qquad dv = \cos\frac{2\pi t}{a}\,dt$$
$$du = dt \qquad v = \frac{a}{2\pi}\sin\frac{2\pi t}{a}$$

$$= \frac{1}{a}\left[t\left(\frac{a}{2\pi}\right)\sin\frac{2\pi t}{a}\Big|_{-a}^{a} - \frac{a}{2\pi}\int_{-a}^{a}\sin\frac{2\pi t}{a}\,dt\right]$$

$$= \frac{1}{a}\left[0 - \frac{a}{2\pi}\left(-\frac{a}{2\pi}\right)\cos\frac{2\pi t}{a}\Big|_{-a}^{a}\right]$$

$$= \frac{a}{2^2\pi^2}(\cos 2\pi - \cos(-2\pi)) = 0.$$

$$a_n = \frac{1}{a}\int_{-a}^{a} t\cos\frac{n\pi t}{a}\,dt$$

$$u = t \qquad dv = \cos\frac{n\pi t}{a}\,dt$$
$$du = dt \qquad v = \frac{a}{n\pi}\sin\frac{n\pi t}{a}$$

$$a_n = \frac{1}{a}\left[t\left(\frac{a}{n\pi}\right)\sin\frac{n\pi t}{a}\Big|_{-a}^{a} - \frac{a}{n\pi}\int_{-a}^{a}\sin\frac{n\pi t}{a}\,dt\right]$$

$$= \frac{1}{a}\left[0 - \frac{a}{n\pi}\left(-\frac{a}{n\pi}\right)\cos\frac{n\pi}{a}\Big|_{-a}^{a}\right] = \frac{a}{n^2\pi^2}\left[\cos\frac{n\pi a}{a} - \cos\left(\frac{-n\pi a}{a}\right)\right]$$

$$= \frac{a}{n^2\pi^2}(\cos n\pi - \cos n\pi) = 0,\ \text{since}\ \cos(-\theta) = \cos\theta.$$

(b) $b_1 = \dfrac{1}{a}\displaystyle\int_{-a}^{a} t\sin\dfrac{1\pi t}{a}\,dt$

$\quad\quad u = t \quad dv = \sin\dfrac{\pi t}{a}\,dt$

$\quad\quad du = dt \quad v = -\dfrac{a}{\pi}\cos\dfrac{\pi t}{a}$

$= \dfrac{1}{a}\left[t\left(-\dfrac{a}{\pi}\right)\cos\dfrac{\pi t}{a}\Big|_{-a}^{a} + \dfrac{a}{\pi}\displaystyle\int_{-a}^{a}\cos\dfrac{\pi t}{a}\,dt\right]$

$= \dfrac{1}{a}\left[-\dfrac{at}{\pi}\cos\dfrac{\pi t}{a} + \dfrac{a^2}{\pi^2}\sin\dfrac{\pi t}{a}\right]\Big|_{-a}^{a}$

$= \dfrac{1}{a}\left[-\dfrac{a^2}{\pi}\cos\pi + \dfrac{a(-a)}{\pi}\cos(-\pi) + 0\right]$

$= \dfrac{2a}{\pi}\quad\text{since }\cos(-\theta) = \cos\theta.$

$b_2 = \dfrac{1}{a}\displaystyle\int_{-a}^{a} t\sin\dfrac{2\pi t}{a}\,dt$

$\quad\quad u = t \quad dv = \sin\dfrac{2\pi t}{a}\,dt$

$\quad\quad du = dt \quad v = -\dfrac{a}{2\pi}\cos\dfrac{2\pi t}{a}$

$= \dfrac{1}{a}\left[t\left(-\dfrac{a}{2\pi}\right)\cos\dfrac{2\pi t}{a}\Big|_{-a}^{a} + \dfrac{a}{2\pi}\displaystyle\int_{-a}^{a}\cos\dfrac{2\pi t}{a}\,dt\right]$

$= \dfrac{1}{a}\left[-\dfrac{at}{2\pi}\cos\dfrac{2\pi t}{a} + \dfrac{a^2}{2^2\pi^2}\sin\dfrac{2\pi t}{a}\right]\Big|_{-a}^{a}$

$= \dfrac{1}{a}\left[-\dfrac{a^2}{2\pi}\cos 2\pi + \dfrac{a(-a)}{2\pi}\cos(-2\pi) + 0\right]$

$= \dfrac{1}{a}\left[-\dfrac{a^2}{2\pi} - \dfrac{a^2}{2\pi}\right] = -\dfrac{2a}{2\pi}.$

$b_3 = \dfrac{1}{a}\displaystyle\int_{-a}^{a} t\sin\dfrac{3\pi t}{a}\,dt$

$\quad\quad u = t \quad dv = \sin\dfrac{3\pi t}{a}\,dt$

$\quad\quad du = dt \quad v = -\dfrac{a}{3\pi}\cos\dfrac{3\pi t}{a}$

$= \dfrac{1}{a}\left[t\left(-\dfrac{a}{3\pi}\right)\cos\dfrac{3\pi t}{a}\Big|_{-a}^{a} + \dfrac{a}{3\pi}\displaystyle\int_{-a}^{a}\cos\dfrac{3\pi t}{a}\,dt\right]$

$= \dfrac{1}{a}\left[-\dfrac{at}{3\pi}\cos\dfrac{3\pi t}{a} + \dfrac{a^2}{3^2\pi^2}\sin\dfrac{3\pi t}{a}\right]\Big|_{-a}^{a}$

$= \dfrac{1}{a}\left[-\dfrac{a^2}{3\pi}\cos 3\pi + \dfrac{a(-a)}{3\pi}\cos(-3\pi) + 0\right]$

$= -\dfrac{a}{3\pi}\cos 3\pi - \dfrac{a}{3\pi}\cos 3\pi \quad\quad \cos(-\theta) = \cos\theta$

$= -\dfrac{a}{3\pi}(-1) - \dfrac{a}{3\pi}(-1) = \dfrac{2a}{3\pi}.$

$b_n = \dfrac{1}{a}\displaystyle\int_{-a}^{a} t\sin\dfrac{n\pi t}{a}\,dt$

$\quad\quad u = t \quad dv = \sin\dfrac{n\pi t}{a}\,dt$

$\quad\quad du = dt \quad v = -\dfrac{a}{n\pi}\cos\dfrac{n\pi t}{a}$

$= \dfrac{1}{a}\left[-\dfrac{at}{n\pi}\cos\dfrac{n\pi t}{a}\Big|_{-a}^{a} + \dfrac{a}{n\pi}\displaystyle\int_{-a}^{a}\cos\dfrac{n\pi t}{a}\,dt\right]$

$= \dfrac{1}{a}\left[-\dfrac{at}{n\pi}\cos\dfrac{n\pi t}{a}\Big|_{-a}^{a} + \dfrac{a^2}{n^2\pi^2}\sin\dfrac{n\pi t}{a}\Big|_{-a}^{a}\right]$

$= \dfrac{1}{a}\left[-\dfrac{a^2}{n\pi}\cos\dfrac{n\pi a}{a} - \dfrac{a^2}{n\pi}\cos\left(\dfrac{-n\pi a}{a}\right) + 0\right]$

$= -\dfrac{a}{n\pi}(\cos n\pi + \cos n\pi) = -\dfrac{2a}{n\pi}\cos n\pi,\ \text{since }\cos(-\theta) = \cos\theta.$

Thus

$$b_n = \begin{cases} -\dfrac{2a}{n\pi}, & n \text{ even} \\[2mm] \dfrac{2a}{n\pi}, & n \text{ odd} \end{cases}$$

$$\begin{aligned} f(t) &= \frac{2a}{1\pi}\sin\frac{1\pi t}{a} - \frac{2a}{2\pi}\sin\frac{2\pi t}{a} + \frac{2a}{3\pi}\sin\frac{3\pi t}{a} - \cdots \\ &= \frac{2a}{\pi}\left(\sin\frac{\pi t}{a} - \frac{1}{2}\sin\frac{2\pi t}{a} + \frac{1}{3}\sin\frac{3\pi t}{a} - \cdots\right). \end{aligned}$$

7.

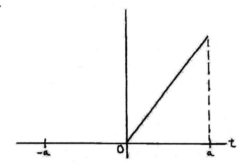

Period: $2p = 2a$, $p = a$.

$$a_0 = \frac{1}{a}\int_{-a}^{a} f(t)\,dt = \frac{1}{a}\int_{-a}^{0} 0\,dt + \frac{1}{a}\int_{0}^{a} t\,dt = \frac{1}{a}\frac{t^2}{2}\Big|_{0}^{a} = \frac{a}{2}$$

$$\frac{a_0}{2} = \frac{a}{4}.$$

$$a_n = \frac{1}{a}\int_{-a}^{a}\cos\frac{n\pi t}{a}\,dt = \frac{1}{a}\int_{-a}^{0} 0\cos\frac{n\pi t}{a}\,dt + \frac{1}{a}\int_{0}^{a} t\cos\frac{n\pi t}{a}\,dt$$

$$= \frac{1}{a}\int_{0}^{a} t\cos\frac{n\pi t}{a}\,dt \qquad\qquad u = t \quad dv = \cos\frac{n\pi t}{a}\,dt$$
$$\qquad\qquad\qquad\qquad\qquad\qquad du = dt \quad v = \frac{a}{n\pi}\sin\frac{n\pi t}{a}$$

$$a_n = \frac{1}{a}\left[\frac{at}{n\pi}\sin\frac{n\pi t}{a}\Big|_{0}^{a} - \frac{a}{n\pi}\int_{0}^{a}\sin\frac{n\pi t}{a}\,dt\right] = \frac{1}{a}\left[0 - \frac{a}{n\pi}\left(-\frac{a}{n\pi}\right)\cos\frac{n\pi t}{a}\Big|_{0}^{a}\right]$$

$$= \frac{a}{n^2\pi^2}(\cos n\pi - 1) = \begin{cases} 0, & n \text{ even} \\[2mm] -\dfrac{2a}{n^2\pi^2}, & n \text{ odd} \end{cases}$$

$$b_n = \frac{1}{a}\int_{-a}^{a}\sin\frac{n\pi t}{a}\,dt = \frac{1}{a}\int_{-a}^{0} 0\sin\frac{n\pi t}{a}\,dt + \frac{1}{a}\int_{0}^{a} t\sin\frac{n\pi t}{a}\,dt$$

$$= \frac{1}{a}\int_{0}^{a} t\sin\frac{n\pi t}{a}\,dt \qquad\qquad u = t \quad dv = \sin\frac{n\pi t}{a}\,dt$$
$$\qquad\qquad\qquad\qquad\qquad\qquad du = dt \quad v = -\frac{a}{n\pi}\cos\frac{n\pi t}{a}$$

$$= \frac{1}{a}\left[-\frac{at}{n\pi}\cos\frac{n\pi t}{a}\Big|_{0}^{a} + \frac{a}{n\pi}\int_{0}^{a}\cos\frac{n\pi t}{a}\,dt\right].$$

Now observe that $\int\cos\dfrac{n\pi t}{a}\,dt = 0$ with the given limits of integration. The first part yields

$$b_n = -\frac{a}{n\pi}\cos n\pi - 0 = \begin{cases} -\dfrac{a}{n\pi}, & n \text{ even} \\[2mm] \dfrac{a}{n\pi}, & n \text{ odd} \end{cases}$$

$$f(t) = \frac{a}{4} - \frac{2a}{1^2\pi^2}\cos\frac{1\pi t}{a} + 0 - \frac{2a}{3^2\pi^2}\cos\frac{3\pi t}{a} + 0 - \frac{2a}{5^2\pi^2}\cos\frac{5\pi t}{a} - \cdots$$

$$+ \frac{a}{1\pi}\sin\frac{1\pi t}{a} - \frac{a}{2\pi}\sin\frac{2\pi t}{a} + \frac{a}{3\pi}\sin\frac{3\pi t}{a} - \frac{a}{4\pi}\sin\frac{4\pi t}{a} + \cdots$$

$$f(t) = \frac{a}{4} - \frac{2a}{\pi^2}\left(\cos\frac{\pi t}{a} + \frac{1}{3^2}\cos\frac{3\pi t}{a} + \frac{1}{5^2}\cos\frac{5\pi t}{a} + \cdots\right)$$

$$+ \frac{a}{\pi}\left(\sin\frac{\pi t}{a} - \frac{1}{2}\sin\frac{2\pi t}{a} + \frac{1}{3}\sin\frac{3\pi t}{a} - \frac{1}{4}\sin\frac{4\pi t}{a} + \cdots\right).$$

9. Since one period is $[-\pi, \pi]$, we have $p = \pi$. Thus

$$\begin{aligned}
a_0 &= \frac{1}{\pi}\int_{-\pi}^{\pi} f(t)\,dt = \frac{1}{\pi}\int_{-\pi}^{0} 0\cdot dt + \frac{1}{\pi}\int_{0}^{\pi}\sin t\,dt \\
&= \frac{1}{\pi}(-\cos t)\Big|_0^{\pi} = \frac{1}{\pi}(1+1) = \frac{2}{\pi}
\end{aligned}$$

and $\dfrac{a_0}{2} = \dfrac{1}{\pi}$.

Next,

$$a_n = \frac{1}{\pi}\int_{-\pi}^{\pi} f(t)\cos\frac{n\pi t}{\pi}\,dt = \frac{1}{\pi}\int_0^{\pi}\sin t\cos nt\,dt.$$

Now, by formula 63, Table 2, with $m = 1$, we get

$$\begin{aligned}
a_n &= \frac{1}{\pi}\left[-\frac{\cos(1+n)t}{2(1+n)} - \frac{\cos(1-n)t}{2(1-n)}\right]\Big|_0^{\pi} \\
&= \frac{1}{2\pi}\left[-\frac{\cos(1+n)\pi}{1+n} - \frac{\cos(1-n)\pi}{1-n} + \frac{1}{1+n} + \frac{1}{1-n}\right]\quad (n \neq 1).
\end{aligned}$$

Observe that this expression is valid for all n except $n = 1$. So if $n \neq 1$, note that

$$\cos(1+n)\pi = \cos(1-n)\pi = \begin{cases} 1, & \text{for } n \text{ odd} \\ -1, & \text{for } n \text{ even} \end{cases}$$

For n <u>odd</u>, we have

$$\frac{1}{2\pi}\left[-\frac{1}{1+n} - \frac{1}{1-n} + \frac{1}{1+n} + \frac{1}{1-n}\right] = 0.$$

For n <u>even</u>, we have

$$\frac{1}{2\pi}\left[\frac{1}{1+n} + \frac{1}{1-n} + \frac{1}{1+n} + \frac{1}{1-n}\right] = \frac{1}{2\pi}\left[\frac{2}{1+n} + \frac{2}{1-n}\right]$$

$$= \frac{1}{\pi}\left[\frac{1}{1+n} + \frac{1}{1-n}\right] = \frac{1}{\pi}\frac{1-n+1+n}{(1+n)(1-n)} = \frac{1}{\pi}\frac{2}{1-n^2} = -\frac{2}{\pi(n^2-1)}.$$ Since these forms are not

valid for $n = 1$, the coefficient a_1 must be evaluated separately:

$$a_1 = \frac{1}{\pi}\int_0^{\pi}\sin t\cos t\,dt = \frac{\sin^2 t}{2\pi}\Big|_0^{\pi} = 0. \qquad u = \sin t;\ du = \cos t\,dt.$$

Thus

$$a_n = \begin{cases} -\dfrac{2}{\pi(n^2-1)}, & n \text{ even} \\[2mm] 0, & n \text{ odd} \end{cases}$$

Next, $b_n = \dfrac{1}{\pi}\displaystyle\int_{-\pi}^{\pi} f(t)\sin\dfrac{n\pi t}{\pi}\,dt = \dfrac{1}{\pi}\int_0^{\pi}\sin t\sin nt\,dt.$

By formula 61, Table 2, with $m = 1$, we get

$$b_n = \frac{1}{\pi}\left[-\frac{\sin(1+n)t}{2(1+n)} + \frac{\sin(1-n)t}{2(1-n)}\right]\Big|_0^{\pi}\quad (n \neq 1)$$

$$= 0, \text{ provided that } n \neq 1.$$

If $n = 1$, we have

$$\begin{aligned}
b_1 &= \frac{1}{\pi}\int_0^{\pi}\sin t\sin t\,dt = \frac{1}{\pi}\int_0^{\pi}\sin^2 t\,dt \\
&= \frac{1}{2\pi}\int_0^{\pi}(1-\cos 2t)\,dt = \frac{1}{2\pi}\left(t - \frac{1}{2}\sin 2t\right)\Big|_0^{\pi} = \frac{1}{2}.
\end{aligned}$$

Substituting the first few values in series (10.33), we obtain

$$f(t) \quad = \quad \frac{1}{\pi} + \frac{1}{2}\sin\frac{1\pi t}{\pi} \qquad\qquad\qquad \frac{a_0}{2} = \frac{1}{\pi},\ b_1 = \frac{1}{2}$$

$$+ 0 - \frac{2}{\pi\cdot 3}\cos\frac{2\pi t}{\pi} + 0 - \frac{2}{\pi\cdot 15}\cos\frac{4\pi t}{\pi}$$

$$+ 0 - \frac{2}{\pi\cdot 35}\cos\frac{6\pi t}{\pi} + 0 - \frac{2}{\pi\cdot 63}\cos\frac{8\pi t}{\pi} + \cdots$$

or

$$f(t) = \frac{1}{\pi} + \frac{1}{2}\sin t - \frac{2}{\pi}\left(\frac{1}{3}\cos 2t + \frac{1}{15}\cos 4t + \frac{1}{35}\cos 6t + \frac{1}{63}\cos 8t + \cdots\right).$$

Chapter 10 Review

1. $r = -\dfrac{1}{3}$, $a = 1$; $S = \dfrac{1}{1 + 1/3} = \dfrac{3}{4}$.

3. $\displaystyle\lim_{n\to\infty}\frac{2n}{4n+3} = \lim_{n\to\infty}\frac{2}{4+3/n} = \frac{1}{2}$

    Since the nth term does not approach 0, the series diverges.

5. Suppose we try the ratio test:

$$\lim_{n\to\infty}\frac{a_{n+1}}{a_n} \quad = \quad \lim_{n\to\infty}\frac{1}{(n+1)\ln^2(n+1)}\frac{n\ln^2 n}{1} = \lim_{n\to\infty}\frac{n}{n+1}\lim_{n\to\infty}\frac{\ln^2 n}{\ln^2(n+1)}$$

$$= \quad 1\cdot\lim_{n\to\infty}\frac{\ln^2 n}{\ln^2(n+1)}.$$

By L'Hospital's rule,

$$\lim_{n\to\infty}\frac{\ln^2 n}{\ln^2(n+1)} \quad = \quad \lim_{n\to\infty}\frac{(2\ln n)\frac{1}{n}}{[2\ln(n+1)]\frac{1}{n+1}} = \lim_{n\to\infty}\frac{n+1}{n}\lim_{n\to\infty}\frac{\ln n}{\ln(n+1)}$$

$$= \quad 1\cdot\lim_{n\to\infty}\frac{1/n}{1/(n+1)} = \lim_{n\to\infty}\frac{n+1}{n} = 1 \quad \text{(test fails)}.$$

By the integral test,

$$\int_2^\infty\frac{dx}{x\ln^2 x} \quad = \quad \lim_{b\to\infty}\int_2^b(\ln x)^{-2}\frac{dx}{x} \qquad u = \ln x;\ du = \frac{1}{x}\,dx.$$

$$= \quad \lim_{b\to\infty}\frac{(\ln x)^{-1}}{-1}\Bigg|_2^b = \lim_{b\to\infty}\left(-\frac{1}{\ln x}\right)\Bigg|_2^b = \lim_{b\to\infty}\left(-\frac{1}{\ln b} + \frac{1}{\ln 2}\right)$$

$$= \quad 0 + \frac{1}{\ln 2} \quad \text{(series converges)}.$$

7. $\dfrac{1}{\ln n} > \dfrac{1}{n}$ \qquad (diverges by comparison to the harmonic series).

9. $\displaystyle\lim_{n\to\infty}\frac{a_{n+1}}{a_n} = \lim_{n\to\infty}\frac{6^{n+1}}{(n+1)!}\frac{n!}{6^n} = \lim_{n\to\infty}\frac{6\cdot 6^n}{(n+1)n!}\frac{n!}{6^n} = \lim_{n\to\infty}\frac{6}{n+1} = 0 < 1$

    The series is therefore convergent by the ratio test.

11.
$$
\begin{aligned}
f(x) &= e^{-x} & f(0) &= 1 \\
f'(x) &= -e^{-x} & f'(0) &= -1 \\
f''(x) &= e^{-x} & f''(0) &= 1 \\
f'''(x) &= -e^{-x} & f'''(0) &= -1 \\
f^{(4)}(x) &= e^{-x} & f^{(4)}(0) &= 1
\end{aligned}
$$

$$
\begin{aligned}
e^{-x} &= 1 - 1x + \frac{1}{2!}x^2 + \frac{-1}{3!}x^3 + \frac{1}{4!}x^4 + \cdots \\
&= 1 - x + \frac{x^2}{2!} - \frac{x^3}{3!} + \frac{x^4}{4!} - \cdots .
\end{aligned}
$$

13. $\sin x = x - \dfrac{x^3}{3!} + \dfrac{x^5}{5!} - \cdots$. Replacing x by x^2,

$$
\sin x^2 = x^2 - \frac{(x^2)^3}{3!} + \frac{(x^2)^5}{5!} - \cdots = x^2 - \frac{x^6}{3!} + \frac{x^{10}}{5!} - \cdots .
$$

15. By (10.14)

$$
e^x = 1 + x + \frac{x^2}{2!} + \frac{x^3}{3!} + \frac{x^4}{4!} + \cdots
$$

$$
\begin{aligned}
f(x) &= \frac{1 - e^x}{x} = \frac{1}{x}\left[1 - \left(1 + x + \frac{x^2}{2!} + \frac{x^3}{3!} + \frac{x^4}{4!} + \cdots\right)\right] \\
&= \frac{1}{x}\left(-x - \frac{x^2}{2!} - \frac{x^3}{3!} - \frac{x^4}{4!} - \cdots\right) = -1 - \frac{x}{2!} - \frac{x^2}{3!} - \frac{x^3}{4!} - \cdots .
\end{aligned}
$$

17. $\cos x = 1 - \dfrac{x^2}{2!} + \dfrac{x^4}{4!} - \dfrac{x^6}{6!} + \cdots$

$$
\cos(0.5) = 1 - \frac{(0.5)^2}{2!} + \frac{(0.5)^4}{4!} \qquad \text{(three terms)}
$$

$$
= 0.877604.
$$

max. error (fourth term): $-\dfrac{(0.5)^6}{6!} = -0.000022.$

(a) $\quad$ 0.877604 $\qquad$ sum of first three terms

$\qquad \dfrac{-0.000022}{}$ $\qquad$ error

(b) $\quad$ 0.877582

The values of (a) and (b) agree to four decimal places (after rounding off): 0.8776.

19.
$$
\cos x = 1 - \frac{x^2}{2!} + \frac{x^4}{4!} - \frac{x^6}{6!} + \frac{x^8}{8!} - \cdots
$$

$$
\cos x^3 = 1 - \frac{x^6}{2!} + \frac{x^{12}}{4!} - \frac{x^{18}}{6!} + \frac{x^{24}}{8!} + \cdots
$$

$$
\begin{aligned}
\int_0^{0.9} \cos x^3\, dx &= \int_0^{0.9}\left(1 - \frac{x^6}{2!} + \frac{x^{12}}{4!} - \frac{x^{18}}{6!} + \frac{x^{24}}{8!} - \cdots\right) dx \\
&= \left. x - \frac{x^7}{7\cdot 2!} + \frac{x^{13}}{13\cdot 4!} - \frac{x^{19}}{19\cdot 6!} + \frac{x^{25}}{25\cdot 8!} + \cdots \right|_0^{0.9} \\
&= 0.9 - \frac{(0.9)^7}{7\cdot 2!} + \frac{(0.9)^{13}}{13\cdot 4!} - \frac{(0.9)^{19}}{19\cdot 6!} \\
&= 0.8666. \qquad \text{(four terms)}
\end{aligned}
$$

error (fifth term): $\dfrac{(0.9)^{25}}{25\cdot 8!} = 7.1 \times 10^{-8}.$

21.

$$e^x = 1 + x + \frac{x^2}{2!} + \frac{x^3}{3!} + \cdots$$

$$e^{-0.20t^2} = 1 - 0.20t^2 + \frac{(-0.20t^2)^2}{2!} + \frac{(-0.20t^2)^3}{3!} + \cdots$$

$$= 1 - 0.20t^2 + \frac{(0.20)^2}{2!}t^4 - \frac{(0.20)^3}{3!}t^6 + \cdots$$

$$\int_0^{0.10} \left(1 - 0.20t^2 + \frac{(0.20)^2}{2!}t^4 - \frac{(0.20)^3}{3!}t^6 + \cdots \right) dt$$

$$= t - 0.20\frac{t^3}{3} + \frac{(0.20)^2}{2!}\frac{t^5}{5} - \frac{(0.20)^3}{3!}\frac{t^7}{7}\Big|_0^{0.10}$$

$$= 0.10 - 0.20\frac{(0.10)^3}{3} + \frac{(0.20)^2}{2!}\frac{(0.10)^5}{5} \qquad \text{(three terms)}$$

$$= 0.0999 \approx 0.10 \text{ coulombs (two significant digits).}$$

error: $-\dfrac{(0.20)^3}{3!}\dfrac{(0.10)^7}{7} = -1.9 \times 10^{-11}$.

23.

$$f(x) = \sin x \qquad f\left(\frac{\pi}{4}\right) = \frac{1}{\sqrt{2}}$$

$$f'(x) = \cos x \qquad f'\left(\frac{\pi}{4}\right) = \frac{1}{\sqrt{2}}$$

$$f''(x) = -\sin x \qquad f''\left(\frac{\pi}{4}\right) = -\frac{1}{\sqrt{2}}$$

$$f'''(x) = -\cos x \qquad f'''\left(\frac{\pi}{4}\right) = -\frac{1}{\sqrt{2}}$$

$$\sin x = \frac{1}{\sqrt{2}} + \frac{1}{\sqrt{2}}\left(x - \frac{\pi}{4}\right) - \frac{1}{\sqrt{2}}\frac{1}{2!}\left(x - \frac{\pi}{4}\right)^2 - \frac{1}{\sqrt{2}}\frac{1}{3!}\left(x - \frac{\pi}{4}\right)^3 + \cdots$$

$$44° = 45° - 1° = \frac{\pi}{4} - \frac{\pi}{180}$$

Thus $x - \dfrac{\pi}{4} = \left(\dfrac{\pi}{4} - \dfrac{\pi}{180}\right) - \dfrac{\pi}{4} = -\dfrac{\pi}{180}$.

$$\sin 44° = \frac{1}{\sqrt{2}} + \frac{1}{\sqrt{2}}\left(-\frac{\pi}{180}\right) - \frac{1}{\sqrt{2}}\frac{1}{2!}\left(-\frac{\pi}{180}\right)^2 - \frac{1}{\sqrt{2}}\frac{1}{3!}\left(-\frac{\pi}{180}\right)^3 = 0.69466.$$

25. $\displaystyle \lim_{x \to 0} \frac{1 - \cos x}{x} = \lim_{x \to 0} \frac{1 - \left(1 - \dfrac{x^2}{2!} + \dfrac{x^4}{4!} - \dfrac{x^6}{6!} + \cdots \right)}{x} = \lim_{x \to 0} \left(\dfrac{x}{2!} - \dfrac{x^3}{4!} + \dfrac{x^5}{6!} - \cdots \right) = 0.$

Chapter 11

First-Order Differential Equations

11.1   What is a Differential Equation?

1. $y = 2e^{3x}$; $\dfrac{dy}{dx} = 6e^{3x}$. Substituting in $\dfrac{dy}{dx} - 3y = 0$, we get
   $6e^{3x} - 3(2e^{3x}) = 0$ so that the solution checks.

3. $y = xe^x + 2e^x$; by the product rule, $\dfrac{dy}{dx} = xe^x + e^x \cdot 1 + 2e^x = xe^x + 3e^x$.
   Substituting in the given equation, we get
   $$\begin{aligned} \dfrac{dy}{dx} - y &= e^x & \text{given equation} \\ (xe^x + 3e^x) - (xe^x + 2e^x) &= e^x. & \text{collecting terms} \end{aligned}$$
   The solution checks.

5. $\begin{aligned} y &= 2\cos 2x + 3\sin 2x \\ y' &= -4\sin 2x + 6\cos 2x \\ y'' &= -8\cos 2x - 12\sin 2x \end{aligned}$
   Substituting in the left side of $y'' + 4y = 0$, we get
   $(-8\cos 2x - 12\sin 2x) + 4(2\cos 2x + 3\sin 2x)$
   $= -8\cos 2x - 12\sin 2x + 8\cos 2x + 12\sin 2x = 0$. The solution checks.

7. $\begin{aligned} y &= c_1 e^{-3x} + c_2 e^{2x} \\ y' &= -3c_1 e^{-3x} + 2c_2 e^{2x} \\ y'' &= 9c_1 e^{-3x} + 4c_2 e^{2x} \end{aligned}$
   Substituting in the left side of the given equation, we get
   $(9c_1 e^{-3x} + 4c_2 e^{2x}) + (-3c_1 e^{-3x} + 2c_2 e^{2x}) - 6(c_1 e^{-3x} + c_2 e^{2x})$
   $= (9c_1 - 3c_1 - 6c_1)e^{-3x} + (4c_2 + 2c_2 - 6c_2)e^{2x} = 0$. The solution checks.

9. $y = x^2 - 4$, $\dfrac{dy}{dx} = 2x$.
   Substituting in the left side of the given equation, we get
   $x\dfrac{dy}{dx} - y = x(2x) - (x^2 - 4) = 2x^2 - x^2 + 4 = x^2 + 4$, which is equal to the right side.

11. $y = x + x \ln x - 1$

$\dfrac{dy}{dx} = 1 + x \cdot \dfrac{1}{x} + \ln x \cdot 1$ product rule

$\qquad = 2 + \ln x.$

Substituting in the left side of the given equation, we have

$x^2(2 + \ln x) - x(x + \ln x - 1) = 2x^2 + x^2 \ln x - x^2 - x^2 \ln x + x = x^2 + x,$ the right side.

13. $y = c \cos 3x - x \cos 3x.$ By the product rule,

$\begin{aligned}
y' &= -3c \sin 3x - x(-3 \sin 3x) - \cos 3x \cdot 1 \\
&= -3c \sin 3x + 3x \sin 3x - \cos 3x \\
y'' &= -9c \cos 3x + 3x(3 \cos 3x) + 3 \sin 3x + 3 \sin 3x \\
&= -9c \cos 3x + 9x \cos 3x + 6 \sin 3x.
\end{aligned}$

Substituting in the left side of the given equation, we get

$\begin{aligned}
y'' + 9y &= (-9c \cos 3x + 9x \cos 3x + 6 \sin 3x) + 9(c \cos 3x - x \cos 3x) \\
&= 6 \sin 3x, \text{ the right side.}
\end{aligned}$

15. $y = c_1 e^x + c_2 e^{-6x} + x e^x$

$\dfrac{dy}{dx} = c_1 e^x - 6c_2 e^{-6x} + x e^x + e^x$

$\dfrac{d^2 y}{dx^2} = c_1 e^x + 36c_2 e^{-6x} + x e^x + 2e^x$

Then

$\dfrac{d^2 y}{dx^2} + 5\dfrac{dy}{dx} - 6y = c_1 e^x + 36c_2 e^{-6x} + x e^x + 2e^x + 5c_1 e^x - 30c_2 e^{-6x} + 5x e^x$

$+ 5e^x - 6c_1 e^x - 6c_2 e^{-x} - 6x e^x = 7e^x,$ the right side.

17. $\dfrac{dy}{dx} = 3x^2$

$y = x^3 + c$ integrating

Now let $x = 2$ and $y = 5$:

$5 = 2^3 + c$ or $c = -3.$

So $y = x^3 - 3.$

19. $\dfrac{dy}{dx} = \sec^2 x$

$y = \tan x + c$ integrating

Now let $y = 1$ and $x = \dfrac{\pi}{4}$:

$1 = \tan \dfrac{\pi}{4} + c$

$1 = 1 + c$ and $c = 0.$

So $y = \tan x.$

21. $\dfrac{d^2 y}{dx^2} = e^x$

$\dfrac{dy}{dx} = e^x + c_1$ integrating

$y = e^x + c_1 x + c_2$ integrating again

Substituting $(0,0)$ and $(1,1)$, respectively, we obtain the system of equations

$0 = 1 + c_2$

$1 = e + c_1 + c_2$

Thus $c_2 = -1.$ From the second equation, we get

$1 = e + c_1 - 1$ or $c_1 = 2 - e.$

The solution is therefore given by $y = e^x + (2 - e)x - 1.$

11.2 Separation of Variables

1.
$$x^2\, dx + y\, dy = 0$$
$$\int x^2\, dx + \int y\, dy = c_1$$
$$\frac{x^3}{3} + \frac{y^2}{2} = c_1$$
$$2x^3 + 3y^2 = 6c_1 \qquad \text{multiplying by 6}$$
$$2x^3 + 3y^2 = c \quad (\text{let } c = 6c_1)$$

3.
$$(1 + x^2)\, dx = 3y\, dy$$
$$\int (1 + x^2)\, dx = \int 3y\, dy$$
$$x + \frac{1}{3}x^3 = 3\frac{y^2}{2} + c_1$$
$$6x + 2x^3 = 9y^2 + 6c_1 \qquad \text{multiplying by 6}$$
$$6x + 2x^3 = 9y^2 + c \qquad \text{letting } c = 6c_1$$

5.
$$2x\, dx + (1 + x^2)\, dy = 0$$
$$\frac{2x\, dx}{1 + x^2} + dy = 0 \qquad \text{dividing by } 1 + x^2$$
$$\int \frac{2x\, dx}{1 + x^2} + \int dy = c$$
$$\int \frac{du}{u} + \int dy = c \qquad u = 1 + x^2; \ du = 2x\, dx.$$
$$\ln(1 + x^2) + y = c$$

7.
$$(1 + x^2)\frac{dy}{dx} + y = 0$$
$$(1 + x^2)\, dy + y\, dx = 0 \qquad \text{differential form}$$
$$\frac{dy}{y} + \frac{dx}{1 + x^2} = 0 \qquad \text{dividing by } y(1 + x^2)$$
$$\int \frac{dy}{y} + \int \frac{dx}{1 + x^2} = c$$
$$\ln |y| + \operatorname{Arctan} x = c$$

9.
$$2y\, dx + 3x\, dy = 0$$
$$\frac{2\, dx}{x} + \frac{3\, dy}{y} = 0 \qquad \text{dividing by } xy$$
$$\int \frac{2\, dx}{x} + \int \frac{3\, dy}{y} = c_2 \qquad \text{Form: } \int \frac{du}{u}$$
$$2\ln |x| + 3\ln |y| = c_2 \qquad \text{Recall: } n\ln A = \ln A^n$$
$$\ln x^2 + \ln |y|^3 = \ln c_1 \qquad (\text{let } \ln c_1 = c_2, \ c_1 > 0)$$
$$\ln |x^2 y^3| = \ln c_1 \qquad \ln A + \ln B = \ln AB$$
$$|x^2 y^3| = c_1, \ c_1 > 0$$
$$x^2 y^3 = \pm c_1$$
$$x^2 y^3 = c, \ c \neq 0 \qquad (\text{let } c = \pm c_1)$$

11.
$$1 + (x^2y - x^2)\frac{dy}{dx} = 0$$
$$dx + (x^2y - x^2)\,dy = 0 \qquad \text{differential form}$$
$$dx + x^2(y - 1)\,dy = 0 \qquad \text{factoring } x^2$$
$$\frac{dx}{x^2} + (y - 1)\,dy = 0 \qquad \text{dividing by } x^2$$
$$\int x^{-2}\,dx + \int (y - 1)\,dy = c_1$$
$$\frac{x^{-1}}{-1} + \frac{1}{2}y^2 - y = c_1$$
$$-\frac{1}{x} + \frac{1}{2}y^2 - y = c_1$$
$$-2 + xy^2 - 2xy = 2xc_1 \qquad \text{multiplying by } 2x$$
$$xy^2 - 2xy - 2 = cx \qquad \text{letting } c = 2c_1$$

13. $dx - y\,dx + x\,dy = 0$
$$(1 - y)\,dx + x\,dy = 0 \qquad \text{factoring } dx$$
$$\frac{dx}{x} + \frac{dy}{1 - y} = 0 \qquad \text{dividing by } x(1 - y)$$
$$\int \frac{dx}{x} + \int \frac{dy}{1 - y} = c_1$$
$$\int \frac{dx}{x} - \int \frac{-dy}{1 - y} = c_1 \qquad u = 1 - y;\ du = -dy.$$
$$\ln|x| - \ln|1 - y| = \ln c_2,\ c_2 > 0 \qquad \text{letting } \ln c_2 = c_1$$
$$\ln\left|\frac{x}{1 - y}\right| = \ln c_2 \qquad \ln A - \ln B = \ln \frac{A}{B}$$
$$\left|\frac{x}{1 - y}\right| = c_2$$
$$\frac{x}{1 - y} = \pm c_2$$
$$\frac{x}{1 - y} = c,\ c \neq 0 \qquad \text{letting } c = \pm c_2$$
$$x = c(1 - y)$$

15.
$$\frac{dV}{dP} = -\frac{V}{P}$$
$$\frac{dV}{V} = -\frac{dP}{P} \qquad \text{separating variables}$$
$$\int \frac{dV}{V} = -\int \frac{dP}{P}$$
$$\ln|V| = -\ln|P| + c_2 \qquad \text{form: } \int \frac{du}{u} = \ln|u| + c$$
$$\ln|P| + \ln|V| = \ln c_1 \qquad \text{letting } \ln c_1 = c_2$$
$$\ln|PV| = \ln c_1 \qquad \ln A + \ln B = \ln AB$$
$$|PV| = c_1$$
$$PV = \pm c_1$$
$$PV = c \qquad \text{letting } c = \pm c_1$$

17. $\quad dx + (2\cos^2 x - y\cos^2 x)\,dy \;=\; 0$

$$dx + \cos^2 x(2 - y)\,dy \;=\; 0 \qquad\qquad \text{factoring } \cos^2 x$$

$$\frac{dx}{\cos^2 x} + (2 - y)\,dy \;=\; 0 \qquad\qquad \text{dividing by } \cos^2 x$$

$$\int \frac{dx}{\cos^2 x} + \int (2 - y)\,dy \;=\; c_1$$

$$\int \sec^2 x\,dx + \int (2 - y)\,dy \;=\; c_1 \qquad\qquad \frac{1}{\cos x} = \sec x$$

$$\tan x + 2y - \frac{1}{2}y^2 \;=\; c_1$$

$$2\tan x + 4y - y^2 \;=\; 2c_1 \qquad\qquad \text{multiplying by 2}$$

$$2\tan x + 4y - y^2 \;=\; c \qquad\qquad \text{letting } c = 2c_1$$

19. $$\cos^2 t + y\csc t\,\frac{dy}{dt} \;=\; 0$$

$$\cos^2 t\,dt + y\csc t\,dy \;=\; 0$$

$$\frac{\cos^2 t\,dt}{\csc t} + y\,dy \;=\; 0 \qquad\qquad \text{dividing by } \csc t$$

$$\int \cos^2 t\sin t\,dt + \int y\,dy \;=\; c_1 \qquad\qquad \frac{1}{\csc t} = \sin t$$

$$-\int (\cos t)^2(-\sin t\,dt) + \int y\,dy \;=\; c_1 \qquad\qquad u = \cos t;\ du = -\sin t\,dt.$$

$$-\int u^2\,du + \int y\,dy \;=\; c_1$$

$$-\frac{1}{3}u^3 + \frac{1}{2}y^2 \;=\; c_1$$

$$-\frac{1}{3}\cos^3 t + \frac{1}{2}y^2 \;=\; c_1$$

$$3y^2 - 2\cos^3 t \;=\; 6c_1 \qquad\qquad \text{multiplying by 6}$$

$$3y^2 - 2\cos^3 t \;=\; c \qquad\qquad \text{letting } c = 6c_1$$

21. $$\sqrt{v^2 + 1}\,dt + vt^2\,dv \;=\; 0$$

$$\frac{dt}{t^2} + \frac{v\,dv}{\sqrt{v^2 + 1}} \;=\; 0 \qquad\qquad \text{dividing by } t^2\sqrt{v^2 + 1}$$

$$\int \frac{dt}{t^2} + \int \frac{v\,dv}{(v^2 + 1)^{1/2}} \;=\; c$$

$$\int t^{-2}\,dt + \int (v^2 + 1)^{-1/2}v\,dv \;=\; c \qquad\qquad u = v^2 + 1;\ du = 2v\,dv.$$

$$\int t^{-2}\,dt + \frac{1}{2}\int (v^2 + 1)^{-1/2}(2v\,dv) \;=\; c$$

$$\frac{t^{-1}}{-1} + \frac{1}{2}\frac{(v^2 + 1)^{1/2}}{1/2} \;=\; c$$

$$-\frac{1}{t} + \sqrt{v^2 + 1} \;=\; c$$

$$-1 + t\sqrt{v^2 + 1} \;=\; ct \qquad\qquad \text{multiplying by } t$$

23. $T_1\, dT_1 + (\csc T_1 + T_2 \csc T_1)\, dT_2 \;=\; 0$

$\qquad T_1\, dT_1 + \csc T_1 (1 + T_2)\, dT_2 \;=\; 0$

$\qquad \dfrac{T_1\, dT_1}{\csc T_1} + (1 + T_2)\, dT_2 \;=\; 0$ $\qquad$ dividing by $\csc T_1$

$\qquad \displaystyle\int T_1 \sin T_1\, dT_1 + \int (1 + T_2)\, dT_2 \;=\; c_1$ $\qquad \dfrac{1}{\csc T_1} = \sin T_1$

$\qquad\qquad\qquad u \;=\; T_1 \qquad\qquad dV \;=\; \sin T_1\, dT_1$

$\qquad\qquad\qquad du \;=\; dT_1 \qquad\qquad v \;=\; -\cos T_1$ $\qquad$ integration by parts

$-T_1 \cos T_1 + \displaystyle\int \cos T_1\, dT_1 + T_2 + \dfrac{1}{2}T_2^2 \;=\; c_1$

$\qquad -T_1 \cos T_1 + \sin T_1 + T_2 + \dfrac{1}{2}T_2^2 \;=\; c_1$

$\qquad 2 \sin T_1 - 2 T_1 \cos T_1 + 2 T_2 + T_2^2 \;=\; c$

25. $(y^2 - 1) \cos x\, dx + 2y \sin x\, dy \;=\; 0$

$\qquad \dfrac{\cos x}{\sin x}\, dx + \dfrac{2y}{y^2 - 1}\, dy \;=\; 0$ $\qquad$ separating variables

$\qquad \ln|\sin x| + \ln|y^2 - 1| \;=\; \ln c_1,\ c_1 > 0$ $\qquad$ integrating

$\qquad \ln|(\sin x)(y^2 - 1)| \;=\; \ln c_1$ $\qquad \ln A + \ln B = \ln AB$

$\qquad |(y^2 - 1)\sin x| \;=\; c_1,\ c_1 > 0$

$\qquad (y^2 - 1)\sin x \;=\; \pm c_1$

$\qquad (y^2 - 1)\sin x \;=\; c,\ c \neq 0$

27. $x e^y\, dx + e^{-x}\, dy \;=\; 0$ $\qquad$ multiply by $e^{-y} e^x$

$\qquad \displaystyle\int x e^x\, dx + \int e^{-y}\, dy \;=\; c$

$\qquad\qquad u \;=\; x \qquad\qquad dv \;=\; e^x\, dx$

$\qquad\qquad du \;=\; dx \qquad\qquad v \;=\; e^x$ $\qquad$ integration by parts

$\qquad x e^x - \displaystyle\int e^x\, dx - e^{-y} \;=\; c$

$\qquad x e^x - e^x - e^{-y} \;=\; c$

29. $(e^x \tan y + \tan y)\dfrac{dy}{dx} + e^x \;=\; 0$

$\qquad (e^x + 1)\tan y\, dy + e^x\, dx \;=\; 0$

$\qquad \displaystyle\int \tan y\, dy + \int \dfrac{e^x}{e^x + 1}\, dx \;=\; \ln c_1$ $\qquad u = e^x + 1;\ du = e^x\, dx.$

$\qquad \ln|\sec y| + \ln(e^x + 1) \;=\; \ln c_1,\ c_1 > 0$ $\qquad$ integrating

$\qquad \ln|(e^x + 1)\sec y| \;=\; \ln c_1$ $\qquad \ln A + \ln B = \ln AB$

$\qquad |(e^x + 1)\sec y| \;=\; c_1,\ c_1 > 0$

$\qquad (e^x + 1)\sec y \;=\; \pm c_1$

$\qquad (e^x + 1)\sec y \;=\; c,\ c \neq 0$

31.
$$y\frac{dy}{dx} + 2x\sec y = 0$$
$$y\,dy + 2x\sec y\,dx = 0$$
$$\frac{y\,dy}{\sec y} + 2x\,dx = 0 \qquad \text{dividing by } \sec y$$
$$\int y\cos y\,dy + \int 2x\,dx = c \qquad \frac{1}{\sec y} = \cos y$$

$$
\begin{array}{cccc}
u &=& y & dv &=& \cos y\,dy \\
du &=& dy & v &=& \sin y
\end{array}
\qquad \text{integration by parts}
$$

$$y\sin y - \int \sin y\,dy + x^2 = c$$
$$y\sin y + \cos y + x^2 = c$$

33. $x\,dy - y\,dx = 0, \; y = 2 \text{ when } x = 1$
$$\frac{dy}{y} - \frac{dx}{x} = 0$$
$$\ln y - \ln x = \ln c \qquad \text{integrating}$$
$$\ln\left(\frac{y}{x}\right) = \ln c$$
$$\frac{y}{x} = c$$

Thus $y = cx$. Substituting the given values, we get $2 = c \cdot 1$ or $y = 2x$.

35. $(y+2)\,dx + (x-3)\,dy = 0$
$$\frac{dx}{x-3} + \frac{dy}{y+2} = 0$$
$$\ln|x-3| + \ln|y+2| = \ln c_1 \qquad \text{form: } \int \frac{du}{u}$$
$$\ln|(x-3)(y+2)| = \ln c_1 \qquad \ln A + \ln B = \ln AB$$
$$|(x-3)(y+2)| = c_1$$
$$(x-3)(y+2) = \pm c_1 = c$$
$$(2-3)(5+2) = c \text{ or } c = -7 \qquad x = 2, \; y = 5$$
$$(y+2)(x-3) = -7$$

37. $dx + x\tan y\,dy = 0, \; y = 0 \text{ when } x = 1$
$$\frac{dx}{x} + \tan y\,dy = 0$$
$$\ln x + \ln \sec y = \ln c \qquad \text{integrating}$$
$$\ln x\sec y = \ln c \qquad \ln A + \ln B = \ln AB$$
$$x\sec y = c$$

If $x = 1$ and $y = 0$, we get $1 \cdot \sec 0 = 1 = c$, so that
$$x\sec y = 1$$
$$x = \frac{1}{\sec y}$$
$$x = \cos y$$

11.3   First-Order Linear Differential Equations

1. Step 1. $\dfrac{dy}{dx} + 1y = 1$ already in standard form

    Step 2. $I.F. = e^{\int 1\,dx} = e^{x}$

    Step 3. $e^{x}\left(\dfrac{dy}{dx} + y\right) = e^{x} \cdot 1$ multiplying by e^{x}

    Step 4. $\dfrac{d}{dx}(ye^{x}) = e^{x}$ by (11.9)

    Step 5. $ye^{x} \;=\; \displaystyle\int e^{x}\,dx = e^{x} + c$

    $\qquad\quad y \;=\; 1 + ce^{-x}$

3. Step 1. $\dfrac{dy}{dx} - 2y = 3e^{3x}$ already in standard form

    Step 2. $I.F. = e^{\int (-2)\,dx} = e^{-2x}$

    Step 3. $e^{-2x}\left(\dfrac{dy}{dx} - 2y\right) = e^{-2x}e^{3x}$ multiplying by $I.F.$

    Step 4. $\dfrac{d}{dx}\left(ye^{-2x}\right) = e^{x}$ $\dfrac{d}{dx}(y \cdot I.F.)$

    Step 5. $ye^{-2x} \;=\; e^{x} + c$ multiplying by e^{2x}

    $\qquad\quad y \;=\; e^{3x} + ce^{2x}$

5. $2\dfrac{dy}{dx} - 8xy = e^{2x^{2}}$

    Step 1. $\dfrac{dy}{dx} - 4xy = \dfrac{1}{2}e^{2x^{2}}$

    Step 2. $I.F. = e^{\int (-4x)\,dx} = e^{-2x^{2}}$

    Step 3. $e^{-2x^{2}}\left(\dfrac{dy}{dx} - 4xy\right) = e^{-2x^{2}} \cdot \dfrac{1}{2}e^{2x^{2}} = \dfrac{1}{2}$

    Step 4. $\dfrac{d}{dx}(ye^{-2x^{2}}) = \dfrac{1}{2}$

    Step 5. $ye^{-2x^{2}} \;=\; \dfrac{1}{2}x + c$

    $\qquad\quad y \;=\; \left(\dfrac{1}{2}x + c\right)e^{2x^{2}}$

7. $\dfrac{1}{2}\dfrac{dy}{dx} + y\cos x = \cos x$

    Step 1. $\dfrac{dy}{dx} + 2y\cos x = 2\cos x$

    Step 2. $I.F. = e^{\int 2\cos x\,dx} = e^{2\sin x}$

    Step 3. $e^{2\sin x}\left(\dfrac{dy}{dx} + 2y\cos x\right) = e^{2\sin x} \cdot 2\cos x$ multiplying by $I.F.$

    Step 4.   The left becomes the derivative of the product of
    y and $I.F.$

    $\qquad \dfrac{d}{dx}\left(ye^{2\sin x}\right) = e^{2\sin x} \cdot 2\cos x$

    Step 5. $ye^{2\sin x} = \displaystyle\int e^{2\sin x} \cdot 2\cos x\,dx = e^{2\sin x} + c$ $u = 2\sin x;\ du = 2\cos x\,dx.$

    $\qquad y = 1 + ce^{-2\sin x}$ multiplying by $e^{-2\sin x}$

9. $x\,dy + (y - x)\,dx = 0$

Step 1. $\quad x\dfrac{dy}{dx} + y - x \;=\; 0$

$\qquad\qquad \dfrac{dy}{dx} + \left(\dfrac{1}{x}\right)y \;=\; 1$

Step 2. $\quad I.F. = e^{\int (1/x)\,dx} = e^{\ln x} = x \qquad$ by(11.12)

Step 3. $\quad x\left(\dfrac{dy}{dx} + \dfrac{y}{x}\right) = x \qquad\qquad$ multiplying by $I.F.$

Step 4. $\quad \dfrac{d}{dx}(xy) = x \qquad\qquad\qquad$ derivative of the product of y and $I.F.$

Step 5. $\quad xy \;=\; \dfrac{1}{2}x^2 + c \qquad\qquad$ integrating

$\qquad\qquad y \;=\; \dfrac{1}{2}x + \dfrac{c}{x}$

11. $y' = e^x - \dfrac{y}{x}$

Step 1. $\quad y' + \dfrac{1}{x}y = e^x$

Step 2. $\quad I.F. = e^{\int (1/x)\,dx} = e^{\ln x} = x \qquad\qquad\qquad$ by (11.12)

Step 3. $\quad x\left(y' + \dfrac{1}{x}y\right) = xe^x \qquad\qquad\qquad$ multiplying by $I.F.$

Step 4. $\quad \dfrac{d}{dx}(xy) = xe^x \qquad\qquad\qquad\qquad \dfrac{d}{dx}(y \cdot I.F.)$

Step 5. $\quad xy = \displaystyle\int xe^x\,dx$

$\qquad$ Ingegrating by parts: $\quad u \;=\; x \qquad dv \;=\; e^x\,dx$

$\qquad\qquad\qquad\qquad\qquad\qquad du \;=\; dx \qquad v \;=\; e^x$

$\qquad xy \;=\; xe^x - \displaystyle\int e^x\,dx$

$\qquad xy \;=\; xe^x - e^x + c$

13. $\dfrac{dy}{dx} - \dfrac{2y}{x} - x^2 \sec^2 x = 0$

Step 1. $\quad \dfrac{dy}{dx} - \dfrac{2}{x}y = x^2 \sec^2 x$

Step 2. $\quad I.F. \;=\; e^{\int (-2/x)\,dx} = e^{-2\ln x}$

$\qquad\qquad\quad =\; e^{\ln x^{-2}} = x^{-2} = \dfrac{1}{x^2}$

Step 3. $\quad \dfrac{1}{x^2}\left(\dfrac{dy}{dx} - \dfrac{2}{x}y\right) = \dfrac{1}{x^2}(x^2 \sec^2 x) \qquad$ multiplying by $\dfrac{1}{x^2}$

Step 4. $\quad \dfrac{d}{dx}\left(\dfrac{1}{x^2}y\right) = \sec^2 x \qquad\qquad$ derivative of the product of y and $I.F.$

Step 5. $\quad \dfrac{1}{x^2}y \;=\; \tan x + c \qquad\qquad$ integrating

$\qquad\qquad y \;=\; x^2 \tan x + cx^2$

15. $xy' = 3y + x^5 \sin x$

Step 1. $y' - \dfrac{3}{x}y = x^4 \sin x$

Step 2. $I.F. = e^{\int (-3/x)\,dx} = e^{-3\ln x} = e^{\ln x^{-3}} = x^{-3} = \dfrac{1}{x^3}$

Step 3. $\dfrac{1}{x^3}\left(y' - \dfrac{3}{x}y\right) = \dfrac{1}{x^3}\cdot x^4 \sin x$ multiplying by $I.F.$

Step 4. $\dfrac{d}{dx}\left(y\cdot\dfrac{1}{x^3}\right) = x\sin x$ $\dfrac{d}{dx}(y\cdot I.F.)$

Step 5. $y\dfrac{1}{x^3} = \displaystyle\int x\sin x\,dx$

Integration by parts: $u = x$ $dv = \sin x\,dx$

                               $du = dx$ $v = -\cos x$

$$\dfrac{y}{x^3} = -x\cos x + \int \cos x\,dx$$

$$= -x\cos x + \sin x + c$$

$$y = x^3 \sin x - x^4 \cos x + cx^3$$

17. $xy' - 2y = x^3 e^x$

Step 1. $\dfrac{dy}{dx} - \dfrac{2}{x}y = x^2 e^x$

Step 2. $I.F. = e^{\int (-2/x)\,dx} = e^{-2\ln x} = e^{\ln x^{-2}} = x^{-2} = \dfrac{1}{x^2}$

Step 3. $\dfrac{1}{x^2}\left(\dfrac{dy}{dx} - \dfrac{2}{x}y\right) = \dfrac{1}{x^2}(x^2 e^x)$ multiplying by $I.F.$

Step 4. $\dfrac{d}{dx}\left(\dfrac{1}{x^2}y\right) = e^x$

Step 5. $\dfrac{1}{x^2}y = e^x + c$ integrating

           $y = x^2 e^x + cx^2$

19. $(y-1)\sin x\,dx + dy = 0$

Step 1. $\dfrac{dy}{dx} + y\sin x = \sin x$

Step 2. $I.F. = e^{\int \sin x\,dx} = e^{-\cos x}$

Step 3. $e^{-\cos x}\left(\dfrac{dy}{dx} + y\sin x\right) = e^{-\cos x}\sin x$ multiplying by $I.F.$

Step 4. $\dfrac{d}{dx}(ye^{-\cos x}) = e^{-\cos x}\sin x$ $\dfrac{d}{dx}(y\cdot I.F.)$

Step 5. $ye^{-\cos x} = \displaystyle\int e^{-\cos x}\sin x\,dx = e^{-\cos x} + c$ $u = -\cos x;\ du = \sin x\,dx.$

       $y = 1 + ce^{\cos x}$

21. $y' - y\tan x - \cos x = 0$

Step 1. $\dfrac{dy}{dx} - (\tan x)y = \cos x$

Step 2. $I.F. = e^{-\int \tan x\,dx} = e^{\ln\cos x} = \cos x$

Step 3. $\cos x\left(\dfrac{dy}{dx} - y\tan x\right) = \cos^2 x$ multiplying by $I.F.$

Step 4. $\dfrac{d}{dx}(y\cos x) = \dfrac{1}{2}(1 + \cos 2x)$ half-angle formula

Step 5. $y\cos x = \dfrac{1}{2}\displaystyle\int(1 + \cos 2x)\,dx$ $u = 2x;\ du = 2\,dx.$

      $y\cos x = \dfrac{1}{2}x + \dfrac{1}{4}\sin 2x + c$

      $4y\cos x = 2x + \sin 2x + c$

23. $(x+1)y' + y = \dfrac{x+1}{x-1}$

   Step 1. $y' + \dfrac{1}{x+1}y = \dfrac{1}{x-1}$

   Step 2. $I.F. = e^{\int dx/(x+1)} = e^{\ln(x+1)} = x + 1$

   Step 3. $(x+1)\left(y' + \dfrac{1}{x+1}y\right) = \dfrac{x+1}{x-1}$

   Step 4. $\dfrac{d}{dx}[y(x+1)] = \dfrac{x+1}{x-1}$

   Step 5. $y(x+1) = \displaystyle\int \dfrac{x+1}{x-1}\,dx = \int \dfrac{x-1+2}{x-1}\,dx = \int \left(1 + \dfrac{2}{x-1}\right)dx$

   $\qquad y(x+1) = x + 2\ln|x-1| + c$

25. $y' - \dfrac{1}{x}y = x^2 \sin x^2$ already in standard form (Step 1).

   Step 2. $I.F. = e^{\int(-1/x)\,dx} = e^{-\ln x} = e^{\ln x^{-1}} = x^{-1} = \dfrac{1}{x}$

   Step 3. $\dfrac{1}{x}\left(\dfrac{dy}{dx} - \dfrac{1}{x}y\right) = x\sin x^2$ $\qquad\qquad$ multiplying by $\dfrac{1}{x}$

   Step 4. $\dfrac{d}{dx}\left(\dfrac{y}{x}\right) = x\sin x^2$

   Step 5. $\dfrac{y}{x} \;=\; \displaystyle\int x\sin x^2\,dx$

   $\qquad\qquad\;=\; \dfrac{1}{2}\displaystyle\int \sin x^2 (2x)\,dx \qquad u = x^2;\ du = 2x\,dx.$

   $\qquad \dfrac{y}{x} \;=\; -\dfrac{1}{2}\cos x^2 + c$

   $\qquad\quad y \;=\; -\dfrac{1}{2}x\cos x^2 + cx$

27. $y' + y\tan t = \sec t$

   Step 1. $y' + y\tan t = \sec t$ $\qquad\qquad$ already in standard form

   Step 2. $I.F. = e^{\int \tan t\,dt} = e^{\ln\sec t} = \sec t$

   Step 3. $\sec t(y' + y\tan t) = \sec^2 t$

   Step 4. $\dfrac{d}{dt}(y\sec t) = \sec^2 t$

   Step 5. $y\sec t = \tan t + c$

29. $t\dfrac{dr}{dt} + r = t\ln t$

   Step 1. $\dfrac{dr}{dt} + \dfrac{1}{t}r = \ln t$ $\qquad\qquad$ dividing by t

   Step 2. $I.F. = e^{\int(1/t)\,dt} = e^{\ln t} = t$

   Step 3. $t\left(\dfrac{dr}{dt} + \dfrac{1}{t}r\right) = t\ln t$ $\qquad\qquad$ multiplying by $I.F.$

   Step 4. $\dfrac{d}{dt}(tr) = t\ln t$

   Step 5.   Integrating by parts: $u = \ln t \quad dv = t\,dt$

   $\qquad\qquad\qquad\qquad\qquad du = \dfrac{1}{t}\,dt \quad v = \dfrac{1}{2}t^2$

   $\qquad tr \;=\; \dfrac{1}{2}t^2\ln t - \displaystyle\int \dfrac{1}{t}\cdot\dfrac{1}{2}t^2\,dt$

   $\qquad tr \;=\; \dfrac{1}{2}t^2\ln t - \displaystyle\int \dfrac{1}{2}t\,dt$

   $\qquad tr \;=\; \dfrac{1}{2}t^2\ln t - \dfrac{1}{4}t^2 + c$

   $\qquad\; r \;=\; \dfrac{1}{2}t\ln t - \dfrac{1}{4}t + \dfrac{c}{t}$ $\qquad\qquad$ dividing by t

31. $s\dfrac{dr}{ds} - r = s^3 e^{3s}$

Step 1. $\dfrac{dr}{ds} - \dfrac{1}{s}r = s^2 e^{3s}$

Step 2. $I.F. = e^{\int (-1/s)\,ds} = e^{-\ln s} = e^{\ln s^{-1}} = s^{-1} = \dfrac{1}{s}$

Step 3. $\dfrac{1}{s}\left(\dfrac{dr}{ds} - \dfrac{1}{s}r\right) = se^{3s}$

Step 4. $\dfrac{d}{ds}\left(r \cdot \dfrac{1}{s}\right) = se^{3s}$

Step 5. $r \cdot \dfrac{1}{s} = \displaystyle\int se^{3s}\,ds$

$$\text{Integration by parts:} \quad u = s \quad dv = e^{3s}\,ds$$
$$du = ds \quad v = \tfrac{1}{3}e^{3s}$$

$$
\begin{aligned}
r \cdot \frac{1}{s} &= \frac{1}{3}se^{3s} - \frac{1}{3}\int e^{3s}\,ds \\
&= \frac{1}{3}se^{3s} - \frac{1}{9}e^{3s} + c \\
r &= \frac{1}{3}s^2 e^{3s} - \frac{1}{9}se^{3s} + cs
\end{aligned}
$$

33. $\dfrac{dy}{dx} + y = 6e^{-x}$, $y = 2$ when $x = 0$.

$I.F. = e^x$

$$
\begin{aligned}
e^x\left(\frac{dy}{dx} + y\right) &= e^x(6e^{-x}) &&\text{multiplying by } I.F. \\
\frac{d}{dx}(ye^x) &= 6 \\
ye^x &= 6x + c &&\text{integrating} \\
y &= 6xe^{-x} + ce^{-x} &&\text{multiplying by } e^{-x} \\
2 &= 0 + c \cdot 1 \text{ or } c = 2 &&y = 2,\ x = 0 \\
y &= 6xe^{-x} + 2e^{-x} \\
y &= 2e^{-x}(3x + 1)
\end{aligned}
$$

35. $dy + x^2 y\,dx = 2x^2\,dx$

Step 1. $\dfrac{dy}{dx} + x^2 y = 2x^2$

Step 2. $I.F. = e^{\int x^2\,dx} = e^{(1/3)x^3}$

Step 3. $e^{(1/3)x^3}\left(\dfrac{dy}{dx} + x^2 y\right) = e^{(1/3)x^3} \cdot 2x^2$

Step 4. $\dfrac{d}{dx}\left(ye^{(1/3)x^3}\right) = e^{(1/3)x^3} \cdot 2x^2$

Step 5. $ye^{(1/3)x^3} = 2\displaystyle\int e^{(1/3)x^3} x^2\,dx$ $\qquad\qquad\qquad u = \dfrac{1}{3}x^3;\ du = x^2\,dx.$

$ye^{(1/3)x^3} = 2e^{(1/3)x^3} + c$

$y = 2 + ce^{(-1/3)x^3}$

Now let $y = 3$ and $x = 0$:

$3 = 2 + c$ and $c = 1$; so $y = 2 + e^{(-1/3)x^3}$.

37.
$$L\frac{di}{dt} + Ri = E$$

$$\frac{di}{dt} + \frac{R}{L}i = \frac{E}{L} \qquad \text{dividing by } L$$

$$e^{(R/L)t}\left(\frac{di}{dt} + \frac{R}{L}i\right) = \frac{E}{L}e^{(R/L)t} \qquad I.F. = e^{\int (R/L)\,dt}$$

$$\frac{d}{dt}\left(ie^{(R/L)t}\right) = \frac{E}{L}e^{(R/L)t}$$

$$ie^{Rt/L} = \frac{E}{L}\cdot\frac{L}{R}e^{(R/L)t} + c \qquad u = \frac{R}{L}t;\ du = \frac{R}{L}\,dt.$$

$$i = \frac{E}{R} + ce^{-Rt/L}$$

From the condition $i = 0$ when $t = 0$, $c = -E/R$. So
$$i = \frac{E}{R}\left(1 - e^{-Rt/L}\right).$$

11.4   Applications of First-Order Differential Equations

1.
$$L\frac{di}{dt} + Ri = e(t)$$

$$0.2\frac{di}{dt} + 5i = 5$$

$$\frac{di}{dt} + \frac{5}{0.2}i = \frac{5}{0.2}$$

$$\frac{di}{dt} + 25i = 25 \qquad I.F. = e^{25t}$$

$$e^{25t}\left(\frac{di}{dt} + 25i\right) = 25e^{25t}$$

$$\frac{d}{dt}(ie^{25t}) = 25e^{25t}$$

$$ie^{25t} = \int 25e^{25t}\,dt \qquad u = 25t;\ du = 25\,dt.$$

$$ie^{25t} = e^{25t} + c$$

$$i = 1 + ce^{-25t}$$

Initial condition: if $t = 0$, $i = 0$. Thus $0 = 1 + c$ or $c = -1$, and $i = 1 - e^{-25t}$.

3. $\dfrac{dN}{dt} = kN$ by (11.13).

Step 1. $\dfrac{dN}{dt} - kN = 0$

Step 2. $I.F. = e^{-kt}$

Step 3. $e^{-kt}\left(\dfrac{dN}{dt} - kN\right) = 0$

Step 4. $\dfrac{d}{dt}\left(Ne^{-kt}\right) = 0$

Step 5. $Ne^{-kt} = c$

$N = ce^{kt}$

Given: when $t = 0$, $N = 100\,\text{g}$; $100 = ce^0 = c$. So $N = 100e^{kt}$. To evaluate k, we use the second pair of values: when $t = 10$, $N = 80\,\text{g}$: $80 = 100e^{10k}$ or $e^{10k} = 0.8$. Taking natural logarithms,

$$\ln e^{10k} = \ln(0.8)$$
$$10k \ln e = \ln(0.8) \qquad\qquad \ln e = 1$$

and
$$k = \frac{\ln(0.8)}{10} = -0.0223.$$

The solution is $N = 100e^{-0.0223t}$.

5. By Example 1, $N = N_0 e^{kt}$. Since the half-life is 5.27 years, we have $N = \dfrac{N_0}{2}$ when $t = 5.27$:

$$\frac{1}{2} N_0 = N_0 e^{5.27k}$$
$$\frac{1}{2} = e^{5.27k}$$
$$\ln \frac{1}{2} = \ln e^{5.27k} = 5.27 k \ln e = 5.27 k \text{ (since } \ln e = 1).$$

So, $k = \dfrac{\ln(1/2)}{5.27} = -0.1315$. Solution: $N = N_0 e^{-0.1315t}$.

If 80% of the initial amount has decayed, then 20% is left. So we need to find t such that $N = 0.20 N_0$:

$$0.20 N_0 = N_0 e^{-0.1315t}$$
$$0.20 = e^{-0.1315t}$$
$$\ln(0.20) = \ln e^{-0.1315t} = -0.1315 t \ln e = -0.1315 t$$
$$t = \frac{\ln(0.20)}{-0.1315} = 12.2 \text{ years.}$$

7. $\dfrac{dN}{dt} = kN$; by Exercise 3, $N = ce^{kt}$. If N_0 is the initial quantity, then $N_0 = ce^0 = c$ and $N = N_0 e^{kt}$.

If 1% of the given quantity has decayed, then 99% is left: $N = 0.99 N_0$ when $t = 20$ years. So

$$0.99 N_0 = N_0 e^{20k}$$
$$0.99 = e^{20k}$$
$$\ln 0.99 = \ln e^{2k} = 20 k \ln e$$
$$k = \frac{\ln 0.99}{20} = -0.0005025 \qquad \ln e = 1$$

and
$$N = N_0 e^{-0.0005025t}.$$

The half-life is the time t corresponding to $N = \dfrac{1}{2} N_0$:

$$\frac{1}{2} N_0 = N_0 e^{-0.0005025t}$$
$$\frac{1}{2} = e^{-0.0005025t}$$
$$\ln \frac{1}{2} = -0.0005025 t \ln e$$
$$t = \frac{\ln(1/2)}{-0.0005025} = 1379.4 \approx 1380 \text{ years.}$$

9. By Example 1, $N = N_0 e^{kt}$. Since the half-life is 5600 years, we have $N = \frac{1}{2}N_0$ when $t = 5600$ years:

$$\frac{1}{2}N_0 = N_0 e^{5600k}$$

$$\frac{1}{2} = e^{5600k}$$

$$\ln\frac{1}{2} = \ln e^{5600k} = 5600k \text{ (since } \ln e = 1)$$

$$k = \frac{\ln(1/2)}{5600} = -0.000124.$$

Solution: $N = N_0 e^{-0.000124t}$. Now find t such that $N = 0.25N_0$:

$$0.25N_0 = N_0 e^{-0.000124t}$$

$$0.25 = e^{-0.000124t} \qquad\qquad \text{dividing by } N_0$$

$$\ln(0.25) = -0.000124t$$

$$t = \frac{\ln(0.25)}{-0.000124} \approx 11,200 \text{ years.}$$

11. If $\dfrac{dS}{dt} = Sr$, then

Step 1. $\dfrac{dS}{dt} - rS = 0$

Step 2. $I.F. = e^{-rt}$

Step 3. $e^{-rt}\left(\dfrac{dS}{dt} - rS\right) = 0$

Step 4. $\dfrac{d}{dt}\left(Se^{-rt}\right) = 0$

Step 5. $Se^{-rt} = c$

             $S = ce^{rt}.$

When $t = 0$, $S = S_0$, so that $c = S_0$ and $S = S_0 e^{rt}$.

(a) $S = 1000e^{0.0775(10)} = \$2170.59$

(b) $S = 100e^{rt}$. $S = \$200$ when $t = 10$; so $200 = 100e^{10r}$ and $2 = e^{10r}$.

$$\ln 2 = \ln e^{10r} = 10r \ln e = 10r \cdot 1$$

$$\ln 2 = 10r$$

$$r = \frac{\ln 2}{10} = 0.069 = 6.9\%$$

(c) $S_0 = Se^{-rt}$

$$S_0 = 500e^{-0.08(5)} = \$335.16$$

13. Since $T_m = 70°$, the equation is

$$\frac{dT}{dt} = -k(T - 70)$$

$$\frac{dT}{dt} + kT = 70k \qquad\qquad I.F. = e^{kt}$$

$$\frac{d}{dt}(Te^{kt}) = 70ke^{kt}$$

$$Te^{kt} = 70\int e^{kt}(k\,dt) \qquad u = kt;\ du = k\,dt.$$

$$Te^{kt} = 70e^{kt} + c$$

$$T = 70 + ce^{-kt}.$$

Initial condition: if $t = 0$, $T = 20$: $20 = 70 + c$ or $c = -50$ and $T = 70 - 50e^{-kt}$.

From the second condition, $T = 35$ when $t = 2$, we have

$$35 = 70 - 50e^{-2k}$$

$$-35 = -50e^{-2k}$$

$$\frac{7}{10} = e^{-2k}$$

$$\ln\left(\frac{7}{10}\right) = -2k \text{ or } k = -\frac{1}{2}\ln\left(\frac{7}{10}\right) = 0.1783.$$

Solution: $T = 70 - 50e^{-0.1783t}$. Now find t such that $T = 69°$:

$$69 = 70 - 50e^{-0.1783t}$$

$$\frac{1}{50} = e^{-0.1783t}$$

$$\ln\left(\frac{1}{50}\right) = -0.1783t$$

$$t = \frac{\ln(1/50)}{-0.1783} = 21.9\,\text{min}.$$

15. Since $T_m = 60°\,\text{F}$, the equation is $\dfrac{dT}{dt} = -k(T - 60)$.

Step 1. $\qquad\qquad\qquad \dfrac{dT}{dt} + kT = 60k$

Step 2. $\qquad\qquad\qquad\quad I.F. = e^{kt}$

Step 3. $\qquad\qquad e^{kt}\left(\dfrac{dT}{dt} + kT\right) = e^{kt} \cdot 60k$

Step 4. $\qquad\qquad\quad \dfrac{d}{dt}\left(Te^{kt}\right) = e^{kt} \cdot 60k$

Step 5. $\quad Te^{kt} = 60\displaystyle\int e^{kt} k\,dt \qquad u = kt;\ du = k\,dt.$

$$= 60e^{kt} + c$$

$$T = 60 + ce^{-kt}$$

When $t = 0$, $T = -15°\,\text{F}$: $-15 = 60 + ce^0$ or $c = -75$.

So $T = 60 - 75e^{-kt}$. From the second condition, $T = -5°$ when $t = 3\,\text{min}$, we obtain

$$-5 = 60 - 75e^{-3k}$$

$$\ln\frac{-5 - 60}{-75} = \ln e^{-3k} = -3k$$

$$k = 0.0477$$

Solution: $T = 60 - 75e^{-0.0477t}$. Now find t such that $T = 50°\,\text{F}$:

$$50 = 60 - 75e^{-0.0477t}$$

$$\frac{50 - 60}{-75} = e^{-0.0477t}$$

$$\ln\frac{50 - 60}{-75} = \ln e^{-0.0477t} = -0.0477t$$

$$t = 42.2\,\text{min}.$$

17. From the equation $m\dfrac{dv}{dt} = mg - kv$ we get

$$10\frac{dv}{dt} = 100 - 0.2v$$

$$\frac{dv}{dt} + \frac{0.2}{10}v = 10$$

$$\frac{dv}{dt} + 0.02v = 10 \qquad\qquad I.F. = e^{0.02t}$$

$$\frac{d}{dt}(ve^{0.02t}) = 10e^{0.02t} \qquad\qquad u = 0.02t; \ du = 0.02\,dt.$$

$$ve^{0.02t} = \frac{10}{0.02}e^{0.02t} + c = 500e^{0.02t} + c$$

$$v = 500 + ce^{-0.02t}.$$

Initial condition: if $t = 0$, $v = 0$, or $0 = 500 + c$, and $c = -500$.

$v = 500 - 500e^{-0.02t} = 500(1 - e^{-0.02t})$

If $t = 10$, then $v = 500(1 - e^{-0.2}) = 91\,\text{m/s}$.

Also, $\lim\limits_{t\to\infty} v = \lim\limits_{t\to\infty} 500(1 - e^{-0.02t}) = 500\,\text{m/s}$.

19. The equation has the form

$$m\frac{dv}{dt} = w - 4v^2$$

$$10\frac{dv}{dt} = 100 - 4v^2$$

$$\frac{10\,dv}{100 - 4v^2} = dt \qquad\qquad \text{separating variabes}$$

$$\frac{10\,dv}{25 - v^2} = 4\,dt \qquad\qquad \text{multiplying both sides by 4}$$

$$\frac{10}{(5 + v)(5 - v)} = 4\,dt \qquad\qquad \text{factoring}$$

The left side is now split into partial fractions:

$$\frac{10}{(5 + v)(5 - v)} = \frac{A}{5 + v} + \frac{B}{5 - v} = \frac{A(5 - v) + B(5 + v)}{(5 + v)(5 - v)}$$

$$A(5 - v) + B(5 + v) = 10$$

$\underline{v = -5}$: $A(10) + 0 = 10$ and $A = 1$

$\underline{v = 5}$: $0 + B(10) = 10$ and $B = 1$

$$\int\left(\frac{1}{5 + v} + \frac{1}{5 - v}\right)dv = \int 4\,dt$$

$$\ln|5 + v| - \ln|5 - v| = 4t + c$$

$$\ln\left|\frac{5 + v}{5 - v}\right| = 4t + c$$

Initial condition: when $t = 0$, $v = 0$; thus $\ln|1| = 0 + c$ or $c = 0$.

Hence $\ln\left|\dfrac{5 + v}{5 - v}\right| = 4t$. Since v starts at 0, it must remain less than 5 to avoid division by 0. As a result, $(5 + v)/(5 - v)$ is always positive, and the solution can be written without absolute value signs: $\ln\dfrac{5 + v}{5 - v} = 4t$.

Another form is obtained by solving for v:

$$e^{4t} = \frac{5 + v}{5 - v} \quad\text{and}\quad \frac{5 - v}{5 + v} = e^{-4t}$$

$$5e^{-4t} + ve^{-4t} = 5 - v$$

$$ve^{-4t} + v = 5 - 5e^{-4t}$$

$$v = \frac{5(1 - e^{-4t})}{1 + e^{-4t}}$$

Finally, as $t \to \infty$, $v \to 5\,\text{m/s}$.

21.
$$\frac{dx}{dt} = kx$$

$$\frac{dx}{dt} - kx = 0 \qquad\qquad I.F. = e^{-kt}$$

$$\frac{d}{dt}(xe^{-kt}) = 0$$

$$xe^{-kt} = c \text{ and } x = ce^{kt}.$$

If $x = x_0$ when $t = 0$, then $c = x_0$. The solution is
$x = x_0 e^{kt}$.

From the given condition, $x = \frac{3}{4}x_0$ when $t = 10$, we have

$$\frac{3}{4}x_0 = x_0 e^{10k}$$

$$\ln\left(\frac{3}{4}\right) = 10k \text{ or } k = \frac{\ln(3/4)}{10} = -0.02877.$$

Thus, $x = x_0 e^{-0.02877t}$. If $x = \frac{1}{10}x_0$, then   one-tenth unconverted

$$\frac{1}{10}x_0 = x_0 e^{-0.02877t}$$

$$\ln\left(\frac{1}{10}\right) = -0.02877t$$

$$t = \frac{\ln(1/10)}{-0.02877} = 80\,\text{s}.$$

23. $x^2 + y^2 = c^2$; differentiating implicitly,

$$2x + 2y\frac{dy}{dx} = 0$$

$$\frac{dy}{dx} = -\frac{x}{y}.$$

The orthogonal trajectories must therefore satisfy the following condition:

$$\frac{dy}{dx} = \frac{y}{x} \qquad\qquad \text{negative reciprocal}$$

$$\frac{dy}{y} = \frac{dx}{x}$$

$$\ln|y| = \ln|x| + \ln k_1$$

$$\ln\left|\frac{y}{x}\right| = \ln k_1 \text{ or } \left|\frac{y}{x}\right| = k_1.$$

It follows that $\frac{y}{x} = \pm k_1$ and $y = kx$.

25. $y^2 = 4px$. Differentiating implicitly, we get $2y\frac{dy}{dx} = 4p$.

From the first equation, $4p = \frac{y^2}{x}$. Substituting in the second equation, we get

$$2y\frac{dy}{dx} = \frac{y^2}{x} \text{ or } \frac{dy}{dx} = \frac{y}{2x}.$$

So the orthogonal trajectories satisfy the condition

$$\frac{dy}{dx} = -\frac{2x}{y} \qquad\qquad \text{negative reciprocal}$$

$$y\,dy = -2x\,dx$$

$$\frac{y^2}{2} = -x^2 + k_1.$$

Thus $2x^2 + y^2 = k$.

27. $y = ce^{-2x}$; differentiating, $\dfrac{dy}{dx} = c(-2e^{-2x})$.

To eliminate c_1 observe that from the first equation, $c = ye^{2x}$. Substituting in the second equation, we get $\dfrac{dy}{dx} = ye^{2x}(-2e^{-2x}) = -2y$.

For the orthogonal trajectories, dy/dx is equal to the negative reciprocal:

$\dfrac{dy}{dx} = \dfrac{1}{2y}$ or $2y\,dy = dx$ and $y^2 = x + k$.

29.
$$\begin{aligned}
xy &= c \qquad \text{given family} \\
y &= \frac{c}{x} \\
\frac{dy}{dx} &= -\frac{c}{x^2}
\end{aligned}$$

Substituting $c = xy$, we get $\dfrac{dy}{dx} = -\dfrac{xy}{x^2} = -\dfrac{y}{x}$. The orthogonal family therefore satisfies the condition

$$\begin{aligned}
\frac{dy}{dx} &= \frac{x}{y} \qquad \text{negative reciprocal} \\
y\,dy &= x\,dx \\
\frac{1}{2}y^2 &= \frac{1}{2}x^2 + k_1 \\
y^2 &= x^2 + 2k_1 \\
y^2 - x^2 &= k.
\end{aligned}$$

31. Since $M = 10000$, the equation is

$$\begin{aligned}
\frac{dN}{dt} &= kN(10000 - N) \\
\frac{dN}{N(10000 - N)} &= k\,dt.
\end{aligned}$$

To integrate, we need to split the left side into partial fractions:

$$\begin{aligned}
\frac{1}{N(10000 - N)} &= \frac{A}{N} + \frac{B}{10000 - N} \\
&= \frac{A(10000 - N) + BN}{N(10000 - N)} \\
A(10000 - N) + BN &= 1.
\end{aligned}$$

$\underline{N = 0}:$ $A(10000) + 0 = 1$ and $A = \dfrac{1}{10000}$.

$\underline{N = 10000}:$ $0 + B(10000) = 1$ and $B = \dfrac{1}{10000}$.

Integrating,

$$\begin{aligned}
\frac{1}{10000}\int \left(\frac{1}{N} + \frac{1}{10000 - N}\right) dN &= \int k\,dt \\
\frac{1}{10000}\left[\ln N - \ln(10000 - N)\right] &= kt + c_1 \\
\ln\frac{N}{10000 - N} &= 10000kt + \ln c.
\end{aligned}$$

(Here $10000c_1$ was replaced by $\ln c$.)

Initial condition: when $t = 0, N = 1000$.

$\ln\dfrac{1000}{10000 - 1000} = 0 + \ln c$ or $\ln c = \ln\dfrac{1}{9}$.

This yields $\ln\dfrac{N}{10000 - N} = 10000kt + \ln\dfrac{1}{9}$, or $\ln\dfrac{9N}{10000 - N} = 10000kt$.

From the second condition (when $t = 1,\ N = 2000$), we get

$\ln\dfrac{9(2000)}{10000 - 2000} = 10000k$ or $k = \dfrac{1}{10000}\ln\dfrac{9}{4}$.

The solution is therefore

$\ln\dfrac{9N}{10000 - N} = 10000\dfrac{1}{10000}\left(\ln\dfrac{9}{4}\right)t$ or $\ln\dfrac{9N}{10000 - N} = \left(\ln\dfrac{9}{4}\right)t$.

This solution can be put into an explicit form by solving for N:

$$e^{[\ln(9/4)]t} = \frac{9N}{10000 - N}$$

$$10000e^{[\ln(9/4)]t} - Ne^{[\ln(9/4)]t} = 9N$$

$$N = \frac{10000e^{[\ln(9/4)]t}}{9 + e^{[\ln(9/4)]t}} \cdot \frac{e^{-[\ln(9/4)]t}}{e^{-[\ln(9/4)]t}}$$

$$N = \frac{10000}{1 + 9e^{-[\ln(9/4)]t}}.$$

33. $\dfrac{dx}{dt}$ = rate of gain−rate of loss.

rate of gain $= \left(0.2\dfrac{\text{lb}}{\text{gal}}\right)\left(4\dfrac{\text{gal}}{\text{min}}\right) = 0.8\dfrac{\text{lb}}{\text{min}}$

rate of loss $= \left(\dfrac{x\,\text{lb}}{200\,\text{gal}}\right)\left(4\dfrac{\text{gal}}{\text{min}}\right) = \dfrac{x}{50}\dfrac{\text{lb}}{\text{min}}$

Thus $\dfrac{dx}{dt} = 0.8 - \dfrac{x}{50} = \dfrac{40 - x}{50}$. Separating variables

$$\frac{dx}{40 - x} = \frac{dt}{50}$$

$$\frac{-dx}{40 - x} = -\frac{dt}{50} \qquad u = 40 - x;\ \ du = -dx.$$

$$\ln(40 - x) = -\frac{1}{50}t + \ln c$$

$$\ln\frac{40 - x}{c} = -\frac{1}{50}t$$

$$e^{-t/50} = \frac{40 - x}{c}$$

$$ce^{-t/50} = 40 - x$$

$$x = 40 - ce^{-t/50}.$$

When $t = 0$, $x = 20$: $20 = 40 - c$, or $c = 20$: $x(t) = 40 - 20e^{-t/50}$.
When $t = 30\,\text{min}$, $x = 29\,\text{lb}$.

35. $\dfrac{dx}{dt}$ = rate of gain − rate of loss.
rate of gain $= 0$

rate of loss $= \left(\dfrac{x\,\text{lb}}{200\,\text{gal}}\right)\left(4\dfrac{\text{gal}}{\text{min}}\right) = \dfrac{x}{50}\dfrac{\text{lb}}{\text{min}}$

The equation is

$$\frac{dx}{dt} = -\frac{x}{50} \quad \text{or} \quad \frac{dx}{x} = -\frac{1}{50}\,dt$$

$$\ln x = -\frac{1}{50}t + \ln c$$

$$\ln\frac{x}{c} = -\frac{1}{50}t$$

$$\frac{x}{c} = e^{-t/50}$$

$$x = ce^{-t/50}.$$

When $t = 0$, $x = 20$: $20 = ce^0 = c$.
$x(t) = 20e^{-t/50}$ and $x(30) = 11.0\,\text{lb}$.

11.5   Numerical Solutions

1. (a) $dy = (1 - y)\, dx$, initial conditions: $x = 0$, $y = 2$, $dx = 0.05$.

$$
\begin{array}{c|l}
x & y \\
\hline
0 & 2 \\
0.05 & (1 - y_0)\, dx + y_0 = (1 - 2)(0.05) + 2 = 1.95 \\
0.10 & (1 - y_1)\, dx + y_1 = (1 - 1.95)(0.05) + 1.95 = 1.9025 \\
0.15 & (1 - y_2)\, dx + y_2 = (1 - 1.9025)(0.05) + 1.9025 = 1.857375 \\
\text{etc.} &
\end{array}
$$

To obtain the solution using a spreadsheet, enter 0 in cell $A1$ and 2 in cell $B1$. To generate the x-values, enter +A1+0.05 in cell $A2$. Now copy the contents of this cell, mark the block $A3 - A41$, and paste.

For the corresponding y-values, enter the following expression in cell $B2$:

    +(1-B1)*0.05+B1

Now copy the contents of this cell, mark the block $B3 - B41$, and paste.

3. (a) $dy = (x - y)\, dx$, initial conditions: $x = 0$, $y = 3$, $dx = 0.01$.

$$
\begin{array}{c|l}
x & y \\
\hline
0 & 3 \\
0.01 & (x_0 - y_0)\, dx + y_0 = (0 - 3)(0.01) + 3 = 2.97 \\
0.02 & (x_1 - y_1)\, dx + y_1 = (0.01 - 2.97)(0.01) + 2.97 = 2.9404 \\
0.03 & (x_2 - y_2)\, dx + y_2 = (0.02 - 2.9404)(0.01) + 2.9404 = 2.911196
\end{array}
$$

5. $dy = (y^2 - x^2)\, dx$, initial conditions: $x = 0$, $y = 1$, $dx = 0.01$.

$$
\begin{array}{c|l}
x & y \\
\hline
0 & 1 \\
0.01 & (y_0^2 - x_0^2)\, dx + y_0 = (1^2 - 0^2)(0.01) + 1 = 1.01 \\
0.02 & (1.01^2 - 0.01^2)(0.01) + 1.01 = 1.0202 \\
0.03 & (1.0202^2 - 0.02^2)(0.01) + 1.0202 = 1.0306041 \\
\text{etc.} &
\end{array}
$$

To obtain the solution using a spreadsheet, enter 0 in cell $A1$ and 1 in cell $B1$. To generate the x-values, enter +A1+0.01 in cell $A2$. Now copy the contents of this cell, mark the block $A3 - A51$, and paste.

For the corresponding y-values, enter the following expression in cell $B2$:

    +(B1^2-A1^2)*0.01+B1

Now copy the contents of this cell, mark the block $B3 - B51$, and paste.

7. (b) $dy = e^{-xy}\, dx$, initial conditions: $x = 0$, $y = 0$, $dx = 0.01$.

$$
\begin{array}{c|l}
x & y \\
\hline
0 & 0 \\
0.01 & e^{-x_0 y_0}\, dx + y_0 = 1(0.01) + 0 = 0.01 \\
0.02 & e^{-x_1 y_1}\, dx + y_1 = e^{-(0.01)(0.01)}(0.01) + 0.01 = 0.019999 \\
0.03 & e^{-x_2 y_2}\, dx + y_2 = e^{-(0.02)(0.019999)}(0.01) + 0.019999 = 0.029995
\end{array}
$$

9. We start with $(x_0, y_0) = (0, 2)$ and calculate y_1:

$$\frac{dy}{dx} = 1 - y$$

$$x_0 = 0$$

$$y_0 = 2$$

$$h = 0.05$$

$$K_1 = f(x_0, y_0) = 1 - 2 = -1$$

$$K_2 = f\left(x_0 + \frac{1}{2}h, y_0 + \frac{1}{2}hK_1\right) = 1 - \left[2 + \frac{1}{2}(0.05)(-1)\right] = -0.975$$

$$K_3 = f\left(x_0 + \frac{1}{2}h, y_0 + \frac{1}{2}hK_2\right) = 1 - \left[2 + \frac{1}{2}(0.05)(-0.975)\right] = -0.975625$$

$$K_4 = f(x_0 + h, y_0 + hK_3) = 1 - [2 + 0.05(-0.975625)] = -0.951219$$

$$y_1 = y_0 + \frac{1}{6}(0.05)(K_1 + 2K_2 + 2K_3 + K_4) = 1.951229$$

11. To illustrate the procedure, we calculate y_1 and y_2, starting with $(x_0, y_0) = (0, 3)$.

$$\frac{dy}{dx} = x - y$$

$$x_0 = 0$$

$$y_0 = 3$$

$$h = 0.01$$

$$K_1 = 0 - 3 = -3$$

$$K_2 = 0 + \frac{1}{2}(0.01) - \left[3 + \frac{1}{2}(0.01)(-3)\right] = -2.98$$

$$K_3 = 0 + \frac{1}{2}(0.01) - \left[3 + \frac{1}{2}(0.01)(-2.98)\right] = -2.9801$$

$$K_4 = 0 + 0.01 - [3 + 0.01(-2.9801)] = -2.960199$$

$$y_1 = y_0 + \frac{1}{6}(0.01)(K_1 + 2K_2 + 2K_3 + K_4) = 2.970199$$

$$x_1 = 0.01$$

$$y_1 = 2.970199$$

$$h = 0.01$$

$$K_1 = 0.01 - 2.970199 = -2.960199$$

$$K_2 = 0.01 + \frac{1}{2}(0.01) - \left[2.970199 + \frac{1}{2}(0.01)(-2.960199)\right] = -2.940398$$

$$K_3 = 0.01 + \frac{1}{2}(0.01) - \left[2.970199 + \frac{1}{2}(0.01)(-2.940398)\right] = -2.940497$$

$$K_4 = 0.01 + 0.01 - [2.970199 + 0.01(-2.940497)] = -2.920794$$

$$y_2 = y_1 + \frac{1}{6}(0.01)(K_1 + 2K_2 + 2K_3 + K_4) = 2.940794$$

13. To illustrate the procedure, we calculate y_1 and y_2, starting with $(x_0, y_0) = (0, 1)$.

$$\frac{dy}{dx} = y^2 - x^2, \; dx = 0.05$$

$$x_0 = 0$$

$$y_0 = 1$$

$$h = 0.05$$

$$K_1 = 1^2 - 0^2 = 1$$

$$K_2 = \left[1 + \frac{1}{2}(0.05)(1)\right]^2 - \left[0 + \frac{1}{2}(0.05)\right]^2 = 1.05$$

$$K_3 = \left[1 + \frac{1}{2}(0.05)(1.05)\right]^2 - \left[0 + \frac{1}{2}(0.05)\right]^2 = 1.05256406$$

$$K_4 = \left[1 + 0.05(1.05256406)\right]^2 - (0 + 0.05)^2 = 1.10552613$$

$$y_1 = y_0 + \frac{1}{6}(0.05)(K_1 + 2K_2 + 2K_3 + K_4) = 1.05258879$$

$$x_1 = 0.05$$

$$y_1 = 1.05258879$$

$$h = 0.05$$

$$K_1 = 1.05258879^2 - 0.05^2 = 1.10544316$$

$$K_2 = \left[1.05258879 + \frac{1}{2}(0.05)(1.10544316)\right]^2 - \left[0.05 + \frac{1}{2}(0.05)\right]^2 = 1.16126077$$

$$K_3 = \left[1.05258879 + \frac{1}{2}(0.05)1.16126077)\right]^2 - \left[0.05 + \frac{1}{2}(0.05)\right]^2 = 1.16427749$$

$$K_4 = \left[1.05258879 + 0.05(1.16427749)\right]^2 - (0.05 + 0.05)^2 = 1.22388256$$

$$y_2 = y_1 + \frac{1}{6}(0.05)(K_1 + 2K_2 + 2K_3 + K_4) = 1.110759$$

Chapter 11 Review

1. $y' = x - y$

   Step 1. $\dfrac{dy}{dx} + y = x$

   Step 2. $I.F. = e^x$

   Step 3. $e^x\left(\dfrac{dy}{dx} + y\right) = xe^x$

   Step 4. $\dfrac{d}{dx}(ye^x) = xe^x$

   Step 5. $ye^x = \displaystyle\int xe^x \, dx$

   (Integration by parts) $\quad \begin{array}{llll} u &=& x & dv &=& e^x \, dx \\ du &=& dx & v &=& e^x \end{array}$

   $$ye^x = xe^x - \int e^x \, dx = xe^x - e^x + c$$

   $$y = x - 1 + ce^{-x}$$

3. $y' = x - 2xy$. As a linear equation:

Step 1. $\dfrac{dy}{dx} + 2xy = x$

Step 2. $I.F. = e^{\int 2x\,dx} = e^{x^2}$

Step 3. $e^{x^2}\left(\dfrac{dy}{dx} + 2xy\right) = xe^{x^2}$

Step 4. $\dfrac{d}{dx}\left(ye^{x^2}\right) = xe^{x^2}$

Step 5.
$$
\begin{aligned}
ye^{x^2} &= \int xe^{x^2}\,dx \\
&= \frac{1}{2}\int e^{x^2}(2x\,dx) \qquad u = x^2;\ du = 2x\,dx.\\
&= \frac{1}{2}e^{x^2} + c \\
y &= \frac{1}{2} + ce^{-x^2}
\end{aligned}
$$

By separation of variables:

$$
\begin{aligned}
\frac{dy}{dx} &= x(1 - 2y) \\
\frac{dy}{1 - 2y} &= x\,dx \qquad u = 1 - 2y;\ du = -2y\,dy.\\
\frac{-2\,dy}{1 - 2y} &= -2x\,dx \qquad \text{multiplying both sides by } -2\\
\ln|1 - 2y| &= -x^2 + c \qquad \text{form: } \int \frac{du}{u}
\end{aligned}
$$

To show that the two solutions agree, we need to solve the latter for y in terms of x. First replace c by $\ln c_1$ and observe that $|1 - 2y| = |2y - 1|$.

$$
\begin{aligned}
\ln|2y - 1| &= -x^2 + \ln c_1 \\
\ln\frac{|2y - 1|}{c_1} &= -x^2 \\
\frac{|2y - 1|}{c_1} &= e^{-x^2} \\
|2y - 1| &= c_1 e^{-x^2} \\
2y - 1 &= \pm c_1 e^{-x^2} \\
y &= \frac{1}{2} \pm \frac{1}{2}c_1 e^{-x^2} \\
y &= \frac{1}{2} + ce^{-x^2} \qquad \text{replacing } \pm\frac{1}{2}c_1 \text{ by } c
\end{aligned}
$$

5.
$$
\begin{aligned}
(1 + y^2)\,dx + (x^2 y + y)\,dy &= 0 \\
(1 + y^2)\,dx + y(x^2 + 1)\,dy &= 0 \\
\frac{dx}{1 + x^2} + \frac{y\,dy}{1 + y^2} &= 0 \qquad u = 1 + y^2;\ du = 2y\,dy.\\
\frac{2\,dx}{1 + x^2} + \frac{2y\,dy}{1 + y^2} &= 0 \qquad \text{multiplying by } 2\\
\int \frac{2\,dx}{1 + x^2} + \int \frac{2y\,dy}{1 + y^2} &= c \\
2\operatorname{Arctan} x + \ln(1 + y^2) &= c
\end{aligned}
$$

7. $$2y\,dx + x\,dy = 0$$
$$\frac{2\,dx}{x} + \frac{dy}{y} = 0$$
$$2\ln|x| + \ln|y| = \ln c_1$$
$$\ln|x^2| + \ln|y| = \ln c_1$$
$$\ln|x^2 y| = \ln c_1$$
$$|x^2 y| = c_1$$
$$x^2 y = \pm c_1 \text{ and } x^2 y = c$$

9. $(x^4 + 2y)\,dx - x\,dy = 0$. This equation is linear:

Step 1. $\quad -x\,dy + 2y\,dx = -x^4\,dx$
$$\frac{dy}{dx} - \left(\frac{2}{x}\right)y = x^3$$

Step 2. $\quad I.F. = e^{-\int (2/x)\,dx} = e^{-2\ln x} = e^{\ln x^{-2}} = x^{-2}$

Step 3. $\quad x^{-2}\left[\frac{dy}{dx} - \left(\frac{2}{x}\right)y\right] = x^{-2}x^3$

Step 4. $\quad \frac{d}{dx}(x^{-2}y) = x^{-2}x^3 = x$

Step 5. $\quad x^{-2}y = \frac{1}{2}x^2 + c$
$$y = \frac{1}{2}x^4 + cx^2$$

11. $$x\sin^2 y\,dx - \cot y\,dy = 0$$
$$x\,dx - \frac{\cot y}{\sin^2 y}\,dy = 0$$
$$x\,dx - \frac{\cos y}{\sin y}\frac{1}{\sin^2 y} = 0$$
$$\int x\,dx - \int (\sin y)^{-3}\cos y\,dy = c_1 \qquad u = \sin y;\ du = \cos y\,dy.$$
$$\int x\,dx - \int u^{-3}\,du = c_1$$
$$\frac{1}{2}x^2 - \frac{u^{-2}}{-2} = c_1$$
$$x^2 + \frac{1}{u^2} = 2c_1$$
$$x^2 + \frac{1}{\sin^2 y} = 2c_1$$
$$x^2 + \csc^2 y = c$$

13. As a linear equation:
$$\frac{dy}{dx} + (\sec x)y = 0 \qquad I.F. = e^{\int \sec x\,dx} = e^{\ln(\sec x + \tan x)} = \sec x + \tan x$$
$$\frac{d}{dx}[y(\sec x + \tan x)] = 0$$
$$y(\sec x + \tan x) = c$$

Separation of variables:
$$dy + y\sec x\,dx = 0$$
$$\frac{dy}{y} + \sec x\,dx = 0$$
$$\ln y + \ln(\sec x + \tan x) = \ln c \qquad \text{integrating}$$
$$\ln y(\sec x + \tan x) = \ln c$$
$$y(\sec x + \tan x) = c.$$

Since $y = 2$ when $x = \dfrac{\pi}{4}$, we get

$$2\left[\sec\left(\frac{\pi}{4}\right) + \tan\left(\frac{\pi}{4}\right)\right] = c$$

$$2\left(\sqrt{2}+1\right) = c.$$

Hence, $y(\sec x + \tan x) = 2\left(\sqrt{2}+1\right).$

15. $\dfrac{dN}{dt} = kN$. By Exercise 7, Section 11.4, $N = N_0 e^{kt}$.

Given: when $t = 2\,\text{h}$, $N = 2N_0$.

$$2N_0 = N_0 e^{2k}$$
$$2 = e^{2k}$$
$$\ln 2 = \ln e^{2k} = 2k \ln e = 2k \cdot 1 = 2k$$
$$k = \frac{1}{2}\ln 2$$

Solution: $N = N_0 e^{[(\ln 2)/2]t}$.

Now find t such that $N = 3N_0$:

$$3N_0 = N_0 e^{[(\ln 2)/2]t}$$
$$3 = e^{[(\ln 2)/2]t}$$
$$\ln 3 = \left(\frac{1}{2}\ln 2\right)t$$

and

$$t = \frac{\ln 3}{\frac{1}{2}\ln 2} \approx 3.2\,\text{h}.$$

17. Since $T_m = 65°\,\text{F}$, the equation is

$$\frac{dT}{dt} = -k(T - 65)$$

$$\frac{dT}{dt} + kT = 65k \qquad\qquad I.F. = e^{kt}$$

$$\frac{d}{dt}(Te^{kt}) = 65ke^{kt}$$

$$Te^{kt} = 65e^{kt} + c$$

$$T = 65 + ce^{-kt}.$$

If $t = 0$, $T = T_0$ (unknown initial temperature). Thus $T_0 = 65 + c$ or $c = T_0 - 65$.

$T = 65 + (T_0 - 65)e^{-kt}$

If $t = 15$, then $T = 0$, and if $t = 30$, then $T = 20$:

$$0 = 65 + (T_0 - 65)e^{-15k}$$

$$20 = 65 + (T_0 - 65)e^{-30k}$$

$$\overline{\qquad\qquad\qquad\qquad\qquad\qquad\qquad}$$

$$-65e^{15k} = T_0 - 65 \qquad\qquad (1)$$

$$-45e^{30k} = T_0 - 65$$

$$\overline{\qquad\qquad\qquad\qquad\qquad\qquad\qquad}$$

$$-65e^{15k} + 45e^{30k} = 0 \qquad\qquad \text{(subtracting)}$$

$$-65 + 45e^{15k} = 0 \qquad\qquad \text{(dividing by } e^{15k})$$

$$e^{15k} = \frac{65}{45}$$

$$15k = \ln\left(\frac{65}{45}\right)$$

$$k = \frac{1}{15}\ln\left(\frac{65}{45}\right) = 0.0245.$$

So from Equation (1), $T_0 = 65 - 65e^{15k} = -29°\,\text{F}.$

19. $m\dfrac{dv}{dt} = w - kv$. Given: $m = 6\,\text{kg}$, $w = 60\,\text{N}$, and $k = 1$.

$6\dfrac{dv}{dt} = 60 - v$

Step 1. $\dfrac{dv}{dt} + \dfrac{1}{6}v = 10$

Step 2. $\text{I.F.} = e^{(1/6)t}$

Step 3. $e^{(1/6)t}\left(\dfrac{dv}{dt} + \dfrac{1}{6}v\right) = 10e^{(1/6)t}$

Step 4. $\dfrac{d}{dt}\left[ve^{(1/6)t}\right] = 10e^{(1/6)t}$

Step 5. $ve^{(1/6)t} = 10\displaystyle\int e^{(1/6)t}\,dt \qquad\qquad u = \dfrac{1}{6}t;\ \ du = \dfrac{1}{6}\,dt.$

$\qquad\qquad ve^{(1/6)t} = 10\cdot 6\displaystyle\int e^{(1/6)t}\left(\dfrac{1}{6}\,dt\right)$

$\qquad\qquad\qquad = 60e^{(1/6)t} + c$

$\qquad\qquad v = 60 + ce^{-(1/6)t}$

Initial condition: $v = 0$ when $t = 0$.

$0 = 60 + c$ or $c = -60$

Solution: $v = 60 - 60e^{-(1/6)t} = 60\left(1 - e^{-(1/6)t}\right)$.

21. Differentiating implicitly, we get $2x - 4y\dfrac{dy}{dx} = 0$ or $\dfrac{dy}{dx} = \dfrac{x}{2y}$.

Condition for orthogonal trajectories:

$\dfrac{dy}{dx} = -\dfrac{2y}{x}$

$\dfrac{dy}{y} = -\dfrac{2\,dx}{x}$

$\ln y = -2\ln x + \ln k$

$\ln y + 2\ln x = \ln k$

$\ln y + \ln x^2 = \ln k$, whence $x^2 y = k$.

Chapter 12

Higher-Order Linear Differential Equations

12.1 Higher-Order Homogeneous Differential Equations

1. $m^2 - 13m + 42 = 0$

   $(m-6)(m-7) = 0$

   $m = 6, 7$

   $y = c_1 e^{6x} + c_2 e^{7x}$

3. $6m^2 - m - 2 = 0$

   $(3m-2)(2m+1) = 0$

   $m = \dfrac{2}{3}, -\dfrac{1}{2}$

   $y = c_1 e^{2x/3} + c_2 e^{-x/2}$

5. $4m^2 + 7m - 2 = 0$

   $(4m-1)(m+2) = 0$

   $m = \dfrac{1}{4}, -2$

   $y = c_1 e^{x/4} + c_2 e^{-2x}$

7. $m^2 - m - 1 = 0$; by the quadratic formula,

$$m = \frac{-(-1) \pm \sqrt{(-1)^2 - 4(1)(-1)}}{2 \cdot 1} = \frac{1 \pm \sqrt{5}}{2}$$

$$= \frac{1}{2} \pm \frac{\sqrt{5}}{2}.$$

$$\begin{aligned} y &= c_1 e^{\left(1/2 + \sqrt{5}/2\right)x} + c_2 e^{\left(1/2 - \sqrt{5}/2\right)x} \\ &= c_1 e^{(1/2)x} e^{\left(\sqrt{5}/2\right)x} + c_2 e^{(1/2)x} e^{-\left(\sqrt{5}/2\right)x} \\ &= e^{x/2}\left(c_1 e^{\sqrt{5}x/2} + c_2 e^{-\sqrt{5}x/2}\right) \end{aligned}$$

9. $2m^2 - 3m + 1 = 0$

$(2m - 1)(m - 1) = 0$

$$m = 1, \frac{1}{2}$$

$y = c_1 e^x + c_2 e^{x/2}$

11. $m^2 - 9 = 0;\ m^2 = 9;\ m = \pm 3$

$$y = c_1 e^{3x} + c_2 e^{-3x}$$

$$Dy = y' = 3c_1 e^{3x} - 3c_2 e^{-3x}$$

Now substitute the initial conditions:

0	$=$	$c_1 + c_2$	$y = 0$ when $x = 0$
6	$=$	$3c_1 - 3c_2$	$Dy = 6$ when $x = 0$

0	$=$	$c_1 + c_2$	first equation
2	$=$	$c_1 - c_2$	second equation
2	$=$	$2c_1$	adding

$c_1 = 1$ and $c_2 = -1$.

Substituting in the general solution: $y = e^{3x} - e^{-3x}$.

13. $m^2 - m - 2 = 0$

$(m - 2)(m + 1) = 0$

$$m = 2, -1$$

$y = c_1 e^{2x} + c_2 e^{-x}$

Substituting $(0, 0)$ and $(1, 1)$, respectively, we get

(1) $0 = c_1 + c_2$ $y = 0,\ x = 0$

(2) $1 = c_1 e^2 + c_2 e^{-1}$ $y = 1,\ x = 1$

From the first equation, $c_1 = -c_2$. Substituting in the second equation, we get

$1 = -c_2 e^2 + c_2 e^{-1}$

$1 = c_2 \left(-e^2 + e^{-1} \right)$

$c_2 = \dfrac{1}{e^{-1} - e^2} = \dfrac{e}{1 - e^3}$

Thus $c_1 = -\dfrac{e}{1 - e^3}$. So the solution becomes

$$y = -\frac{e}{1 - e^3} e^{2x} + \frac{e}{1 - e^3} e^{-x} = \frac{e}{1 - e^3} \left(e^{-x} - e^{2x} \right).$$

15. $m^3 - 7m + 6 = 0$. Possible rational roots: $\pm 1,\ \pm 2,\ \pm 3,\ \pm 6$.

$1 + 0 - 7 + 6 \underline{)1}$

$\underline{+1 + 1 - 6}$

$1 + 1 - 6 + 0$ $x = 1$ is a root.

$m^2 + m - 6 = 0$

$(m - 2)(m + 3) = 0$

$$m = 2, -3 \quad \text{(remaining roots)}$$

$y = c_1 e^x + c_2 e^{2x} + c_3 e^{-3x}$

17. $m^3 - m^2 - 4m - 2 = 0$. Possible rational roots: ± 1, ± 2.

$$1 - 1 - 4 - 2)\underline{-1}$$
$$\frac{-1 + 2 + 2}{1 - 2 - 2 + 0} \qquad x = -1 \text{ is a root.}$$
$$m^2 - 2m - 2 = 0$$

$$
\begin{aligned}
m &= \frac{-(-2) \pm \sqrt{(-2)^2 - 4(1)(-2)}}{2 \cdot 1} \\
&= \frac{2 \pm \sqrt{4 + 8}}{2} = \frac{2 \pm 2\sqrt{3}}{2} = 1 \pm \sqrt{3}
\end{aligned}
$$

So the roots are: -1, $1 \pm \sqrt{3}$.

$$y = c_1 e^{-x} + c_2 e^{(1+\sqrt{3})x} + c_2 e^{(1-\sqrt{3})x}$$

19. $m^2 + m - 1 = 0$

$$m = \frac{-1 \pm \sqrt{1^2 - 4(1)(-1)}}{2} = \frac{-1 \pm \sqrt{5}}{2} = -\frac{1}{2} \pm \frac{\sqrt{5}}{2}$$

$$
\begin{aligned}
y &= c_1 e^{(-1/2 + \sqrt{5}/2)x} + c_2 e^{(-1/2 - \sqrt{5}/2)x} \\
&= c_1 e^{(-1/2)x} e^{(\sqrt{5}/2)x} + c_2 e^{(-1/2)x} e^{-(\sqrt{5}/2)x} \\
&= e^{-x/2} \left(c_1 e^{\sqrt{5}\,x/2} + c_2 e^{-\sqrt{5}\,x/2} \right)
\end{aligned}
$$

21. $m^2 - 2m - 2 = 0$

$$
\begin{aligned}
m &= \frac{-(-2) \pm \sqrt{(-2)^2 - 4(1)(-2)}}{2 \cdot 1} \\
&= \frac{2 \pm \sqrt{4 + 8}}{2} = \frac{2 \pm 2\sqrt{3}}{2} = \frac{2(1 \pm \sqrt{3})}{2} = 1 \pm \sqrt{3}
\end{aligned}
$$

$$
\begin{aligned}
y &= c_1 e^{(1+\sqrt{3})x} + c_2 e^{(1-\sqrt{3})x} \\
y &= c_1 e^{x + \sqrt{3}\,x} + c_2 e^{x - \sqrt{3}\,x} \\
y &= c_1 e^{x} e^{\sqrt{3}\,x} + c_2 e^{x} e^{-\sqrt{3}\,x} \\
y &= e^{x} \left(c_1 e^{\sqrt{3}\,x} + c_2 e^{-\sqrt{3}\,x} \right)
\end{aligned}
$$

23. $m^2 - 4m - 2 = 0$

$$
\begin{aligned}
m &= \frac{-(-4) \pm \sqrt{(-4)^2 - 4(1)(-2)}}{2} = \frac{4 \pm \sqrt{24}}{2} = \frac{4 \pm \sqrt{4 \cdot 6}}{2} \\
&= \frac{4 \pm 2\sqrt{6}}{2} = \frac{2\left(2 \pm \sqrt{6}\right)}{2} = 2 \pm \sqrt{6}
\end{aligned}
$$

$$
\begin{aligned}
y &= c_1 e^{(2+\sqrt{6})x} + c_2 e^{(2-\sqrt{6})x} \\
&= c_1 e^{2x} e^{\sqrt{6}\,x} + c_2 e^{2x} e^{-\sqrt{6}\,x} \\
&= e^{2x} \left(c_1 e^{\sqrt{6}\,x} + c_2 e^{-\sqrt{6}\,x} \right)
\end{aligned}
$$

25. $m^2 + 6m - 6 = 0$

$$m = \frac{-6 \pm \sqrt{6^2 - 4(1)(-6)}}{2 \cdot 1}$$

$$= \frac{-6 \pm \sqrt{36 + 24}}{2} = \frac{-6 \pm \sqrt{4 \cdot 15}}{2} = \frac{-6 \pm 2\sqrt{15}}{2} = \frac{2(-3 \pm \sqrt{15})}{2}$$

$$= -3 \pm \sqrt{15}$$

$$\begin{aligned}
y &= c_1 e^{(-3+\sqrt{15})x} + c_2 e^{(-3-\sqrt{15})x} \\
&= c_1 e^{-3x} e^{\sqrt{15}\,x} + c_2 e^{-3x} e^{-\sqrt{15}\,x} \\
&= e^{-3x} \left(c_1 e^{\sqrt{15}\,x} + c_2 e^{-\sqrt{15}\,x} \right)
\end{aligned}$$

27. $2m^2 + 4m + 1 = 0$

$$m = \frac{-4 \pm \sqrt{4^2 - 4(2)(1)}}{2 \cdot 2} = \frac{-4 \pm \sqrt{8}}{4} = \frac{-4 \pm \sqrt{4 \cdot 2}}{4}$$

$$= \frac{-4 \pm 2\sqrt{2}}{4} = \frac{2\left(-2 \pm \sqrt{2}\right)}{4} = \frac{-2 \pm \sqrt{2}}{2} = -1 \pm \frac{\sqrt{2}}{2}$$

$$\begin{aligned}
y &= c_1 e^{(-1+\sqrt{2}/2)x} + c_2 e^{(-1-\sqrt{2}/2)x} \\
&= c_1 e^{-x} e^{(\sqrt{2}/2)x} + c_2 e^{-x} e^{-(\sqrt{2}/2)x} \\
&= e^{-x} \left(c_1 e^{(\sqrt{2}/2)x} + c_2 e^{-(\sqrt{2}/2)x} \right)
\end{aligned}$$

29.
$$\begin{aligned}
3m^2 - m - 2 &= 0 \\
(3m + 2)(m - 1) &= 0 \\
m &= 1,\ -\frac{2}{3}
\end{aligned}$$
$$y = c_1 e^x + c_2 e^{-(2/3)x}$$

12.2   Auxiliary Equations with Repeating or Complex Roots

1.
$$\begin{aligned}
m^2 + 6m + 9 &= 0 \\
(m + 3)^2 &= 0 \\
m &= -3,\ -3 \qquad \text{(repeating root)}
\end{aligned}$$
$$y = c_1 e^{-3x} + c_2 x e^{-3x}$$

3.
$$4m^2 - 4m + 1 = 0$$
$$\begin{aligned}
(2m - 1)^2 &= 0 \\
m &= \frac{1}{2},\ \frac{1}{2} \qquad \text{(repeating root)}
\end{aligned}$$
$$y = c_1 e^{(1/2)x} + c_2 x e^{(1/2)x}$$

5.
$$\begin{aligned}
9m^2 + 12m + 4 &= 0 \\
(3m + 2)^2 &= 0 \\
m &= -\frac{2}{3},\ -\frac{2}{3} \qquad \text{(repeating root)}
\end{aligned}$$
$$y = c_1 e^{-(2/3)x} + c_2 x e^{-(2/3)x}$$

7. $4m^2 - 20m + 25 \;=\; 0$

$\quad\quad\quad (2m - 5)^2 \;=\; 0$

$\quad\quad\quad\quad\quad\quad m \;=\; \dfrac{5}{2}, \dfrac{5}{2} \quad\quad$ (repeating root)

$\quad y = c_1 e^{(5/2)x} + c_2 x e^{(5/2)x}$

9. $m^2 - 4m + 5 = 0$

$$m \;=\; \frac{-(-4) \pm \sqrt{(-4)^2 - 4(1)(5)}}{2 \cdot 1} = \frac{4 \pm \sqrt{16 - 20}}{2}$$

$$=\; \frac{4 \pm \sqrt{-4}}{2} = \frac{4 \pm 2j}{4} = \frac{2(2 \pm j)}{2} = 2 \pm j$$

$\quad y = e^{2x}(c_1 \cos x + c_2 \sin x)$

11. $m^2 + 4m + 8 = 0$

$$m = \frac{-4 \pm \sqrt{4^2 - 4(1)(8)}}{2} = \frac{-4 \pm \sqrt{-16}}{2} = \frac{-4 \pm 4j}{2} = -2 \pm 2j$$

$\quad y = e^{-2x}(c_1 \cos 2x + c_2 \sin 2x)$

13. $2m^2 - 2m + 1 = 0$

$$m = \frac{2 \pm \sqrt{4 - 8}}{4} = \frac{2 \pm \sqrt{-4}}{4} = \frac{2 \pm 2j}{4} = \frac{1}{2} \pm \frac{1}{2}j$$

$\quad y = e^{(1/2)x}\left(c_1 \cos \dfrac{1}{2}x + c_2 \sin \dfrac{1}{2}x \right)$

15. $m^2 + 25 = 0; \;\; m^2 = -25; \;\; m = \pm 5j$

$\quad y = c_1 \cos 5x + c_2 \sin 5x$

17. $m^2 - 6m + 9 \;=\; 0$

$\quad\quad\quad (m - 3)^2 \;=\; 0$

$\quad\quad\quad\quad\quad\quad m \;=\; 3, \, 3$

$\quad y = c_1 e^{3x} + c_2 x e^{3x}$

19. $m^4 + 2m^3 \;=\; 0$

$\quad\quad m^3(m + 2) \;=\; 0$

$\quad\quad\quad\quad\quad m \;=\; 0, \, 0, \, 0, \, -2$

Here we have a triple root and a single root:

$\quad y \;=\; c_1 e^{0x} + c_2 x e^{0x} + c_3 x^2 e^{0x} + c_4 e^{-2x}$

$\quad\quad =\; c_1 + c_2 x + c_3 x^2 + c_4 e^{-2x}$

21. $m^2 + m + 2 = 0$

$$m = \frac{-1 \pm \sqrt{1 - 8}}{2} = \frac{-1 \pm \sqrt{7}j}{2} = -\frac{1}{2} \pm \frac{\sqrt{7}}{2}j$$

$\quad y = e^{-x/2}\left(c_1 \cos \dfrac{\sqrt{7}}{2}x + c_2 \sin \dfrac{\sqrt{7}}{2}x \right)$

23. $m^2 - 3m + 5 = 0$

$$m = \frac{-(-3) \pm \sqrt{(-3)^2 - 4(1)(5)}}{2} = \frac{3 \pm \sqrt{-11}}{2} = \frac{3 \pm \sqrt{11}\,j}{2} = \frac{3}{2} \pm \frac{\sqrt{11}}{2}j$$

$$y = e^{3x/2}\left(c_1 \cos \frac{\sqrt{11}}{2}x + c_2 \sin \frac{\sqrt{11}}{2}x\right)$$

25. $2m^2 - 4m + 5 = 0$

$$m = \frac{4 \pm \sqrt{16 - 40}}{4} = \frac{4 \pm \sqrt{-24}}{4} = \frac{4 \pm \sqrt{4 \cdot (-6)}}{4} = \frac{4 \pm 2\sqrt{6}\,j}{4} = 1 \pm \frac{1}{2}\sqrt{6}\,j$$

$$y = e^x\left(c_1 \cos \frac{1}{2}\sqrt{6}\,x + c_2 \sin \frac{1}{2}\sqrt{6}\,x\right)$$

27. $2m^2 + 4m - 1 = 0$

$$
\begin{aligned}
m &= \frac{-4 \pm \sqrt{4^2 - 4(2)(-1)}}{2 \cdot 2} = \frac{-4 \pm \sqrt{24}}{4} = \frac{-4 \pm \sqrt{4 \cdot 6}}{4} \\
&= \frac{-4 \pm 2\sqrt{6}}{4} = \frac{2\left(-2 \pm \sqrt{6}\right)}{4} = \frac{-2 \pm \sqrt{6}}{2} = -1 \pm \frac{\sqrt{6}}{2}
\end{aligned}
$$

$$
\begin{aligned}
y &= c_1 e^{\left(-1+\sqrt{6}/2\right)x} + c_2 e^{\left(-1-\sqrt{6}/2\right)x} \\
&= c_1 e^{-x} e^{\left(\sqrt{6}/2\right)x} + c_2 e^{-x} e^{-\left(\sqrt{6}/2\right)x} \\
&= e^{-x}\left(c_1 e^{\left(\sqrt{6}/2\right)x} + c_2 e^{-\left(\sqrt{6}/2\right)x}\right)
\end{aligned}
$$

29. $\quad m^2 - 100 \;=\; 0$

$$m \;=\; \pm 10$$

$$y = c_1 e^{10x} + c_2 e^{-10x}$$

31. $m^2 + 100 = 0;\; m^2 = -100;\; m = \pm 10j$

$$y = c_1 \cos 10x + c_2 \sin 10x$$

33. $\quad 3m^3 - 2m^2 + m \;=\; 0$

$$m(3m^2 - 2m + 1) \;=\; 0$$

$$m \;=\; 0,$$

$$m \;=\; \frac{2 \pm \sqrt{4 - 12}}{6} = \frac{2 \pm \sqrt{2(4)(-1)}}{6} = \frac{2 \pm 2\sqrt{2}\,j}{6}$$

$$\;=\; \frac{1}{3} \pm \frac{1}{3}\sqrt{2}\,j$$

$$
\begin{aligned}
y &= c_1 e^{0x} + e^{(1/3)x}\left(c_2 \cos \frac{\sqrt{2}}{3}x + c_3 \sin \frac{\sqrt{2}}{3}x\right) \\
&= c_1 + e^{x/3}\left(c_2 \cos \frac{\sqrt{2}}{3}x + c_3 \sin \frac{\sqrt{2}}{3}x\right)
\end{aligned}
$$

35. $m^2 - 4m + 2 = 0$

$$m = \frac{-(-4) \pm \sqrt{(-4)^2 - 4(1)(2)}}{2} = \frac{4 \pm \sqrt{8}}{2} = \frac{4 \pm \sqrt{4 \cdot 2}}{2} = \frac{4 \pm 2\sqrt{2}}{2} = \frac{2\left(2 \pm \sqrt{2}\right)}{2} = 2 \pm \sqrt{2}$$

$$\begin{aligned} y &= c_1 e^{(2+\sqrt{2})x} + c_2 e^{(2-\sqrt{2})x} \\ &= c_1 e^{2x} e^{\sqrt{2}\,x} + c_2 e^{2x} e^{-\sqrt{2}\,x} \\ &= e^{2x}\left(c_1 e^{\sqrt{2}\,x} + c_2 e^{-\sqrt{2}\,x}\right) \end{aligned}$$

37. $\begin{aligned} m^2 + 4 &= 0 \\ m^2 &= -4 \\ m &= \pm 2j = 0 \pm 2j \end{aligned}$

$y = e^{0x}(c_1 \cos 2x + c_2 \sin 2x) = c_1 \cos 2x + c_2 \sin 2x$

39. $\begin{aligned} m^3 - 4m^2 + 4m &= 0 \\ m(m^2 - 4m + 4) &= 0 \\ m(m - 2)^2 &= 0 \\ m &= 0,\ 2,\ 2 \end{aligned}$

$y = c_1 e^{0x} + c_2 e^{2x} + c_3 x e^{2x} = c_1 + c_2 e^{2x} + c_3 x e^{2x}$

41. $\begin{aligned} m^3 - 9m = m(m^2 - 9) &= 0 \\ m &= 0,\ \pm 3 \end{aligned}$

$\begin{aligned} y &= c_1 e^{0x} + c_2 e^{-3x} + c_3 e^{3x} \\ y &= c_1 + c_2 e^{-3x} + c_3 e^{3x} \end{aligned}$

43. $\begin{aligned} (m - 2)^4 &= 0 \\ m &= 2,\ 2,\ 2,\ 2 \end{aligned}$

$y = c_1 e^{2x} + c_2 x e^{2x} + c_3 x^2 e^{2x} + c_4 x^3 e^{2x}$

45. $\begin{aligned} m^2 + 1 &= 0 \\ m &= \pm j \end{aligned}$

$y = c_1 \cos x + c_2 \sin x$

Substituting $(0,0)$ and $\left(\dfrac{\pi}{2}, 1\right)$, respectively, we get

(1) $0 = c_1 \cdot 1 + c_2 \cdot 0$ or $c_1 = 0$; $y = 0,\ x = 0$

(2) $1 = c_1 \cdot 0 + c_2 \cdot 1$ or $c_2 = 1$. $y = 1,\ x = \dfrac{\pi}{2}$

Thus $y = \sin x$ is the solution.

47. $m^2 - 2m + 2 = 0$

$$m = \frac{-(-2) \pm \sqrt{(-2)^2 - 4(1)(2)}}{2} = \frac{2 \pm \sqrt{-4}}{2} = \frac{2 \pm 2j}{2} = 1 \pm j$$

$y = e^x(c_1 \cos x + c_2 \sin x)$

By the product rule,

$Dy = y' = e^x(-c_1 \sin x + c_2 \cos x) + e^x(c_1 \cos x + c_2 \sin x)$

Now substitute the initial conditions:

$\begin{aligned} 0 &= 1(c_1 + 0) & y = 0 \text{ when } x = 0 \\ -1 &= 1(0 + c_2) + 1(c_1 + 0) & Dy = -1 \text{ when } x = 0 \end{aligned}$

From the first equation, $c_1 = 0$. Substituting in the second equation, we obtain

$-1 = 1(c_2) + 1(0)$ or $c_2 = -1$.

Finally, we substitute $c_1 = 0$ and $c_2 = -1$ in the general solution:

$y = e^x [0 + (-1) \sin x] = -e^x \sin x.$

49.
$$\begin{aligned} m^4 + 18m^2 + 81 &= 0 \\ (m^2 + 9)^2 = (m^2 + 9)(m^2 + 9) &= 0 \\ m &= 3j,\ 3j,\ -3j,\ -3j \end{aligned}$$

$$\begin{aligned} y &= c_1 e^{3jx} + c_2 x e^{3jx} + c_3 e^{-3jx} + c_4 x e^{-3jx} \\ &= c_1(\cos 3x + j \sin 3x) + c_2 x(\cos 3x + j \sin 3x) \\ &\quad + c_3(\cos 3x - j \sin 3x) + c_4 x(\cos 3x - j \sin 3x) \\ &= (c_1 + c_3)\cos 3x + j(c_1 - c_3)\sin 3x + (c_2 + c_4)x \cos 3x + j(c_2 - c_4)x \sin 3x \end{aligned}$$

So y has the following form:

$y = c_1 \cos 3x + c_2 \sin 3x + c_3 x \cos 3x + c_4 x \sin 3x$

51. $-1,\ \dfrac{1}{2} \pm \dfrac{\sqrt{2}}{2}j$

53. $6,\ -4,\ -4$

55. $-1,\ -1,\ 1 \pm \dfrac{\sqrt{5}}{2}j$

12.3   Nonhomogeneous Equations

1. $(D^2 - 6D + 9)y = e^x$

$\begin{aligned} y_c:\ m^2 - 6m + 9 &= 0 \\ (m - 3)^2 &= 0 \\ m &= 3,\ 3 \end{aligned}$

$y_c = c_1 e^{3x} + c_2 x e^{3x}$

$y_p:\ y_p = Ae^x;\ y_p' = Ae^x;\ y_p'' = Ae^x$

Substituting in $(D^2 - 6D + 9)y = e^x$:

$$\begin{aligned} y_p'' - 6y_p' + 9y_p &= e^x \\ Ae^x - 6Ae^x + 9Ae^x &= e^x \\ 4Ae^x &= e^x \\ A &= \frac{1}{4}. \end{aligned}$$

So $y_p = Ae^x = \dfrac{1}{4}e^x$ and $y = y_c + y_p = c_1 e^{3x} + c_2 x e^{3x} + \dfrac{1}{4}e^x.$

3. $(D^2 - 6D + 9)y = 9x$

$y_c : y_c = c_1e^{3x} + c_2xe^{3x}$ (same as in Exercise 1)

$y_p : y_p = Ax + B; \ y_p' = A; \ y_p'' = 0$

Substituting in $(D^2 - 6D + 9)y = 9x$:

$$y_p'' - 6y_p' + 9y_p \ = \ 9x$$
$$0 - 6A + 9(Ax + B) \ = \ 9x$$
$$9Ax - 6A + 9B \ = \ 9x + 0.$$

Now compare coefficients:

$$9A \ = \ 9 \qquad x\text{-coefficients}$$
$$-6A + 9B \ = \ 0. \qquad \text{constants}$$

We obtain $A = 1$ and $B = \dfrac{6}{9}A = \dfrac{2}{3}$. So $y_p = x + \dfrac{2}{3}$ and $y = y_c + y_p = c_1e^{3x} + c_2xe^{3x} + x + \dfrac{2}{3}$.

5. $(D^2 - D - 2)y = 2x^2$

$y_c : \ m^2 - m - 2 = (m - 2)(m + 1) \ = \ 0$
$$m \ = \ 2, -1$$

$y_c = c_1e^{2x} + c_2e^{-x}$

$y_p : \ y_p = Ax^2 + Bx + C, \ y_p' = 2Ax + B, \ y_p'' = 2A$

Substituting in $(D^2 - D - 2)y = 2x^2$, we get

$$2A - (2Ax + B) - 2(Ax^2 + Bx + C) \ = \ 2x^2$$
$$2A - 2Ax - B - 2Ax^2 - 2Bx - 2C \ = \ 2x^2$$
$$-2Ax^2 + (-2A - 2B)x + (2A - B - 2C) \ = \ 2x^2 + 0x + 0.$$

$$-2A \ = \ 2 \qquad x^2\text{-coefficients}$$
$$-2A - 2B \ = \ 0 \qquad x\text{-coefficients}$$
$$2A - B - 2C \ = \ 0 \qquad \text{constants}$$

We obtain: $A = -1$, $B = 1$, and $C = -\dfrac{3}{2}$.

So $y_p = Ax^2 + Bx + C = -x^2 + x - \dfrac{3}{2}$ and $y = y_c + y_p = c_1e^{2x} + c_2e^{-x} - x^2 + x - \dfrac{3}{2}$.

7. $(D^2 - D + 1)y = 1 - x^2$

$y_c : \ m^2 - m + 1 \ = \ 0$
$$m \ = \ \frac{-(-1) \pm \sqrt{(-1)^2 - 4}}{2} = \frac{1 \pm \sqrt{3}\,\mathrm{j}}{2} = \frac{1}{2} \pm \frac{\sqrt{3}}{2}\mathrm{j}$$

$y_c = e^{(1/2)x}\left(c_1 \cos\dfrac{\sqrt{3}}{2}x + c_2 \sin\dfrac{\sqrt{3}}{2}x\right)$

$y_p : \ y_p = Ax^2 + Bx + C; \ y_p' = 2Ax + B; \ y_p'' = 2A$

Substituting in $(D^2 - D + 1)y = 1 - x^2$:

$$y_p'' - y_p' + y_p \ = \ 1 - x^2$$
$$2A - (2Ax + B) + (Ax^2 + Bx + C) \ = \ 1 - x^2$$
$$Ax^2 + (-2A + B)x + (2A - B + C) \ = \ -x^2 + 0x + 1.$$

Comparing coefficients:

$$A \ = \ -1 \qquad x^2\text{-coefficients}$$
$$-2A + B \ = \ 0 \qquad x\text{-coefficients}$$
$$2A - B + C \ = \ 1. \qquad \text{constants}$$

We obtain: $A = -1$, $B = -2$, and $C = 1$. So $y_p = -x^2 - 2x + 1$

and $y = y_c + y_p = e^{(1/2)x}\left(c_1 \cos\dfrac{\sqrt{3}}{2}x + c_2 \sin\dfrac{\sqrt{3}}{2}x\right) - x^2 - 2x + 1.$

9. $(D^2 - D + 2)y = 4e^{3x}$

$y_c: \quad m^2 - m + 2 \;=\; 0$

$$m \;=\; \frac{-(-1) \pm \sqrt{(-1)^2 - 4(1)(2)}}{2 \cdot 1}$$

$$\;=\; \frac{1 \pm \sqrt{-7}}{2} = \frac{1 \pm \sqrt{7}\,\mathrm{j}}{2} = \frac{1}{2} \pm \frac{\sqrt{7}}{2}\mathrm{j}$$

$$y_c = e^{x/2}\left(c_1 \cos \frac{\sqrt{7}}{2}x + c_2 \sin \frac{\sqrt{7}}{2}x\right)$$

$y_p: y_p = Ae^{3x}; \; y_p' = 3Ae^{3x}; \; y_p'' = 9Ae^{3x}$

Substituting in $(D^2 - D + 2)y = 4e^{3x}$:

$$y_p'' - y_p' + 2y_p \;=\; 4e^{3x}$$

$$9Ae^{3x} - 3Ae^{3x} + 2Ae^{3x} \;=\; 4e^{3x}$$

$$8Ae^{3x} \;=\; 4e^{3x}$$

$$A \;=\; \frac{1}{2}.$$

So $y_p = \dfrac{1}{2}e^{3x}$ and $y = e^{x/2}\left(c_1 \cos \dfrac{\sqrt{7}}{2}x + c_2 \sin \dfrac{\sqrt{7}}{2}x\right) + \dfrac{1}{2}e^{3x}$.

11. $(D^2 - 6D + 9)y = 9\cos 3x$

$y_c: m^2 - 6m + 9 = 0; \; (m-3)^2 = 0; \; m = 3, \; 3$

$y_c = c_1 e^{3x} + c_2 x e^{3x}$

$y_p: y_p = A\cos 3x + B\sin 3x; \; y_p' = -3A\sin 3x + 3B\cos 3x; \; y_p'' = -9A\cos 3x - 9B\sin 3x$

Substituting in $(D^2 - 6D + 9)y = 9\cos 3x$,

$$y_p'' - 6y_p' + 9y_p \;=\; 9\cos 3x$$

$$-9A\cos 3x - 9B\sin 3x - 6(-3A\sin 3x + 3B\cos 3x)$$
$$+9(A\cos 3x + B\sin 3x) \;=\; 9\cos 3x$$

$$-18B\cos 3x + 18A\sin 3x \;=\; 9\cos 3x + 0\sin 3x$$

So $A = 0$ and $B = -\dfrac{1}{2}$, and $y_p = -\dfrac{1}{2}\sin 3x$.

$$y = c_1 e^{3x} + c_2 x e^{3x} - \frac{1}{2}\sin 3x$$

13. $y_c: \quad m^2 + 4m + 3 \;=\; 0$

$$(m+1)(m+3) \;=\; 0$$

$$m \;=\; -1, \; -3$$

$y_c = c_1 e^{-x} + c_2 e^{-3x}$

$y_p: y_p = A + Be^x; \; y_p' = Be^x; \; y_p'' = Be^x$

Substituting in the equation:

$$y'' + 4y' + 3y \;=\; 6 + e^x$$

$$Be^x + 4Be^x + 3(A + Be^x) \;=\; 6 + e^x$$

$$8Be^x + 3A \;=\; 6 + e^x$$

$$A \;=\; 2, \; B = \frac{1}{8}$$

So $y_p = 2 + \dfrac{1}{8}e^x$ and $y = c_1 e^{-x} + c_2 e^{-3x} + \dfrac{1}{8}e^x + 2$.

15. $(D^2 + 1)y = 6\sin 2x$

$y_c: m^2 + 1 = 0; m = \pm j$

$y_c = c_1 \cos x + c_2 \sin x$

$y_p: y_p = A\cos 2x + B\sin 2x; y_p' = -2A\sin 2x + 2B\cos 2x; y_p'' = -4A\cos 2x - 4B\sin 2x$

Substituting in $(D^2 + 1)y = 6\sin 2x$, we get

$$(-4A\cos 2x - 4B\sin 2x) + (A\cos 2x + B\sin 2x) = 6\sin 2x$$
$$-3A\cos 2x - 3B\sin 2x = 0\cos 2x + 6\sin 2x.$$

We obtain $A = 0$ and $B = -2$, so that $y_p = -2\sin 2x$ and $y = c_1\cos x + c_2\sin x - 2\sin 2x$.

17. $y_c: \quad m^2 + 5m + 6 = 0$

$\quad\quad (m+2)(m+3) = 0$

$\quad\quad\quad\quad\quad\quad m = -2, -3$

$y_c = c_1 e^{-2x} + c_2 e^{-3x}$

$y_p: y_p = A\cos x + B\sin x; y_p' = -A\sin x + B\cos x; y_p'' = -A\cos x - B\sin x$

Substituting in the equation $(D^2 + 5D + 6)y = 4\cos x + 6\sin x$, we get

$$(-A\cos x - B\sin x) + 5(-A\sin x + B\cos x) + 6(A\cos x + B\sin x) = 4\cos x + 6\sin x$$
$$(5A + 5B)\cos x + (-5A + 5B)\sin x = 4\cos x + 6\sin x.$$

Comparing coefficients:

$\quad 5A + 5B = 4$

$\quad \underline{-5A + 5B = 6}$

$\quad\quad 10B = 10, \quad B = 1$

$5A + 5 = 4, \quad A = -\dfrac{1}{5} \quad$ first equation

So $y_p = -\dfrac{1}{5}\cos x + \sin x$ and $y = c_1 e^{-2x} + c_2 e^{-3x} - \dfrac{1}{5}\cos x + \sin x.$

19. $(D^2 - 2D + 1)y = 3e^{2x}$

$y_c: \quad m^2 - 2m + 1 = 0$

$\quad\quad (m-1)^2 = 0$

$\quad\quad\quad\quad\quad m = 1, 1$

$y_c = c_1 e^x + c_2 x e^x$

$y_p: y_p = Ae^{2x}; y_p' = 2Ae^{2x}; y_p'' = 4Ae^{2x}$

Substituting in $(D^2 - 2D + 1)y = 3e^{2x}$, we get

$\quad 4Ae^{2x} - 2(2Ae^{2x}) + Ae^{2x} = 3e^{2x}$

$\quad\quad\quad\quad\quad\quad Ae^{2x} = 3e^{2x}$ and $A = 3$.

So $y_p = 3e^{2x}$ and $y = c_1 e^x + c_2 x e^x + 3e^{2x}.$

21. $y_c: m^3 - 2m^2 - m + 2 = 0$

Possible rational roots: $\pm 1, \pm 2$; one root is $m = 1$.

$1 - 2 - 1 + 2)\underline{1}$

$\quad\quad \underline{1 - 1 - 2}$

$\overline{1 - 1 - 2 + 0}$

Since the resulting factor is $m^2 - m - 2 = (m-2)(m+1)$, the other roots are 2 and -1.

$y_c = c_1 e^x + c_2 e^{-x} + c_3 e^{2x}$

$y_p: y_p = Ae^{3x}; y_p' = 3Ae^{3x}; y_p'' = 9Ae^{3x}; y_p''' = 27Ae^{3x}$

Substituting in the equation: $27Ae^{3x} - 2\left(9Ae^{3x}\right) - 3Ae^{3x} + 2Ae^{3x} = 8e^{3x}.$

$8A = 8, \quad A = 1$ and $y_p = e^{3x}.$

$y = c_1 e^x + c_2 e^{-x} + c_3 e^{2x} + e^{3x}$

23. $(D^2 - 2D + 5)y = 4xe^x$

$y_c : m^2 - 2m + 5 = 0$

$$m = \frac{-(-2) \pm \sqrt{(-2)^2 - 4(1)(5)}}{2} = \frac{2 \pm \sqrt{-16}}{2} = 1 \pm 2j$$

$y_c = e^x(c_1 \cos 2x + c_2 \sin 2x)$

$y_p : y_p = Axe^x + Be^x;$

$\qquad y_p' = Axe^x + Ae^x + Be^x \qquad$ product rule

$\qquad y_p'' = Axe^x + 2Ae^x + Be^x$

Substituting in $(D^2 - 2D + 5)y = 4xe^x$,

$(Axe^x + 2Ae^x + Be^x) - 2(Axe^x + Ae^x + Be^x) + 5(Axe^x + Be^x) = 4xe^x.$

Collecting terms, $4Axe^x + 4Be^x = 4xe^x.$

We obtain $A = 1$ and $B = 0$, so that $y_p = xe^x$ and $y = e^x(c_1 \cos 2x + c_2 \sin 2x) + xe^x.$

25. $y_c : m^2 + 9 = 0$

$\qquad m = \pm 3j$

$y_c = c_1 \cos 3x + c_2 \sin 3x$

$y_p : y_p = Ae^{3x}; \ y_p' = 3Ae^{3x}; \ y_p'' = 9Ae^{3x}$

Substituting in the equation:

$9Ae^{3x} + 9Ae^{3x} = 9e^{3x}$

$\qquad 18A = 9, \ A = \frac{1}{2}$ and $y_p = \frac{1}{2}e^{3x}.$

$y = c_1 \cos 3x + c_2 \sin 3x + \frac{1}{2}e^{3x}$

$Dy = -3c_1 \sin 3x + 3c_2 \cos 3x + \frac{3}{2}e^{3x}$

Substituting in the given conditions, we have

(1) $\quad 1 = c_1 \cdot 1 + c_2 \cdot 0 + \frac{1}{2} \qquad y = 1, \ x = 0$

(2) $\frac{3}{2} = -3c_1 \cdot 0 + 3c_2 \cdot 1 + \frac{3}{2} \qquad Dy = \frac{3}{2}, \ x = 0$

$\qquad c_1 = \frac{1}{2}, \ c_2 = 0$

The solution is $y = \frac{1}{2} \cos 3x + \frac{1}{2}e^{3x} = \frac{1}{2}\left(\cos 3x + e^{3x}\right).$

27. $(D^2 + 1)y = 6 \cos 2x$

$y_c : m^2 + 1 = 0; \ m = \pm j$

$y_c = c_1 \cos x + c_2 \sin x$

$y_p : y_p = A \cos 2x + B \sin 2x; \ y_p' = -2A \sin 2x + 2B \cos 2x; \ y_p'' = -4A \cos 2x - 4B \sin 2x$

Substituting,

$(-4A \cos 2x - 4B \sin 2x) + (A \cos 2x + B \sin 2x) = 6 \cos 2x$

$\qquad\qquad -3A \cos 2x - 3B \sin 2x = 6 \cos 2x.$

We obtain $A = -2$ and $B = 0$, so that $y_p = -2 \cos 2x$ and

$\qquad y = c_1 \cos x + c_2 \sin x - 2 \cos 2x$

$Dy = y' = -c_1 \sin x + c_2 \cos x + 4 \sin 2x.$

Substituting the initial conditions,

$3 = c_1 + 0 - 2; \qquad y = 3$ when $x = 0$

$1 = 0 + c_2 + 0. \qquad Dy = 1$ when $x = 0$

So $c_1 = 5$ and $c_2 = 1$; the solution is $y = 5 \cos x + \sin x - 2 \cos 2x.$

29. $(D^2 + 2D + 5)y = 10\cos x$

$y_c : m^2 + 2m + 5 = 0$

$$m = \frac{-2 \pm \sqrt{4 - 20}}{2} = \frac{-2 \pm 4j}{2} = -1 \pm 2j$$

$y_c = e^{-x}(c_1 \cos 2x + c_2 \sin 2x)$

$y_p : y_p = A\cos x + B\sin x; \ y_p' = -A\sin x + B\cos x; \ y_p'' = -A\cos x - B\sin x$

Substituting in the given equation:

$(-A\cos x - B\sin x) + 2(-A\sin x + B\cos x) + 5(A\cos x + B\sin x) = 10\cos x$

$\qquad\qquad (4A + 2B)\cos x + (-2A + 4B)\sin x = 10\cos x$

$\qquad\qquad (4A + 2B)\cos x + (-2A + 4B)\sin x = 10\cos x + 0\sin x.$

$\quad 4A + 2B = 10$

$\underline{-2A + 4B = 0}$

$\quad 4A + 2B = 10$

$\underline{-4A + 8B = 0}$

$\qquad 10B = 10, B = 1$

$\quad -2A + 4 = 0, \ A = 2 \qquad y_p = 2\cos x + \sin x$

General solution:

$\quad y = e^{-x}(c_1 \cos 2x + c_2 \sin 2x) + 2\cos x + \sin x$

$Dy = e^{-x}(-2c_1 \sin 2x + 2c_2 \cos 2x) - e^{-x}(c_1 \cos 2x + c_2 \sin 2x) - 2\sin x + \cos x.$

Now substitute the given conditions:

$5 = c_1 + 2 \qquad\qquad x = 0, \ y = 5$

$6 = 2c_2 - c_1 + 1 \qquad x = 0, \ Dy = 6$

Thus $c_1 = 3$ and $c_2 = 4$, so that the solution is

$y = e^{-x}(3\cos 2x + 4\sin 2x) + 2\cos x + \sin x.$

31. $(D^2 + 4)y = 10\sin 3x$

$y_p = A\cos 3x + B\sin 3x; \ y_p' = -3A\sin 3x + 3B\cos 3x; \ y_p'' = -9A\cos 3x - 9B\sin 3x$

Substituting in the given equation,

$(-9A\cos 3x - 9B\sin 3x) + 4(A\cos 3x + B\sin 3x) = 10\sin 3x$

$\qquad\qquad -5A\cos 3x - 5B\sin 3x = 0\cos 3x + 10\sin 3x.$

It follows directly that $A = 0$ and $B = -2$; so $y_p = -2\sin 3x.$

33. $(D^2 - D + 2)y = \cos x$

$y_p = A\cos x + B\sin x; \ y_p' = -A\sin x + B\cos x; \ y_p'' = -A\cos x - B\sin x$

Substituting:

$(-A\cos x - B\sin x) - (-A\sin x + B\cos x) + 2(A\cos x + B\sin x) = \cos x$

$\qquad\qquad (A - B)\cos x + (A + B)\sin x = \cos x + 0\sin x.$

$\quad A - B = 1$

$\underline{A + B = 0}$

$\quad 2A \quad = 1, \quad A = \frac{1}{2}$

$\qquad B = -A, \ B = -\frac{1}{2}$

$y_p = \frac{1}{2}\cos x - \frac{1}{2}\sin x$

35. $(D^2 + D - 2)y = 2x^2 + 1$

$y_p = Ax^2 + Bx + C; \ y_p' = 2Ax + B; \ y_p'' = 2A$

Substituting,

$2A + (2Ax + B) - 2(Ax^2 + Bx + C) = 2x^2 + 1$

$-2Ax^2 + (2A - 2B)x + (2A + B - 2C) = 2x^2 + 0x + 1.$

$\quad -2A = 2 \qquad x^2\text{-coefficients}$

$\quad 2A - 2B = 0 \qquad x\text{-coefficients}$

$2A + B - 2C = 1 \qquad \text{constants}$

We obtain $A = -1, \ B = -1,$ and $C = -2; \ y_p = -x^2 - x - 2.$

37. $(D^2 + 2)y = 2e^{2x} + 2$

$y_p = Ae^{2x} + B; \ y_p' = 2Ae^{2x}; \ y_p'' = 4Ae^{2x}$

Substituting:

$4Ae^{2x} + 2(Ae^{2x} + B) = 2e^{2x} + 2$

$\qquad 6Ae^{2x} + 2B = 2e^{2x} + 2$

$\qquad\qquad A = \dfrac{1}{3}, \ B = 1.$

$y_p = \dfrac{1}{3}e^{2x} + 1$

39. $(D^2 - D - 4)y = x + 2e^{3x}$

$y_p = Ax + B + Ce^{3x}; \ y_p' = A + 3Ce^{3x}; \ y_p'' = 9Ce^{3x}$

Substituting:

$9Ce^{3x} - (A + 3Ce^{3x}) - 4(Ax + B + Ce^{3x}) = x + 2e^{3x}$

$\qquad -4Ax + (-A - 4B) + 2Ce^{3x} = x + 0 + 2e^{3x}.$

$\quad -4A = 1 \qquad x\text{-coefficients}$

$-A - 4B = 0 \qquad \text{constants}$

$\quad 2C = 2 \qquad \text{coefficients of } e^{3x}$

We obtain: $A = -\dfrac{1}{4}, \ B = \dfrac{1}{16},$ and $C = 1.$

$y_p = -\dfrac{1}{4}x + \dfrac{1}{16} + e^{3x}.$

41. $(D^2 + 3)y = 2x + \cos x$

$y_p = Ax + B + C\cos x + D\sin x; \ y_p' = A - C\sin x + D\cos x; \ y_p'' = -C\cos x - D\sin x$

Substituting:

$(-C\cos x - D\sin x) + 3(Ax + B + C\cos x + D\sin x) = 2x + \cos x$

$\qquad\qquad 3Ax + 3B + 2C\cos x + 2D\sin x = 2x + \cos x$

$\qquad\qquad\qquad\qquad\qquad\qquad = 2x + 0 + \cos x + 0\sin x.$

$A = \dfrac{2}{3}, \ B = 0, \ C = \dfrac{1}{2}, \ D = 0$

$y_p = \dfrac{2}{3}x + \dfrac{1}{2}\cos x$

43. $(D^2 - D - 3)y = 6xe^x$

$\dfrac{d}{dx}(6xe^x) = 6xe^x + 6e^x$

Since the right side is a linear combination of y_p and its derivatives, the form of y_p is the following:

$y_p = Axe^x + Be^x$; $y_p' = Axe^x + Ae^x + Be^x$; $y_p'' = Axe^x + 2Ae^x + Be^x$.

Substituting:

$Axe^x + 2Ae^x + Be^x - (Axe^x + Ae^x + Be^x) - 3(Axe^x + Be^x) = 6xe^x$

$-3Axe^x + (A - 3B)e^x = 6xe^x + 0e^x$.

We obtain: $A = -2$, $A - 3B = 0$, so $B = -\dfrac{2}{3}$.

$y_p = -2xe^x - \dfrac{2}{3}e^x$

45. $(D^2 - 4)y = 8e^{2x}$

y_c : $m^2 - 4 = 0$, $m = \pm 2$

$y_c = c_1e^{-2x} + c_2e^{2x}$

Since the right side of the given equation, $8e^{2x}$, is a term in y_c, we need the annihilator. (In other words, y_p is not of the form Ae^{2x}.)

Right side $= 8e^{2x}$

$m' = 2$

$m' - 2 = 0$

Annihilator: $D - 2$.

Applying the annihilator to the equation, we get

$(D - 2)(D^2 - 4)y = (D - 2)(8e^{2x}) = 0$

$(m - 2)(m^2 - 4) = 0$

$(m - 2)(m - 2)(m + 2) = 0$

$m = 2,\ 2,\ -2$.

So y has the form: $y = c_1e^{-2x} + c_2e^{2x} + c_3xe^{2x}$.

Since $y_c = c_1e^{-2x} + c_2e^{2x}$, we conclude that

$y_p = Axe^{2x}$;

$y_p' = 2Axe^{2x} + Ae^{2x}$;

$y_p'' = 4Axe^{2x} + 2Ae^{2x} + 2Ae^{2x} = 4Axe^{2x} + 4Ae^{2x}$.

Substituting in the given equation, we get

$(4Axe^{2x} + 4Ae^{2x}) - 4Axe^{2x} = 8e^{2x}$

$4Ae^{2x} = 8e^{2x}$

$A = 2$ and $y_p = 2xe^{2x}$.

$y = y_c + y_p = c_1e^{-2x} + c_2e^{2x} + 2xe^{2x}$

47. $(D^2 - D - 6)y = 10e^{-2x}$

$y_c:$ $\quad m^2 - m - 6 = 0$

$\qquad (m-3)(m+2) = 0$

$\qquad\qquad m = 3, -2$

$y_c = c_1 e^{3x} + c_2 e^{-2x}$

Since the right side of the given equation, $10e^{-2x}$, is of the form $c_2 e^{-2x}$, y_p cannot be Ae^{-2x}.

So we need the annihilator:

$\quad$ Right side $= 10e^{-2x}$

$\qquad m' = -2$

$\qquad m' + 2 = 0$

Annihilator: $D + 2$.

Applying the annihilator to the given equation, we get

$\quad (D+2)(D^2 - D - 6)y = (D+2)(10e^{-2x}) = 0$

$(m+2)(m+2)(m-3) = 0$

$\qquad\qquad m = -2, -2, 3.$

So y has the form: $y = c_1 e^{3x} + c_2 e^{-2x} + c_3 x e^{-2x}$.

Since $y_c = c_1 e^{3x} + c_2 e^{-2x}$, we conclude that

$y_p = Axe^{-2x};\ y_p' = -2Axe^{-2x} + Ae^{-2x};\ y_p'' = 4Axe^{-2x} - 2Ae^{-2x} - 2Ae^{-2x} = 4Axe^{-2x} - 4Ae^{-2x}.$

Substituting in $(D^2 - D - 6)y = 10e^{-2x}$, we get

$(4Axe^{-2x} - 4Ae^{-2x}) - (-2Axe^{-2x} + Ae^{-2x}) - 6Axe^{-2x} = 10e^{-2x}$

$\qquad\qquad\qquad\qquad -5Ae^{-2x} = 10e^{-2x}$ and $A = -2$.

So $y_p = -2xe^{-2x}$ and $y = c_1 e^{3x} + c_2 e^{-2x} - 2xe^{-2x}$.

49. $(D^2 + 3D - 28)y = 11e^{4x}$

$y_c: m^2 + 3m - 28 = (m+7)(m-4) = 0$

$\qquad\qquad m = 4, -7$

$y_c = c_1 e^{-7x} + c_2 e^{4x}$

(Note that the right side, $11e^{4x}$, is one of the terms in y_c.)

$y_p:$ $\quad$ Right side $= 11e^{4x}$

$\qquad\qquad m' = 4$

$\qquad\quad m' - 4 = 0$

$\quad$ Annihilator: $D - 4$.

Applying the annihilator to the equation, we get

$(D-4)(D^2 + 3D - 28)y = (D-4)(11e^{4x}) = 0$

$\quad (m-4)(m-4)(m+7) = 0$

$\qquad\qquad m = 4, 4, -7.$

So y has the following form: $y = c_1 e^{-7x} + c_2 e^{4x} + c_3 x e^{4x}$.

Since $y_c = c_1 e^{-7x} + c_2 e^{4x}$, we conclude that

$y_p = Axe^{4x};\ y_p' = 4Axe^{4x} + Ae^{4x};$

$y_p'' = 16Axe^{4x} + 4Ae^{4x} + 4Ae^{4x} = 16Axe^{4x} + 8Ae^{4x}.$

Substituting in the given equation,

$(16Axe^{4x} + 8Ae^{4x}) + 3(4Axe^{4x} + Ae^{4x}) - 28Axe^{4x} = 11e^{4x}$

$\quad 16Axe^{4x} + 8Ae^{4x} + 12Axe^{4x} + 3Ae^{4x} - 28Axe^{4x} = 11e^{4x}$

$\qquad\qquad\qquad\qquad\qquad 11Ae^{4x} = 11e^{4x}$

$\qquad\qquad\qquad\qquad\qquad\qquad A = 1.$

So $y_p = xe^{4x}$ and $y = y_c + y_p = c_1 e^{-7x} + c_2 e^{4x} + xe^{4x}$.

51. $(D^2 - 1)y = 2e^x$

$y_c: \ m^2 - 1 = 0; \ m = \pm 1$

$y_c = c_1 e^{-x} + c_2 e^x$

(Observe that the right side is one of the terms in y_c.)

$y_p: \quad$ Right side$=2e^x$

$$m' = 1$$

$$m' - 1 = 0$$

Annihilator: $D - 1$.

Applying the annihilator to the equation, we get

$$(D - 1)(D^2 - 1)y = (D - 1)(2e^x) = 0$$

$$(m - 1)(m - 1)(m + 1) = 0$$

$$m = 1, \ 1, \ -1.$$

So y has the form: $y = c_1 e^{-x} + c_2 e^x + c_3 x e^x$.

Since $y_c = c_1 e^{-x} + c_2 e^x$, we conclude that

$y_p = Axe^x; \ y_p' = Axe^x + Ae^x; \ y_p'' = Axe^x + 2Ae^x$.

Substituting in the given equation, we get

$Axe^x + 2Ae^x - Axe^x = 2e^x$

$$A = 1.$$

So $y_p = xe^x$ and $y = c_1 e^{-x} + c_2 e^x + xe^x = c_1 e^{-x} + (c_2 + x)e^x$.

53. $(D^2 - 4)y = 4 + e^{2x}$

$y_c: \ m^2 - 4 = 0, \ m = \pm 2$

$y_c = c_1 e^{-2x} + c_2 e^{2x}$

$y_p: \quad$ Right side$=4 + e^{2x}$

$$m' = 0, \ 2$$

$$m'(m' - 2) = 0$$

Annihilator: $D(D - 2)$.

Applying the annihilator to the equation:

$$D(D - 2)(D^2 - 4)y = D(D - 2)(4 + e^{2x}) = 0$$

$$m(m - 2)(m^2 - 4) = 0$$

$$m = 0, \ 2, \ 2, \ -2.$$

So y has the form: $y = c_1 e^{-2x} + c_2 e^{2x} + c_3 x e^{2x} + c_4 e^{0x}$.

Since $y_c = c_1 e^{-2x} + c_2 e^{2x}$, we conclude that

$y_p = Axe^{2x} + B; \ y_p' = 2Axe^{2x} + Ae^{2x}; \ y_p'' = 4Axe^{2x} + 2Ae^{2x} + 2Ae^{2x} = 4Axe^{2x} + 4Ae^{2x}$.

Substituting in the given equation,

$$(4Axe^{2x} + 4Ae^{2x}) - 4(Axe^{2x} + B) = 4 + e^{2x}$$

$$4Axe^{2x} + 4Ae^{2x} - 4Axe^{2x} - 4B = 4 + e^{2x}$$

$$-4B + 4Ae^{2x} = 4 + e^{2x}$$

$$-4B = 4$$

$$4A = 1.$$

So $A = \dfrac{1}{4}$ and $B = -1$ and $y_p = \dfrac{1}{4}xe^{2x} - 1$.

$y = y_c + y_p = c_1 e^{-2x} + c_2 e^{2x} + \dfrac{1}{4}xe^{2x} - 1$

55. $(D^2 + 1)y = 2\sin x$

$y_c :\ m^2 + 1 = 0;\ m = \pm j$

$y_c = c_1 \cos x + c_2 \sin x$

Since the right side is one of the terms in y_c, we need the annihilator.

$y_p :$ Right side $= 2\sin x$

$$m' = \pm j$$
$$(m' - j)(m' + j) = 0$$
$$(m')^2 + 1 = 0$$

Annihilator: $D^2 + 1$.

Applying the annihilator to the given equation, we get

$$(D^2 + 1)(D^2 + 1)y = (D^2 + 1)(2\sin x) = 0$$
$$(m^2 + 1)(m^2 + 1) = 0$$
$$m = \pm j,\ \pm j.$$

By Exercise 49, Section 12.2, $y = c_1 \cos x + c_2 \sin x + c_3 x \cos x + c_4 x \sin x$.

It follows that

$y_p = Ax \cos x + Bx \sin x;$

$y_p' = -Ax \sin x + A \cos x + Bx \cos x + B \sin x;$

$y_p'' = -Ax \cos x - A \sin x - A \sin x - Bx \sin x + B \cos x + B \cos x$

     $= -Ax \cos x - 2A \sin x - Bx \sin x + 2B \cos x.$

Substituting in the given equation:

$-Ax \cos x - 2A \sin x - Bx \sin x + 2B \cos x + Ax \cos x + Bx \sin x = 2\sin x$

$$-2A \sin x + 2B \cos x = 2\sin x.$$

We obtain $A = -1$ and $B = 0$, so that $y_p = -x \cos x$ and $y = c_1 \cos x + c_2 \sin x - x \cos x$.

12.4   Applications of Second-Order Equations

1. By Hooke's law

$$F = kx$$
$$4 = k \cdot \frac{1}{2}\ \text{or}\ k = 8 \qquad \text{spring constant}$$
$$\text{mass} = \frac{4}{32} = \frac{1}{8}\ \text{slug} \qquad \text{mass}$$
$$m\frac{d^2x}{dt^2} + kx = 0 \qquad b = 0$$
$$\frac{1}{8}\frac{d^2x}{dt^2} + 8x = 0$$
$$\frac{d^2x}{dt^2} + 64x = 0 \qquad \text{equation}$$

Conditions: (1)   If $t = 0$, $x = \dfrac{1}{4}$ ft   initial position

            (2)   If $t = 0$, $\dfrac{dx}{dt} = 0$ intital velocity

Auxiliary equation: $m^2 + 64 = 0$, $m = \pm 8j$.

Thus $x(t) = c_1 \cos 8t + c_2 \sin 8t;$

        $x'(t) = -8c_1 \sin 8t + 8c_2 \cos 8t.$

From the initial conditions: $\dfrac{1}{4} = c_1 + c_2(0)$ or $c_1 = \dfrac{1}{4};$

                        $0 = 0 + 8c_2$ or $c_2 = 0.$

Substituting c_1 and c_2, we get $x(t) = \dfrac{1}{4} \cos 8t.$

3. By Hooke's law,

$$F = kx$$

$$4 = k \cdot \frac{1}{2} \text{ or } k = 8 \qquad \text{spring constant}$$

$$m = \frac{4}{32} = \frac{1}{8} \text{ slug} \qquad \text{mass}$$

$$m\frac{d^2x}{dt^2} + kx = f(t) \qquad b = 0$$

$$\frac{1}{8}\frac{d^2x}{dt^2} + 8x = \frac{1}{4}\cos 6t$$

$$\frac{d^2x}{dt^2} + 64x = 2\cos 6t \qquad \text{multiplying by 8}$$

Conditions: (1) If $t = 0$, $x = \frac{1}{4}$ ft $\qquad$ intitial position

$\qquad\qquad\qquad$ (2) If $t = 0$, $\frac{dx}{dt} = 0$ $\qquad$ initial velocity

x_c : $m^2 + 64 = 0$; $m = \pm 8j$

$x_c = c_1 \cos 8t + c_2 \sin 8t$

x_p : $x_p = A\cos 6t + B\sin 6t$; $x_p' = -6A\sin 6t + 6B\cos 6t$; $x_p'' = -36A\cos 6t - 36B\sin 6t$

Substituting,

$$(-36A\cos 6t - 36B\sin 6t) + 64(A\cos 6t + B\sin 6t) = 2\cos 6t$$

$$28A\cos 6t + 28B\sin 6t = 2\cos 6t + 0\sin 6t.$$

We obtain: $A = \frac{2}{28} = \frac{1}{14}$ and $B = 0$. So $x_p = \frac{1}{14}\cos 6t$ and

$$x(t) = c_1 \cos 8t + c_2 \sin 8t + \frac{1}{14}\cos 6t;$$

$$x'(t) = -8c_1 \sin 8t + 8c_2 \cos 8t - \frac{6}{14}\sin 6t.$$

From the initial conditions,

$$\frac{1}{4} = c_1 + 0 + \frac{1}{14}; \qquad x = \frac{1}{4} \text{ when } t = 0$$

$$0 = 0 + 8c_2 - 0. \qquad \frac{dx}{dt} = 0 \text{ when } t = 0$$

So $c_2 = 0$ and $c_1 = \frac{1}{4} - \frac{1}{14} = \frac{5}{28}$. The solution is $x(t) = \frac{5}{28}\cos 8t + \frac{1}{14}\cos 6t.$

5. By Hooke's law,

$$F = kx$$

$$12 = k \cdot 2 \text{ or } k = 6 \qquad \text{spring constant}$$

$$\text{mass: } \frac{12}{32} = \frac{3}{8} \text{ slug} \qquad \text{mass}$$

$$\frac{3}{8}\frac{d^2x}{dt^2} + 6x = 0 \qquad b = 0$$

$$\frac{d^2x}{dt^2} + 16x = 0 \qquad \text{equation}$$

Initial conditions: (1) If $t = 0$, $x = \frac{8}{12} = \frac{2}{3}$ ft $\qquad$ initial position

$\qquad\qquad\qquad$ (2) If $t = 0$, $\frac{dx}{dt} = 3$ ft/s $\qquad$ initial velocity

Auxiliary equation: $m^2 + 16 = 0$ or $m = \pm 4j$.

Thus $\quad x(t) = c_1 \cos 4t + c_2 \sin 4t$;

$$x'(t) = -4c_1 \sin 4t + 4c_2 \cos 4t.$$

From initial conditions: $\quad \frac{2}{3} = c_1 + c_2(0) \quad \text{or} \quad c_1 = \frac{2}{3};$

$$3 = 0 + 4c_2 \quad \text{or} \quad c_2 = \frac{3}{4}.$$

Substituting c_1 and c_2: $x(t) = \frac{2}{3}\cos 4t + \frac{3}{4}\sin 4t.$

7. From Exercise 5, $\dfrac{d^2x}{dt^2} + 16x = 0$.

Initial conditions: (1)   If $t = 0$, $x = \dfrac{2}{3}$ ft   initial position

(2)   If $t = 0$, $\dfrac{dx}{dt} = -4$ ft/s   initial velocity

From $m^2 + 16 = 0$, we get $m = \pm 4j$ and

$$x(t) \;=\; c_1 \cos 4t + c_2 \sin 4t;$$
$$x'(t) \;=\; -4c_1 \sin 4t + 4c_2 \cos 4t.$$

$$\dfrac{2}{3} \;=\; c_1 + 0 \qquad x = \dfrac{2}{3} \text{ when } t = 0$$
$$-4 \;=\; 0 + 4c_2 \qquad \dfrac{dx}{dt} = -4 \text{ when } t = 0$$

So $c_1 = \dfrac{2}{3}$ and $c_2 = -1$ and $x(t) = \dfrac{2}{3}\cos 4t - \sin 4t$.

9. From Hooke's law:
$$F \;=\; kx$$
$$4 \;=\; k \cdot 2 \text{ or } k = 2 \qquad \text{spring constant}$$
$$\text{mass: } \dfrac{4}{32} \;=\; \dfrac{1}{8} \text{ slug} \qquad \text{mass}$$
$$\dfrac{1}{8}\dfrac{d^2x}{dt^2} + \dfrac{1}{8}\dfrac{dx}{dt} + 2x \;=\; 0 \qquad b = \dfrac{1}{8}$$
$$\dfrac{d^2x}{dt^2} + \dfrac{dx}{dt} + 16x \;=\; 0 \qquad \text{equation}$$

Initial conditions: (1)   If $t = 0$, $x = \dfrac{1}{2}$ ft   intitial position

(2)   If $t = 0$, $\dfrac{dx}{dt} = 0$ initial velocity

Auxiliary equation:

$$m^2 + m + 16 \;=\; 0$$
$$m \;=\; \dfrac{-1 \pm \sqrt{1-64}}{2} = \dfrac{-1 \pm \sqrt{63}\,j}{2}$$
$$\;=\; \dfrac{-1 \pm 3\sqrt{7}\,j}{2} = -\dfrac{1}{2} \pm \dfrac{3}{2}\sqrt{7}\,j.$$

Thus $x(t) = e^{-t/2}\left(c_1 \cos \dfrac{3}{2}\sqrt{7}\,t + c_2 \sin \dfrac{3}{2}\sqrt{7}\,t \right)$.

By the product rule,

$$x'(t) = e^{-t/2}\left(-\dfrac{3}{2}\sqrt{7}\,c_1 \sin \dfrac{3}{2}\sqrt{7}\,t + \dfrac{3}{2}\sqrt{7}\,c_2 \cos \dfrac{3}{2}\sqrt{7}\,t \right) - \dfrac{1}{2}e^{-t/2}\left(c_1 \cos \dfrac{3}{2}\sqrt{7}\,t + c_2 \sin \dfrac{3}{2}\sqrt{7}\,t \right).$$

In the first equation, using $t = 0$ and $x = \dfrac{1}{2}$:

$\dfrac{1}{2} = 1(c_1 + c_2 \cdot 0)$, so that $c_1 = \dfrac{1}{2}$.

In the second equation, let $t = 0$ and $x'(t) = 0$:

$0 = 1\left(0 + \dfrac{3}{2}\sqrt{7}\,c_2 \right) - \dfrac{1}{2}(c_1 + 0)$, so that $0 = \dfrac{3}{2}\sqrt{7}\,c_2 - \dfrac{1}{2}c_1$. Since $c_1 = \dfrac{1}{2}$,

$\dfrac{3}{2}\sqrt{7}\,c_2 - \dfrac{1}{4} = 0$ and $c_2 = \dfrac{1}{6\sqrt{7}}$.

It follows that $x(t) = e^{-t/2}\left(\dfrac{1}{2}\cos \dfrac{3}{2}\sqrt{7}\,t + \dfrac{1}{6\sqrt{7}}\sin \dfrac{3}{2}\sqrt{7}\,t \right)$.

11. Given: $k = 3$, $m = \dfrac{8}{32} = \dfrac{1}{4}$ slug, and $b = \dfrac{1}{4}$.

$$m\frac{d^2x}{dt^2} + b\frac{dx}{dt} + kx = 0$$
$$\frac{1}{4}\frac{d^2x}{dt^2} + \frac{1}{4}\frac{dx}{dt} + 3x = 0$$
$$\frac{d^2x}{dt^2} + \frac{dx}{dt} + 12x = 0$$

(1) If $t = 0$, $x = -\dfrac{3}{12} = -\dfrac{1}{4}$ ft   initital position

(2) If $t = 0$, $\dfrac{dx}{dt} = 0$ initial velocity

Auxiliary equation:

$$m^2 + m + 12 = 0$$
$$m = \frac{-1 \pm \sqrt{1^2 - 4(1)(12)}}{2} = \frac{-1 \pm \sqrt{-47}}{2} = -\frac{1}{2} \pm \frac{\sqrt{47}}{2}j.$$

Thus $x(t) = e^{-t/2}\left(c_1 \cos \dfrac{\sqrt{47}}{2}t + c_2 \sin \dfrac{\sqrt{47}}{2}t\right).$

By the product rule,

$$x'(t) = e^{-t/2}\left(-\frac{\sqrt{47}}{2}c_1 \sin \frac{\sqrt{47}}{2}t + \frac{\sqrt{47}}{2}c_2 \cos \frac{\sqrt{47}}{2}t\right) - \frac{1}{2}e^{-t/2}\left(c_1 \cos \frac{\sqrt{47}}{2}t + c_2 \sin \frac{\sqrt{47}}{2}t\right).$$

In the first equation, let $t = 0$ and $x = -\dfrac{1}{4}$:

$-\dfrac{1}{4} = 1(c_1 + 0)$ or $c_1 = -\dfrac{1}{4}$.

In the second equation, let $t = 0$ and $x'(t) = 0$:

$$0 = 1\left(0 + \frac{\sqrt{47}}{2}c_2\right) - \frac{1}{2}(c_1 + 0)$$
$$0 = \frac{\sqrt{47}}{2}c_2 - \frac{1}{2}c_1$$
$$0 = \frac{\sqrt{47}}{2}c_2 - \frac{1}{2}\left(-\frac{1}{4}\right) \text{ and } c_2 = -\frac{1}{4\sqrt{47}}.$$

The solution is $x(t) = e^{-t/2}\left(-\dfrac{1}{4}\cos\dfrac{\sqrt{47}}{2}t - \dfrac{1}{4\sqrt{47}}\sin\dfrac{\sqrt{47}}{2}t\right).$

13. We are given that $k = 10$, $b = \dfrac{5}{8}$, and $m = \dfrac{10}{32} = \dfrac{5}{16}$.

$$\frac{5}{16}\frac{d^2x}{dt^2} + \frac{5}{8}\frac{dx}{dt} + 10x = 0$$
$$\frac{d^2x}{dt^2} + 2\frac{dx}{dt} + 32x = 0 \qquad \text{equation}$$

(1) If $t = 0$, $x = \dfrac{1}{2}$ ft   initial position

(2) If $t = 0$, $\dfrac{dx}{dt} = -3\,\text{ft/s}$ initial velocity

Auxiliary equation:

$$m^2 + 2m + 32 = 0$$
$$m = \frac{-2 \pm \sqrt{4 - 4(32)}}{2} = \frac{-2 \pm \sqrt{-124}}{2} = \frac{-2 \pm 2\sqrt{31}j}{2} = -1 \pm \sqrt{31}j.$$

Thus $x(t) = e^{-t}\left(c_1 \cos \sqrt{31}\,t + c_2 \sin \sqrt{31}\,t\right).$

By the product rule,

$$x'(t) = e^{-t}\left(-\sqrt{31}\,c_1 \sin \sqrt{31}\,t + \sqrt{31}\,c_2 \cos \sqrt{31}\,t\right) - e^{-t}\left(c_1 \cos \sqrt{31}\,t + c_2 \sin \sqrt{31}\,t\right).$$

In the first equation, let $t = 0$ and $x = \frac{1}{2}$:

$\frac{1}{2} = 1(c_1 + 0)$ or $c_1 = \frac{1}{2}$.

In the second equation, let $t = 0$ and $x'(t) = -3$:

$-3 = 1\left(0 + \sqrt{31}\, c_2\right) - 1(c_1 + 0)$.

Since $c_1 = \frac{1}{2}$, we get $-3 = \sqrt{31}\, c_2 - \frac{1}{2}$, or $c_2 = -\dfrac{5}{2\sqrt{31}}$.

Finally, $x(t) = e^{-t}\left(\dfrac{1}{2}\cos\sqrt{31}\,t - \dfrac{5}{2\sqrt{31}}\sin\sqrt{31}\,t\right)$.

15. By Hooke's law,

$$F = kx$$

$$2 = k \cdot \frac{1}{2} \text{ or } k = 4 \qquad \text{spring constant}$$

$$m = \frac{2}{32} = \frac{1}{16} \text{ slug} \qquad \text{mass}$$

damping force: $1\dfrac{dx}{dx}$ so that $b = 1$.

$$\frac{1}{16}\frac{d^2x}{dt^2} + 1\frac{dx}{dt} + 4x = 2\sin 8t$$

$$\frac{d^2x}{dt^2} + 16\frac{dx}{dt} + 64x = 32\sin 8t$$

Initial conditions: (1)   If $t = 0$, $x = \dfrac{3}{12} = \dfrac{1}{4}$ ft   initial position

                   (2)   If $t = 0$, $\dfrac{dx}{dt} = 0$ initial velocity

$$x_c: \quad m^2 + 16m + 64 = 0$$

$$(m + 8)^2 = 0$$

$$m = -8, -8$$

$x_c = c_1 e^{-8t} + c_2 t e^{-8t}$

$x_p: x_p = A\cos 8t + B\sin 8t$; $x_p' = -8A\sin 8t + 8B\cos 8t$; $x_p'' = -64A\cos 8t - 64B\sin 8t$

Substituting,

$$(-64A\cos 8t - 64B\sin 8t) + 16(-8A\sin 8t + 8B\cos 8t) + 64(A\cos 8t + B\sin 8t) = 32\sin 8t.$$

Collecting terms, $-128A\sin 8t + 128B\cos 8t = 32\sin 8t + 0\cos 8t$.

We obtain: $A = -\dfrac{32}{128} = -\dfrac{1}{4}$ and $B = 0$. So $x_p = -\dfrac{1}{4}\cos 8t$ and

$$x(t) = c_1 e^{-8t} + c_2 t e^{-8t} - \frac{1}{4}\cos 8t;$$

$$x'(t) = -8c_1 e^{-8t} - 8c_2 t e^{-8t} + c_2 e^{-8t} + 2\sin 8t.$$

Substituting the initial conditions,

$$\frac{1}{4} = c_1 + 0 - \frac{1}{4} \qquad\qquad x = \frac{1}{4} \text{ when } t = 0$$

$$0 = -8c_1 + 0 + c_2 + 0. \qquad\quad \frac{dx}{dt} = 0 \text{ when } t = 0$$

We obtain: $c_1 = \dfrac{1}{2}$ and $0 = -8\left(\dfrac{1}{2}\right) + c_2$ or $c_2 = 4$; thus

$$x(t) = \frac{1}{2}e^{-8t} + 4te^{-8t} - \frac{1}{4}\cos 8t.$$

17. By Hooke's law

$$F = kx$$

$$19.6 = k \cdot 0.098 \text{ or } k = 200.$$

$m\dfrac{d^2x}{dt^2} + kx = 0$	$b = 0$
$2.0\dfrac{d^2x}{dt^2} + 200x = 0$	
$\dfrac{d^2x}{dt^2} + 100x = 0$	equation
$m^2 + 100 = 0$	auxiliary equation

$$m = \pm 10j$$

$$x = c_1 \cos 10t + c_2 \sin 10t$$

$$\frac{dx}{dt} = -10c_1 \sin 10t + 10c_2 \cos 10t$$

From the description in the problem, we have the following intitial conditions: if $t = 0$, then $x = 0.25$ and $\dfrac{dx}{dt} = 0$. It follows that

(1) $0.25 = c_1$;

(2) $0 = 10c_2$, or $c_2 = 0$.

Thus $x(t) = 0.25 \cos 10t$.

19. From Exercise 17,

$2.0\dfrac{d^2x}{dt^2} + 4\dfrac{dx}{dt} + 200x = 0$	$b = 4$
$\dfrac{d^2x}{dt^2} + 2\dfrac{dx}{dt} + 100x = 0.$	

(1) If $t = 0$, $x = 0.25$ initial position

(2) If $t = 0$, $\dfrac{dx}{dt} = 0$ initial velocity

Auxiliary equation:

$$m^2 + 2m + 100 = 0$$

$$m = \frac{-2 \pm \sqrt{4 - 400}}{2} = \frac{-2 \pm \sqrt{396}\,j}{2} = \frac{-2 \pm 2\sqrt{99}\,j}{2} = -1 \pm \sqrt{99}\,j.$$

$$x(t) = e^{-t}\left(c_1 \cos \sqrt{99}\,t + c_2 \sin \sqrt{99}\,t\right)$$

$$x'(t) = e^{-t}\left(-\sqrt{99}\,c_1 \sin \sqrt{99}\,t + \sqrt{99}\,c_2 \cos \sqrt{99}\,t\right) - e^{-t}\left(c_1 \cos \sqrt{99}\,t + c_2 \sin \sqrt{99}\,t\right)$$

Using the initial conditions,

$$0.25 = 1(c_1 + 0) \text{ or } c_1 = 0.25;$$

$$0 = 1\left(0 + \sqrt{99}\,c_2\right) - 1(c_1 + 0)$$

$$0 = \sqrt{99}\,c_2 - 0.25 \text{ and } c_2 = \frac{0.25}{\sqrt{99}}.$$

Thus $x(t) = e^{-t}\left(0.25 \cos \sqrt{99}\,t + \dfrac{0.25}{\sqrt{99}} \sin \sqrt{99}\,t\right)$.

21. From Exercise 17, $k = 200$. We are also given that $m = 2.0\,\text{kg}$, $b = 4$, and $f(t) = 20 \sin 5t$.

$$m\frac{d^2x}{dt^2} + b\frac{dx}{dt} + kx = f(t)$$

$$2.0\frac{d^2x}{dt^2} + 4\frac{dx}{dt} + 200x = 20\sin 5t$$

$$\text{and}$$

$$\frac{d^2x}{dt^2} + 2\frac{dx}{dt} + 100x = 10\sin 5t \qquad (*)$$

$x_c:\ m^2 + 2m + 100 = 0$

$$m = \frac{-2 \pm \sqrt{4 - 400}}{2} = \frac{-2 \pm \sqrt{-396}}{2}$$

$$= -1 \pm \frac{1}{2}\sqrt{396}\,j = -1 \pm \sqrt{99}\,j$$

$x_c = e^{-t}\left(c_1 \cos\sqrt{99}\,t + c_2 \sin\sqrt{99}\,t\right)$

$x_p:\ x_p = A\cos 5t + B\sin 5t;\ x_p' = -5A\sin 5t + 5B\cos 5t;\ x_p'' = -25A\cos 5t - 25B\sin 5t$

Substituting in equation (*):

$(-25A\cos 5t - 25B\sin 5t) + 2(-5A\sin 5t + 5B\cos 5t)$

$$+100(A\cos 5t + B\sin 5t) = 10\sin 5t$$

$$(75A + 10B)\cos 5t + (-10A + 75B)\sin 5t = 0\cos 5t + 10\sin 5t$$

$$75A + 10B = 0$$

$$-10A + 75B = 10$$

$$A = \frac{\begin{vmatrix} 0 & 10 \\ 10 & 75 \end{vmatrix}}{\begin{vmatrix} 75 & 10 \\ -10 & 75 \end{vmatrix}} = \frac{0 - 100}{75^2 + 10^2} = \frac{-100}{5725} = -0.01747$$

$$B = \frac{\begin{vmatrix} 75 & 0 \\ -10 & 10 \end{vmatrix}}{\begin{vmatrix} 75 & 10 \\ -10 & 75 \end{vmatrix}} = \frac{750}{5725} = 0.131$$

General solution:

$$x(t) = e^{-t}\left(c_1\cos\sqrt{99}\,t + c_2\sin\sqrt{99}\,t\right) - 0.01747\cos 5t + 0.131\sin 5t;$$

$$x'(t) = e^{-t}\left(-\sqrt{99}\,c_1\sin\sqrt{99}\,t + \sqrt{99}\,c_2\cos\sqrt{99}\,t\right) - e^{-t}\left(c_1\cos\sqrt{99}\,t + c_2\sin\sqrt{99}\,t\right)$$

$$+0.0874\sin 5t + 0.655\cos 5t.$$

Initial conditions: if $t = 0$, $x = 0.25$ and $\dfrac{dx}{dt} = 0$.

(1) $0.25 = c_1 - 0.01747$, or $c_1 = 0.267$

(2) $0 = \sqrt{99}\,c_2 - c_1 + 0.655$

$0 = \sqrt{99}\,c_2 - 0.267 + 0.655$, or $c_2 = -0.0390$

Thus $x_c = e^{-t}\left(0.267\cos\sqrt{99}\,t - 0.0390\sin\sqrt{99}\,t\right);$

$\quad\quad x_p = -0.01747\cos 5t + 0.131\sin 5t.$

Using two significant digits,

$x_c = e^{-1.0t}\left(0.27\cos 9.9t - 0.039\sin 9.9t\right);$

$x_p = -0.017\cos 5t + 0.13\sin 5t.$

23.
$$L\frac{d^2q}{dt^2} + \frac{1}{C}q = 0$$

$$\frac{d^2q}{dt^2} + \frac{1}{1.0 \times 10^{-4}}q = 0$$

$$\frac{d^2q}{dt^2} + 10^4 q = 0$$

Initial conditions: (1) If $t = 0$, $q = 0$.

(2) If $t = 0$, $i = \dfrac{dq}{dt} = 10$.

Auxiliary equation: $m^2 + 10^4 = 0$; $m = \pm 10^2 j = \pm 100j$.

$$q(t) = c_1 \cos 100t + c_2 \sin 100t$$

$$q'(t) = -100c_1 \sin 100t + 100c_2 \cos 100t$$

$$0 = c_1 + 0 \text{ or } c_1 = 0$$

$$10 = 0 + 100c_2 \text{ or } c_2 = \frac{1}{10}$$

$$q(t) = \frac{1}{10} \sin 100t$$

$$i = \frac{dq}{dt} = 10 \cos 100t$$

25.
$$L\frac{d^2q}{dt^2} + \frac{1}{C}q = e(t)$$

$$0.5\frac{d^2q}{dt^2} + \frac{1}{8 \times 10^{-4}}q = 50 \sin 100t$$

$$\frac{d^2q}{dt^2} + 2500q = 100 \sin 100t \qquad (*)$$

$q_c :$ $\quad q_c = c_1 \cos 50t + c_2 \sin 50t$

$q_p :$ $\quad q_p = A \cos 100t + B \sin 100t$

$$q_p' = -100A \sin 100t + 100B \cos 100t$$

$$q_p'' = -10000A \cos 100t - 10000B \sin 100t$$

Substituting in equation (*):

$$(-10000A \cos 100t - 10000B \sin 100t) + 2500(A \cos 100t + B \sin 100t) = 100 \sin 100t$$

$$(-10000 + 2500)B \sin 100t + (-10000 + 2500)A \cos 100t = 100 \sin 100t.$$

$$-7500B = 100, \ B = -\frac{1}{75}$$

$$A = 0$$

General solution: $\quad q = c_1 \cos 50t + c_2 \sin 50t - \left(\frac{1}{75}\right) \sin 100t;$

$$\frac{dq}{dt} = -50c_1 \sin 50t + 50c_2 \cos 50t - \left(\frac{4}{3}\right) \cos 100t.$$

Initial conditions: if $t = 0$, then $q = 0$ and $\dfrac{dq}{dt} = 0$.

(1) $0 = c_1$

(2) $0 = 50c_2 - \dfrac{4}{3}$, or $c_2 = \dfrac{2}{75}$

Thus $q(t) = \dfrac{2}{75} \sin 50t - \dfrac{1}{75} \sin 100t.$

27. $$m\frac{d^2x}{dt^2} + b\frac{dx}{dt} + kx = f(t)$$

$$1.2\frac{d^2x}{dt^2} + 1.5\frac{dx}{dt} + 80x = 10\sin 5t$$

Steady-state solution:
$$\begin{aligned} x_p &= A\cos 5t + B\sin 5t; \\ x_p' &= -5A\sin 5t + 5B\cos 5t; \\ x_p'' &= -25A\cos 5t - 25B\sin 5t. \end{aligned}$$

Substituting:

$$1.2(-25A\cos 5t - 25B\sin 5t) + 1.5(-5A\sin 5t + 5B\cos 5t)$$
$$+80(A\cos 5t + B\sin 5t) = 10\sin 5t$$

$$(50A + 7.5B)\cos 5t + (-7.5A + 50B)\sin 5t = 0\cos 5t + 10\sin 5t.$$

$$50A + 7.5B = 0$$
$$-7.5A + 50B = 10$$

$$A = \frac{\begin{vmatrix} 0 & 7.5 \\ 10 & 50 \end{vmatrix}}{\begin{vmatrix} 50 & 7.5 \\ -7.5 & 50 \end{vmatrix}} = \frac{0 - 7.5(10)}{50^2 + 7.5^2} = \frac{-75}{2556.25} = -0.029$$

$$B = \frac{\begin{vmatrix} 50 & 0 \\ -7.5 & 10 \end{vmatrix}}{\begin{vmatrix} 50 & 7.5 \\ -7.5 & 50 \end{vmatrix}} = \frac{(50)(10) - 0}{2556.25} = 0.20$$

$$x_p = -0.029\cos 5t + 0.20\sin 5t$$

29. $$\frac{d^2q}{dt^2} + 10\frac{dq}{dt} + 100q = 50\cos 10t$$

$$\begin{aligned} q_p &= A\cos 10t + B\sin 10t \\ q_p' &= -10A\sin 10t + 10B\cos 10t \\ q_p'' &= -100A\cos 10t - 100B\sin 10t \end{aligned}$$

Substituting:

$$(-100A\cos 10t - 100B\sin 10t) + 10(-10A\sin 10t + 10B\cos 10t)$$
$$+100(A\cos 10t + B\sin 10t) = 50\cos 10t$$

or

$$100B\cos 10t - 100A\sin 10t = 50\cos 10t$$
$$B = \frac{1}{2},\ A = 0.$$

Thus $q_p = \dfrac{1}{2}\sin 10t$ and $i_p = 5\cos 10t$.

Chapter 12 Review

1. $D^4y = 0$, $m^4 = 0$, so that $m = 0,\ 0,\ 0,\ 0$.

$$y = c_1 + c_2x + c_3x^2 + c_4x^3$$

3. $(D^2 + 4)y = 0$

$m^2 + 4 = 0$; $m^2 = -4$; $m = \pm 2j$

$$y = c_1\cos 2x + c_2\sin 2x$$

5. $(D^2 - 2D - 2)y = 0;$

$$m^2 - 2m - 2 = 0$$

$$m = \frac{2 \pm \sqrt{4 + 8}}{2} = \frac{2 \pm 2\sqrt{3}}{2} = 1 \pm \sqrt{3}$$

$$y = c_1 e^{(1+\sqrt{3})x} + c_2 e^{(1-\sqrt{3})x} = e^x \left(c_1 e^{\sqrt{3}\,x} + c_2 e^{-\sqrt{3}\,x} \right)$$

7. $(3D^2 - D + 1)y = 0$

$$3m^2 - m + 1 = 0$$

$$m = \frac{-(-1) \pm \sqrt{(-1)^2 - 4(3)(1)}}{2 \cdot 3} = \frac{1 \pm \sqrt{-11}}{6} = \frac{1}{6} \pm \frac{\sqrt{11}}{6}j$$

$$y = e^{(1/6)x} \left(c_1 \cos \frac{\sqrt{11}}{6}x + c_2 \sin \frac{\sqrt{11}}{6}x \right)$$

9. $(m - 2)^2 (m^2 + 1) = 0$

$$m = 2,\ 2,\ \pm j$$
$$y = c_1 e^{2x} + c_2 x e^{2x} + c_3 \cos x + c_4 \sin x$$

11. $2m^2 - m + 1 = 0$

$$m = \frac{-(-1) \pm \sqrt{(-1)^2 - 4(2)(1)}}{2 \cdot 2} = \frac{1 \pm \sqrt{-7}}{4} = \frac{1}{4} \pm \frac{\sqrt{7}}{4}j$$

$$y = e^{(1/4)x} \left(c_1 \cos \frac{\sqrt{7}}{4}x + c_2 \sin \frac{\sqrt{7}}{4}x \right)$$

13. $(3D^2 - 2D - 2)y = 0$

$$3m^2 - 2m - 2 = 0$$

$$m = \frac{2 \pm \sqrt{4 + 24}}{6} = \frac{2 \pm 2\sqrt{7}}{6} = \frac{1}{3} \pm \frac{1}{3}\sqrt{7}$$

$$y = c_1 e^{(1/3 + \sqrt{7}/3)x} + c_2 e^{(1/3 - \sqrt{7}/3)x} = e^{(1/3)x} \left(c_1 e^{(\sqrt{7}/3)x} + c_2 e^{-(\sqrt{7}/3)x} \right)$$

15. $(D^2 - 3D - 4)y = 6e^x$

$y_c : \quad m^2 - 3m - 4 = 0$

$\qquad (m - 4)(m + 1) = 0$

$\qquad\qquad\qquad m = 4,\ -1$

$y_c = c_1 e^{4x} + c_2 e^{-x}$

$y_p : \ y_p = Ae^x;\ y_p' = Ae^x;\ y_p'' = Ae^x$

Substituting in $(D^2 - 3D - 4)y = 6e^x$, we get

$\qquad Ae^x - 3Ae^x - 4Ae^x = 6e^x$

$\qquad\qquad\qquad -6Ae^x = 6e^x$

$\qquad\qquad\qquad\qquad A = -1.$

So $y_p = -e^x$ and $y = y_c - e^x.$

17. $(D^2 - 3D - 4)y = 2\sin x$

$y_c:$ $\quad m^2 - 3m - 4 \;=\; 0$

$\qquad (m-4)(m+1) \;=\; 0$

$\qquad\qquad\qquad m \;=\; 4,\ -1$

$y_c = c_1 e^{4x} + c_2 e^{-x}$

$y_p:$ $\; y_p = A\cos x + B\sin x;\; y_p' = -A\sin x + B\cos x;\; y_p'' = -A\cos x - B\sin x$

Substituting in the equation:

$(-A\cos x - B\sin x) - 3(-A\sin x + B\cos x) - 4(A\cos x + B\sin x) \;=\; 2\sin x$

$(-5A - 3B)\cos x + (3A - 5B)\sin x \;=\; 0\cos x + 2\sin x$

$-5A - 3B \;=\; 0$

$3A - 5B \;=\; 2.$

Solution: $B = -\dfrac{5}{17},\ A = \dfrac{3}{17}.$

Thus $y = c_1 e^{4x} + c_2 e^{-x} + \dfrac{3}{17}\cos x - \dfrac{5}{17}\sin x.$

19. $(D^2 - 3D - 4)y = 10xe^{-x}$

$y_c:$ $\quad m^2 - 3m - 4 \;=\; 0$

$\qquad (m-4)(m+1) \;=\; 0$

$\qquad\qquad\qquad m \;=\; 4,\ -1$

$y_c = c_1 e^{4x} + c_2 e^{-x}$

(Observe that the root $m = -1$ of the auxiliary equation coincides with a root associated with the annihilator.)

$y_p:$ $\quad$ Right side $\;=\; 10xe^{-x}$

$\qquad\qquad\quad m' \;=\; -1,\ -1 \qquad$ repeating root

$\qquad\quad (m' + 1)^2 \;=\; 0$

$\qquad$ Annihilator: $\quad (D+1)^2.$

Now apply the annihilator to the given equation:

$(D+1)^2(D^2 - 3D - 4)y \;=\; (D+1)^2(10xe^{-x}) = 0$

$(m+1)^2(m+1)(m-4) \;=\; 0$

$\qquad\qquad\qquad m \;=\; 4,\ -1,\ -1,\ -1.$

It follows that y has the form: $y = c_1 e^{4x} + c_2 e^{-x} + c_3 x e^{-x} + c_4 x^2 e^{-x}.$

Since $y_c = c_1 e^{4x} + c_2 e^{-x}$, y_p must have the following form:

$y_p \;=\; Axe^{-x} + Bx^2 e^{-x};$

$y_p' \;=\; -Axe^{-x} + Ae^{-x} - Bx^2 e^{-x} + 2Bxe^{-x};$

$y_p'' \;=\; Axe^{-x} - Ae^{-x} - Ae^{-x} + Bx^2 e^{-x} - 2Bxe^{-x} - 2Bxe^{-x} + 2Be^{-x}$

$\qquad =\; Axe^{-x} - 2Ae^{-x} + Bx^2 e^{-x} - 4Bxe^{-x} + 2Be^{-x}.$

Substituting in $(D^2 - 3D - 4)y = 10xe^{-x}$, we get

$(Axe^{-x} - 2Ae^{-x} + Bx^2 e^{-x} - 4Bxe^{-x} + 2Be^{-x})$

$-3(-Axe^{-x} + Ae^{-x} - Bx^2 e^{-x} + 2Bxe^{-x}) - 4(Axe^{-x} + Bx^2 e^{-x}) = 10xe^{-x}.$

After collecting terms, we get $(-5A + 2B)e^{-x} - 10Bxe^{-x} = 10xe^{-x}.$

We obtain: $\qquad -10B \;=\; 10$

$\qquad\qquad -5A + 2B \;=\; 0.$

Thus $B = -1$ and $A = -\dfrac{2}{5}$; so that $y_p = -\dfrac{2}{5}xe^{-x} - x^2 e^{-x}.$

21. $(D^2 - 2D - 3)y = 0$

$$m^2 - 2m - 3 = 0$$
$$(m - 3)(m + 1) = 0$$
$$m = 3, -1$$
$$y = c_1 e^{3x} + c_2 e^{-x}$$
$$Dy = 3c_1 e^{3x} - c_2 e^{-x}$$

If $x = 0$, then $y = 0$ and $Dy = -4$:

(1) $\quad 0 = c_1 + c_2$

(2) $\dfrac{-4 = 3c_1 - c_2}{-4 = 4c_1}$ $\quad$ or $c_1 = -1$; $c_2 = 1$.

Thus $y = e^{-x} - e^{3x}$.

23. $0.100\dfrac{d^2q}{dt^2} + 40.0\dfrac{dq}{dt} + \dfrac{1}{2.00 \times 10^{-3}}q = 100.0\cos 20.0t$

$$\dfrac{d^2q}{dt^2} + 400\dfrac{dq}{dt} + 5000q = 1000\cos 20t$$

$q_p = A\cos 20t + B\sin 20t;$

$q_p' = -20A\sin 20t + 20B\cos 20t;$

$q_p'' = -400A\cos 20t - 400B\sin 20t$

Substituting,

$(-400A\cos 20t - 400B\sin 20t) + 400(-20A\sin 20t + 20B\cos 20t)$

$\qquad\qquad\qquad\qquad +5000(A\cos 20t + B\sin 20t) = 1000\cos 20t$

$(4600A + 8000B)\cos 20t + (-8000A + 4600B)\sin 20t = 1000\cos 20t$

$$4600A + 8000B = 1000$$
$$\dfrac{-8000A + 4600B = 0}{}$$
$$46A + 80B = 10$$
$$-80A + 46B = 0$$

$$A = \dfrac{\begin{vmatrix} 10 & 80 \\ 0 & 46 \end{vmatrix}}{\begin{vmatrix} 46 & 80 \\ -80 & 46 \end{vmatrix}} = \dfrac{10(46)}{46^2 + 80^2} = \dfrac{460}{8516} = 0.054$$

$$B = \dfrac{\begin{vmatrix} 46 & 10 \\ -80 & 0 \end{vmatrix}}{\begin{vmatrix} 46 & 80 \\ -80 & 46 \end{vmatrix}} = \dfrac{0 + 10(80)}{8516} = 0.0939$$

$q(t) = 0.054\cos 20.0t + 0.0939\sin 20.0t$

$i(t) = q'(t) = -1.08\sin 20.0t + 1.88\cos 20.0t$

25. By Hooke's law,

$$F = kx$$

$$5 = k \cdot \frac{1}{2} \text{ or } k = 10 \qquad \text{spring constant}$$

$$\text{mass} = \frac{5}{32} \text{ slug} \qquad \text{mass}$$

$$m\frac{d^2x}{dt^2} + kx = f(t) \qquad b = 0$$

$$\frac{5}{32}\frac{d^2x}{dt^2} + 10x = \frac{1}{8}\cos 4t \qquad f(t) = \frac{1}{8}\cos 4t$$

$$(*) \qquad \frac{d^2x}{dt^2} + 64x = \frac{4}{5}\cos 4t \qquad \text{multiplying by } \frac{32}{5}$$

$x_c: \ m^2 + 64 = 0, \ m = \pm 8j$

$x_c = c_1 \cos 8t + c_2 \sin 8t$

$x_p: \ x_p = A\cos 4t + B\sin 4t; \ x_p' = -4A\sin 4t + 4B\cos 4t; \ x_p'' = -16A\cos 4t - 16B\sin 4t$

Substituting in (*):

$$(-16A\cos 4t - 16B\sin 4t) + 64(A\cos 4t + B\sin 4t) = \frac{4}{5}\cos 4t$$

or

$$48A\cos 4t + 48B\sin 4t = \frac{4}{5}\cos 4t.$$

It follows that $B = 0$ and $A = \dfrac{4}{5} \cdot \dfrac{1}{48} = \dfrac{1}{60}$.

General solution:

$$x(t) = x_c + x_p = c_1\cos 8t + c_2\sin 8t + \frac{1}{60}\cos 4t;$$

$$x'(t) = -8c_1\sin 8t + 8c_2\cos 8t - \frac{1}{15}\sin 4t.$$

Initial conditions: (1) If $t = 0$, $x = \dfrac{1}{4}$ ft \qquad initial position

(2) If $t = 0$, $\dfrac{dx}{dt} = -5$ ft/s \qquad initial velocity

Substituting:

$\dfrac{1}{4} = c_1 + 0 + \dfrac{1}{60}$, or $c_1 = \dfrac{1}{4} - \dfrac{1}{60} = \dfrac{15}{60} - \dfrac{1}{60} = \dfrac{14}{60} = \dfrac{7}{30}$;

$-5 = 0 + 8c_2 + 0$, or $c_2 = -\dfrac{5}{8}$.

Substituting the values of c_1 and c_2, we get

$$x(t) = \frac{7}{30}\cos 8t - \frac{5}{8}\sin 8t + \frac{1}{60}\cos 4t.$$

Chapter 13

The Laplace Transform

Sections 13.1-13.3

1. Transform 5:

$$\mathcal{L}\{\sin at\} = \int_0^\infty e^{-st} \sin at\, dt$$

Integration by parts: $\quad u = e^{-st} \qquad dv = \sin at\, dt$

$$du = -se^{-st} \qquad v = -\frac{1}{a}\cos at$$

$$\lim_{b\to\infty} \left(-\frac{1}{a}e^{-st}\cos at \Big|_0^b \right) - \frac{s}{a}\int_0^\infty e^{-st}\cos at\, dt$$

$$= \lim_{b\to\infty} \left(-\frac{1}{a}e^{-sb}\cos ab + \frac{1}{a} \right) - \frac{s}{a}\int_0^\infty e^{-st}\cos at\, dt = 0 + \frac{1}{a} - \frac{s}{a}\int_0^\infty e^{-st}\cos at\, dt$$

Integrating by parts again: $\quad u = e^{-st} \qquad dv = \cos at\, dt$

$$du = -se^{-st} \qquad v = \frac{1}{a}\sin at$$

$$\int_0^\infty e^{-st}\sin at\, dt = \frac{1}{a} - \frac{s}{a}\left[\frac{1}{a}e^{-st}\sin at \Big|_0^\infty + \frac{s}{a}\int_0^\infty e^{-st}\sin at\, dt \right]$$

$$= \frac{1}{a} - \frac{s}{a}(0) - \frac{s^2}{a^2}\int_0^\infty e^{-st}\sin at\, dt$$

Solving for the integral,

$$\left(1 + \frac{s^2}{a^2} \right) \int_0^\infty e^{-st}\sin at\, dt = \frac{1}{a}$$

$$\int_0^\infty e^{-st}\sin at\, dt = \frac{1}{a}\cdot\frac{a^2}{s^2 + a^2} = \frac{a}{s^2 + a^2}.$$

3. $f(t) = 2 + 3e^{-t} = 2\cdot 1 + 3e^{-t}$

$$F(s) = 2\mathcal{L}\{1\} + 3\mathcal{L}\{e^{-t}\} = 2\cdot\frac{1}{s} + 3\frac{1}{s+1} \quad \text{(by transforms 1 and 4, respectively) or}$$

$$F(s) = \frac{2}{s}\frac{s+1}{s+1} + \frac{3}{s+1}\frac{s}{s} = \frac{2s+2+3s}{s(s+1)} = \frac{5s+2}{s(s+1)}.$$

5. $f(t) = t + \cos 2t$

$$F(s) = \frac{1}{s^2} + \frac{s}{s^2+4} \quad \text{by transforms 2 and 6, respectively.}$$

Thus, $F(s) = \dfrac{s^2 + 4 + s^3}{s^2(s^2+4)} = \dfrac{s^3 + s^2 + 4}{s^2(s^2+4)}.$

7. $f(t) = e^{2t}\sin 5t$. By transform 7 with $a = 2$ and $b = 5$:

$$F(s) = \frac{5}{(s-2)^2 + 5^2} = \frac{5}{(s-2)^2 + 25}.$$

9. $f(t) = t^3 e^{-4t}$. By transform 9 with $n = 3$ and $a = -4$:
$$F(s) = \frac{3!}{(s+4)^4} = \frac{6}{(s+4)^4}.$$

11. $f(t) = 2t^4 e^{-t}$. By transform 9 with $n = 4$ and $a = -1$:
$$F(s) = 2 \cdot \frac{4!}{(s+1)^{4+1}} = \frac{48}{(s+1)^5}.$$

13. $f(t) = 4 - 5\sin 2t$. By transforms 1 and 5, respectively:
$$F(s) = \frac{4}{s} - 5\frac{2}{s^2+4} = \frac{4}{s} - \frac{10}{s^2+4}.$$

15. $F(s) = \dfrac{10}{s^2+4} = 5\dfrac{2}{s^2+2^2}$
    $f(t) = 5\sin 2t$ by transform 5.

17. $F(s) = \dfrac{1}{(s+2)^3}$. Transform 9 with $n = 2$ and $a = -2$. The form needs to be adjusted:

    insert 2! and place $\dfrac{1}{2!}$ in front.
    $$F(s) = \frac{1}{2!}\frac{2!}{(s+2)^3} \text{ or } F(s) = \frac{1}{2}\frac{2!}{(s+2)^3}.$$
    It follows that $f(t) = \dfrac{1}{2}t^2 e^{-2t}$.

19. $F(s) = \dfrac{4}{(s^2+4)^2}$. This form can be made to fit transform 12 with $a = 2$: the numerator must

    be $2a^3 = 2 \cdot 2^3 = 16$. One way to make this adjustment is to insert the factor 4 and place $\dfrac{1}{4}$ in
    front. So
    $$F(s) = \frac{1}{4}\frac{4 \cdot 4}{(s+4)^2} = \frac{1}{4}\frac{2 \cdot 2^3}{(s+2^2)^2};$$
    $$f(t) = \frac{1}{4}(\sin 2t - 2t\cos 2t).$$

21. $F(s) = \dfrac{2s}{(s^2+9)^2}$. This form can be made to fit transform 13 with $a = 3$: insert the required

    3 and place $\dfrac{1}{3}$ in front. $F(s) = \dfrac{1}{3}\dfrac{2 \cdot 3s}{(s^2+3^2)^2}$ so $f(t) = \dfrac{1}{3}t\sin 3t$.

23. $F(s) = \dfrac{5s}{s^2+6} = 5\dfrac{s}{s^2+\left(\sqrt{6}\right)^2}$
    $f(t) = 5\cos\sqrt{6}\,t$ by transform 6 with $a = \sqrt{6}$.

25. $F(s) = \dfrac{2}{s^2(s^2+4)}$. Transform 11 with $a = 2$. Insert the missing 2^2 and place $\dfrac{1}{2^2} = \dfrac{1}{4}$ in front:
    $$F(s) = \frac{1}{2^2} \cdot \frac{2 \cdot 2^2}{s^2(s^2+4)}.$$
    So $F(s) = \dfrac{1}{4}\dfrac{2^3}{s^2(s^2+4)}$ and $f(t) = \dfrac{1}{4}(2t - \sin 2t) = \dfrac{1}{2}t - \dfrac{1}{4}\sin 2t$.

27. $F(s) = \dfrac{2s+4}{(s+2)^2+4} = 2\dfrac{s+2}{(s+2)^2+2^2}$
    $f(t) = 2e^{-2t}\cos 2t$ by transform 8 with $a = -2$ and $b = 2$.

29. $F(s) = \dfrac{1}{(s+3)^2 + 5}$. Transform 7 with $a = -3$ and $b = \sqrt{5}$. To fit this form, we need to insert

$\sqrt{5}$ and place $\dfrac{1}{\sqrt{5}}$ in front:

$$F(s) = \frac{1}{\sqrt{5}} \frac{\sqrt{5}}{(s+3)^2 + 5} \text{ and } f(t) = \frac{1}{\sqrt{5}} e^{-3t} \sin \sqrt{5}\, t.$$

31. Here we need to complete the square in the denominator and use transform 7:

$$\begin{aligned}
F(s) &= \frac{\sqrt{10}}{s^2 - 2s + 11} = \frac{\sqrt{10}}{(s-1)^2 + 10}; \\
f(t) &= e^t \sin \sqrt{10}\, t. \qquad (a = 1,\ b = \sqrt{10})
\end{aligned}$$

33. Here we complete the square in the denominator and then use transforms 8 and 7 (after splitting the fraction).

$$\begin{aligned}
\mathcal{L}^{-1}\left\{ \frac{s}{s^2 - 6s + 10} \right\} &= \mathcal{L}^{-1}\left\{ \frac{s}{(s-3)^2 + 1} \right\} = \mathcal{L}^{-1}\left\{ \frac{(s - \underline{3} + \underline{3})}{(s-3)^2 + 1} \right\} \\
&= \mathcal{L}^{-1}\left\{ \frac{s-3}{(s-3)^2 + 1} \right\} + \mathcal{L}^{-1}\left\{ \frac{3}{(s-3)^2 + 1} \right\} \\
&= \mathcal{L}^{-1}\left\{ \frac{s-3}{(s-3)^2 + 1} \right\} + 3\mathcal{L}^{-1}\left\{ \frac{1}{(s-3)^2 + 1} \right\} \\
&= e^{3t} \cos t + 3e^{3t} \sin t = e^{3t}(\cos t + 3 \sin t)
\end{aligned}$$

35. $$\begin{aligned}
F(s) &= \frac{s}{s^2 - 2s + 6} = \frac{s}{(s-1)^2 + 5} = \frac{(s-1)+1}{(s-1)^2 + 5} \\
&= \frac{s-1}{(s-1)^2 + 5} + \frac{1}{(s-1)^2 + 5} = \frac{s-1}{(s-1)^2 + 5} + \frac{1}{\sqrt{5}} \frac{\sqrt{5}}{(s-1)^2 + 5}
\end{aligned}$$

$f(t) = e^t \cos \sqrt{5}\, t + \dfrac{1}{\sqrt{5}} e^t \sin \sqrt{5}\, t$ by transforms 8 and 7, respectively. Factoring e^t, we also

have $f(t) = e^t \left(\cos \sqrt{5}\, t + \dfrac{1}{\sqrt{5}} \sin \sqrt{5}\, t \right)$.

37. $\dfrac{1}{s(s+1)}$. Rule I, distinct linear factors:

$$\begin{aligned}
\frac{A}{s} + \frac{B}{s+1} &= \frac{A(s+1) + Bs}{s(s+1)} \\
A(s+1) + Bs &= 1.
\end{aligned}$$

Let $s = 0$: $\quad A(1) + 0 = 1$

$$A = 1.$$

Let $s = -1$: $\quad 0 + B(-1) = 1$

$$B = -1.$$

$$\mathcal{L}^{-1}\left\{ \frac{1}{s(s+1)} \right\} = \mathcal{L}^{-1}\left\{ \frac{1}{s} - \frac{1}{s+1} \right\} = 1 - e^{-t} \text{ by transforms 1 and 4, respectively.}$$

39. $\dfrac{2s+1}{(s-2)(s+3)}$. Distinct linear factors:

$$\frac{2s+1}{(s-2)(s+3)} = \frac{A}{s-2} + \frac{B}{s+3} = \frac{A(s+3)+B(s-2)}{(s-2)(s+3)}$$

$$A(s+3)+B(s-2) = 2s+1.$$

Let $s=-3$: $\;0+B(-5) = -5.\;\;$ Let $s=2$: $\;A(5)+0 = 5$

$$B = 1. \hspace{5.5cm} A = 1.$$

$$\mathcal{L}^{-1}\left\{\frac{2s+1}{(s-2)(s+3)}\right\} = \mathcal{L}^{-1}\left\{\frac{1}{s-2} + \frac{1}{s+3}\right\} = e^{2t} + e^{-3t} \text{ by transform 4.}$$

41. $\dfrac{s^2}{(s-2)(s+2)(s-4)}$. Rule I, distinct linear factors:

$$\frac{A}{s-2} + \frac{B}{s+2} + \frac{C}{s-4} = \frac{A(s+2)(s-4)+B(s-2)(s-4)+C(s-2)(s+2)}{(s-2)(s+2)(s-4)}$$

$$A(s+2)(s-4)+B(s-2)(s-4)+C(s-2)(s+2) = s^2.$$

Let $s=-2$: $\;0+B(-4)(-6)+0 = (-2)^2$

$$24B = 4$$

$$B = \frac{1}{6}.$$

Let $s=4$: $\;0+0+C(2)(6) = 4^2$

$$12C = 16$$

$$C = \frac{4}{3}.$$

Let $s=2$: $\;A(4)(-2)+0+0 = 2^2$

$$-8A = 4$$

$$A = -\frac{1}{2}.$$

$$\mathcal{L}^{-1}\left\{-\frac{1}{2}\frac{1}{s-2} + \frac{1}{6}\frac{1}{s+2} + \frac{4}{3}\frac{1}{s-4}\right\} = -\frac{1}{2}e^{2t} + \frac{1}{6}e^{-2t} + \frac{4}{3}e^{4t} \text{ by transform 4.}$$

43. $\dfrac{3s^2}{(s+2)^2(s-1)}$. Rule I, linear factors, one repeating and one distinct:

$$\frac{3s^2}{(s+2)^2(s-1)} = \frac{A}{s+2} + \frac{B}{(s+2)^2} + \frac{C}{s-1}$$

$$= \frac{A}{s+2}\frac{(s+2)(s-1)}{(s+2)(s-1)} + \frac{B}{(s+2)^2}\frac{s-1}{s-1} + \frac{C}{s-1}\frac{(s+2)^2}{(s+2)^2}$$

$$= \frac{A(s+2)(s-1)+B(s-1)+C(s+2)^2}{(s+2)^2(s-1)}$$

$$A(s+2)(s-1)+B(s-1)+C(s+2)^2 = 3s^2.$$

Let $s=-2$: $\;0+B(-3)+0=12$ or $B=-4$. Let $s=1$: $\;0+0+C(3)^2=3$ or $C=\dfrac{1}{3}$.

Since there are only two distinct factors, we seem to have run out of values to substitute. So we use the values already obtained for B and C and then let $s=$ any value:

$$A(s+2)(s-1) - 4(s-1) + \frac{1}{3}(s+2)^2 = 3s^2.$$

Let $s=$ any value (such as $s=0$):

$$A(2)(-1) - 4(-1) + \frac{1}{3}(2)^2 = 0$$

$$-2A + 4 + \frac{4}{3} = 0$$

$$A = \frac{8}{3}.$$

$$F(s) = \frac{8}{3}\frac{1}{s+2} + \frac{-4}{(s+2)^2} + \frac{1}{3}\frac{1}{s-1};$$

$$f(t) = \frac{8}{3}e^{-2t} - 4te^{-2t} + \frac{1}{3}e^t \text{ by transforms 4 and 9, respectively.}$$

45. $\dfrac{1}{(s+1)(s^2+1)}$. (Distinct factors, one linear, one quadratic.) By Rules I and II:

$$\frac{A}{s+1} + \frac{Bs+C}{s^2+1} = \frac{A(s^2+1)+(Bs+C)(s+1)}{(s+1)(s^2+1)}$$

$$A(s^2+1)+(Bs+C)(s+1) = 1.$$

Let $s = -1$: $\quad A(2) + 0 = 1$

$$A = \frac{1}{2}.$$

We have no more real values to substitute, since s^2+1 is positive for all real s. So we use the value for A already found

$$\frac{1}{2}(s^2+1)+(Bs+C)(s+1)=1$$

and let $s = 0$, yielding C:

$$\frac{1}{2}(1)+C(1)=1 \text{ and } C = \frac{1}{2}.$$

We now have $\dfrac{1}{2}(s^2+1)+\left(Bs+\dfrac{1}{2}\right)(s+1)=1$. To find B, let $s=$ any value (such as $s=1$).

Let $s = 1$: $\quad \dfrac{1}{2}(2)+\left(B+\dfrac{1}{2}\right)(2) = 1 \qquad\qquad A = \dfrac{1}{2},\ C = \dfrac{1}{2}$

$$1 + \left(B+\frac{1}{2}\right)(2) = 1$$

$$\left(B+\frac{1}{2}\right)(2) = 0$$

$$B = -\frac{1}{2}.$$

$$\mathcal{L}^{-1}\left\{\frac{1}{2}\frac{1}{s+1}-\frac{1}{2}\frac{s}{s^2+1}+\frac{1}{2}\frac{1}{s^2+1}\right\}=\frac{1}{2}e^{-t}-\frac{1}{2}\cos t+\frac{1}{2}\sin t$$

47. $\dfrac{s}{(s+1)(s^2+1)}$. Distinct factors, one linear and one quadratic:

$$\frac{s}{(s+1)(s^2+1)} = \frac{A}{s+1}+\frac{Bs+C}{s^2+1} = \frac{A(s^2+1)+(Bs+C)(s+1)}{(s+1)(s^2+1)}$$

$$A(s^2+1)+(Bs+C)(s+1) = s.$$

Let $s = -1$: $\quad A(2) + 0 = -1$ or $A = -\dfrac{1}{2}$.

We have no more real values to substitute since s^2+1 is positive for all real s. So we use the value for A already found, that is,

$$-\frac{1}{2}(s^2+1)+(Bs+C)(s+1)=s$$

and let $s = 0$ to find C:

$$-\frac{1}{2}(1)+C(1)=0 \text{ or } C=\frac{1}{2}.$$

We now have $-\dfrac{1}{2}(s^2+1)+\left(Bs+\dfrac{1}{2}\right)(s+1)=s.$

To find B, let $s=$ any value (such as $s=1$):

$$-\frac{1}{2}(2)+\left(B+\frac{1}{2}\right)(2) = 1$$

$$-1+2B+1 = 1$$

$$B = \frac{1}{2}.$$

$$F(s) = -\frac{1}{2}\frac{1}{s+1}+\frac{\frac{1}{2}s+\frac{1}{2}}{s^2+1}=-\frac{1}{2}\frac{1}{s+1}+\frac{1}{2}\frac{s}{s^2+1}+\frac{1}{2}\frac{1}{s^2+1};$$

$$f(t) = -\frac{1}{2}e^{-t}+\frac{1}{2}\cos t+\frac{1}{2}\sin t.$$

49. $\dfrac{9s}{(s+2)^2(s-1)}$. Rule I, repeating linear factors:

$$\frac{9s}{(s+2)^2(s-2)} = \frac{A}{s+2} + \frac{B}{(s+2)^2} + \frac{C}{s-1}$$
$$= \frac{A}{s+2}\cdot\frac{(s+2)(s-1)}{(s+2)(s-1)} + \frac{B}{(s+2)^2}\cdot\frac{s-1}{s-1} + \frac{C}{s-1}\frac{(s+2)^2}{(s+2)^2}.$$

Equating numerators: $A(s+2)(s-1) + B(s-1) + C(s+2)^2 = 9s$.

We start by substituting convenient values.

Let $s = -2$: $B(-3) = -18 \qquad B = 6$.

Let $s = 1$: $C(9) = 9 \qquad C = 1$.

Because of the repeating factor, we have run out of values. So let's use the values already found,

$$A(s+2)(s-1) + 6(s-1) + 1(s+2)^2 = 9s$$

and let $s = $ any value (such as $s = 0$) to find A:

$$A(2)(-1) + 6(-1) + 2^2 = 0$$
$$A = -1.$$

We now have
$$F(s) = \frac{-1}{s+2} + \frac{6}{(s+2)^2} + \frac{1}{s-1}$$
$$f(t) = -e^{-2t} + 6te^{-2t} + e^t$$

by transforms 4 and 9, respectively.

51. $\dfrac{s}{(s+1)^2}$. Repeating linear factor:

$$\frac{s}{(s+1)^2} = \frac{A}{s+1} + \frac{B}{(s+1)^2} = \frac{A(s+1)+B}{(s+1)^2}$$
$$A(s+1) + B = s.$$

Let $s = -1$: $0 + B = -1$ or $B = -1$.

Using this value, we get $A(s+1) - 1 = s$. Now let $s = $ any value (such as $s = 0$)

$s = 0$: $A(1) - 1 = 0$ or $A = 1$.

$$F(s) = \frac{1}{s+1} + \frac{-1}{(s+1)^2};$$
$$f(t) = e^{-t} - te^{-t}.$$

A simple alternative is to write
$$\frac{s}{(s+1)^2} = \frac{s+1-1}{(s+1)^2} = \frac{s+1}{(s+1)^2} - \frac{1}{(s+1)^2} = \frac{1}{s+1} - \frac{1}{(s+1)^2}.$$

53.
$$\frac{2s^2+2s+1}{(s^2+2s+2)(s-1)} = \frac{As+B}{s^2+2s+2} + \frac{C}{s-1}$$
$$= \frac{(As+B)(s-1)+C(s^2+2s+2)}{(s^2+2s+2)(s-1)}$$

$$(As+B)(s-1) + C(s^2+2s+2) = 2s^2+2s+1.$$

Let $s = 1$: $0 + C(5) = 5$

$$C = 1.$$

Because of the quadratic factor, let us use the value of C already found,

$$(As+B)(s-1) + 1(s^2+2s+2) = 2s^2+2s+1$$

and let $s = 0$ to find B.

Let $s = 0$: $B(-1) + 1(2) = 1 \qquad$ (since $C = 1$)

$$B = 1.$$

Let $s = 2$: $(2A+1)(1) + 1(10) = 13$ $\qquad B = 1, \; C = 1$

$$2A + 1 = 3$$

$$A = 1.$$

$$\mathcal{L}^{-1} \left\{ \frac{s+1}{(s+1)^2 + 1} + \frac{1}{s-1} \right\} = e^{-t} \cos t + e^t$$

55.
$$\frac{1}{(s^2 + 4s + 7)(s+4)} = \frac{As+B}{s^2 + 4s + 7} + \frac{C}{s+4}$$
$$= \frac{(As+B)(s+4) + C(s^2 + 4s + 7)}{(s^2 + 4s + 7)(s+4)}$$

$$(As+B)(s+4) + C(s^2 + 4s + 7) = 1.$$

Let $s = -4$: $0 + C(7) = 1$ or $C = \dfrac{1}{7}$.

Because of the quadratic factor, let us use the value of C already found and then let $s = 0$ to determine B:

$$(As+B)(s+4) + \frac{1}{7}(s^2 + 4s + 7) = 1.$$

$s = 0$: $B(4) + \dfrac{1}{7}(7) = 1$ or $B = 0$.

We now have $(As+0)(s+4) + \dfrac{1}{7}(s^2 + 4s + 7) = 1$.

To find A, let $s =$ any value (such as $s = 1$):

$A(5) + \dfrac{1}{7}(12) = 1$ or $A = -\dfrac{1}{7}$.

$$\begin{aligned}
F(s) &= -\frac{1}{7}\frac{s}{s^2 + 4s + 7} + \frac{1}{7}\frac{1}{s+4} \\
&= -\frac{1}{7}\frac{s}{(s+2)^2 + 3} + \frac{1}{7}\frac{1}{s+4} \\
&= -\frac{1}{7}\frac{s+2-2}{(s+2)^2 + 3} + \frac{1}{7}\frac{1}{s+4} \\
&= -\frac{1}{7}\left(\frac{s+2}{(s+2)^2 + 3} - 2\frac{1}{(s+2)^2 + 3} \right) + \frac{1}{7}\frac{1}{s+4} \\
&= -\frac{1}{7}\left(\frac{s+2}{(s+2)^2 + 3} - \frac{2}{\sqrt{3}}\frac{\sqrt{3}}{(s+2)^2 + 3} \right) + \frac{1}{7}\frac{1}{s+4} \\
&= -\frac{1}{7}\frac{s+2}{(s+2)^2 + 3} + \frac{2}{7\sqrt{3}}\frac{\sqrt{3}}{(s+2)^2 + 3} + \frac{1}{7}\frac{1}{s+4} \\
&= \frac{1}{7}\frac{1}{s+4} - \frac{1}{7}\frac{s+2}{(s+2)^2 + 3} + \frac{2\sqrt{3}}{21}\frac{\sqrt{3}}{(s+2)^2 + 3}; \\
f(t) &= \frac{1}{7}e^{-4t} - \frac{1}{7}e^{-2t}\cos\sqrt{3}\,t + \frac{2\sqrt{3}}{21}e^{-2t}\sin\sqrt{3}\,t.
\end{aligned}$$

13.4 Solution of Linear Equations by Laplace Transforms

1. $y' - y = 0$, $y(0) = 1$

Step 1. $sY(s) - y(0) - Y(s) = 0$ $\qquad$ by (13.11)

Step 2. $sY(s) - 1 - Y(s) = 0$ $\qquad$ since $y(0) = 1$

Step 3. $sY(s) - Y(s) = 1$

$\qquad (s-1)Y(s) = 1$ $\qquad$ factoring $Y(s)$

$\qquad\qquad Y(s) = \dfrac{1}{s-1}$ $\qquad$ dividing by $s-1$

Step 4. $y = e^t$ $\qquad$ by transform 4

3. $y' - 2y = 4$, $y(0) = 0$

Step 1. $sY(s) - y(0) - 2Y(s) = \dfrac{4}{s}$ by (13.11)

Step 2. $sY(s) - 0 - 2Y(s) = \dfrac{4}{s}$ $y(0) = 0$

Step 3. $Y(s)(s - 2) = \dfrac{4}{s}$ factoring $Y(s)$

$Y(s) = \dfrac{4}{s(s - 2)}$ dividing by $s - 2$

Step 4. $Y(s) = \dfrac{A}{s} + \dfrac{B}{s - 2} = \dfrac{A(s - 2) + Bs}{s(s - 2)}$

$A(s - 2) + Bs = 4$

$s = 2:\ 0 + 2B = 4$ or $B = 2$.

$s = 0:\ A(-2) + 0 = 4$ or $A = -2$.

$Y(s) = \dfrac{-2}{s} + \dfrac{2}{s - 1}$

$y = 2e^t - 2$

5. $y' - 2y = e^{2t}$, $y(0) = 0$

Step 1. $sY(s) - y(0) - 2Y(s) = \dfrac{1}{s - 2}$

Step 2. $sY(s) - 2Y(s) = \dfrac{1}{s - 2}$ $y(0) = 0$

Step 3. $(s - 2)Y(s) = \dfrac{1}{s - 2}$ factoring $Y(s)$

$Y(s) = \dfrac{1}{(s - 2)^2}$ dividing by $s - 2$

Step 4. $y = te^{2t}$ by transform 9

7. $y' + 4y = te^{-4t}$, $y(0) = 3$

Step 1. $sY(s) - y(0) + 4Y(s) = \dfrac{1}{(s + 4)^2}$ transform 9

Step 2. $sY(s) - 3 + 4Y(s) = \dfrac{1}{(s + 4)^2}$ $y(0) = 3$

Step 3. $sY(s) + 4Y(s) = \dfrac{1}{(s + 4)^2} + 3$ solving for $Y(s)$

$Y(s) = \dfrac{1}{(s + 4)^3} + \dfrac{3}{s + 4}$

Step 4. $Y(s) = \dfrac{1}{2!}\dfrac{2!}{(s + 4)^3} + \dfrac{3}{s + 4}$

$y = \dfrac{1}{2}t^2 e^{-4t} + 3e^{-4t}$ (by transforms 9 and 4, respectively).

9. $y'' + 9y = 0$, $y(0) = 1$, $y'(0) = -2$

   Step 1. $s^2 Y(s) - sy(0) - y'(0) + 9Y(s) = 0$ by (13.12)

   Step 2. $s^2 Y(s) - s + 2 + 9Y(s) = 0$ $y(0) = 1$, $y'(0) = -2$

   Step 3. $s^2 Y(s) + 9Y(s) = s - 2$

           $(s^2 + 9)Y(s) = s - 2$

                $Y(s) = \dfrac{s - 2}{s^2 + 9}$

   Step 4. Find the inverse transform:

   $$\begin{aligned}
   Y(s) &= \frac{s - 2}{s^2 + 9} = \frac{s}{s^2 + 9} - \frac{2}{s^2 + 9} \\
   &= \frac{s}{s^2 + 9} - 2\frac{1}{s^2 + 9} \\
   &= \frac{s}{s^2 + 9} - \frac{2}{3}\frac{3}{s^2 + 9} \\
   y &= \cos 3t - \frac{2}{3}\sin 3t
   \end{aligned}$$

11. $y'' + y = 2\sin t$, $y(0) = 0$, $y'(0) = 0$

    Step 1. $s^2 Y(s) - sy(0) - y'(0) + Y(s) = \dfrac{2}{s^2 + 1}$ by (13.12)

    Step 2. $s^2 Y(s) + Y(s) = \dfrac{2}{s^2 + 1}$

    Step 3. $Y(s)(s^2 + 1) = \dfrac{2}{s^2 + 1}$

                 $Y(s) = \dfrac{2}{(s^2 + 1)^2}$

    Step 4. By transform 12 with $a = 1$, $y = \sin t - t\cos t$.

13. $y'' + 4y = 2\cos 2t$, $y(0) = -2$, $y'(0) = 0$

    Step 1. $s^2 Y(s) - sy(0) - y'(0) + 4Y(s) = \dfrac{2s}{s^2 + 4}$

    Step 2. $s^2 Y(s) + 2s - 0 + 4Y(s) = \dfrac{2s}{s^2 + 4}$

    Step 3. $s^2 Y(s) + 4Y(s) = -2s + \dfrac{2s}{s^2 + 4}$

            $Y(s)(s^2 + 4) = -2s + \dfrac{2s}{s^2 + 4}$

                 $Y(s) = -\dfrac{2s}{s^2 + 4} + \dfrac{2s}{(s^2 + 4)^2}$

    Step 4. (Transforms 6 and 13 with $a = 2$.) For the second term on the right, we need to insert a 2 and place $\dfrac{1}{2}$ in front:

    $$\begin{aligned}
    Y(s) &= -\frac{2s}{s^2 + 4} + \frac{1}{2}\frac{2 \cdot 2s}{(s^2 + 4)^2} \\
    y &= -2\cos 2t + \frac{1}{2}t\sin 2t.
    \end{aligned}$$

15. $y' - y = \cos 2t$, $y(0) = 0$

   Step 1. $sY(s) - sy(0) - Y(s) = \dfrac{s}{s^2 + 4}$

   Step 2. $sY(s) - Y(s) = \dfrac{s}{s^2 + 4}$ $y(0) = 0$

   Step 3. $Y(s)(s - 1) = \dfrac{\frac{s}{s^2 + 4}}{}$

$$Y(s) = \dfrac{s}{(s - 1)(s^2 + 4)}$$

   Step 4.
$$Y(s) = \dfrac{s}{(s - 1)(s^2 + 4)} = \dfrac{A}{s - 1} + \dfrac{Bs + C}{s^2 + 4}$$
$$= \dfrac{A(s^2 + 4) + (Bs + C)(s - 1)}{(s - 1)(s^2 + 4)}$$

$$A(s^2 + 4) + (Bs + C)(s - 1) = s$$

$s = 1:$ $A(5) + 0 = 1$ or $A = \dfrac{1}{5}$.

$\dfrac{1}{5}(s^2 + 4) + (Bs + C)(s - 1) = s$ $A = \dfrac{1}{5}$

Now let $s = 0$ to find C:

$\dfrac{1}{5}(4) + C(-1) = 0$ or $C = \dfrac{4}{5}$.

Using this value, we get

$\dfrac{1}{5}(s^2 + 4) + \left(Bs + \dfrac{4}{5}\right)(s - 1) = s$.

To find B, let $s = $ any value (such as $s = -1$):

$$\dfrac{1}{5}(5) + \left(-B + \dfrac{4}{5}\right)(-2) = -1$$
$$1 + 2B - \dfrac{8}{5} = -1$$
$$B = -\dfrac{1}{5}$$

$$Y(s) = \dfrac{1}{5}\dfrac{1}{s - 1} - \dfrac{1}{5}\dfrac{s}{s^2 + 4} + \dfrac{4}{5}\dfrac{1}{s^2 + 4} = \dfrac{1}{5}\dfrac{1}{s - 1} - \dfrac{1}{5}\dfrac{s}{s^2 + 4} + \dfrac{2}{5}\dfrac{2}{s^2 + 4}.$$
$$y = \dfrac{1}{5}e^t - \dfrac{1}{5}\cos 2t + \dfrac{2}{5}\sin 2t.$$

17. $y'' - 4y' + 4y = e^{3t}$, $y(0) = 0$, $y'(0) = -2$

   Step 1. $s^2 Y(s) - sy(0) - y'(0) - 4[sY(s) - y(0)] + 4Y(s) = \dfrac{1}{s - 3}$

   Step 2. $s^2 Y(s) + 2 - 4sY(s) + 4Y(s) = \dfrac{1}{s - 3}$

   Step 3. $s^2 Y(s) - 4sY(s) + 4Y(s) = -2 + \dfrac{1}{s - 3}$

$$Y(s)(s^2 - 4s + 4) = -2 + \dfrac{1}{s - 3}$$
$$Y(s)(s - 2)^2 = -2 + \dfrac{1}{s - 3}$$
$$Y(s) = -\dfrac{2}{(s - 2)^2} + \dfrac{1}{(s - 3)(s - 2)^2}$$

   Step 4.   Find the inverse transform. The second fraction on the right needs to be expanded by Rule I (repeating linear factors):

$$\dfrac{1}{(s - 3)(s - 2)^2} = \dfrac{A}{s - 3} + \dfrac{B}{s - 2} + \dfrac{C}{(s - 2)^2}$$
$$= \dfrac{A}{s - 3} \cdot \dfrac{(s - 2)^2}{(s - 2)^2} + \dfrac{B}{s - 2}\dfrac{(s - 2)(s - 3)}{(s - 2)(s - 3)} + \dfrac{C}{(s - 2)^2}\dfrac{s - 3}{s - 3}$$

Equating numerators: $A(s - 2)^2 + B(s - 2)(s - 3) + C(s - 3) = 1$.

Let $s = 2:$ $-C = 1$ or $C = -1$.

Let $s = 3:$ $A(1) = 1$ or $A = 1$.

Because of the repeating factor, we have run out of values to substitute. So let's use the values of A and C already found,

$$1(s-2)^2 + B(s-2)(s-3) + (-1)(s-3) = 1$$

and let $s =$ any value (such as $s = 0$) to find B.

Let $s = 0$: $\quad (-2)^2 + B(-2)(-3) + (-1)(-3) \quad = \quad 1$

$$B \quad = \quad -1.$$

We now have

$$Y(s) \quad = \quad -\frac{2}{(s-2)^2} + \frac{1}{s-3} + \frac{-1}{s-2} + \frac{-1}{(s-2)^2}$$

$$= \quad -\frac{3}{(s-2)^2} - \frac{1}{s-2} + \frac{1}{s-3}.$$

Step 4. $y = -3te^{2t} - e^{2t} + e^{3t}$ by transforms 9 and 4, respectively.

19. $y'' - 6y' + 9y = 12t^2 e^{3t}, \ y(0) = y'(0) = 0$

Step 1. $\quad s^2 Y(s) - sy(0) - y'(0) - 6\left[sY(s) - y(0)\right] + 9Y(s) = 12\dfrac{2!}{(s-3)^3} \qquad$ (by transform 9)

Step 2. $\quad s^2 Y(s) - 6sY(s) + 9Y(s) = \dfrac{24}{(s-3)^3}$

Step 3. $\quad Y(s)(s^2 - 6s + 9) \quad = \quad \dfrac{24}{(s-3)^3}$

$$Y(s)(s-3)^2 \quad = \quad \dfrac{24}{(s-3)^3}$$

$$Y(s) \quad = \quad \dfrac{24}{(s-3)^5}$$

Step 4. By transform 9 with $n = 4$,

$$Y(s) = \dfrac{4!}{(s-3)^5} \text{ and } y = t^4 e^{3t}.$$

21. $y'' - 4y' + 10y = 0, \ y(0) = -3, \ y'(0) = 0$

Step 1. $\quad s^2 Y(s) - sy(0) - y'(0) - 4[sY(s) - y(0)] + 10Y(s) = 0$

Step 2. $\quad s^2 Y(s) + 3s - 4sY(s) - 12 + 10Y(s) = 0$

Step 3. $\quad s^2 Y(s) - 4sY(s) + 10Y(s) \quad = \quad -3s + 12$

$$Y(s)(s^2 - 4s + 10) \quad = \quad -3s + 12$$

$$Y(s) \quad = \quad \dfrac{-3s + 12}{s^2 - 4s + 10}$$

Step 4. To find the inverse transform, we need to complete the square in the denominator and use transforms 7 and 8:

$$Y(s) \quad = \quad -3\frac{s-4}{(s-2)^2 + 6}$$

$$= \quad -3\frac{s-2-2}{(s-2)^2 + 6}$$

$$= \quad -3\left[\frac{s-2}{(s-2)^2 + 6} - 2\frac{1}{(s-2)^2 + 6}\right]$$

$$= \quad -3\left[\frac{s-2}{(s-2)^2 + 6} - \frac{2}{\sqrt{6}}\frac{\sqrt{6}}{(s-2)^2 + 6}\right]$$

$$= \quad -3\frac{s-2}{(s-2)^2 + 6} + \frac{6}{\sqrt{6}}\frac{\sqrt{6}}{(s-2)^2 + 6};$$

$$y \quad = \quad -3e^{2t}\cos\sqrt{6}\,t + \frac{6}{\sqrt{6}}e^{2t}\sin\sqrt{6}\,t$$

$$= \quad -3e^{2t}\cos\sqrt{6}\,t + \sqrt{6}e^{2t}\sin\sqrt{6}\,t$$

$$= \quad e^{2t}\left(-3\cos\sqrt{6}\,t + \sqrt{6}\sin\sqrt{6}\,t\right).$$

23. $y'' + y = 4e^t,\ y(0) = y'(0) = 0$

Step 1. $s^2 Y(s) - sy(0) - y'(0) + Y(s) = \dfrac{4}{s-1}$

Step 2. $s^2 Y(s) + Y(s) = \dfrac{4}{s-1}$

Step 3. $Y(s) = \dfrac{4}{(s-1)(s^2+1)}$

Step 4. $Y(s) = \dfrac{A}{s-1} + \dfrac{Bs+C}{s^2+1} = \dfrac{A(s^2+1)+(Bs+C)(s-1)}{(s-1)(s^2+1)}$

$A(s^2+1) + (Bs+C)(s-1) = 4$

$s = 1:\ A(2) + 0 = 4$ or $A = 2$.

$2(s^2+1) + (Bs+C)(s-1) = 4$ (since $A = 2$).

Now let $s = 0$ to find C: $2(1) + C(-1) = 4$ or $C = -2$.

Using this value, we now have: $2(s^2+1) + (Bs-2)(s-1) = 4$.

Finally, to find B, we let $s = $ any value (such as $s = -1$): $2(2) + (-B-2)(-2) = 4$ or $B = -2$.

So $Y(s) = \dfrac{2}{s-1} + \dfrac{-2s-2}{s^2+1} = 2\left(\dfrac{1}{s-1} - \dfrac{s}{s^2+1} - \dfrac{1}{s^2+1}\right)$;

$\qquad y = 2\left(e^t - \cos t - \sin t\right).$

25. $y'' - 4y = 3\cos t,\ y(0) = y'(0) = 0$

$$s^2 Y(s) - sy(0) - y'(0) - 4Y(s) = \dfrac{3s}{s^2+1}$$

$$(s^2 - 4)Y(s) = \dfrac{3s}{s^2+1}$$

$$Y(s) = \dfrac{3s}{(s^2-4)(s^2+1)}$$

$$Y(s) = \dfrac{3s}{(s-2)(s+2)(s^2+1)}$$

$$\dfrac{3s}{(s-2)(s+2)(s^2+1)} = \dfrac{A}{s-2} + \dfrac{B}{s+2} + \dfrac{Cs+D}{s^2+1}$$

$$= \dfrac{A(s+2)(s^2+1) + B(s-2)(s^2+1) + (Cs+D)(s-2)(s+2)}{(s-2)(s+2)(s^2+1)}$$

$A(s+2)(s^2+1) + B(s-2)(s^2+1) + (Cs+D)(s-2)(s+2) = 3s$

Let $s = 2:\quad A(4)(5) + 0 + 0 = 6$

$$A = \dfrac{3}{10}.$$

Let $s = -2:\quad 0 + B(-4)(5) + 0 = -6$

$$B = \dfrac{3}{10}.$$

Now use the values of A and B already found. Then let $s = 0$ to find D.

Let $s = 0:\quad \dfrac{3}{10}(2)(1) + \dfrac{3}{10}(-2)(1) + D(-2)(2) = 0 \qquad A = B = \dfrac{3}{10}$

$$D = 0.$$

Let $s = 1:\quad \dfrac{3}{10}(3)(2) + \dfrac{3}{10}(-1)(2) + C(-1)(3) = 3$

$$\dfrac{9}{5} - \dfrac{3}{5} - 3C = 3$$

$$-3C = \dfrac{15}{5} - \dfrac{9}{5} + \dfrac{3}{5} = \dfrac{9}{5}$$

$$C = -\dfrac{3}{5}.$$

$$Y(s) = \dfrac{3}{10}\dfrac{1}{s-2} + \dfrac{3}{10}\dfrac{1}{s+2} - \dfrac{3}{5}\dfrac{s}{s^2+1}$$

$$y = \dfrac{3}{10}e^{2t} + \dfrac{3}{10}e^{-2t} - \dfrac{3}{5}\cos t.$$

27. $y'' + 2y' + 5y = 8e^t$, $y(0) = 0$, $y'(0) = 0$

Step 1. $s^2 Y(s) - sy(0) - y'(0) + 2(sY(s) - y(0)) + 5Y(s) = \dfrac{8}{s-1}$

Step 2. $s^2 Y(s) + 2sY(s) + 5Y(s) = \dfrac{8}{s-1}$

Step 3. $Y(s)(s^2 + 2s + 5) = \dfrac{8}{s-1}$

$$Y(s) = \frac{8}{(s-1)(s^2 + 2s + 5)}$$

Step 4. $\dfrac{A}{s-1} + \dfrac{Bs + C}{s^2 + 2s + 5}$

$A(s^2 + 2s + 5) + (Bs + C)(s - 1) = 8$

$s = 1:$ $A(8) = 8$ or $A = 1$.

$1(s^2 + 2s + 5) + (Bs + C)(s - 1) = 8$ since $A = 1$

Now let $s = 0$ to find C:

$1(5) + C(-1) = 8$ or $C = -3$.

Using this value, we now have: $1(s^2 + 2s + 5) + (Bs - 3)(s - 1) = 8$.

Finally, to find B, we let $s = $ any value (such as $s = -1$):

$4 + (-B - 3)(-2) = 8$ or $B = -1$.

So $Y(s) = \dfrac{1}{s-1} + \dfrac{-s-3}{s^2 + 2s + 5}$

$\qquad = \dfrac{1}{s-1} - \dfrac{s+3}{(s+1)^2 + 4} = \dfrac{1}{s-1} - \dfrac{s+1+2}{(s+1)^2 + 4}$

$\qquad = \dfrac{1}{s-1} - \dfrac{s+1}{(s+1)^2 + 4} - \dfrac{2}{(s+1)^2 + 4};$

$\quad y = e^t - e^{-t}\cos 2t - e^{-t}\sin 2t.$

29. $\qquad L\dfrac{d^2 q}{dt^2} + R\dfrac{dq}{dt} + \dfrac{1}{C}q = e(t)$

$0.1\dfrac{d^2 q}{dt^2} + 6.0\dfrac{dq}{dt} + \dfrac{1}{0.02}q = 6.0$

Omitting final zeros, we get

$0.1\dfrac{d^2 q}{dt^2} + 6\dfrac{dq}{dt} + 50q = 6$

$\dfrac{d^2 q}{dt^2} + 60\dfrac{dq}{dt} + 500q = 60.$

$s^2 Q(s) - sq(0) - q'(0) + 60(sQ(s) - q(0)) + 500Q(s) = \dfrac{60}{s}$

$s^2 Q(s) + 60sQ(s) + 500Q(s) = \dfrac{60}{s}$

$Q(s) = \dfrac{60}{s(s^2 + 60s + 500)} = \dfrac{60}{s(s + 50)(s + 10)}$

$\qquad = \dfrac{A}{s} + \dfrac{B}{s + 50} + \dfrac{C}{s + 10} = \dfrac{3}{25}\dfrac{1}{s} + \dfrac{3}{100}\dfrac{1}{s + 50} - \dfrac{3}{20}\dfrac{1}{s + 10}$

$q(t) = \dfrac{3}{25} + \dfrac{3}{100}e^{-50t} - \dfrac{3}{20}e^{-10t}$

$\qquad = 0.12 + 0.03e^{-50t} - 0.15e^{-10t}.$

31. (Exercise 1)

$F = kx$

$4 = k \cdot \dfrac{1}{2}$ or $k = 8$; mass$= \dfrac{4}{32} = \dfrac{1}{8}$ slug.

Equation: $\dfrac{1}{8}\dfrac{d^2x}{dt^2} + 8x = 0.$

$\dfrac{d^2x}{dt^2} + 64x = 0,\ x(0) = \dfrac{1}{4},\ x'(0) = 0.$

$$
\begin{aligned}
s^2 X(s) - sx(0) - x'(0) + 64X(s) &= 0 \\
X(s)(s^2 + 64) &= \frac{1}{4}s \\
X(s) &= \frac{1}{4}\frac{s}{s^2 + 64} \\
x(t) &= \frac{1}{4}\cos 8t.
\end{aligned}
$$

(Exercise 9)

$F = kx$

$4 = k \cdot 2$ or $k = 2$; mass$= \dfrac{4}{32} = \dfrac{1}{8}$ slug; $b = \dfrac{1}{8}.$

Equation: $\dfrac{1}{8}\dfrac{d^2x}{dt^2} + \dfrac{1}{8}\dfrac{dx}{dt} + 2x = 0.$

$\dfrac{d^2x}{dt^2} + \dfrac{dx}{dt} + 16x = 0,\ x(0) = \dfrac{1}{2},\ x'(0) = 0.$

$$
\begin{aligned}
s^2 X(s) - sx(0) - x'(0) + sX(s) - x(0) + 16X(s) &= 0 \\
s^2 X(s) - \frac{1}{2}s - 0 + sX(s) - \frac{1}{2} + 16X(s) &= 0 \\
X(s)(s^2 + s + 16) &= \frac{1}{2}s + \frac{1}{2}
\end{aligned}
$$

$$
\begin{aligned}
X(s) &= \frac{1}{2}\frac{s+1}{s^2+s+16} = \frac{1}{2}\frac{s+1}{\left(s+\frac{1}{2}\right)^2 + \frac{63}{4}} \\
&= \frac{1}{2}\frac{s+\frac{1}{2}+\frac{1}{2}}{\left(s+\frac{1}{2}\right)^2 + \frac{63}{4}} \\
&= \frac{1}{2}\frac{s+\frac{1}{2}}{\left(s+\frac{1}{2}\right)^2 + \frac{63}{4}} + \frac{1}{4}\frac{1}{\left(s+\frac{1}{2}\right)^2 + \frac{63}{4}} \\
&= \frac{1}{2}\frac{s+\frac{1}{2}}{\left(s+\frac{1}{2}\right)^2 + \frac{63}{4}} + \frac{1}{4}\frac{2}{\sqrt{63}}\frac{\sqrt{63}/2}{\left(s+\frac{1}{2}\right)^2 + \frac{63}{4}}
\end{aligned}
$$

$$
x(t) = \frac{1}{2}e^{-t/2}\cos\frac{\sqrt{63}}{2}t + \frac{1}{2\sqrt{63}}e^{-t/2}\sin\frac{\sqrt{63}}{2}t
$$

or

$$
x(t) = e^{-t/2}\left(\frac{1}{2}\cos\frac{3}{2}\sqrt{7}\,t + \frac{1}{6\sqrt{7}}\sin\frac{3}{2}\sqrt{7}\,t\right).
$$

33. $y'' + 4y = 5\cos 3t,\ y(0) = y'(0) = 0$

$$
\begin{aligned}
s^2 Y(s) - sy(0) - y'(0) + 4Y(s) &= \frac{5s}{s^2+9} \\
Y(s)(s^2+4) &= \frac{5s}{s^2+9} \\
Y(s) &= \frac{5s}{(s^2+9)(s^2+4)} = \frac{s}{s^2+4} - \frac{s}{s^2+9} \\
y &= \cos 2t - \cos 3t.
\end{aligned}
$$

35. $y'' + 9y = 2\sin 2t$, $y(0) = 0$, $y'(0) = 2$

$$s^2 Y(s) - sy(0) - y'(0) + 9Y(s) = \frac{4}{s^2 + 4}$$

$$s^2 Y(s) - 2 + 9Y(s) = \frac{4}{s^2 + 4}$$

$$\begin{aligned}
Y(s) &= \frac{2}{s^2 + 9} + \frac{4}{(s^2 + 4)(s^2 + 9)} = \frac{2}{s^2 + 9} + \frac{4}{5}\frac{1}{s^2 + 4} - \frac{4}{5}\frac{1}{s^2 + 9} \\
&= \frac{6}{5}\frac{1}{s^2 + 9} + \frac{4}{5}\frac{1}{s^2 + 4} = \frac{2}{5}\frac{3}{s^2 + 9} + \frac{2}{5}\frac{2}{s^2 + 4}. \\
y &= \frac{2}{5}\sin 3t + \frac{2}{5}\sin 2t.
\end{aligned}$$

37. $y'' + 4y' + 6y = 3\cos\sqrt{6}\,t$, $y(0) = -2$, $y'(0) = 0$

$$s^2 Y(s) - sy(0) - y'(0) + 4\left[sY(s) - y(0)\right] + 6Y(s) = \frac{3s}{s^2 + 6}$$

$$s^2 Y(s) + 2s + 4sY(s) + 8 + 6Y(s) = \frac{3s}{s^2 + 6}$$

$$(s^2 + 4s + 6)Y(s) = -2s - 8 + \frac{3s}{s^2 + 6}$$

$$\begin{aligned}
Y(s) &= \frac{-2s - 8}{s^2 + 4s + 6} + \frac{3s}{(s^2 + 6)(s^2 + 4s + 6)} \\
&= \frac{-2s - 8}{s^2 + 4s + 6} + \frac{\frac{3}{4}}{s^2 + 6} - \frac{\frac{3}{4}}{s^2 + 4s + 6} \\
&= \frac{-2s - \frac{32}{4} - \frac{3}{4}}{s^2 + 4s + 6} + \frac{\frac{3}{4}}{s^2 + 6} \\
&= -2 \cdot \frac{s + \frac{35}{8}}{(s + 2)^2 + 2} + \frac{\frac{3}{4}}{s^2 + 6} \\
&= -2\left[\frac{s + 2 - 2 + \frac{35}{8}}{(s + 2)^2 + 2}\right] + \frac{\frac{3}{4}}{s^2 + 6} \\
&= -2\left[\frac{s + 2}{(s + 2)^2 + 2} + \frac{\frac{19}{8}}{(s + 2)^2 + 2}\right] + \frac{\frac{3}{4}}{s^2 + 6} \\
&= -2 \cdot \frac{s + 2}{(s + 2)^2 + 2} - \frac{19}{4}\frac{1}{\sqrt{2}}\frac{\sqrt{2}}{(s + 2)^2 + 2} + \frac{\frac{3}{4}}{s^2 + 6}. \\
y &= -2e^{-2t}\cos\sqrt{2}\,t - \frac{19}{4\sqrt{2}}e^{-2t}\sin\sqrt{2}\,t + \frac{3}{4\sqrt{6}}\sin\sqrt{6}\,t.
\end{aligned}$$

39. $F = kx$

$12 = k \cdot 2$ or $k = 6$; mass $= \frac{12}{32} = \frac{3}{8}$ slug; $b = 0$.

$$\frac{3}{8}\frac{d^2 x}{dt^2} + 6x = 12\sin t$$

$$\frac{d^2 x}{dt^2} + 16x = 32\sin t, \quad x(0) = 1, \quad x'(0) = 0$$

$$s^2 X(s) - sx(0) - x'(0) + 16X(s) = \frac{32}{s^2 + 1}$$

$$s^2 X(s) - s + 16X(s) = \frac{32}{s^2 + 1}$$

$$\begin{aligned}
X(s) &= \frac{s}{s^2 + 16} + \frac{32}{(s^2 + 16)(s^2 + 1)} = \frac{s}{s^2 + 16} + \frac{32}{15}\frac{1}{s^2 + 1} - \frac{32}{15}\frac{1}{s^2 + 16} \\
&= \frac{s}{s^2 + 16} + \frac{32}{15}\frac{1}{s^2 + 1} - \frac{8}{15}\frac{4}{s^2 + 16}. \\
x(t) &= \cos 4t + \frac{32}{15}\sin t - \frac{8}{15}\sin 4t.
\end{aligned}$$

41. $0.100\dfrac{d^2q}{dt^2} + 10.0\dfrac{dq}{dt} + \dfrac{1}{1.00 \times 10^{-3}}q = 10.0\sin 20.0t$

Multiplying by 10 and leaving out final zeros:

$$\dfrac{d^2q}{dt^2} + 100\dfrac{dq}{dt} + 10000q = 100\sin 20t.$$

$$Q(s) \;=\; \dfrac{2000}{(s^2 + 400)(s^2 + 100s + 10000)}$$

$$=\; \dfrac{\frac{5}{2404}s}{s^2 + 100s + 10000} + \dfrac{\frac{5}{601}}{s^2 + 100s + 10000} - \dfrac{\frac{5}{2404}s}{s^2 + 400} + \dfrac{\frac{120}{601}}{s^2 + 400}$$

The last two terms lead to: $\dfrac{-5}{2404}\cos 20.0t + \dfrac{6}{601}\sin 20.0t.$

The first two terms need to be rewritten as follows:

$$\dfrac{5}{2404}\dfrac{s + \frac{2404}{5}\cdot\frac{5}{601}}{(s+50)^2 + 7500} \;=\; \dfrac{5}{2404}\dfrac{s+4}{(s+50)^2 + 7500} \;=\; \dfrac{5}{2404}\dfrac{s + 50 - 50 + 4}{(s+50)^2 + 7500}$$

$$=\; \dfrac{5}{2404}\dfrac{s+50}{(s+50)^2 + 7500} + \dfrac{5}{2404}(-46)\dfrac{1}{\sqrt{7500}}\dfrac{\sqrt{7500}}{(s+50)^2 + 7500}.$$

Inverse transform: $0.00208e^{-50t}\cos 50\sqrt{3}\,t - 0.00110e^{-50t}\sin 50\sqrt{3}\,t.$

Chapter 13 Review

1. $f(t) \;=\; 2e^{-3t}$

$F(s) \;=\; 2\dfrac{1}{s+3} = \dfrac{2}{s+3} \qquad$ by transform 4

3. $f(t) = 2t^3 + \sin 3t$

$F(s) = 2\mathcal{L}\left\{t^3\right\} + \mathcal{L}\left\{\sin 3t\right\} = 2\cdot\dfrac{3!}{s^{3+1}} + \dfrac{3}{s^2 + 3^2}$ by transforms 3 and 5, respectively, or

$F(s) = \dfrac{12}{s^4} + \dfrac{3}{s^2 + 9}.$

5. $f(t) = 2t - \sin 2t$

$F(s) = \dfrac{2}{s^2} - \dfrac{2}{s^2 + 4}$ by transforms 2 and 5, respectively.

Thus $F(s) = \dfrac{2(s^2 + 4) - 2s^2}{s^2(s^2 + 4)} = \dfrac{8}{s^2(s^2 + 4)}.$

7. $F(s) = \dfrac{s-2}{(s-2)^2 + 5}.$ By transform 8 with $a = 2$ and $b = \sqrt{5}$,

$f(t) = e^{2t}\cos\sqrt{5}\,t.$

9. $F(s) \;=\; \dfrac{s}{s^2 - 2s + 5} = \dfrac{s}{(s-1)^2 + 4} = \dfrac{(s-1) + 1}{(s-1)^2 + 4}$

$=\; \dfrac{s-1}{(s-1)^2 + 4} + \dfrac{1}{2}\dfrac{2}{(s-1)^2 + 4}$

$f(t) \;=\; e^t\cos 2t + \dfrac{1}{2}e^t\sin 2t \qquad$ transforms 8 and 7, respectively

$=\; e^t\left(\cos 2t + \dfrac{1}{2}\sin 2t\right).$

11. $F(s) = \dfrac{s}{(s+1)(s-2)} = \dfrac{A}{s+1} + \dfrac{B}{s-2} = \dfrac{A(s-2) + B(s+1)}{(s+1)(s-2)}$

$A(s-2) + B(s+1) = s$

Let $s = 2$: $0 + B(3) = 2$ or $B = \dfrac{2}{3}.$ Let $s = -1$: $A(-3) + 0 = -1$ or $A = \dfrac{1}{3}.$

$F(s) = \dfrac{1}{3}\dfrac{1}{s+1} + \dfrac{2}{3}\dfrac{1}{s-2}.$

$f(t) = \dfrac{1}{3}e^{-t} + \dfrac{2}{3}e^{2t}.$

13. $F(s) = \dfrac{1}{(s+2)(s-3)(s-4)} = \dfrac{A}{s+2} + \dfrac{B}{s-3} + \dfrac{C}{s-4}$

$\qquad = \dfrac{A(s-3)(s-4) + B(s+2)(s-4) + C(s+2)(s-3)}{(s+2)(s-3)(s-4)}$

$A(s-3)(s-4) + B(s+2)(s-4) + C(s+2)(s-3) = 1$

Let $s = 3$: $B(5)(-1) = 1$. Let $s = 4$: $C(6)(1) = 1$

$\qquad\qquad\qquad B = -\dfrac{1}{5}.$ $\qquad\qquad\qquad C = \dfrac{1}{6}.$

Let $s = -2$: $A(-5)(-6) = 1$

$\qquad\qquad\qquad A = \dfrac{1}{30}.$

$F(s) = \dfrac{1}{30}\dfrac{1}{s+2} - \dfrac{1}{5}\dfrac{1}{s-3} + \dfrac{1}{6}\dfrac{1}{s-4}.$

$f(t) = \dfrac{1}{30}e^{-2t} - \dfrac{1}{5}e^{3t} + \dfrac{1}{6}e^{4t}.$

15. $y' + 2y = 0,\ y(0) = 1$

Step 1. $sY(s) - y(0) + 2Y(s) = 0$

Step 2. $sY(s) - 1 + 2Y(s) = 0$ $\qquad\qquad y(0) = 1$

Step 3. $Y(s)(s+2) = 1$

$\qquad\qquad Y(s) = \dfrac{1}{s+2}$

Step 4. $y = e^{-2t}$

17. $y' + 2y = te^{-2t},\ y(0) = -1$

$sY(s) - y(0) + 2Y(s) = \dfrac{1}{(s+2)^2}$ $\qquad$ by transform 9

$sY(s) + 1 + 2Y(s) = \dfrac{1}{(s+2)^2}$ $\qquad$ $y(0) = -1$

$(s+2)Y(s) = \dfrac{1}{(s+2)^2} - 1$

$Y(s) = \dfrac{1}{(s+2)^3} - \dfrac{1}{s+2}$

$\qquad = \dfrac{1}{2}\dfrac{2!}{(s+2)^3} - \dfrac{1}{s+2}.$

$y = \dfrac{1}{2}t^2e^{-2t} - e^{-2t}$

$\qquad = e^{-2t}\left(\dfrac{1}{2}t^2 - 1\right)$ $\qquad$ by transforms 9 and 4, respectively

19. $y'' - 2y' - 3y = 0,\ y(0) = 0,\ y'(0) = -4$

Step 1. $s^2Y(s) - sy(0) - y'(0) - 2[sY(s) - y(0)] - 3Y(s) = 0$

Step 2. $s^2Y(s) - 0 - (-4) - 2sY(s) + 0 - 3Y(s) = 0$

Step 3. $s^2Y(s) - 2sY(s) - 3Y(s) = -4$

$\qquad\qquad Y(s) = \dfrac{-4}{s^2 - 2s - 3} = \dfrac{-4}{(s+1)(s-3)}$

Step 4. $Y(s) = \dfrac{A}{s+1} + \dfrac{B}{s-3} = \dfrac{A(s-3) + B(s+1)}{(s+1)(s-3)}$

$A(s-3) + B(s+1) = -4$

Let $s = 3$: $0 + B(4) = -4$ or $B = -1$.

Let $s = -1$: $A(-4) + 0 = -4$ or $A = 1$.

$Y(s) = \dfrac{1}{s+1} - \dfrac{1}{s-3}$

$y = e^{-t} - e^{3t}$

21. $y'' + 2y' + 5y = 0$, $y(0) = 1$, $y'(0) = 0$

$$
\begin{aligned}
s^2 Y(s) - s y(0) - y'(0) + 2(s Y(s) - y(0)) + 5Y(s) &= 0 \\
s^2 Y(s) - s + 2(sY(s) - 1) + 5Y(s) &= 0 \\
s^2 Y(s) - s + 2sY(s) - 2 + 5Y(s) &= 0 \\
(s^2 + 2s + 5)Y(s) &= s + 2 \\
Y(s) &= \frac{s+2}{s^2 + 2s + 5}
\end{aligned}
$$

$$
\begin{aligned}
Y(s) &= \frac{s+2}{(s+1)^2 + 4} = \frac{(s+1)+1}{(s+1)^2 + 4} \\
&= \frac{s+1}{(s+1)^2 + 4} + \frac{1}{2}\frac{2}{(s+1)^2 + 4} \\
y &= e^{-t}\cos 2t + \frac{1}{2}e^{-t}\sin 2t = e^{-t}\left(\cos 2t + \frac{1}{2}\sin 2t\right)
\end{aligned}
$$

23. $y'' + 2y' + 5y = 3e^{-2t}$, $y(0) = 1$, $y'(0) = 1$

Step 1. $s^2 Y(s) - s y(0) - y'(0) + 2[sY(s) - y(0)] + 5Y(s) = \dfrac{3}{s+2}$

Step 2. $s^2 Y(s) - s - 1 + 2sY(s) - 2 + 5Y(s) = \dfrac{3}{s+2}$

$$
s^2 Y(s) + 2sY(s) + 5Y(s) = s + 3 + \frac{3}{s+2}
$$

$$
Y(s) = \frac{s+3}{s^2 + 2s + 5} + \frac{3}{(s+2)(s^2 + 2s + 5)}
$$

We decompose the second fraction:

$$
\frac{3}{(s+2)(s^2 + 2s + 5)} = \frac{A}{s+2} + \frac{Bs+C}{s^2 + 2s + 5}
$$

$A(s^2 + 2s + 5) + (Bs + C)(s+2) = 3$

Let $s = -2$: $A(5) = 3$ or $A = \dfrac{3}{5}$.

So $\dfrac{3}{5}(s^2 + 2s + 5) + (Bs + C)(s+2) = 3$.

Now let $s = 0$ to find C:

$\dfrac{3}{5}(5) + C(2) = 3$ or $C = 0$.

Using this value, we get: $\dfrac{3}{5}(s^2 + 2s + 5) + (Bs + 0)(s+2) = 3$.

To find B, let $s =$ any value (such as $s = 1$):

$\dfrac{3}{5}(8) + B(3) = 3$ or $B = -\dfrac{3}{5}$.

We now have

$$
\begin{aligned}
Y(s) &= \frac{s+3}{s^2 + 2s + 5} + \frac{3}{5}\frac{1}{s+2} - \frac{3}{5}\frac{s}{s^2 + 2s + 5} \\
&= \frac{3}{5}\frac{1}{s+2} + \frac{(2/5)s + 3}{s^2 + 2s + 5} \\
&= \frac{3}{5}\frac{1}{s+2} + \frac{2}{5}\frac{s + 15/2}{(s+1)^2 + 4} \\
&= \frac{3}{5}\frac{1}{s+2} + \frac{2}{5}\frac{s + 1 - 1 + 15/2}{(s+1)^2 + 4} \\
&= \frac{3}{5}\frac{1}{s+2} + \frac{2}{5}\frac{s+1}{(s+1)^2 + 4} + \frac{13}{5}\frac{1}{(s+1)^2 + 4} \\
&= \frac{3}{5}\frac{1}{s+2} + \frac{2}{5}\frac{s+1}{(s+1)^2 + 4} + \frac{13}{10}\frac{2}{(s+1)^2 + 4};
\end{aligned}
$$

$$
y = \frac{3}{5}e^{-2t} + \frac{2}{5}e^{-t}\cos 2t + \frac{13}{10}e^{-t}\sin 2t
$$

25. By Hooke's law,

$$F = kx$$

$$4 = k \cdot \frac{1}{2} \text{ or } k = 8$$

mass: $\dfrac{4}{32} = \dfrac{1}{8}$ slug.

Equation: $\dfrac{1}{8}\dfrac{d^2x}{dt^2} + 2\dfrac{dx}{dt} + 8x = 0$ since $b = 2$

and

$$\frac{d^2x}{dt^2} + 16\frac{dx}{dt} + 64x = 0.$$

Initial conditions: (1)   If $t = 0$, $x = 0$ initial position

(2)   If $t = 0$, $\dfrac{dx}{dt} = -4$ initial velocity

$$s^2 X(s) - sx(0) - x'(0) + 16(sX(s) - x(0)) + 64X(s) = 0$$

$$s^2 X(s) - (-4) + 16sX(s) + 64X(s) = 0$$

$$X(s) = \frac{-4}{s^2 + 16s + 64} = -\frac{4}{(s+8)^2}$$

$$x(t) = -4te^{-8t} \text{ by transform 9.}$$